Lecture Notes in Physics

The Editorial Policy for Proceedings

The series Lecture Notes in Physics reports new developments in physical research and teaching – quickly, informally, and at a high level. The proceedings to be considered for publication in this series should be limited to only a few areas of research, and these should be closely related to each other. The contributions should be of a high standard and should avoid lengthy redraftings of papers already published or about to be published elsewhere. As a whole, the proceedings should aim for a balanced presentation of the theme of the conference including a description of the techniques used and enough motivation for a broad readership. It should not be assumed that the published proceedings must reflect the conference in its entirety. (A listing or abstracts of papers presented at the meeting but not included in the proceedings could be added as an appendix.)
When applying for publication in the series Lecture Notes in Physics the volume's editor(s) should submit sufficient material to enable the series editors and their referees to make a fairly accurate evaluation (e.g. a complete list of speakers and titles of papers to be presented and abstracts). If, based on this information, the proceedings are (tentatively) accepted, the volume's editor(s), whose name(s) will appear on the title pages, should select the papers suitable for publication and have them refereed (as for a journal) when appropriate. As a rule discussions will not be accepted. The series editors and Springer-Verlag will normally not interfere with the detailed editing except in fairly obvious cases or on technical matters.
Final acceptance is expressed by the series editor in charge, in consultation with Springer-Verlag only after receiving the complete manuscript. It might help to send a copy of the authors' manuscripts in advance to the editor in charge to discuss possible revisions with him. As a general rule, the series editor will confirm his tentative acceptance if the final manuscript corresponds to the original concept discussed, if the quality of the contribution meets the requirements of the series, and if the final size of the manuscript does not greatly exceed the number of pages originally agreed upon. The manuscript should be forwarded to Springer-Verlag shortly after the meeting. In cases of extreme delay (more than six months after the conference) the series editors will check once more the timeliness of the papers. Therefore, the volume's editor(s) should establish strict deadlines, or collect the articles during the conference and have them revised on the spot. If a delay is unavoidable, one should encourage the authors to update their contributions if appropriate. The editors of proceedings are strongly advised to inform contributors about these points at an early stage.
The final manuscript should contain a table of contents and an informative introduction accessible also to readers not particularly familiar with the topic of the conference. The contributions should be in English. The volume's editor(s) should check the contributions for the correct use of language. At Springer-Verlag only the prefaces will be checked by a copy-editor for language and style. Grave linguistic or technical shortcomings may lead to the rejection of contributions by the series editors. A conference report should not exceed a total of 500 pages. Keeping the size within this bound should be achieved by a stricter selection of articles and not by imposing an upper limit to the length of the individual papers. Editors receive jointly 30 complimentary copies of their book. They are entitled to purchase further copies of their book at a reduced rate. As a rule no reprints of individual contributions can be supplied. No royalty is paid on Lecture Notes in Physics volumes. Commitment to publish is made by letter of interest rather than by signing a formal contract. Springer-Verlag secures the copyright for each volume.

The Production Process

The books are hardbound, and the publisher will select quality paper appropriate to the needs of the author(s). Publication time is about ten weeks. More than twenty years of experience guarantee authors the best possible service. To reach the goal of rapid publication at a low price the technique of photographic reproduction from a camera-ready manuscript was chosen. This process shifts the main responsibility for the technical quality considerably from the publisher to the authors. We therefore urge all authors and editors of proceedings to observe very carefully the essentials for the preparation of camera-ready manuscripts, which we will supply on request. This applies especially to the quality of figures and halftones submitted for publication. In addition, it might be useful to look at some of the volumes already published. As a special service, we offer free of charge LATEX and TEX macro packages to format the text according to Springer-Verlag's quality requirements. We strongly recommend that you make use of this offer, since the result will be a book of considerably improved technical quality. To avoid mistakes and time-consuming correspondence during the production period the conference editors should request special instructions from the publisher well before the beginning of the conference. Manuscripts not meeting the technical standard of the series will have to be returned for improvement.

For further information please contact Springer-Verlag, Physics Editorial Department V, Tiergartenstrasse 17, D-69121 Heidelberg, FRG

L. Mathelitsch W. Plessas (Eds.)

Substructures of Matter as Revealed with Electroweak Probes

Proceedings of the
32. Internationale Universitätswochen
für Kern- und Teilchenphysik
Schladming, Austria, 24 February - 5 March 1993

Springer-Verlag
Berlin Heidelberg New York
London Paris Tokyo
Hong Kong Barcelona
Budapest

Editors

Leopold Mathelitsch
Willibald Plessas
Institut für Theoretische Physik, Universität Graz
Universitätsplatz 5, A-8010 Graz, Austria

ISBN 3-540-57575-8 Springer-Verlag Berlin Heidelberg New York
ISBN 0-387-57575-8 Springer-Verlag New York Berlin Heidelberg

Printed in Germany

Typesetting: Camera ready by author/editor using the TEX/LATEX macro package from Springer-Verlag

58/3140-543210 - Printed on acid-free paper

Preface

This volume presents the written versions of the lectures delivered at the 1993 Schladming Winter School. This year the School was devoted to the theme **"Substructures of Matter as Revealed with Electroweak Probes"**. The aim was to review the evidences that have been gained about the various layers of matter using electroweak probes from low to high (i.e. the highest presently available) energy/momentum transfers. A particular intention was to comprehensively discuss the properties of the constituents that make up matter at each level and also their interactions.

The field of probing matter with electroweak particles has a long and bright tradition. As early as in the first part of this century the structure of atoms was explored by electron scattering. Similar techniques, though at increased energies, were subsequently used to investigate the structures of nuclei, and they finally led to the detection of the partons inside the nucleons and the other hadrons. It is not exaggerating to state that it was electroweak probes that at every stage provided the first insight into new (sub)structures, each time revealing more fundamental constituents. Also in the future, lepton experiments will possibly have the potential to show whether the particles that nowadays appear as point-like actually have structure or are themselves composite.

It is well known that, by accelerating electroweak projectiles to higher and higher energies, smaller and smaller regions of space can be resolved. Consequently these particles, when directed onto material targets, can be used just as a microscope to look at small objects. In recent years, this has led to the construction of larger and larger accelerators, also requiring bigger and more expensive installations. In this respect the race still goes on, and one may wonder which accelerator centre will come up with the discovery of so-called new physics.

However, our present understanding of the fundamental building blocks of matter and their basic forces, which are essentially 6 quarks and 6 leptons interacting via forces described by the standard model, has still to be further tested in more conventional (non-asymptotic) regimes. Specifically, the basic gauge theory of the strong-interaction sector, quantum chromodynamics (QCD), has to be carried through down to the domain of nuclei and nuclear processes, i.e. low- and intermediate-energy hadronic physics. It is certainly a challenging aim to explain the whole subatomic world starting from the fundamentals of QCD. Of course, achieving this aim has already turned out to be extremely demanding, and no real breakthrough has yet been made. In this situation it seems highly desirable to collect further, more detailed and more precise, insight into the properties of objects that have long been known, like nuclei and nucleons. Indeed, many new experimental and theoretical efforts are underway to investigate low- and intermediate-energy electroweak reactions. We mention, as an example, the many contemporary attempts to explore the precise structure and the specific electromagnetic properties of the neutron at accelerators such as MAMI, ELSA, and CEBAF.

In view of these arguments we tried to make the School interdisciplinary in the sense that we tried to bring together scientists dealing with similar problems at different energies. We believe that fostering such contacts is extremely instructive for all those who are usually restricted to a limited regime of their own research. In fact, we are quite confident that to solve the remaining tough problems of lepton-hadron physics, a broad understanding of all its facets in the different regimes of energy and momentum transfers will be a prerequisite.

The School started off with a review of electromagnetic interactions in the nuclear domain by H. Arenhövel, which also served as an introduction to the general (theoretical) treatment of electromagnetic processes. The extension to the regime of nucleon excitation and meson production was presented by R.M. Davidson, while the present experimental situation at low and intermediate energies (up to the GeV region) was summarized by B.H. Schoch. These three lectures, which may together be regarded as an up-to-date review of the more or less conventional field of electromagnetic processes in terms of hadrons, are placed first in these proceedings.

An introduction to the subhadronic world was given by W.J. Stirling. He presented an overview of the current evidence on the properties of partons and discussed the fundamentals of QCD. U.-G. Meißner explained how non-perturbative QCD can be formulated and how far it can actually be carried out. The very detailed knowledge of the behaviour of the constituents of the proton, gained recently from polarized structure functions, was treated by E. Reya. Small-x physics, a field that has likewise attracted big interest recently, was summarized by J. Kwieciński. A comprehensive overview of neutrino and muon scattering was given by A. Milsztajn; his lectures, co-authored by M. Virchaux, are particularly timely, since the corresponding experimental situation has now emerged into a well-balanced shape.

An introduction to electroweak theory was presented by G. Altarelli, giving also an insight into the current status of precision tests and a model-independent analysis of the data. Finally, G. Wolf described the HERA collider and reported on first experimental results measured both at ZEUS and H1.

At the School R.E. Taylor (Stanford University) delivered a distinguished lecture on "Deep Inelastic Scattering and the Early Observation of Partons", mainly dealing with the SLAC experiments that led to the observation of quarks. As the contents of this lecture are largely covered by the Nobel articles published in Reviews of Modern Physics **63** (1991) 573-629, no written account is included here.

Many excellent seminars were also presented at the School, but due to space limitations could not be included in these proceedings. However, a list of the seminar speakers and the topics addressed is given at the end of this volume, so that interested readers may request information or pertinent material themselves.

Finally, at this place, we would like to express our sincere thanks to the lecturers for investing such a lot of effort in their contributions, to the sponsors of the School, above all the Austrian Ministry for Science and Research and the Styrian Government, for providing generous support, to our colleagues in the organizing committee for cooperative assistance, and to Mrs. E. Neuhold for processing the files prepared by the authors in TEX or LATEX into the final form.

Graz, October 1993

W. Plessas (Director of the School)
L. Mathelitsch (Scientific Secretary)

Contents

Small x Physics

By J. Kwieciński

Nucleon Structure Functions

By A. Milsztajn and M.Virchaux

The Standard Electroweak Theory and Its Experimental Tests

By G. Altarelli

First Results from HERA

By G. Wolf

Contributors

Altarelli, G.
Theory Division, CERN
CH-1211 Genève 23, Switzerland

Arenhövel, H.
Institut für Kernphysik, Johannes Gutenberg-Universität
J.-J.-Becher-Weg 45, D-55099 Mainz, Germany

Davidson, R.M.
Institut für Kernphysik, Johannes Gutenberg-Universität
J.-J.-Becher-Weg 45, D-55099 Mainz, Germany

Kwieciński, J.
Department of Theoretical Physics, H. Niewodniczański Institute of Nuclear Physics
ul. Radzikowskiego 152, PL-31-342 Kraków, Poland

Meißner, U.-G.
Institut für Theoretische Physik, Universität Bern
Sidlerstraße 5, CH-3012 Bern, Switzerland

Milsztajn, A.
DAPNIA/SPP, CEN Saclay
F-91191 Gif-sur-Yvette, France

Reya, E.
Institut für Physik, Universität Dortmund
D-44221 Dortmund, Germany

Schoch, B.H.
Physikalisches Institut, Friedrich-Wilhelms-Universität
D-53090 Bonn, Germany

Stirling, W.J.

Department of Mathematical Sciences and Physics
University of Durham, Durham DH1 3LE, United Kingdom

Virchaux, M.

DAPNIA/SPP, CEN Saclay
F-91191 Gif-sur-Yvette, France

Wolf, G.

DESY
Notkestraße 52, D-22603 Hamburg, Germany

Electromagnetic Interactions of Nucleons and Nuclei at Low Energy and Momentum Transfer *

Hartmuth Arenhövel

Institut für Kernphysik, Johannes Gutenberg-Universität, J.-J.-Becher-Weg 45, D-55099 Mainz, Germany

1 Introduction

Looking back at the historical development of physics it is fascinating to see the extraordinary rôle the electromagnetic interaction has played. It was the first theory which incorporated special relativity though not immediately recognized and it was also the first and simplest example of a gauge theory which is now a basic feature of fundamental interactions. Its mathematical structure as displayed in the famous Maxwell equations has such a high aesthetical value that it led hundred years ago Ludwig Boltzmann to use as motto for his lectures on Maxwell theory "War es ein Gott, der diese Zeichen schrieb...?" [Bol 93], a liberal quote from Goethe's "Faust". All our visual recognition of the world around us is based on the electromagnetic interaction. And it was also the key for revealing the microstructure of matter as for example through the characteristic spectra of atoms which led combined with the problem of black body radiation - a merge of thermodynamics with Maxwell theory - to the recognition of the quantum structure of matter.

Thus it is not surprising that also in nuclear physics since its beginning the electromagnetic probe has always been at the forefront of nucleon and nuclear structure investigations. In order to underline this fact, let me just mention a few important milestones of the history of photonuclear reactions and electron nucleus scattering.

1909: The discovery of the nucleus in α-particle scattering by a gold foil by Geiger and Marsden [GeM 09]. The experimental results were in agreement with the cross section of a point scattering center (Rutherford cross section) except for very small impact parameters demonstrating that the interior of an atom contains a charge which is concentrated in an almost pointlike region in the center of the atom, called the nucleus. From the deviations of the experimental data from the Rutherford cross section for very small impact parameters corresponding to large scattering angles, Rutherford estimated the size of the atomic nucleus to be of the order of a few 10^{-15}m [Rut 11]. Thus this experiment pioneered future electron scattering experiments. It furthermore marks also the beginning of nuclear physics.

1934: The photodisintegration of the deuteron by Chadwick and Goldhaber which may be considered the first photonuclear experiment [ChG 34]. These authors also emphasized the special role of the deuteron being "as important in nuclear physics as

* Supported by the Deutsche Forschungsgemeinschaft (SFB 201)

the hydrogen atom is in atomic physics". In fact, in the course of the development of nuclear physics this fundamental process has received continued interest until today [ArS 91].

1947: The discovery of the giant dipole resonance in nuclear photoabsorption by Baldwin and Klaiber [BaK 47] whose integrated strength almost exhausts the classical dipole sum rule. It is a very important collective mode of motion common to all nuclei and has found generalizations to other types of collectivity.

1951: The first electron nucleus scattering experiment by Lyman, Hanson and Scott [LyH 51] opening up a new era of nuclear structure investigations by elastic and inelastic electron nucleus scattering which has supplied us with a wealth of information on static and dynamic properties of nuclei [DFW 66,CdA 80].

1955: First indication of a finite proton size in an electron proton scattering experiment by Hofstadter and McAllister [HoM 55]. From the deviations of the Rosenbluth cross section for a point Dirac particle these authors estimated the proton size to be of the order $(0.7 \pm 0.2) \cdot 10^{-15}$ m.

1967: Begin of deep inelastic scattering experiments at Stanford by Friedman, Kendall and Taylor which led to the discovery of the parton substructure of nucleons, i.e., pointlike scattering centers [Fri 91,Ken 91,Tay 91]. It gave the experimental basis to the quark model of hadrons and led finally to modern quantum chromodynamics as a non-abelian gauge theory of strong interactions.

1972: The absolute measurement of total nuclear photoabsorption cross sections by Ziegler and collaborators [AhB 72,AhB 75], which showed for the first time experimentally the strong enhancement of the - up to the π-threshold - integrated total absorption cross section over the classical Thomas-Reiche-Kuhn (TRK) sum rule which has been attributed to meson exchange current effects.

1973: The recognition of strong meson exchange current effects in an isovector $M1$-transition in $d(e,e')pn$ near threshold at high momentum transfer [HoR 73].

The special role of the electromagnetic interaction for unraveling the microstructure of the world is due to the fact that its properties are well known from classical physics and that it is weak enough to allow lowest order perturbation treatment resulting in simple and unique interpretations of experimental results. In the course of the history of our understanding of nuclear structure one may distinguish different stages of development.

The first stage is what is now commonly called the classical or conventional framework. Its basic assumption is that the nucleus is built from pointlike, i.e., structureless particles, protons and neutrons, commonly named nucleons. These nucleons are weakly bound so that one may assume a non-relativistic description. The interaction between the nucleons is taken in the simplest form of a two-body interaction, namely a phenomenological NN potential, which has to be determined from nucleon-nucleon scattering. The main theoretical task then is to solve the many-body problem.

As we know, this classical framework has been very successful in understanding many phenomena in nuclear structure and nuclear reactions. However, it was clear from the beginning that a few important questions had been left unanswered. One of them is concerned with the nature of the strong interaction. What is the true physical mechanism of the strong force? Another is related to the role of internal nucleon structure as first of all manifest in the static properties of the nucleons, the anomalous magnetic moments of proton and neutron and more general in their electromagnetic form factors and, furthermore, in the appearance of nucleon resonances, called isobars.

With respect to the electromagnetic properties of nuclei a new problem showed up related to the presence of an exchange part in the NN interaction, which was a necessary, though phenomenological ingredient in order to obtain saturation of nuclear binding. It has the effect of exchanging a proton with a neutron. This process is obviously connected with a flow of electric charge (see Fig. 1.1), in other words, the exchange interaction induces an additional current, the so-called exchange current of two-body character. Given the exchange part of the potential, the question is how to construct the corresponding two-body current operator. It is clear that one has to go beyond a phenomenological description towards a dynamical model for the underlying degrees of freedom.

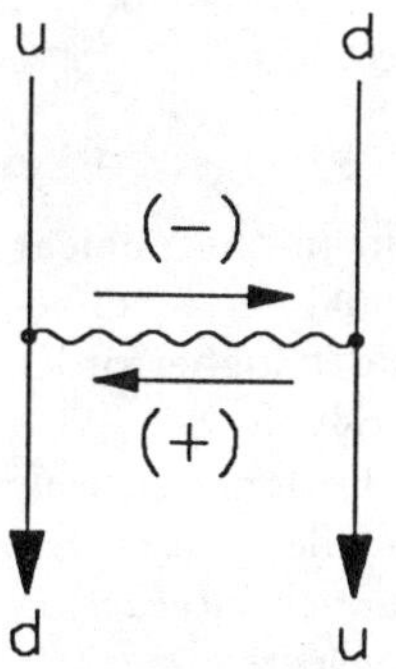

Fig. 1.1 Exchange interaction with indication of charge flow.

A first important answer was given in 1935 by Yukawa [Yuk 35]. Motivated by the fact that the electromagnetic interaction between two charged particles is mediated by the exchange of a virtual photon, he hypothesized a particle with finite mass - called meson - as mediator of the strong force. In contrast to the photon, the meson should have a finite mass since the strong force is short ranged according to experimental evidence. It needs also to have a charge in order to describe a charge exchange process. In fact, twelve years later the pion - the meson of lowest mass - had been discovered. Since then a whole family of mesons have been found acting as mediators of the strong force.

In principle, the appearance of mesons as mediators of the interaction has drastically changed our picture of the nucleus. Because now the nucleus is populated also by exchanged mesons which, however, are virtual since they are far from the mass shell. Due to their short range according to the time-energy uncertainty relation and because of the low nuclear density one expects, however, a small probability for these subnuclear components where mesons are present. Therefore, in order to keep the original simple picture of the nucleus containing nuleons only, one can formally eliminate these meson degrees of freedom by introducing effective operators which describe their action. This leads to models of the NN interaction in terms of an effective "One-Boson-Exchange-Potential" (OBEP) as diagrammatically depicted in Fig. 1.2. They have become quite successful although also rather involved.

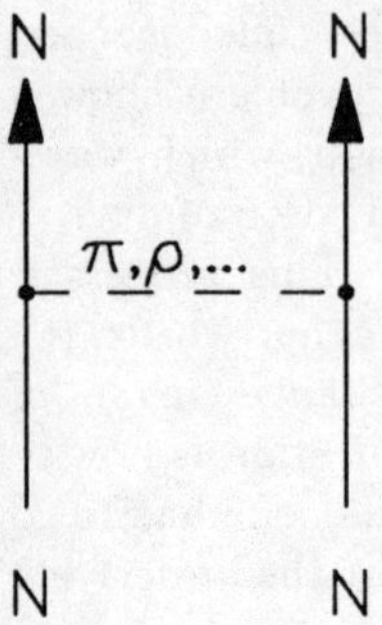

Fig. 1.2 One-boson-exchange interaction.

The prize one pays for keeping only nucleon degrees of freedom in the nuclear wave function is the appearance of energy dependence and non-locality of the effective operators. Furthermore, in the NN interaction one has to consider higher order contributions from many-boson exchange and in addition many-body forces. Also for the electromagnetic interaction one finds, as mentioned above, additional current contributions of two- and in principle many-body character, the so-called meson exchange currents (MEC). These are intimately connected to the exchange interaction of the nuclear force via current conservation [Sie 37] and describe the electromagnetic interaction during the exchange of mesons.

Another modification of the traditional picture arises from the above mentioned fact that the nucleons themselves possess internal dynamics as is manifest in their spatial extension and in a rich internal excitation spectrum as seen, e.g., in photoabsorption or pion reactions (see Fig. 1.3). Thus the nucleus contains now besides

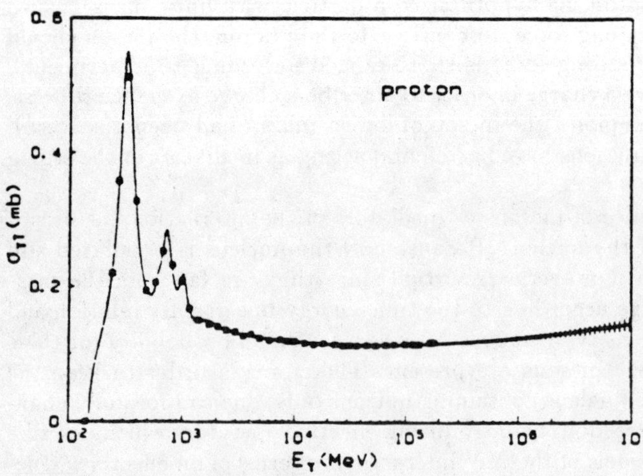

Fig. 1.3 Total photoabsorption cross section of the proton (from [Ahr 85]).

protons, neutrons and mesons also virtually excited nucleons or isobars (Δ, N^*,...). Also in this case, one expects a small probability for those components where isobars are present and thus these internal degrees of freedom can be eliminated again by introducing effective operators. This then leads to additional contributions to the NN interaction and to the electromagnetic current as shown in Fig. 1.4.

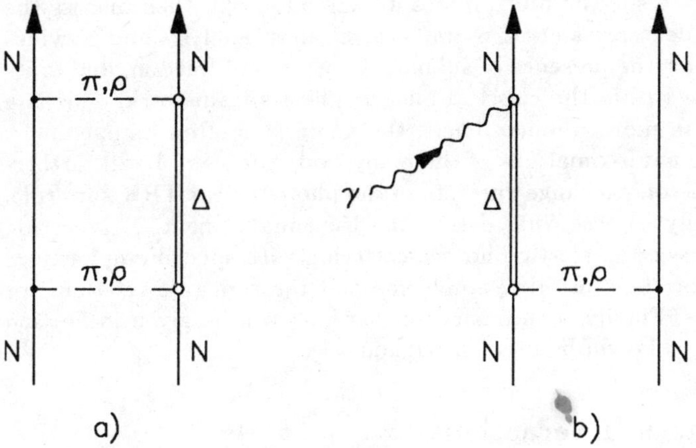

Fig. 1.4 Isobar contribution to the NN interaction (a), and to the electromagnetic current (b).

But even this already quite complicated picture of the nucleus again had to be modified by the advent of quantum chromodynamics (QCD) where quarks and gluons constitute the fundamental degrees of freedom for the structure and interaction of hadrons. For example, in the simple constituent quark model the nucleons are built from three quarks in a colour singlet coupled to spin and isospin 1/2 while the Δ is then obtained by a spin-isospin flip to $S = T = 3/2$. Correspondingly, the mesons appear as bound quark-antiquark systems. Consequently, the NN interaction and the exchange currents have to be constructed as effective operators from these new degrees of freedom which, however, should go over into the traditional meson exchange picture for sufficiently large internucleon distances.

Thus, one of the interesting questions we face at present is the question of possible evidence of quark-gluon dynamics in nuclei. In other words: Where will the conventional effective theory fail? However, since any evidence will be indirect, i.e., will appear as deviations of experimental results from the traditional theoretical description, one has to assure oneself that the deviations are not caused by some approximations in the conventional framework.

In these lectures I will concentrate on the manifestation of subnuclear degrees of freedom in terms of meson and isobar degrees of freedom in electromagnetic processes where their presence usually is described in terms of so-called exchange or interaction currents. In Section 2 I will first discuss the general properties of the electromagnetic interaction, the gauge conditions and low-energy theorems which follow from gauge invariance, the charge and current density operators for a non-relativistic system of nucleons and the Siegert theorem. In Section 3 I will sketch the basic ideas and construction methods for the exchange current operators as effective operators and in Section 4 the model of nuclear isobar configurations introducing explicitly isobar degrees of freedom into the nuclear wave function.

The general features of one- and two-photon processes will be discussed in Section 5. First the expressions for the cross sections of photoabsorption and electron scattering will be reviewed. As a specific but important example, I will then discuss the two-body break-up of the deuteron since it permits the cleanest analysis and provides one of the best evidences for the presence of subnuclear degrees of freedom due to its simple two-body structure within the classical nuclear physics framework. This is a unique situation because in more complex nuclei the analysis is often hampered by presently still unavoidable approximations of the many-body problem. I will furthermore discuss the role of meson exchange currents in the photonuclear TRK sum rule, in particular, I will carefully analyse what determines the enhancement. This section will end with a brief discussion of elastic photon scattering with special emphasis on the low-energy theorem for the scattering amplitude and the sum rule relations for the low-energy parameters. Finally, a summary and outlook will be given in Section 6. Natural units ($c = 1$, $\hbar = 1$) will be used throughout.

2 The Electromagnetic Interaction

I will begin with a brief review of the Maxwell equations for the electromagnetic field. The dynamical variables are given by the electric field $\mathbf{E}(\mathbf{x}, t)$ and the magnetic induction $\mathbf{B}(\mathbf{x}, t)$. They obey the field equations which fix their divergence and curl

$$\boldsymbol{\nabla} \cdot \mathbf{B} = 0\,, \qquad \boldsymbol{\nabla} \cdot \mathbf{E} = 4\pi\rho \tag{2.1}$$

$$\boldsymbol{\nabla} \times \mathbf{E} + \partial_t \mathbf{B} = 0\,, \qquad \boldsymbol{\nabla} \times \mathbf{B} - \partial_t \mathbf{E} = 4\pi \mathbf{j}\,. \tag{2.2}$$

Here I have used the notation $\partial_t := \partial/\partial t$. The two equations on the left side are homogeneous while the other two on the right side are inhomogeneous because of their coupling to matter as described by the external charge and current densities ρ and $\mathbf{j}$, respectively. Taking the divergence of the second inhomogeneous equation one finds current conservation

$$\boldsymbol{\nabla} \cdot \mathbf{j} + \partial_t \rho = 0\,. \tag{2.3}$$

By introducing scalar and vector potentials $\phi(\mathbf{x}, t)$ and $\mathbf{A}(\mathbf{x}, t)$, respectively, by

$$\mathbf{B} = \boldsymbol{\nabla} \times \mathbf{A}\,, \qquad \mathbf{E} = -\boldsymbol{\nabla}\phi - \partial_t \mathbf{A}\,, \tag{2.4}$$

the two homogeneous Maxwell equations are automatically fulfilled. The potentials are determined from the following set of coupled equations which follow from the inhomogeneous Maxwell equations by inserting the expressions of (2.4)

$$\Delta\phi - \partial_t^2 \phi = -4\pi\rho - \partial_t(\boldsymbol{\nabla} \cdot \mathbf{A} + \partial_t \phi) \tag{2.5}$$

$$\Delta\mathbf{A} - \partial_t^2 \mathbf{A} = -4\pi \mathbf{j} + \boldsymbol{\nabla}(\boldsymbol{\nabla} \cdot \mathbf{A} + \partial_t \phi)\,. \tag{2.6}$$

Thus, instead of the field strengths $\mathbf{E}$ and $\mathbf{B}$, one can describe the electromagnetic field in terms of the potentials ϕ and $\mathbf{A}$ as dynamical variables. However, one has to keep in mind that the potentials are not completely fixed by (2.4). One may change them by a so-called gauge transformation

$$\mathbf{A} \rightarrow \mathbf{A} + \boldsymbol{\nabla}\lambda\,, \tag{2.7}$$

$$\phi \rightarrow \phi - \partial_t \lambda\,, \tag{2.8}$$

where λ is an arbitrary scalar function, without changing the field strengths **E** and **B**. This gauge freedom can be used to simplify (2.5, 2.6). It has furthermore the important consequence that any physical observable which is expressed in terms of the electromagnetic potentials should remain invariant under such a gauge transformation ("gauge invariance").

Let me now introduce the covariant notation for convenience. The 4-vectors are denoted by

$$x^\mu = (t, \mathbf{x})\,, \qquad x_\mu = g_{\mu\nu} x^\nu\,, \tag{2.9}$$

where I use the metric

$$g_{\mu\nu} = \begin{pmatrix} 1 & 0 & 0 & 0 \\ 0 & -1 & 0 & 0 \\ 0 & 0 & -1 & 0 \\ 0 & 0 & 0 & -1 \end{pmatrix}\,. \tag{2.10}$$

For convenience, I will use the notation $\partial_\mu := \partial / \partial x^\mu$.

Furthermore, the 4-potential is then given by $A^\mu = (\phi, \mathbf{A})$ and the field strengths are the components of the field tensor $F_{\mu\nu}$ which is defined by

$$\begin{aligned} F_{\mu\nu} &= \partial_\mu A_\nu - \partial_\nu A_\mu \\ &= \begin{pmatrix} 0 & E_1 & E_2 & E_3 \\ -E_1 & 0 & -B_3 & B_2 \\ -E_2 & B_3 & 0 & -B_1 \\ -E_3 & -B_2 & B_1 & 0 \end{pmatrix}\,. \end{aligned} \tag{2.11}$$

This field tensor is completely antisymmetric, i.e., $F_{\mu\nu} = -F_{\nu\mu}$. Then one obtains the covariant form of the Maxwell equations in terms of the field tensor

$$\varepsilon_{\rho\mu\nu\sigma} \partial^\rho F^{\mu\nu} = 0\,, \tag{2.12}$$

$$\partial_\mu F^{\mu\nu} = 4\pi j^\nu\,, \tag{2.13}$$

where the 4-current is given by $j^\mu = (\rho, \mathbf{j})$ and $\varepsilon_{\rho\mu\nu\sigma}$ denotes the totally antisymmetric unit tensor. The set of equations for the potentials (2.5, 2.6) becomes

$$\partial_\nu \partial^\nu A^\mu = 4\pi j^\mu + \partial^\mu (\partial_\nu A^\nu)\,, \tag{2.14}$$

and the general gauge transformation is given by

$$A_\mu \rightarrow A_\mu - \partial_\mu \lambda\,. \tag{2.15}$$

Obviously, the field tensor is gauge invariant. It is the coupling of the electromagnetic field to the current which provides the means to explore the structure of hadrons with the electromagnetic probe.

Now I will discuss the general properties of the electromagnetic interaction operators which follow from the requirement of invariance under gauge transformations.

As already mentioned, the electromagnetic probe is weak and hence we can expect that a given hadronic system responds weakly to a weak electromagnetic field. Thus it is reasonable to assume that the Hamiltonian of the system has a Taylor expansion with respect to the electromagnetic potential $A_\mu(\mathbf{x}, t)$ which reads up to second order

$$H(A) = H_0 + H_{e.m.}(A)\,, \tag{2.16}$$

with

$$\begin{aligned} H_{e.m.}(A) = &\int d^3x j_\mu(\mathbf{x}, t) A^\mu(\mathbf{x}, t) \\ &+ \frac{1}{2}\int d^3x d^3y A^\mu(\mathbf{x}, t) B_{\mu\nu}(\mathbf{x}, \mathbf{y}, t) A^\nu(\mathbf{y}, t)\,. \end{aligned} \tag{2.17}$$

This expansion is the basis for a perturbative treatment of the e.m. interaction. Its specific form is determined by the requirement of Lorentz covariance.

The expansion coefficients in (2.17) define the current density $j_\mu(\mathbf{x}, t)$ and the two-photon operator $B_{\mu\nu}(\mathbf{x}, \mathbf{y}, t)$. They are obtained from the full Hamiltonian $H(A)$ by taking the first and second functional derivatives with respect to A at $A \equiv 0$. In detail, one has for the current operator

$$j_\mu(\mathbf{x}, t) = \left.\frac{\delta H(A)}{\delta A^\mu(\mathbf{x}, t)}\right|_{A\equiv 0}\,, \tag{2.18}$$

and for the two-photon operator

$$B_{\mu\nu}(\mathbf{x}, \mathbf{y}, t) = \left.\frac{\delta^2 H(A)}{\delta A^\mu(\mathbf{x}, t)\delta A^\nu(\mathbf{y}, t)}\right|_{A\equiv 0}\,. \tag{2.19}$$

It is obvious from (2.19) that the two-photon operator is symmetric, i.e.,

$$B_{\mu\nu}(\mathbf{x}, \mathbf{y}, t) = B_{\nu\mu}(\mathbf{y}, \mathbf{x}, t)\,. \tag{2.20}$$

The corresponding interaction diagrams for the current and the two-photon operators are shown in Fig. 2.1.

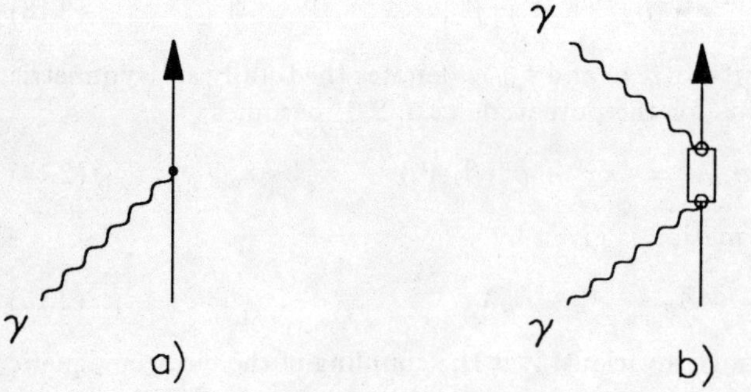

Fig. 2.1 Diagrams for current operator (a) and two-photon operator (b).

2.1 Gauge Conditions for the Electromagnetic Operators

Let me now exploit the consequences of gauge invariance. The transformation of the electromagnetic potential

$$A_\mu \to A'_\mu = A_\mu - \partial_\mu \lambda\,, \tag{2.21}$$

induces a U(1)-gauge transformation of the hadronic field

$$\psi \to \psi' = e^{i\chi(\lambda)}\psi\,, \tag{2.22}$$

where $\lambda(\mathbf{x},t)$ is an arbitrary scalar field and $\chi(\lambda)$ is a local, hermitean phase operator which acts on the hadronic field ψ. It is a linear functional in λ, i.e.,

$$\chi(\lambda) = \int d^3x\,\Omega(\mathbf{x},t)\lambda(\mathbf{x},t)\,. \tag{2.23}$$

The local property of the phase operator is contained in the dependence of the operator $\Omega(\mathbf{x},t)$ on the dynamical variables of the system. Its specific form has still to be determined.

While the original field ψ obeys the Schrödinger equation

$$H(A)\psi = i\partial_t\psi\,, \tag{2.24}$$

the transformed field ψ' fulfills the same equation with the transformed Hamilton operator

$$H' = H(A') = H(A-\partial\lambda)\,. \tag{2.25}$$

Inserting H' from (2.25) and ψ' from (2.22) into the Schrödinger equation one finds

$$\begin{aligned} H(A-\partial\lambda)e^{i\chi}\psi &= e^{i\chi}(ie^{-i\chi}\partial_t e^{i\chi} + i\partial_t)\psi \\ &= e^{i\chi}(ie^{-i\chi}\partial_t e^{i\chi} + H(A))\psi\,. \end{aligned} \tag{2.26}$$

Comparison with the original Schrödinger equation for ψ yields then as condition for invariance

$$e^{-i\chi}(H(A-\partial\lambda) - i\partial_t)e^{i\chi} = H(A)\,. \tag{2.27}$$

It is sufficient to consider infinitesimal transformations only which amounts to keeping terms only linear in the gauge function λ. Using (2.17), one then finds the following condition

$$\begin{aligned} i[H(A),\chi] + \partial_t\chi - \int d^3x\, j_\mu(\mathbf{x},t)\partial^\mu\lambda(\mathbf{x},t) = \\ \int d^3x \int d^3y\,\partial^\mu\lambda(\mathbf{x},t)B_{\mu\nu}(\mathbf{x},\mathbf{y},t)A^\nu(\mathbf{y},t)\,. \end{aligned} \tag{2.28}$$

Now I will first exploit this condition for the case $A\equiv 0$ which gives

$$\begin{aligned} i[H_0,\chi] + \partial_t\chi &= \int d^3x(j_0\partial_t\lambda + \mathbf{j}\cdot\boldsymbol{\nabla}\lambda) \\ &= \int d^3x(j_0\partial_t\lambda - \boldsymbol{\nabla}\cdot\mathbf{j}\lambda)\,, \end{aligned} \tag{2.29}$$

where the second line is obtained by partial integration. Because χ is a linear functional in λ one obtains

$$\int d^3x\,(i[H_0,\Omega(\mathbf{x},t)]+\partial_t\chi+\boldsymbol{\nabla}\cdot\mathbf{j})\,\lambda(\mathbf{x},t)=\int d^3x\,(\Omega(\mathbf{x},t)-j_0(\mathbf{x},t))\,\partial_t\lambda(\mathbf{x},t)\,. \tag{2.30}$$

Since $\lambda(\mathbf{x},t)$ may be chosen arbitrarily, one can consider $\lambda(\mathbf{x},t)$ and $\partial_t\lambda(\mathbf{x},t)$ as independent. This then gives two equations, namely first by choosing $\partial_t\lambda\equiv 0$

$$\boldsymbol{\nabla}\cdot\mathbf{j}\,(\mathbf{x},t)+i[H_0,\Omega(\mathbf{x},t)]+\partial_t\Omega(\mathbf{x},t)=0\,, \tag{2.31}$$

and then for arbitrary $\partial_t\lambda$

$$\Omega(\mathbf{x},t)=j_0(\mathbf{x},t)\,, \tag{2.32}$$

which determines the unknown operator. Combining these two equations, one finds

$$\boldsymbol{\nabla}\cdot\mathbf{j}\,(\mathbf{x},t)+i[H_0,\rho(\mathbf{x},t)]+\partial_t\rho(\mathbf{x},t)=0\,, \tag{2.33}$$

which is the well-known continuity equation of the electromagnetic current. It implies conservation of the total charge of the system, i.e.,

$$\frac{d}{dt}Q=i[H_0,Q]+\partial_tQ=0\,, \tag{2.34}$$

where

$$Q(t)=\int d^3x\,j_0(\mathbf{x},t) \tag{2.35}$$

is the electric charge operator.

In order to derive the gauge conditions for the two-photon operator, I will consider those terms of the general gauge condition (2.28) which are linear in A. One finds

$$i\left[\int d^3y\,j_\mu(\mathbf{y},t)A^\mu(\mathbf{y},t),\int d^3x\,\rho(\mathbf{x},t)\lambda(\mathbf{x},t)\right]=\int d^3x\int d^3y\,\partial_x^\nu\lambda(\mathbf{x},t)B_{\nu\mu}(\mathbf{x},\mathbf{y},t)A^\mu(\mathbf{y},t)\,, \tag{2.36}$$

which should again hold for any gauge function λ and any field A. Taking the functional derivative with respect to A, one finds first

$$\begin{aligned} i[j_\mu(\mathbf{y},t),\int d^3x\,\rho(\mathbf{x},t)\lambda(\mathbf{x},t)] &= \int d^3x\,\partial_x^\nu\lambda(\mathbf{x},t)B_{\nu\mu}(\mathbf{x},\mathbf{y},t)\\ &= \int d^3x\,\left(\partial_t\lambda(\mathbf{x},t)B_{0\mu}(\mathbf{x},\mathbf{y},t)-\nabla_{x_j}B_{j\mu}(\mathbf{x},\mathbf{y},t)\lambda(\mathbf{x},t)\right)\,, \end{aligned} \tag{2.37}$$

where again in the last step a partial integration has been done. Finally, the independent variation of λ and $\partial_t\lambda$ leads to the following two conditions

$$B_{\nu 0}(\mathbf{x},\mathbf{y},t)=B_{0\nu}(\mathbf{y},\mathbf{x},t)\equiv 0\,, \tag{2.38}$$

$$\partial_{x_k}B_{kl}(\mathbf{x},\mathbf{y},t)=i[\rho(\mathbf{x},t),j_l(\mathbf{y},t)]\,, \tag{2.39}$$

where k and l run over spatial indices. The last equation constitutes the gauge condition for the two-photon operator setting a constraint with respect to the charge and current operators. Note that the lhs of (2.39) is called a Schwinger term in current algebra [Sch 59]. Furthermore, in a relativistic theory (2.38) does not hold in general and instead of (2.39) one has [Fri 77]

$$\partial_x^\mu B_{\mu\nu}(x,y)=i[\rho(x),j_\nu(y)]\,. \tag{2.40}$$

2.2 Low-Energy Theorems

An important consequence of gauge invariance are the so-called low-energy theorems, which allow to derive very simple relations for the current and two-photon operators in the limit that the photon momenta go to zero. Such theorems have been derived by Sachs and Austern [SaA 51] and later by Low [Low 54] and Gell-Mann and Goldberger [GeG54] on more general grounds for the photon scattering amplitude. Here, I will consider the low-energy theorems for the electromagnetic operators.

To this end I will first introduce the Fourier transforms of the charge and current densities and the two-photon operator, which I assume to be time independent for convenience, by

$$\tilde{\rho}(\mathbf{q}) = \int d^3x \rho(\mathbf{x}) e^{i\mathbf{q}\cdot\mathbf{x}}, \tag{2.41}$$

$$\mathbf{J}(\mathbf{q}) = \int d^3x \mathbf{j}(\mathbf{x}) e^{i\mathbf{q}\cdot\mathbf{x}}, \tag{2.42}$$

$$\tilde{B}_{kl}(\mathbf{q}', \mathbf{q}) = \int d^3x d^3y e^{i\mathbf{q}'\cdot\mathbf{x}} B_{kl}(\mathbf{x}, \mathbf{y}) e^{i\mathbf{q}\cdot\mathbf{y}}. \tag{2.43}$$

The gauge conditions (2.33) and (2.39) then become

$$\mathbf{q}\cdot\mathbf{J}(\mathbf{q}) = [H_0, \tilde{\rho}(\mathbf{q})], \tag{2.44}$$

$$q_k' \tilde{B}_{kl}(\mathbf{q}', \mathbf{q}) = -[\tilde{\rho}(\mathbf{q}'), J_l(\mathbf{q})]. \tag{2.45}$$

Assuming the existence of a Taylor expansion with respect to $\mathbf{q}$ one finds for (2.44) up to second order

$$\mathbf{q}\cdot(\mathbf{J}(0) + (\mathbf{q}\cdot\boldsymbol{\nabla}')\mathbf{J}(\mathbf{q}')|_{\mathbf{q}'\equiv 0}) = [H_0, i\mathbf{q}\cdot\mathbf{D} - \frac{1}{2} q_{l'} Q_{l'l} q_l], \tag{2.46}$$

where I have used for $\tilde{\rho}(\mathbf{q})$ the expansion up to second order

$$\tilde{\rho}(\mathbf{q}) = Z + i\mathbf{q}\cdot\mathbf{D} - \frac{1}{2} q_{l'} Q_{l'l} q_l \tag{2.47}$$

with the total charge Z, the dipole and quadrupole operators, respectively,

$$\mathbf{D} = \int d^3x \rho(\mathbf{x}) \mathbf{x}, \tag{2.48}$$

$$Q_{l'l} = \int d^3x \rho(\mathbf{x}) x_{l'} x_l. \tag{2.49}$$

Now, comparing the first order terms, one finds

$$\mathbf{J}(0) = i[H_0, \mathbf{D}], \tag{2.50}$$

which is called Siegert-Theorem. Its virtue lies in the fact that it allows to calculate the current in the low energy regime from the charge dipole operator, a quantity in general much better known in approximate theories. It can be generalized to electric multipoles of higher order as will be shown in Section 2.5.

For the second order terms one finds correspondingly

$$\nabla_{l'} J_l(0) + \nabla_l J_{l'}(0) = -[H_0, Q_{l'l}], \tag{2.51}$$

which determines only the symmetric part of $\nabla_l J_{l'}(0)$ with respect to the interchange of l with l'. The antisymmetric part, which is related to the magnetic moment contribution

$$\nabla_{l'} J_l(0) - \nabla_l J_{l'}(0) = 2i\varepsilon_{l'lk} M_k, \tag{2.52}$$

remains unrestricted. Collecting the various pieces, one finds for the current up to first order in $\mathbf{q}$

$$J_l(\mathbf{q}) = i[H_0, \mathbf{D}] - \frac{1}{2}[H_0, q_{l'} Q_{l'l}] - i(\mathbf{q} \times \mathbf{M})_l . \tag{2.53}$$

Correspondingly, one finds from the second gauge condition (2.45) for the two-photon operator by expanding both sides up to first order in $\mathbf{q}'$

$$\tilde{B}_{kl}(0, \mathbf{q}) = -i[D_k, J_l(\mathbf{q})] . \tag{2.54}$$

Then, setting first $\mathbf{q} = 0$ and using (2.49) one finds the low-energy limit of the two-photon operator

$$\tilde{B}_{kl}(0, 0) = [D_k, [H_0, D_l]] . \tag{2.55}$$

These limits are the basic ingredients for the low energy theorem of the photon scattering amplitude (see e.g. [Are 86,Are 92a]).

Also the first order terms are completely determined by the gauge condition. To this end one expands (2.54) up to first order in $\mathbf{q}$ and finds

$$\nabla_{l'} \tilde{B}_{kl}(0, \mathbf{q})\Big|_{\mathbf{q}\equiv 0} = i[D_k, \frac{1}{2}[H_0, Q_{l'l}] - i\varepsilon_{ll'n} M_n] . \tag{2.56}$$

The other linear term is obtained from crossing symmetry which follows from (2.20)

$$\tilde{B}_{lk}(\mathbf{q}', \mathbf{q}) = \tilde{B}_{kl}(\mathbf{q}, \mathbf{q}'), \tag{2.57}$$

and thus

$$\nabla'_{l'} \tilde{B}_{lk}(\mathbf{q}', 0)\Big|_{\mathbf{q}'\equiv 0} = \nabla'_{l'} \tilde{B}_{kl}(0, \mathbf{q}')\Big|_{\mathbf{q}'\equiv 0} . \tag{2.58}$$

Altogether one obtains up to first order in $\mathbf{q}$ and $\mathbf{q}'$

$$\begin{aligned} \tilde{B}_{lk}(\mathbf{q}', \mathbf{q}) = {} & [D_k, [H_0, D_l]] \\ & + i[D_k, \frac{1}{2} q_{l'} [H_0, Q_{l'l}] - i(\mathbf{q} \times \mathbf{M})_l] \\ & + i[D_l, \frac{1}{2} q'_{l'} [H_0, Q_{l'k}] - i(\mathbf{q}' \times \mathbf{M})_k] . \end{aligned} \tag{2.59}$$

Even most of the second order terms except for the asymmetric part of the mixed gradient term with respect to $\mathbf{q}$ and $\mathbf{q}'$ are determined by the gauge condition as has been pointed out by Friar [Fri 75a].

2.3 Electromagnetic Operators for a Non-Relativistic System of Nucleons

As next topic I will consider the explicit construction of the electromagnetic operators for a non-relativistic system of interacting nucleons. The Hamiltonian H_0 of (2.16) is assumed to have the simple form

$$\begin{aligned} H_0 &= \sum_i \frac{\mathbf{p}_i^{\,2}}{2M} + \frac{1}{2}\sum_{i,j} V_{ij} \\ &= T + V\,, \end{aligned} \tag{2.60}$$

and for the electromagnetic interaction up to second order in A as given in (2.17) one needs $H(A)$.

The method of minimal substitution allows to derive the Hamiltonian $H(A)$ in a gauge invariant way from the Hamiltonian H_0 in the absence of an electromagnetic field. It consists of the substitution

$$\mathbf{p}_j \to \mathbf{p}_j - e_j \mathbf{A}(\mathbf{x}_j, t)\,, \tag{2.61}$$

$$H_0 \to H_0 + \sum_j e_j A_0(\mathbf{x}_j, t)\,, \tag{2.62}$$

where e_j denotes the charge of the "j-th" particle. With the help of (2.18) and (2.19), one finds for the one-body charge and convection current densities

$$\rho_{[1]}(\mathbf{x}) = \sum_j e_j \delta(\mathbf{x} - \mathbf{r}_j)\,, \tag{2.63}$$

$$\mathbf{j}^c_{[1]}(\mathbf{x}) = \frac{1}{2M}\sum_j e_j \{\mathbf{p}_j, \delta(\mathbf{x} - \mathbf{r}_j)\}\,, \tag{2.64}$$

and for the two-photon operator

$$B_{[1]\,kl}(\mathbf{x}, \mathbf{y}) = \frac{\delta_{kl}}{M}\sum_j e_j^2 \delta(\mathbf{x} - \mathbf{r}_j)\delta(\mathbf{y} - \mathbf{r}_j)\,. \tag{2.65}$$

The suffix "[1]" indicates the one-body character of an operator. All these operators have no explicit time dependence.

Obviously, the spin current is absent in (2.64) because H_0 does not depend on the spin. However, it can also be generated for the charged particles by adding to the Hamiltonian H_0 a spin-dependent term which vanishes identically for $A \equiv 0$

$$\begin{aligned} H_0 &= H_0 + \frac{1}{2M}\sum_j \mu_j ((\boldsymbol{\sigma}_j \cdot \mathbf{p}_j)^2 - (\mathbf{p}_j)^2) \\ &= H_0 + \frac{i}{2M}\sum_j \mu_j \boldsymbol{\sigma}_j \cdot (\mathbf{p}_j \times \mathbf{p}_j)\,, \end{aligned} \tag{2.66}$$

where μ_j determines the magnetic moment $e_j\mu_j/2M$ of the "j-th" particle. Such a term, for example, appears for the kinetic energy in the non-relativistic reduction of the Dirac equation (with $\mu_j = 1$). Applying now the minimal substitution, one obtains the one-body spin current

$$\mathbf{j}^s_{[1]}(\mathbf{x}) = \frac{i}{2M} \sum_j e_j \mu_j \boldsymbol{\sigma}_j \times [\mathbf{p}_j, \delta(\mathbf{x} - \mathbf{r}_j)]$$
$$= \boldsymbol{\nabla}_x \times \sum_j \frac{e_j \mu_j}{2M} \boldsymbol{\sigma}_j \delta(\mathbf{x} - \mathbf{r}_j) . \tag{2.67}$$

This also illustrates nicely that the method of minimal substitution is not a unique procedure. But the difference of the currents is purely transverse, which means that its divergence vanishes identically as is the case for this spin current

$$\boldsymbol{\nabla}_x \cdot \mathbf{j}^s_{[1]}(\mathbf{x}) \equiv 0 . \tag{2.68}$$

I will show another interesting example of this non-uniqueness when I will discuss the phenomenological method of constructing meson exchange currents in Section 3.5. Furthermore, I would like to point out that this procedure works for charged particles only and not for neutral ones, like the neutron. In the latter case, one has to add by hand the corresponding spin current.

It is useful to separate the center-of-mass current from the intrinsic current since only the latter leads to internal excitations

$$\mathbf{j}_{[1]}(\mathbf{x}) = \mathbf{j}_{cm[1]}(\mathbf{x}) + \mathbf{j}_{int[1]}(\mathbf{x}) , \tag{2.69}$$

where

$$\mathbf{j}_{cm[1]}(\mathbf{x}) = \frac{1}{2AM} \{\mathbf{P}, \rho_{[1]}(\mathbf{x})\} , \tag{2.70}$$
$$\mathbf{j}_{int[1]}(\mathbf{x}) = \frac{1}{2M} \sum_j e_j \{\mathbf{p}'_j, \delta(\mathbf{x} - \mathbf{r}_j)\} + \mathbf{j}^s_{[1]}(\mathbf{x}) . \tag{2.71}$$

Here, $\mathbf{P}$ denotes the total momentum, A the mass number and $\mathbf{p}'_j$ the momentum of the "j-th" particle with respect to the c.m. system.

In view of the importance of gauge invariance, let me now consider the gauge conditions for these one-body operators. It is a simple exercise to show that the following relations hold

$$\boldsymbol{\nabla}_x \cdot \mathbf{j}^c_{[1]}(\mathbf{x}) + i[T, \rho_{[1]}(\mathbf{x})] = 0 , \tag{2.72}$$
$$\partial_{x_k} B_{[1]\,kl}(\mathbf{x}, \mathbf{y}) = i[\rho_{[1]}(\mathbf{x}), j^c_{[1]\,l}(\mathbf{y})] , \tag{2.73}$$

where T is defined in (2.60). With respect to the separation of c.m. and intrinsic current one has separate current conservation relations

$$\boldsymbol{\nabla}_x \cdot \mathbf{j}^c_{cm[1]}(\mathbf{x}) + i[\frac{\mathbf{P}^2}{2AM}, \rho_{[1]}(\mathbf{x})] = 0 , \tag{2.74}$$
$$\boldsymbol{\nabla}_x \cdot \mathbf{j}^c_{int[1]}(\mathbf{x}) + i[T^{int}, \rho_{[1]}(\mathbf{x})] = 0 , \tag{2.75}$$

where the intrinsic kinetic energy is denoted by

$$T^{int} = \sum_k \frac{\mathbf{p}'^2_k}{2M} . \tag{2.76}$$

This means, that for a strictly local potential V, i.e.,

$$[V, \rho_{[1]}(\mathbf{x})] = 0\,, \tag{2.77}$$

one can replace in (2.72) the kinetic energy operator T by the total Hamiltonian H_0 and thus the gauge conditions (2.33) and (2.39) are fulfilled.

But for an exchange potential of the form $V^{ex}(r_{12}) = \boldsymbol{\tau}_1 \cdot \boldsymbol{\tau}_2 V_0(r_{12})$, where $\boldsymbol{\tau}_j$ denotes the isospin of a nucleon, and with $e_j = \frac{e}{2}(1+\tau_{j,3})$ one has

$$[V^{ex}, \rho_{[1]}(\mathbf{x})] = ieV_0(r_{12})(\boldsymbol{\tau}_1 \times \boldsymbol{\tau}_2)_3(\delta(\mathbf{x}-\mathbf{r}_1) - \delta(\mathbf{x}-\mathbf{r}_2)) \neq 0\,, \tag{2.78}$$

as is easily verified. Also for a non-local or momentum dependent potential this commutator does not vanish. Therefore, one needs in addition at least two-body exchange operators $\mathbf{j}^{ex}_{[2]}(\mathbf{x})$ and $B^{ex}_{[2]\,kl}(\mathbf{x},\mathbf{y})$ for the current and the two-photon operator, respectively, in order to satisfy the gauge conditions. This can be seen in a very general way by expanding the Hamiltonian, the charge and current density operators in one-, two- and many-body operators indicated by a suffix "$[\nu]$"

$$\begin{aligned} H_0 &= H_{0\,[1]} + H_{0\,[2]} + \dots \\ &= \sum_{\nu\geq 1} H_{0\,[\nu]}\,, \end{aligned} \tag{2.79}$$

$$\rho(\mathbf{x}) = \sum_{\nu\geq 1} \rho_{[\nu]}(\mathbf{x})\,, \tag{2.80}$$

$$\mathbf{j}\,(\mathbf{x}) = \sum_{\nu\geq 1} \mathbf{j}_{[\nu]}(\mathbf{x})\,, \tag{2.81}$$

$$B_{kl}(\mathbf{x},\mathbf{y}) = \sum_{\nu\geq 1} B_{[\nu]kl}(\mathbf{x},\mathbf{y})\,. \tag{2.82}$$

The commutator of an m-body operator $A_{[m]}$ with an n-body operator $B_{[n]}$ can uniquely be expressed as a sum of k-body operators where k runs between $\max(m,n)$ and $m+n-1$

$$\left[A_{[m]}, B_{[n]}\right] = \sum_{k=\max(m,n)}^{m+n-1} \left[A_{[m]}, B_{[n]}\right]_{[k]} \tag{2.83}$$

in an obvious notation. Then one finds from (2.33) and (2.39) by separating the various ν-body operators the following hierarchy of gauge conditions

$$\boldsymbol{\nabla}_x \cdot \mathbf{j}_{[\nu]}(\mathbf{x}) + i\sum_{\mu=1}^{\nu} \sum_{\mu'=\nu+1-\mu}^{\nu} [H_{0\,[\mu]}, \rho_{[\mu']}(\mathbf{x})]_{[\nu]} = 0\,, \tag{2.84}$$

$$\partial_{x_k} B^{ex}_{[\nu]\,kl}(\mathbf{x},\mathbf{y}) - i\sum_{\mu=1}^{\nu} \sum_{\mu'=\nu+1-\mu}^{\nu} [\rho_{[\mu]}(\mathbf{x}), j_{[\mu']\,l}(\mathbf{y})]_{[\nu]} = 0\,. \tag{2.85}$$

In the case of the Hamiltonian H_0 of (2.60), where

$$H_{0[1]} = T\,, \quad H_{0[2]} = V\,, \quad H_{0[\nu]} = 0\,,\ \nu \geq 3\,, \tag{2.86}$$

one obtains

$$\nabla_x \cdot \mathbf{j}_{[\nu]}(\mathbf{x}) + i[T, \rho_{[\nu]}(\mathbf{x})]_{[\nu]} + i[V, \rho_{[\nu-1]}(\mathbf{x})]_{[\nu]} + i[V, \rho_{[\nu]}(\mathbf{x})]_{[\nu]} = 0\,, \tag{2.87}$$

and an analogous equation holds for the two-photon operator. Under the assumption that the charge density is not affected by the exchange interaction, i.e., $\rho^{ex}_{[2]}(\mathbf{x}) \equiv 0$ (Siegert's hypothesis [Sie 37]) or more general $\rho_{[\nu]}(\mathbf{x}) \equiv 0$ for $\nu \geq 2$, one needs only additional two-body operators for $\mathbf{j}$ and for the two-photon operator B_{kl}

$$\mathbf{j}(\mathbf{x}) = \mathbf{j}_{[1]}(\mathbf{x}) + \mathbf{j}^{ex}_{[2]}(\mathbf{x})\,, \tag{2.88}$$

$$B_{kl}(\mathbf{x}, \mathbf{y}) = B_{[1]\,kl}(\mathbf{x}, \mathbf{y}) + B^{ex}_{[2]\,kl}(\mathbf{x}, \mathbf{y})\,, \tag{2.89}$$

and finds that the two conditions (2.33) and (2.39) are fulfilled provided the exchange operators obey

$$\nabla_x \cdot \mathbf{j}^{ex}_{[2]}(\mathbf{x}) + i[V^{ex}, \rho_{[1]}(\mathbf{x})] = 0\,, \tag{2.90}$$

$$\partial_{x_k} B^{ex}_{[2]\,kl}(\mathbf{x}, \mathbf{y}) = i[\rho_{[1]}(\mathbf{x}), j^{ex}_{[2]\,l}(\mathbf{y})]\,. \tag{2.91}$$

The problem one faces, however, is that these two conditions are not sufficient to determine uniquely the required exchange operators because the gauge condition constrains only the divergence while the curl remains unrestricted. For this reason one needs an underlying dynamical model where the original degrees of freedom, which are buried in the effective operators, can be recovered. This will be discussed in Section 3.

2.4 Multipole Decomposition

Since the intrinsic states of hadrons and nuclei can be classified according to the total angular momentum, it is very useful to decompose the charge and current operators in terms of operators which transfer a definite angular momentum. In this way one can take advantage of the angular momentum selection rules and thus can separate the geometrical aspects from the dynamical properties of the system contained in the remaining strengths parameters, the so-called reduced matrix elements [Ros 57].

The starting point for the multipole decomposition of the charge density is the expansion of the plane wave in terms of spherical harmonics $Y_{LM}(\Omega)$ ($\Omega = (\theta, \phi)$)

$$e^{i\mathbf{q}\cdot\mathbf{x}} = 4\pi \sum_{LM} i^L j_L(qx) Y_{LM}(\Omega_x) Y^*_{LM}(\Omega_q)\,, \tag{2.92}$$

where $j_L(z)$ denotes a spherical Bessel function, and Ω_x and Ω_q characterize the spherical angles of $\mathbf{x}$ and $\mathbf{q}$, respectively. Then one obtains for the Fourier transform of the charge density

$$\begin{aligned}\tilde{\rho}(\mathbf{q}) &= \int d^3x \rho(\mathbf{x}) e^{i\mathbf{q}\cdot\mathbf{x}} \\ &= 4\pi \sum_{LM} i^L \left[\int d^3x \rho(\mathbf{x}) j_L(qx) Y_{LM}(\Omega_x)\right] Y^*_{LM}(\Omega_q) \\ &= 4\pi \sum_{LM} i^L C_{LM}(q) Y^*_{LM}(\Omega_q) \\ &= \sqrt{4\pi} \sum_{LM} i^L \hat{L} C_{LM}(q) D^L_{M0}(\phi_q, \theta_q, 0)\,, \end{aligned} \tag{2.93}$$

where I have introduced the charge multipole operators

$$C_{LM} = \int d^3x \rho(\mathbf{x}) j_L(qx) Y_{LM}(\Omega_x) , \tag{2.94}$$

and the D-matrices are taken in the convention of [Ros 57].

In complete analogy one finds a similar decomposition for the transverse current operators. With a corresponding expansion of a transverse plane wave ($\mathbf{e}_\lambda \perp \mathbf{q}$)

$$\mathbf{e}_\lambda e^{i\mathbf{q}\cdot\mathbf{x}} = -\sqrt{2\pi} \sum_{LM} i^L \hat{L} \left(\mathbf{A}_{LM}(e) + \lambda \mathbf{A}_{LM}(m)\right) D^L_{M\lambda}(\phi_q, \theta_q, 0) , \tag{2.95}$$

with electric and magnetic multipole fields

$$\begin{aligned} \mathbf{A}_{LM}(e) &= \frac{1}{q} \boldsymbol{\nabla} \times \mathbf{A}_{LM}(m) \\ &= i \left(\frac{1}{q} \sqrt{\frac{L+1}{L}} \boldsymbol{\nabla} j_L(qx) Y_{LM} - \sqrt{\frac{2L+1}{L}} j_{L+1}(qx) \mathbf{Y}_{LL+1M} \right) \end{aligned} \tag{2.96}$$

$$\begin{aligned} \mathbf{A}_{LM}(m) &= -\frac{i}{L(L+1)} \mathbf{x} \times \boldsymbol{\nabla} j_L(qx) Y_{LM} \\ &= j_L(qx) \mathbf{Y}_{LLM} , \end{aligned} \tag{2.97}$$

where the vector spherical harmonics are given by

$$\mathbf{Y}_{L\ell M} = \sum_{m\lambda} (\ell m 1 \lambda | LM) Y_{\ell m} \mathbf{e}_\lambda , \tag{2.98}$$

one obtains finally

$$\begin{aligned} J_\lambda(\mathbf{q}) &= \int d^3x \mathbf{j}(\mathbf{x}) \cdot \mathbf{e}_\lambda e^{i\mathbf{q}\cdot\mathbf{x}} \\ &= -\sqrt{2\pi} \sum_{LM} i^L \hat{L} (T_{LM}(e) + \lambda T_{LM}(m)) D^L_{M\lambda}(\phi_q, \theta_q, 0) , \end{aligned} \tag{2.99}$$

with the transverse electric and magnetic multipole operators

$$T^{(e)}_{LM} = \frac{1}{q} \int d^3x \mathbf{j}(\mathbf{x}) \cdot (\boldsymbol{\nabla} \times (j_L(qx) \mathbf{Y}_{LLM})) , \tag{2.100}$$

$$T^{(m)}_{LM} = \int d^3x \mathbf{j}(\mathbf{x}) \cdot (j_L(qx) \mathbf{Y}_{LLM}) . \tag{2.101}$$

It is interesting to note that for $\mathbf{A}_{LM}(e)$ according to (2.96) the leading order term $(qx)^{L-1}$ is given as the gradient of a scalar function. This is important for the introduction of the so-called Siegert operators in the next Section 2.5.

Thus one can summarize the multipole decomposition of charge and current density operators by ($\lambda = 0, \pm 1$) [DFW 66,CdA 80]

$$J_\lambda = (-)^\lambda \sqrt{2\pi(1 + \delta_{\lambda 0})} \sum_{LM} i^L \hat{L} \mathcal{O}^\lambda_{LM} D^\lambda_{M\lambda}(\phi_q, \theta_q, 0) , \tag{2.102}$$

$$\mathcal{O}^\lambda_{LM} = \delta_{\lambda 0} C_{LM} + \delta_{|\lambda| 1} (T^{(e)}_{LM} + \lambda T^{(m)}_{LM}) . \tag{2.103}$$

The charge and transverse multipole operators are irreducible spherical tensors of rank L. Therefore, in evaluating them between states of definite angular momentum $|JM\rangle$, one can make use of the Wigner-Eckart theorem and finds for the matrix elements of the charge and current operators [Ros 57]

$$\langle J_f M_f | J_\lambda(\mathbf{q}) | J_i M_i \rangle = (-)^{\lambda + J_f - M_f} \sqrt{2\pi(1+\delta_{\lambda 0})}$$
$$\sum_{LM} i^L \hat{L} \begin{pmatrix} J_f & L & J_i \\ -M_f & M & M_i \end{pmatrix} \langle J_f || \mathcal{O}_L^\lambda || J_i \rangle D^L_{M\lambda}(\phi_q, \theta_q, 0) . \quad (2.104)$$

The 3j-symbol contains the geometrical aspects, i.e., angular momentum selection rules, while the multipole strength is given by the reduced matrix element $\langle J_f || \mathcal{O}_L^\lambda || J_i \rangle$ as determined from the dynamics of the system.

2.5 Generalized Siegert Theorem and Siegert Operators

The generalized Siegert theorem states that in the low energy limit the transverse electric multipoles can be related to the charge multipoles. This is a very important theorem for photo- and electronuclear theory since it allows to calculate reliably within the limits of the theorem the electric transition matrix elements from the charge density without explicit knowledge of exchange currents. I will now briefly derive and discuss this theorem.

The relevant quantities for electric transitions are the electric multipole matrix elements of the intrinsic current operator $\mathbf{j}_{int}(\mathbf{x})$

$$\langle f | T^{(e)}_{LM} | i \rangle = \int d^3x \langle f | \mathbf{j}_{int}(\mathbf{x}) | i \rangle \cdot \mathbf{A}_{LM}(e) , \quad (2.105)$$

where the electric multipole field is given in (2.96). Since the term of leading order in qx is given as a gradient of a scalar function, whose higher order terms, however, are not uniquely determined, it is customary to separate a gradient term from $\mathbf{A}_{LM}(e)$, i.e.,

$$\mathbf{A}_{LM}(e) = \boldsymbol{\nabla} \Phi_{LM}(\mathbf{x}, q) + \mathbf{A}'_{LM}(e; \Phi) , \quad (2.106)$$

where the specific form of $\mathbf{A}'_{LM}(e;\Phi)$ depends on the choice of Φ_{LM}.

This separation can be viewed as a gauge transformation with a gauge function Φ_{LM}. In principle, Φ_{LM} can be chosen arbitrarily, if the current is conserved. However, if that is not the case, e.g., if one uses a simple π-exchange current in conjunction with a phenomenological potential, a considerable uncertainty may be introduced by different choices of Φ_{LM}. On the other hand, this uncertainty is largely reduced if one requires that the lowest order term of $\mathbf{A}'_{LM}(e;\Phi)$ is two orders higher in the leading order, i.e., $(qx)^{L+1}$, than the gradient term [ScW 90]. For this reason, the scalar function Φ_{LM} is chosen in the form

$$\Phi_{LM}(\mathbf{x}, q) = \frac{i}{q} \frac{L+1}{\sqrt{L(L+1)}} \varphi_L(qx) Y_{LM}(\Omega_x) , \quad (2.107)$$

where for φ_L only the leading order is fixed

$$\varphi_L(z) = \frac{z^L}{(2L+1)!!} + \mathcal{O}(z^{L+2}) . \quad (2.108)$$

By partial integration in the gradient term and using current conservation, one arrives at the well known expression

$$\begin{aligned}\langle f|T^{(e)}_{LM}|i\rangle &= i\int d^3x\langle f|[H^{int},\rho(\mathbf{x})]|i\rangle\Phi_{LM}(\mathbf{x},q)\\&\quad+\int d^3x\langle f|\mathbf{j}_{int}(\mathbf{x})|i\rangle\cdot\mathbf{A}'_{LM}(e;\Phi)\\&= i(\epsilon_f-\epsilon_i)\int d^3x\langle f|\rho(\mathbf{x})|i\rangle\Phi_{LM}(\mathbf{x},q)\\&\quad+\int d^3x\langle f|\mathbf{j}_{int}(\mathbf{x})|i\rangle\cdot\mathbf{A}'_{LM}(e;\Phi)\,,\end{aligned}\tag{2.109}$$

where the first term defines a Siegert operator for the general electric multipole operator

$$T^{(e,\,sieg)}_{LM}(\varphi_L)=\frac{i}{q}\frac{L+1}{\sqrt{L(L+1)}}\left[H^{int},\int d^3x\rho(\mathbf{x})\varphi_L(qx)Y_{LM}(\Omega_x)\right].\tag{2.110}$$

Again its specific form depends on the choice of φ_L. It is important to note that the energy difference $(\epsilon_f-\epsilon_i)$ is given by the energy eigenvalues ϵ_λ of the intrinsic Hamiltonian H^{int}. For the special choice $\varphi_L(z)=j_L(z)$ the Siegert operator is simply given by the commutator of the corresponding charge multipole with the intrinsic Hamiltonian

$$T^{(e,\,sieg)}_{LM}(j_L)=\frac{i}{q}\frac{L+1}{\sqrt{L(L+1)}}[H^{int},C_{LM}]\,.\tag{2.111}$$

The Siegert operator is particularly useful in the long-wave-length limit, i.e., for $qx\ll 1$, since then the additional term in (2.109) beyond the Siegert operator is a tiny correction only.

As already said, the Siegert operators contain implicitly exchange current contributions even though the charge operator is a one-body operator in the static limit (see Section 3). This is easy to see, remembering that the contributions of the kinetic and potential energies to the commutator in (2.110) describe just the one- and two-body contributions to the current according to (2.72) and (2.90). Particularly simple is the low energy limit of the $E1$ operator. Because the low-energy limit of $\mathbf{J}(\mathbf{q})$ as given in (2.49)

$$\mathbf{J}_{int}(0)=i[H_0^{int},\mathbf{D}]\tag{2.112}$$

is essentially the $E1$ Siegert operator of (2.110) in this limit. Now, according to (2.72) and (2.90) one has

$$\mathbf{J}_{int[1]}(0)=i[T^{int},\mathbf{D}]\,,\tag{2.113}$$

$$\mathbf{J}_{int[2]}(0)=i[V,\mathbf{D}]\,.\tag{2.114}$$

The latter equation shows how the two-body current operator is generated from the commutator of the potential with the one-body charge operator. Therefore, using the Siegert operators for the electric transitions means implicit inclusion of part of the exchange currents, in fact, the dominant ones [Are 81a,GaH 81]. This fact one has to keep in mind if one includes in addition explicit exchange currents in (2.105) because then one has to apply a Siegert correction for that part which is already included in

(2.109) [FaA 76]. It is also obvious from the form of the Siegert operators (2.110), that the largest effects from exchange currents are to be expected in $E1$ since for the higher multipoles the short range part of the potential contribution to the commutator is largely suppressed due to the $(qx)^L$-behaviour at small distances.

3 Exchange Currents as Effective Operators

In the previous section we have seen that exchange forces demand the existence of corresponding meson exchange currents. A variety of methods have been developed for the construction of effective operators in order to eliminate unwanted degrees of freedom. I will here describe briefly the FST-method of Fukuda, Sawada and Taketani [FuS 54] of how to incorporate meson degrees of freedom in terms of effective operators. Furthermore, I will discuss a phenomenological method using the concept of minimal coupling. Other methods are described in [BlH 58,Fes 62,Che 79,AdT 89, Mat 89,Ris 89].

In order to be specific, I will choose a simple model of a system consisting of nucleons and pions only. The corresponding Hamiltonian is assumed to have the form

$$H = H_N^0 + H_\pi^0 + H_{\pi N}\,, \tag{3.1}$$

with free nucleon and pion Hamiltonians H_N^0 and H_π^0, respectively, and an interaction term $H_{\pi N}$. One divides the total Fock space which comprises all possible pion and nucleon states into two orthogonal subspaces, described by appropriate projection operators P and Q. The P-space consists of all states which contain only nucleons, i.e., no pions are present, while the Q-space, the orthogonal complement of the P-space, consists of all states which contain at least one pion. The projection operators P and Q have the following properties

$$P^2 = P \;,\; Q^2 = Q\,, \tag{3.2}$$

$$PQ = 0 \;,\; QP = 0\,, \tag{3.3}$$

$$P + Q = 1\,. \tag{3.4}$$

For convenience I will introduce the following notation for any operator Ω

$$\Omega_{PP} := P\Omega P\,,\quad \Omega_{PQ} := P\Omega Q \tag{3.5}$$

$$\Omega_{QP} := Q\Omega P\,,\quad \Omega_{QQ} := Q\Omega Q\,, \tag{3.6}$$

so that Ω can be written in the following symbolic matrix notation

$$\Omega = \begin{pmatrix} \Omega_{PP} & \Omega_{PQ} \\ \Omega_{QP} & \Omega_{QQ} \end{pmatrix}. \tag{3.7}$$

The aim now is to arrive at a description in P-space alone for those states of the total space which have a non-vanishing component in P-space.

3.1 The FST Method

The essential idea of the FST method is to find a unitary transformation U which has the property of diagonalizing the total Hamiltonian with respect to the P- and Q-space, so that the transformed Hamiltonian has no overlap matrix elements between P- and Q-space, i.e.,

$$
\begin{aligned}
H' &= U^\dagger H U \\
&= \begin{pmatrix} H'_{PP} & 0 \\ 0 & H'_{QQ} \end{pmatrix} .
\end{aligned} \tag{3.8}
$$

This is the same basic idea as is used in the Foldy-Wouthuysen transformation of the Dirac-Hamiltonian in order to separate the upper and lower components of the Dirac spinors [FoW 50]. In other words, the transformed Hamiltonian has to fulfill the conditions

$$
H'_{PQ} = 0 , \qquad H'_{QP} = 0 , \tag{3.9}
$$

from which U has to be determined.

Noting that any unitary 2×2-matrix can always be written in the form

$$
\gamma \begin{pmatrix} 1 & -\alpha^* \\ \alpha & 1 \end{pmatrix} , \tag{3.10}
$$

where α is a complex number and $\gamma = (1 + \alpha\alpha^*)^{-\frac{1}{2}}$, one obtains a formal solution by the following ansatz for U

$$
\begin{aligned}
U &= \begin{pmatrix} (J^\dagger J)_{PP}^{-\frac{1}{2}} & -F_{PQ}^\dagger (J J^\dagger)_{QQ}^{-\frac{1}{2}} \\ F_{QP} (J^\dagger J)_{PP}^{-\frac{1}{2}} & (J J^\dagger)_{QQ}^{-\frac{1}{2}} \end{pmatrix} \\
&= \begin{pmatrix} 1_{PP} & -F_{PQ}^\dagger \\ F_{QP} & 1_{QQ} \end{pmatrix} \begin{pmatrix} (J^\dagger J)_{PP}^{-\frac{1}{2}} & 0 \\ 0 & (J J^\dagger)_{QQ}^{-\frac{1}{2}} \end{pmatrix} ,
\end{aligned} \tag{3.11}
$$

where

$$
J = 1 + F \tag{3.12}
$$

and F connects only P- with Q-space, i.e.,

$$
F = \begin{pmatrix} 0 & 0 \\ F_{QP} & 0 \end{pmatrix} \tag{3.13}
$$

and $F_{QP} \neq 0$. Then one can write (3.11) in the form

$$
U = (1 + F - F^\dagger) \begin{pmatrix} (J^\dagger J)_{PP}^{-\frac{1}{2}} & 0 \\ 0 & (J J^\dagger)_{QQ}^{-\frac{1}{2}} \end{pmatrix} , \tag{3.14}
$$

from which the unitarity can be proven using the relations

$$
(J J^\dagger F)_{QP} = (F J J^\dagger)_{QP} , \qquad (J^\dagger J F^\dagger)_{PQ} = (F^\dagger J^\dagger J)_{PQ} , \tag{3.15}
$$

which follow easily from the properties of F.

The condition (3.9) then reads

$$(JJ^\dagger)^{-\frac{1}{2}}_{QQ} Q(1+J^\dagger - J)H(1+J-J^\dagger)P(J^\dagger J)^{-\frac{1}{2}}_{PP} = 0\,, \tag{3.16}$$

and one obtains a non-linear equation for the unknown operator F

$$(H + [H,F] - FHF)_{QP} = 0\,. \tag{3.17}$$

It cannot be solved in general but an approximate solution using perturbation theory will be described below.

Assuming that a solution for F has been found, then for any operator Ω the corresponding effective operator acting in P-space only is obtained from the P-space part of the transformed operator Ω'

$$\begin{aligned} \Omega_{eff} &:= PU^\dagger \Omega UP \\ &= \Omega'_{PP} \\ &= \Omega_{PP} + \Omega^{ex}\,, \end{aligned} \tag{3.18}$$

where the last equation defines the exchange operator which effectively takes into account the action of the operator Ω on the Q-space components. As an example, the effective Hamiltonian in P-space has the following form

$$\begin{aligned} H_{eff} &= H'_{PP} \\ &= U_{PP}(J^\dagger HJ)_{PP}U_{PP} \\ &= H_{PP} + V^{ex}\,, \end{aligned} \tag{3.19}$$

and V^{ex} describes the effective NN interaction as mediated by the exchange of mesons in this simple model.

Any eigenfunction $|\phi\rangle$ of H_{eff} will generate an eigenfunction $|\psi\rangle$ of H with the same energy by

$$\begin{aligned} |\psi\rangle &= UP|\phi\rangle \\ &= JU_{PP}|\phi\rangle \\ &= U_{PP}|\phi\rangle + FU_{PP}|\phi\rangle\,. \end{aligned} \tag{3.20}$$

The second part is just the Q-space component of $|\psi\rangle$. U_{PP} takes into account the wave function renormalization since the P-space component of $|\psi\rangle$ is in general not normalized to unity. Obviously, if $|\psi\rangle$ has no component in P-space, it cannot be generated this way.

3.2 Perturbative Solution

For the explicit construction of the effective operators one has to solve (3.17). As already mentioned above, this is in general not possible. However, using perturbation theory one can find an approximate solution [GaH 76]. It rests on the fact that the operator F is related to the only non-diagonal part of the Hamiltonian H_{QP} which in our case is given by the πN coupling $H_{\pi N}$. The basic assumption for the perturbative treatment is that F is analytic in the πN coupling constant, so that one can expand F in a power series with respect to this coupling constant

$$F = \sum_{\nu=1}^{\infty} F^{(\nu)} . \tag{3.21}$$

Inserting this expansion into (3.17), treating H_{PP} and H^0_{QQ} as order zero, H_{QP} and $H^{(1)}_{QQ} = QH_{\pi N}Q$ as order one, one finds successively an equation for the ν-th order contribution $F^{(\nu)}$ of F. In detail, one has for $\nu = 1$

$$H_{QP} + [H, F^{(1)}]_{QP} = 0 , \tag{3.22}$$

and for $\nu \geq 2$

$$[H^0, F^{(\nu)}]_{QP} = \sum_{\mu=1}^{\nu-1} (F^{(\mu)} H F^{(\nu-\mu)})_{QP} - H^{(1)}_{QQ} F^{(\nu-1)} . \tag{3.23}$$

The structure of each equation of this set is of the form

$$[H^0, X]_{QP} = B_{QP} , \quad X = X_{QP} . \tag{3.24}$$

A formal solution can be constructed by using a set of eigenfunctions $\{|n_0\rangle\}$ of H_{PP}, i.e.,

$$H_{PP}|n_0\rangle = E_{n_0}|n_0\rangle , \tag{3.25}$$

assumed to be complete in P-space. It has the form

$$X_{QP} = \sum_{n_0} (H^0_{QQ} - E_{n_0})^{-1} B_{QP}|n_0\rangle\langle n_0| . \tag{3.26}$$

In detail, one finds for the first two orders

$$F^{(1)} = \sum_{n_0} (E_{n_0} - H^0_{QQ})^{-1} H_{QP}|n_0\rangle\langle n_0| , \tag{3.27}$$

$$F^{(2)} = \sum_{n_0} (E_{n_0} - H^0_{QQ})^{-1} H^{(1)}_{QQ} (E_{n_0} - H^0_{QQ})^{-1} H_{QP}|n_0\rangle\langle n_0| . \tag{3.28}$$

A graphical representation in terms of diagrams for $F^{(1)}$ and $F^{(2)}$ is shown in Fig. 3.1 for our model noting that H_{QP} describes the creation of a pion.

Then, with

$$\begin{aligned} U_{PP} &= (1 + F^\dagger F)^{-1/2}_{PP} \\ &= (1 - \frac{1}{2} F^{(1)\dagger} F^{(1)})_{PP} , \end{aligned} \tag{3.29}$$

up to second order, one finds for the exchange part of the effective operator of (3.18)

$$\begin{aligned} \Omega^{ex} = {} & F^{(1)\dagger}\Omega_{QP} + \Omega_{PQ}F^{(1)} + F^{(1)\dagger}\Omega_{QQ}F^{(1)} + F^{(2)\dagger}\Omega_{QP} + \Omega_{PQ}F^{(2)} \\ & - \frac{1}{2}(F^{(1)\dagger}F^{(1)}\Omega_{PP} + \Omega_{PP}F^{(1)\dagger}F^{(1)}) . \end{aligned} \tag{3.30}$$

The various terms in (3.30) are diagrammatically represented in Fig. 3.2. The last one describes the effect of the wave function renormalization due to the restriction on P-space alone.

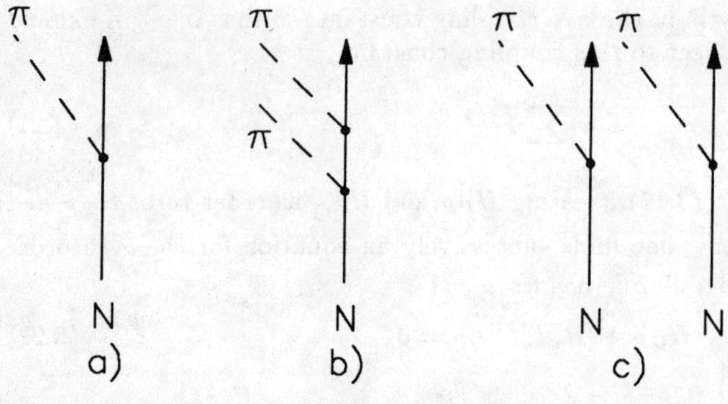

Fig. 3.1 Creation of one pion by $F^{(1)}$ (a), and of two pions by $F^{(2)}$ at one nucleon (b) and at two nucleons (c).

3.3 Explicit Construction of π-Exchange Current

These expressions will now be used for the explicit construction of the pion exchange current. In second quantization the free Hamiltonians of the nucleon and pion fields are taken in the simple form

$$H_N^0 = \frac{1}{2M}\int d^3x (\boldsymbol{\nabla}\psi)^\dagger \cdot \boldsymbol{\nabla}\psi\,, \tag{3.31}$$

$$H_\pi^0 = \frac{1}{2}\sum_\mu \int d^3x((\partial_t\phi_\mu)^\dagger\partial_t\phi_\mu + (\boldsymbol{\nabla}\phi_\mu)^\dagger \cdot \boldsymbol{\nabla}\phi_\mu + m^2(\phi_\mu)^\dagger\phi_\mu)\,, \tag{3.32}$$

where the nucleon field operator ψ is a non-relativistic spinor in spin and isospin space, and ϕ_μ denotes the field operator of a pion with charge μ. The interaction Hamiltonian is given by

$$H_{\pi N} = \frac{f_\pi}{m_\pi}\sum_\mu \int d^3x\psi^\dagger(-)^\mu\tau_{-\mu}\boldsymbol{\sigma}\psi \cdot \boldsymbol{\nabla}\phi_\mu\,. \tag{3.33}$$

The free π-field can be expanded in terms of creation and annihilation operators $a^\dagger_{\mathbf{q},\mu}$ and $a_{\mathbf{q},\mu}$, respectively, of a pion with momentum $\mathbf{q}$ and charge μ

$$\phi_\mu(\mathbf{x},t) = \frac{1}{(2\pi)^{3/2}}\int \frac{d^3q}{\sqrt{2\omega_q}}(a^\dagger_{\mathbf{q},\mu}e^{-i(\mathbf{q}\cdot\mathbf{x}-\omega_q t)} + (-)^\mu a_{\mathbf{q},-\mu}e^{i(\mathbf{q}\cdot\mathbf{x}-\omega_q t)})\,, \tag{3.34}$$

where the pion energy is given by $\omega_q = \sqrt{\mathbf{q}^2 + m_\pi^2}$.

The operators obey the boson commutation relations

$$[a_{\mathbf{q}',\mu'}, a^\dagger_{\mathbf{q},\mu}] = \delta_{\mu'\mu}\delta(\mathbf{q}' - \mathbf{q})\,, \tag{3.35}$$

while all other commutator combinations vanish. Then one obtains the interaction in the form

$$H_{\pi N} = -\frac{i}{(2\pi)^{3/2}}\frac{f_\pi}{m_\pi}\sum_\mu \int d^3x \int \frac{d^3q}{\sqrt{2\omega_q}}\psi^\dagger(\tau_\mu)^\dagger\boldsymbol{\sigma}\cdot\mathbf{q}\psi(a^\dagger_{\mathbf{q},\mu}e^{-i(\mathbf{q}\cdot\mathbf{x}-\omega_q t)} - h.c.)\,. \tag{3.36}$$

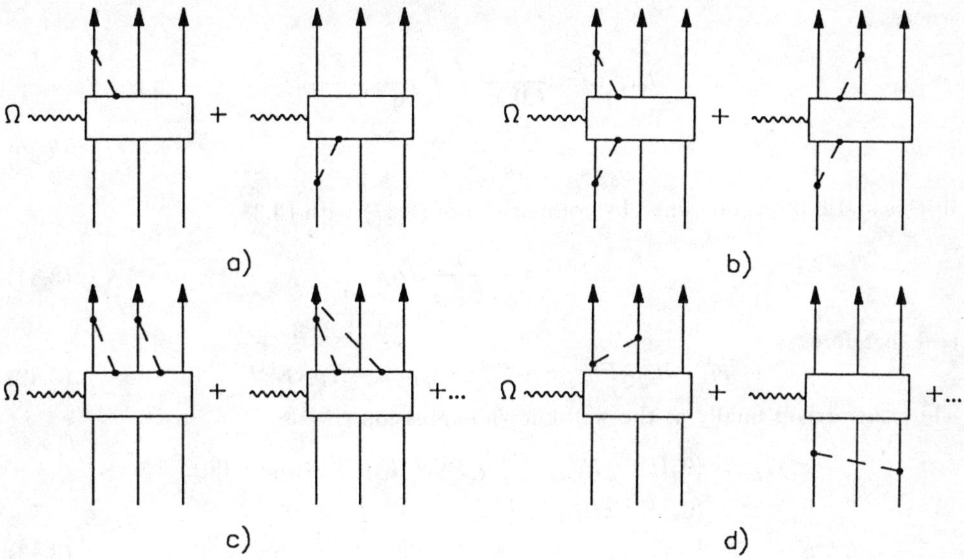

Fig. 3.2 Representation of the various contributions to the exchange operator Ω^{ex}: (a) $F^{(1)\dagger}\Omega_{QP}+\Omega_{PQ}F^{(1)}$, (b) $F^{(1)\dagger}\Omega_{QQ}F^{(1)}$, (c) $F^{(2)\dagger}\Omega_{QP}+\Omega_{PQ}F^{(2)}$, (d) $F^{(1)\dagger}F^{(1)}\Omega_{PP}+\Omega_{PP}F^{(1)\dagger}F^{(1)}$.

In order to further simplify the discussion, I will adopt a strict static model with respect to the nucleons by neglecting their kinetic energy in the intermediate states where pions are present (adiabatic model). This leads then for $F^{(1)}$ to the expression

$$F^{(1)} = \frac{i}{(2\pi)^{3/2}}\frac{f_\pi}{m_\pi}\int\frac{d^3q}{\sqrt{2\omega_q}}\frac{1}{\omega_q}\sum_\mu\int d^3x e^{-i(\mathbf{q}\cdot\mathbf{x}-\omega_q t)}\psi^\dagger(\mathbf{x})(\tau_\mu)^\dagger\boldsymbol{\sigma}\cdot\mathbf{q}\psi(\mathbf{x})a^\dagger_{\mathbf{q},\mu}P \tag{3.37}$$

from (3.27) with $E_{n_0}=0$ and therefore $\sum_{n_0}|n_0\rangle\langle n_0|=1$. $P=|0_\pi\rangle\langle 0_\pi|$ is the projection operator onto the pion vacuum denoted by $|0_\pi\rangle$. Obviously, $F^{(1)}$ describes the creation of a pion with momentum $\mathbf{q}$ and charge μ with a corresponding momentum and charge transfer to the nucleon field. An analogous expression is obtained for $F^{(2)}$ describing the creation of two pions which I will not list here.

From (3.36) one obtains now

$$H_{QP} = -\frac{i}{(2\pi)^{3/2}}\frac{f_\pi}{m_\pi}\int\frac{d^3q}{\sqrt{2\omega_q}}\sum_\mu\int d^3x e^{-i(\mathbf{q}\cdot\mathbf{x}-\omega_q t)}\psi^\dagger(\mathbf{x})(\tau_\mu)^\dagger\boldsymbol{\sigma}\cdot\mathbf{q}\psi(\mathbf{x})a^\dagger_{\mathbf{q},\mu}P\,, \tag{3.38}$$

and

$$\begin{aligned}\langle 0_\pi|F^{(1)\dagger}H_{QP}|0_\pi\rangle = -\frac{f_\pi^2}{2m_\pi^2}\int d^3x\int d^3y\psi^\dagger(\mathbf{x})\psi^\dagger(\mathbf{y})\boldsymbol{\tau}(\mathbf{x})\cdot\boldsymbol{\tau}(\mathbf{y})\\ \boldsymbol{\sigma}(\mathbf{x})\cdot\boldsymbol{\nabla}_x\boldsymbol{\sigma}(\mathbf{y})\cdot\boldsymbol{\nabla}_y J_{m_\pi}(\mathbf{x}-\mathbf{y})\psi(\mathbf{x})\psi(\mathbf{y})\,,\end{aligned} \tag{3.39}$$

where

$$J_{m_\pi}(r) = \frac{1}{(2\pi)^3}\int d^3q \frac{e^{i\mathbf{q}\cdot\mathbf{r}}}{\mathbf{q}^2 + m_\pi^2} = \frac{e^{-m_\pi r}}{4\pi r}\,. \tag{3.40}$$

In this static limit, one finds by comparison of (3.37) with (3.38)

$$F^{(1)} = -\frac{1}{H^0_{QQ}} H_{QP} \tag{3.41}$$

and therefore

$$F^{(1)\dagger} H^0_{QQ} F^{(1)} = -F^{(1)\dagger} H_{QP} = -H_{PQ} F^{(1)}\,. \tag{3.42}$$

Thus one arrives finally at the well-known expression

$$\begin{aligned} V_{eff} &= \langle 0_\pi | F^{(1)\dagger} H^0_{QQ} F^{(1)} + F^{(1)\dagger} H_{QP} + H_{PQ} F^{(1)} | 0_\pi \rangle \\ &= \langle 0_\pi | F^{(1)\dagger} H_{QP} | 0_\pi \rangle \\ &= V_\pi\,, \end{aligned} \tag{3.43}$$

of the static one-pion-exchange potential.

Now I will derive the corresponding effective two-body current operator. According to the general structure of an effective operator, I need the various current components $\mathbf{j}_{PP}$, $\mathbf{j}^0_{QQ}$ and $\mathbf{j}_{QP}$ which are obtained from the original Hamiltonian by minimal substitution yielding

$$\mathbf{j}_{PP}(\mathbf{x}) = P\mathbf{j}_N(\mathbf{x})P\,, \tag{3.44}$$

$$\mathbf{j}^0_{QQ}(\mathbf{x}) = Q\left(\mathbf{j}_N(\mathbf{x}) + \mathbf{j}_\pi(\mathbf{x})\right)Q\,, \tag{3.45}$$

where

$$\mathbf{j}_N(\mathbf{x}) = \frac{e}{4iM}\psi^\dagger(\mathbf{x})(1+\tau_3)\overleftrightarrow{\nabla}\psi(\mathbf{x})\,, \tag{3.46}$$

$$\mathbf{j}_\pi(\mathbf{x}) = \frac{e}{2i}\sum_\mu \mu(\phi_\mu)^\dagger \overleftrightarrow{\nabla}\phi_\mu\,, \tag{3.47}$$

and for the transition current

$$\mathbf{j}_{QP}(\mathbf{x}) = \frac{1}{(2\pi)^{3/2}}\frac{f_\pi}{im_\pi}\sum_\mu \psi^\dagger(\mathbf{x})(\tau_\mu)^\dagger \boldsymbol{\sigma}\psi(\mathbf{x})\int \frac{d^3q}{\sqrt{2\omega_q}} e^{-i(\mathbf{q}\cdot\mathbf{x}-\omega_q t)} a^\dagger_{\mathbf{q},\mu} P\,. \tag{3.48}$$

Then one finds

$$\begin{aligned} \langle 0_\pi | F^{(1)\dagger}\mathbf{j}_{QP}(\mathbf{z})|0_\pi\rangle = -\frac{f_\pi^2}{2m_\pi^2}\int d^3x \int d^3y\, \psi^\dagger(\mathbf{x})\psi^\dagger(\mathbf{y})(\boldsymbol{\tau}(\mathbf{x})\times\boldsymbol{\tau}(\mathbf{y}))_3 \\ \boldsymbol{\sigma}(\mathbf{x})\boldsymbol{\sigma}(\mathbf{y})\cdot\boldsymbol{\nabla}_y J_{m_\pi}(\mathbf{x}-\mathbf{y})\delta(\mathbf{x}-\mathbf{z})\psi(\mathbf{x})\psi(\mathbf{y})\,. \end{aligned} \tag{3.49}$$

This term and the hermitian conjugate constitute the so-called pair or contact current because they arise in pseudoscalar coupling from an intermediate nucleon-antinucleon pair while in pseudovector coupling they stem from the contact e.m. interaction.

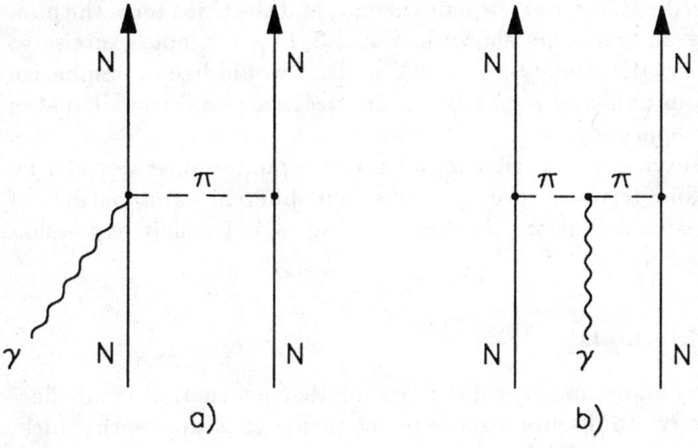

Fig. 3.3 π-exchange current diagrams: (a) pair current and (b) π-current.

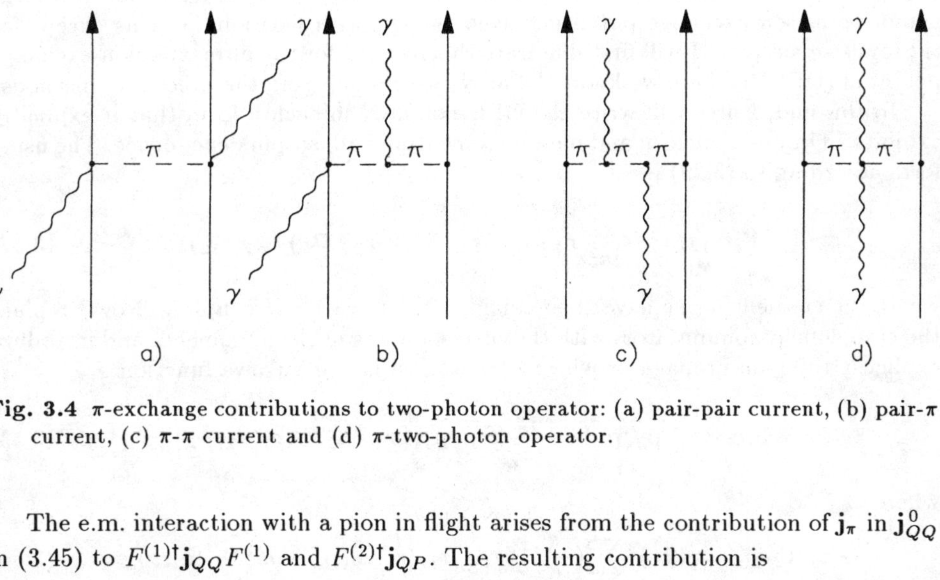

Fig. 3.4 π-exchange contributions to two-photon operator: (a) pair-pair current, (b) pair-π current, (c) π-π current and (d) π-two-photon operator.

The e.m. interaction with a pion in flight arises from the contribution of $\mathbf{j}_\pi$ in $\mathbf{j}^0_{QQ}$ in (3.45) to $F^{(1)\dagger}\mathbf{j}_{QQ}F^{(1)}$ and $F^{(2)\dagger}\mathbf{j}_{QP}$. The resulting contribution is

$$-\frac{f_\pi^2}{2m_\pi^2}\int d^3x \int d^3y \psi^\dagger(\mathbf{x})\psi^\dagger(\mathbf{y})(\boldsymbol{\tau}(\mathbf{x})\times\boldsymbol{\tau}(\mathbf{y}))_3$$
$$\boldsymbol{\sigma}(\mathbf{x})\cdot\boldsymbol{\nabla}_x\boldsymbol{\sigma}(\mathbf{y})\cdot\boldsymbol{\nabla}_y J_{m_\pi}(\mathbf{x}-\mathbf{z})\overleftrightarrow{\nabla} J_{m_\pi}(\mathbf{z}-\mathbf{y})\delta(\mathbf{x}-\mathbf{z})\psi(\mathbf{x})\psi(\mathbf{y})\,. \quad (3.50)$$

Collecting all pieces, one finds for the complete π-exchange current

$$\mathbf{j}^\pi_{[2]}(\mathbf{x}) = -\frac{f_\pi^2}{m_\pi^2}(\boldsymbol{\tau}_1\times\boldsymbol{\tau}_2)_3\,(\delta(\mathbf{x}-\mathbf{r}_1)\boldsymbol{\sigma}_1(\boldsymbol{\sigma}_2\cdot\boldsymbol{\nabla}_2)J_{m_\pi}(r_{12}) - (1\leftrightarrow 2)$$
$$+(\boldsymbol{\sigma}_1\cdot\boldsymbol{\nabla}_1)(\boldsymbol{\sigma}_2\cdot\boldsymbol{\nabla}_2)J_{m_\pi}(\mathbf{r}_1-\mathbf{x})\overleftrightarrow{\nabla}_x J_{m_\pi}(\mathbf{x}-\mathbf{r}_2)\Big)\,. \quad (3.51)$$

The first two terms describe the contact or pair current, and the third term the pion current. The corresponding diagrams are shown in Fig. 3.3. It is a simple exercise to prove that the current conservation relation of (2.90) holds. I would like to emphasize that both contributions from pair and pion current are necessary and that it makes no sense to consider them separately.

In the same way the π-exchange contribution to the two-photon operator can be obtained. Here, one finds four terms according to the four different combinations of vertex and pion propagator contributions as shown in Fig. 3.4. Explicit expression can be found in [Fri 76,Are 80].

3.4 Phenomenological Methods

While the preceding method allows in general the systematic construction of all effective operators, it is instructive to discuss a more phenomenological approach which, even though not unique, is quite useful for the construction of a consistent exchange current for a realistic NN potential [Are 79a]. It is based on one hand on the method of minimal substitution and on the other hand on the observation that most realistic NN potentials are formally constructed in such a way that they appear to a large extent as meson exchange potentials even though with fictitious mesons except for the lowest-mass pion. I will first illustrate this method for the pure one-pion-exchange potential (OPEP) where we know the answer from applying the preceding methods.

To this end, I first will write the OPE potential in such a form that it explicitly exhibits all implicit nucleon and meson momentum and isospin dependence. The usual form according to (3.39) is

$$V_\pi(r_{12}) = -\frac{f_\pi^2}{m_\pi^2}\boldsymbol{\tau}_1\cdot\boldsymbol{\tau}_2\,(\boldsymbol{\sigma}_1\cdot\boldsymbol{\nabla}_1)(\boldsymbol{\sigma}_2\cdot\boldsymbol{\nabla}_2)J_{m_\pi}(r_{12})\,, \qquad (3.52)$$

where the gradient terms have their origin in the p-wave πN-coupling. Now I replace the gradients by commutators with the corresponding nucleon momenta and introduce explicitly the pion propagator which is contained in the Yukawa function J_{m_π}

$$V_\pi(r_{12};\mathbf{p}_1,\mathbf{p}_2,\mathbf{p}_\pi) = \frac{f_\pi^2}{m_\pi^2}[\boldsymbol{\sigma}_1\cdot\mathbf{p}_1,[\boldsymbol{\sigma}_2\cdot\mathbf{p}_2,G(\mathbf{r}_1,\mathbf{r}_2,\mathbf{p}_\pi)]]\,, \qquad (3.53)$$

where

$$G(\mathbf{r}_1,\mathbf{r}_2,\mathbf{p}_\pi) = (2\pi)^3\sum_{\mu\nu}(-)^{\mu+\nu}\tau_{1-\mu}\tau_{2\nu}\langle\mathbf{r}_1,\mu|(\mathbf{p}_\pi^2+m_\pi^2)^{-1}|\mathbf{r}_2,\nu\rangle \qquad (3.54)$$

describes the pion propagation from $\mathbf{r}_2$ to $\mathbf{r}_1$ and includes the isospin operators of the pion-nucleon vertex. Furthermore, I have introduced in (3.54) the wave function $|\mathbf{r},\mu\rangle$ of a pion of charge μ created at the point $\mathbf{r}$. The pion momentum operator is denoted by $\mathbf{p}_\pi$.

The corresponding exchange current is then obtained by minimal substitution in V_π and functional derivative

$$\mathbf{j}_{[2]}^{\pi}(\mathbf{x}) = -\frac{\delta}{\delta\mathbf{A}(\mathbf{x})}V_\pi(r_{12};\mathbf{p}_1-e_1\mathbf{A}(\mathbf{r}_1),\mathbf{p}_2-e_2\mathbf{A}(\mathbf{r}_2),\mathbf{p}_\pi-e_\pi\mathbf{A}(\mathbf{r}_\pi))\,|_{\mathbf{A}\equiv 0}\,. \qquad (3.55)$$

Since the minimal substitution acts in three places - the two vertices and the pion propagator - one obtains the three above mentioned contributions as given in (3.51). In an analogous manner, one obtains the exchange two-photon amplitude from

$$B^{\pi}_{[2]kl}(\mathbf{x},\mathbf{y}) = \frac{\delta^2}{\delta A_k(\mathbf{x})\delta A_l(\mathbf{y})} V_\pi(r_{12}, \mathbf{p}_1 - e_1\mathbf{A}(\mathbf{r}_1), \mathbf{p}_2 - e_2\mathbf{A}(\mathbf{r}_2), \mathbf{p}_\pi - e_\pi \mathbf{A}(\mathbf{r}_\pi)) \,|_{\mathbf{A}\equiv 0} \quad (3.56)$$

Also hadronic form factors can be incorporated in a consistent manner [Are 79a, Oht 89,ScA 89a]. The simplest case is a monopole form factor where in (3.40) J_{m_π} is replaced by

$$\frac{\lambda^2}{(2\pi)^3}\int d^3q \frac{e^{i\mathbf{q}\cdot\mathbf{r}}}{(\mathbf{q}^2+m_\pi^2)(\mathbf{q}^2+\lambda^2)} = \frac{\lambda^2}{\lambda^2 - m_\pi^2}(J_{m_\pi}(r) - J_\lambda(r)), \quad (3.57)$$

where λ is a regularization or cut-off parameter. Higher order regularization can be generated by differentiation with respect to λ^2.

Furthermore, for a phenomenological potential which formally resembles a boson exchange potential but with fictitious mesons, one can construct in the same way a consistent exchange current satisfying (2.90) [BuL 85,Ris 85]. For example, the general structure of the Paris potential in its parametrized form is [LaL 80]

$$V = V^0 + \boldsymbol{\tau}_1 \cdot \boldsymbol{\tau}_2 V^\tau, \quad (3.58)$$

where

$$V^{0/\tau} = \sum_\alpha V_\alpha^{0/\tau}(r, \mathbf{p}^2)\Omega_\alpha. \quad (3.59)$$

Here $\alpha = 0$, σ, T, LS, and $SO2$ enumerates the various spin operators $\Omega_\alpha = 1$, $\boldsymbol{\sigma}_1 \cdot \boldsymbol{\sigma}_2$, S_{12}, $\mathbf{L}\cdot\mathbf{S}$, and Ω_{SO2}, the quadratic spin-orbit operator, respectively. The $\mathbf{p}^2$-dependence is present in the central and the V_σ-part. One has two sources for exchange currents: (i) the p-dependence, and (ii) the isospin dependence. For the explicit p-dependence the method of minimal substitution leads directly to the corresponding exchange current while for the isospin-dependence one can make use of the fact that each term has the formal resemblance of a fictitious meson exchange for which one knows the corresponding gauge invariant exchange current. For example, one may write - for details see [BuL 85] -

$$\boldsymbol{\tau}_1 \cdot \boldsymbol{\tau}_2 \, (\boldsymbol{\sigma}_1 \cdot \boldsymbol{\sigma}_2 V_\sigma + S_{12} V_T) = V(\pi - like) + V(\rho - like), \quad (3.60)$$

and then use the well known expressions for π- and ρ-exchange currents. In this manner one obtains an exchange current which is consistent with the Paris potential, i.e., fulfills (2.90) for this potential.

However, one should be aware that the procedure is not uniquely defined. I will illustrate this for the OPE potential. Instead of using (3.52) one could have used by recoupling

$$(\boldsymbol{\sigma}_1 \cdot \boldsymbol{\nabla}_1)(\boldsymbol{\sigma}_2 \cdot \boldsymbol{\nabla}_2) = \frac{1}{3}(\boldsymbol{\sigma}_1 \cdot \boldsymbol{\sigma}_2)(\boldsymbol{\nabla}_1 \cdot \boldsymbol{\nabla}_2) + \frac{1}{2}(\boldsymbol{\sigma}_1 \times \boldsymbol{\sigma}_2)\cdot(\boldsymbol{\nabla}_1 \times \boldsymbol{\nabla}_2) + \frac{1}{3} S_{12}(\boldsymbol{\sigma}_1, \boldsymbol{\nabla}_1, \boldsymbol{\sigma}_2, \boldsymbol{\nabla}_2) \quad (3.61)$$

the equivalent form in terms of a central and a tensor part

$$V_\pi(r_{12}) = -\frac{f_\pi^2}{3m_\pi^2}\boldsymbol{\tau}_1\cdot\boldsymbol{\tau}_2 \;\{(\boldsymbol{\sigma}_1\cdot\boldsymbol{\sigma}_2)(\boldsymbol{\nabla}_1\cdot\boldsymbol{\nabla}_2)+S_{12}(\boldsymbol{\sigma}_1,\boldsymbol{\nabla}_1,\boldsymbol{\sigma}_2,\boldsymbol{\nabla}_2)\}J_{m_\pi}(r_{12}), \quad (3.62)$$

since the vector operator $((\boldsymbol{\sigma}_1\times\boldsymbol{\sigma}_2)\cdot(\boldsymbol{\nabla}_1\times\boldsymbol{\nabla}_2))$ gives a vanishing contribution. The tensor operator when acting on $J_{m_\pi}(r_{12})$ gives just the usual tensor force.

For the minimal coupling in the tensor operator one may use the following operator identity for the tensor operator

$$\begin{aligned}\frac{1}{3}S_{12}(\boldsymbol{\sigma}_1,\boldsymbol{\nabla}_1,\boldsymbol{\sigma}_2,\boldsymbol{\nabla}_2) &= (\boldsymbol{\sigma}_1\cdot\boldsymbol{\nabla}_1)(\boldsymbol{\sigma}_2\cdot\boldsymbol{\nabla}_2)-\frac{1}{3}(\boldsymbol{\sigma}_1\cdot\boldsymbol{\sigma}_2)(\boldsymbol{\nabla}_1\cdot\boldsymbol{\nabla}_2)\\ &\quad -\frac{1}{2}(\boldsymbol{\sigma}_1\times\boldsymbol{\sigma}_2)\cdot(\boldsymbol{\nabla}_1\times\boldsymbol{\nabla}_2)\\ &= \frac{1}{2}(\boldsymbol{\sigma}_1\cdot\boldsymbol{\nabla}_1)(\boldsymbol{\sigma}_2\cdot\boldsymbol{\nabla}_2)-\frac{1}{3}(\boldsymbol{\sigma}_1\cdot\boldsymbol{\sigma}_2)(\boldsymbol{\nabla}_1\cdot\boldsymbol{\nabla}_2)\\ &\quad +\frac{1}{2}(\boldsymbol{\sigma}_1\cdot\boldsymbol{\nabla}_2)(\boldsymbol{\sigma}_2\cdot\boldsymbol{\nabla}_1). \qquad (3.63)\end{aligned}$$

One could leave out the vector operator in (3.63) because it gives a vanishing contribution and then minimal coupling would yield again the π-exchange current of (3.51). But if one keeps it formally, it gives in minimal substitution an additional exchange current piece, namely

$$\begin{aligned}\mathbf{j}_{[2]}^{\pi\prime}(\mathbf{x}) = -\frac{f_\pi^2}{2m_\pi^2}(\boldsymbol{\tau}_1\times\boldsymbol{\tau}_2)_3\,\Big(\delta(\mathbf{x}-\mathbf{r}_1)(\boldsymbol{\sigma}_1\times\boldsymbol{\sigma}_2)\times\boldsymbol{\nabla}_2 J_{m_\pi}(r_{12})-(1\leftrightarrow 2)\\ -(\boldsymbol{\sigma}_1\times\boldsymbol{\sigma}_2)\cdot(\boldsymbol{\nabla}_1\times\boldsymbol{\nabla}_2)J_{m_\pi}(\mathbf{r}_1-\mathbf{x})\,\overleftrightarrow{\nabla}_x\,J_{m_\pi}(\mathbf{x}-\mathbf{r}_2)\Big), \qquad (3.64)\end{aligned}$$

which, however, is completely transverse, i.e., its divergence vanishes identically

$$\boldsymbol{\nabla}\cdot\mathbf{j}_{[2]}^{\pi\prime}(\mathbf{x}) = 0\,. \qquad (3.65)$$

It would just cancel the current arising from the vector operator in (3.61). This ambiguity is very similar to what I have discussed in connection with the spin current in Section 2.3.

A final remark is concerned with the introduction of electromagnetic form factors for the nucleons. In principle, they should be a natural consequence of a complete dynamical theory including the internal structure of the participating particles. Since such a theory is still lacking, one has to use more phenomenological methods, the simplest of which is to put in by hand the nucleon e.m. factors into the one-body charge and current operators. That means, the Fourier transforms of the operators in (2.63), (2.64) and (2.67) are modified according to

$$\rho_{[1]}(\mathbf{q}) = \sum_l G_{E,l}(\mathbf{q})e^{i\mathbf{q}\cdot\mathbf{r}_l}\,, \qquad (3.66)$$

$$\mathbf{J}_{[1]}(\mathbf{q}) = \frac{1}{2M}\sum_l \left(G_{E,l}(\mathbf{q})\{\mathbf{p}_l, e^{i\mathbf{q}\cdot\mathbf{r}_l}\} + iG_{M,l}(\mathbf{q})\boldsymbol{\sigma}_l\times\mathbf{q}e^{i\mathbf{q}\cdot\mathbf{r}_l}\right). \qquad (3.67)$$

It is important that in the charge and convection current density operators of (3.66) and (3.67) the same form factor G_E occurs, otherwise the current conservation relation (2.72) would not be fulfilled. By the same argument, it is easy to see that for the exchange current the isovector part of the nucleon charge form factor G_E^V has to be taken for both the pair and exchange part as long as one restricts oneself to the strictly non-relativistic domain [FaA 76,FrF 76]. However, at higher momentum transfer one leaves this non-relativistic regime and there a controversy has developed around the question which form factors should appear in the exchange current, e.g., whether the Dirac form factor F_1^V is more appropriate and the interested reader is referred to [Are 87,SiL 88], where a critical discussion of the various arguments is presented. Furthermore, one could always add a completely transverse two-body current like, e.g., the one in (3.64) with a different e.m. form factor without violating current conservation (see also [GrR 87]). But then the theory looses its predictive power due to this arbitrariness.

4 The Model of Nuclear Isobar Configurations

With respect to the intrinsic nucleon degrees of freedom (d.o.f.), one can in principle include them in the same manner by effective operators as has been discussed in the preceding section for the meson d.o.f. Here I will describe an alternative method where one includes the internal d.o.f. in the nuclear wave function by introducing new configurations, called nuclear isobar configurations (IC), where one or several nucleons are internally excited [ArD 71,ArW 72,Gre 76,WeA 78]. The basic idea behind this approach is that nucleons inside a nucleus tend to deform or polarize each other in a close encounter. This polarization can then be described as internal excitation of the nucleons, i.e., isobars. In view of the large internal excitation energy of at least 300 MeV, a nucleon will remain only a short time in the excited isobar state and, therefore, the corresponding wave function component will have a short range structure and a small probability.

Compared to the effective operator method, this model has some advantage since the non-locality of the effective operators due to the intermediate propagation of the internally excited nucleons is automatically included in the wave function description. In both approaches, the internal nucleon d.o.f. are still treated in a phenomenological manner by taking the isobars as particles given by experiment and making simple models of their coupling to mesons.

4.1 Formal Developments

I will illustrate this model by considering for convenience the two-nucleon system. The generalization to more complex nuclei is then obvious. Thus in the model of nuclear isobar configurations the wave function of the two-nucleon system has the general form

$$\Phi(1,2) = \sum_{N_1 N_2} \Phi_{N_1 N_2}(1,2)\,. \tag{4.1}$$

Here, $\Phi_{N_1 N_2}$ denotes the wave function component of Φ for the configuration $(N_1 N_2)$, where the various intrinsic nucleon states (isobars) are represented by N_i running over $N = N(939)$, $\Delta(1232)$, $N'(1440)\ldots$ Such a component can be written in more detail

$$\Phi_{N_1 N_2}(1,2) = \left\{ \begin{array}{ll} \psi_{N_1 N_1}(1,2)\phi_{N_1 N_1}(1,2)\,, & N_2 = N_1 \\ \mathcal{A}_{N_1 N_2}(\psi_{N_1 N_2}(1,2)\phi_{N_1 N_2}(1,2))\,, & N_2 \neq N_1 \end{array} \right\}\,, \tag{4.2}$$

where $\psi_{N_1N_2}$ denotes the orbital, spin and isospin part of the isobar configuration. It is totally antisymmetric under exchange of two particles having the same intrinsic state. For nucleons in different intrinsic states, $\mathcal{A}_{N_1N_2}$ takes care of the proper antisymmetrization (generalized Pauli principle). The remaining intrinsic quantum numbers beyond spin and isospin are contained in $\phi_{N_1N_2}$ which is a product of the intrinsic states $|N_i\rangle$. For $N_1 = N_2$ it is symmetric.

The total Hamiltonian of the two-nucleon system has the form

$$H = \sum_{i=1}^{2}(T(i) + H_{in}(i)) + V(1,2)\,. \tag{4.3}$$

It includes the kinetic energy T, a potential $V(1,2)$ and an intrinsic Hamiltonian H_{in} describing the internal nucleon dynamics. All three terms depend on intrinsic d.o.f., which we do not treat explicitly but describe them by the matrix elements between intrinsic nucleon states $|N_i\rangle$, i.e.,

$$\langle N_i|T|N_j\rangle = \frac{\mathbf{p}^{\,2}}{2M_{N_i}}\delta_{N_iN_j}\,, \tag{4.4}$$

$$\langle N_i|H_{in}|N_j\rangle = (M_{N_i} - M)\delta_{N_iN_j}\,, \tag{4.5}$$

$$\langle N_iN_{i'}|V(1,2)|N_jN_{j'}\rangle = V_{N_iN_{i'}N_jN_{j'}}\,, \tag{4.6}$$

where M_{N_i} denotes the mass of the isobar N_i and M the nucleon mass.

The center-of-mass motion can be separated out. Projecting onto a specific isobar configuration, one obtains a system of coupled equations for the various components of the relative motion

$$\langle \Phi_{N_1N_2}|H_{N_1N_2} - E|\Phi_{N_1N_2}\rangle = -\sum_{N_1'N_2'\neq N_1N_2} \langle \Phi_{N_1N_2}|V|\Phi_{N_1'N_2'}\rangle\,, \tag{4.7}$$

where E denotes the energy of the two-nucleon system in the c.m. system and

$$H_{N_1N_2} = \frac{\mathbf{p}^{\,2}}{2\mu_{N_1N_2}} + \Delta_{N_1N_2} + V_{N_1N_2} \tag{4.8}$$

the diagonal part of the Hamiltonian with respect to the isobar configuration N_1N_2. The relative momentum is denoted by $\mathbf{p}$. Furthermore, the following abbreviations have been used

$$\mu_{N_1N_2} = \frac{M_{N_1}M_{N_2}}{M_{N_1} + M_{N_2}}\,, \tag{4.9}$$

$$\Delta_{N_1N_2} = M_{N_1} + M_{N_2} - 2M\,; \tag{4.10}$$

$$V_{N_1N_2} = V_{N_1N_2N_1N_2}\,. \tag{4.11}$$

The various channels are coupled by the non-diagonal transition potentials $V_{N_1'N_2'N_1N_2}$ which are taken from a meson exchange model. For details see [WeA 78].

4.2 The Impulse Approximation

A complete solution of the system of coupled equations (4.7) is in general rather involved. However, at low excitation energies - low compared to the pion mass - one expects a small admixture probability for those configurations where one or several nucleons are in an internally excited state. In this case, an approximate solution - usually called impulse approximation (IA) - has been found to give quite reliable results [ArD 71]. I will first discuss this impulse approximation.

In IA one first generates the NN-channel wave function from the conventional two-nucleon Schrödinger equation

$$\left(\frac{\mathbf{p}^2}{2\mu_{NN}}+\overline{V}_{NN}-E\right)\Phi_{NN}=0\,, \tag{4.12}$$

where $\overline{V}_{NN}$ is a realistic NN potential fitted to NN scattering data. Thus it includes implicitly the dispersive contribution from other isobar channels, i.e., intermediate isobar excitation, while it is not included in V_{NN}, the diagonal interaction which enters in the equation for the NN channel in the coupled channel (CC) approach

$$\left(\frac{\mathbf{p}^2}{2\mu_{NN}}+V_{NN}-E\right)\Phi_{NN}=-\sum_{N_1'N_2'\neq NN}V_{NN,N_1N_2}\Phi_{N_1N_2}\,. \tag{4.13}$$

One can formally eliminate from (4.7) the other channels by retaining only their coupling to the NN channel and finds then the approximate relation

$$\overline{V}_{NN}=V_{NN}-\sum_{N_1'N_2'\neq NN}V_{NN,N_1N_2}(H_{N_1N_2}-E)^{-1}V_{N_1N_2,NN}\,. \tag{4.14}$$

It clearly shows the above mentioned fact, that $\overline{V}_{NN}$ contains implicitly a dispersive part from intermediate IC.

Once the NN wave function has been determined from (4.12), one can find an approximate expression for the component $\Phi_{N_1N_2}$ ($N_1N_2\neq NN$) neglecting in (4.7) the diagonal interaction $V_{N_1N_2}$ and keeping on the right side the dominant NN component only, i.e.,

$$\Phi_{N_1N_2}=-(H_{N_1N_2}-E)^{-1}V_{N_1N_2,NN}\Phi_{NN}\,. \tag{4.15}$$

This is called the impulse approximation (IA).

4.3 The Coupled Channel Approach

With increasing energy, in particular when approaching the Δ excitation region, one cannot expect the IA to be valid any more. Then one has to use the CC approach. As an example, I will consider the coupled system for the configurations NN, $N\Delta$ and $\Delta\Delta$. In matrix form, (4.7) reads then

$$\begin{pmatrix} H_{NN}-E & V_{NN,N\Delta} & V_{NN,\Delta\Delta}\\ V_{N\Delta,NN} & H_{N\Delta}-E & V_{N\Delta,\Delta\Delta}\\ V_{\Delta\Delta,NN} & V_{\Delta\Delta,N\Delta} & H_{\Delta\Delta}-E\end{pmatrix}\begin{pmatrix}|NN\rangle\\ |N\Delta\rangle\\ |\Delta\Delta\rangle\end{pmatrix}=0\,. \tag{4.16}$$

For the diagonal and transition potentials one may use π- and ρ-exchange potentials. For example, the $NN \to N\Delta$ transition potential has a form analogous to (3.52)

$$V^{\pi}_{N\Delta,NN}(r_{12}) = -\frac{f_{NN\pi} f_{\Delta N\pi}}{m_\pi^2} \boldsymbol{\tau}_{\Delta N} \cdot \boldsymbol{\tau}_{NN} (\boldsymbol{\sigma}_{\Delta N} \cdot \boldsymbol{\nabla}_2)(\boldsymbol{\sigma}_{NN} \cdot \boldsymbol{\nabla}_1) J_{m_\pi}(r_{12}), \quad (4.17)$$

where $\boldsymbol{\tau}_{\Delta N}$ and $\boldsymbol{\sigma}_{\Delta N}$ denote the isospin and spin transition operators from N to Δ. The $\Delta N\pi$-coupling constant can be determined from the decay $\Delta \to N + \pi$ or from quark model predictions. For further details one may consult [Gre 76,WeA 78,LeA 87].

Only for the NN channel such a model would be too simplistic. On the other hand, one cannot just take a realistic potential because it contains already dispersive contributions as mentioned above. One first has to remove these, and several methods have been devised. Details can be found in [GrS 79,LeA 87]. A more ambitious approach is to start from a general semi-phenomenological ansatz for the various diagonal and transition potentials and to fix then the open parameters by fitting NN scattering data. The Argonne V_{28} potential has been constructed in this way for the NN, $N\Delta$ and $\Delta\Delta$ channels [WiS 84].

4.4 Electromagnetic Current Operators

With respect to electromagnetic reactions, one then has to allow also for the interaction with isobar d.o.f. in the electromagnetic current operators in order to describe diagrams of the type shown in Fig. 4.1. The first diagram describes the electromagnetic excitation of a Δ isobar and subsequent one-pion-exchange. It is the most important one and describes above π-production threshold the resonance pion photoproduction. In terms of an effective operator description it is a two-body one-pion-exchange current diagram while the second one describes a two-pion-exchange contribution where the photon interacts with the intermediate Δ isobar. In the framework of nuclear isobar configurations, both diagrams are described by one-body contributions of the N-Δ-transition and the diagonal Δ-current, respectively.

In order to derive the formal expressions for these currents, one can start from the most general relativistic forms using a Rarita-Schwinger description for the 3/2-spin Δ-particle. I will not go through the details of the derivation which may be found in [WeA 78] but will only give the relevant non-relativistic expressions

$$\mathbf{J}_{\Delta N}(\mathbf{q}) = \frac{i}{2M} G^{\Delta N}_{M1}(\mathbf{q}) \boldsymbol{\sigma}_{\Delta N} \times \mathbf{q} e^{i\mathbf{q}\cdot\mathbf{r}} \quad (4.18)$$

$$\mathbf{J}_{\Delta}(\mathbf{q}) = \frac{1}{2M_\Delta} \left(G^{\Delta}_{E}(\mathbf{q}) \{\mathbf{p}, e^{i\mathbf{q}\cdot\mathbf{r}}\} + i G^{\Delta}_{M}(\mathbf{q}) \boldsymbol{\sigma}_{\Delta} \times \mathbf{q} e^{i\mathbf{q}\cdot\mathbf{r}} \right), \quad (4.19)$$

with appropriate e.m. form factors. Here, I have neglected small contributions from $E2$ and have furthermore neglected the mass difference between the nucleon and the Δ relative to the nucleon mass. These expressions will allow to calculate the isobar contributions to the electromagnetic processes to be discussed later.

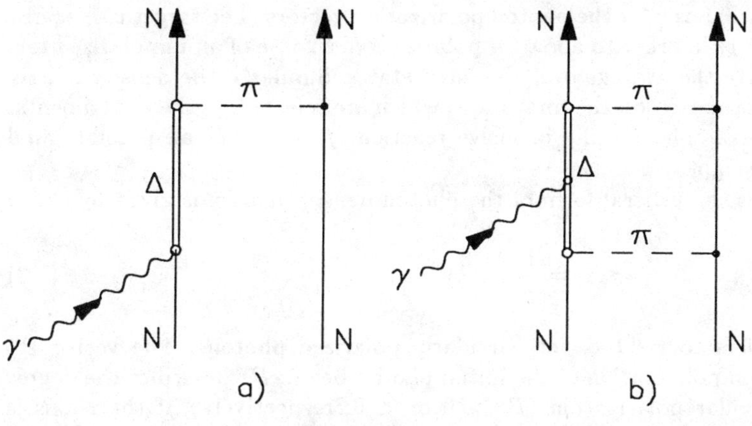

Fig. 4.1 Two-body current contributions from intermediate Δ-excitation: (a) photoexcitation of a Δ isobar, (b) diagonal Δ-current contribution.

5 Electromagnetic Processes

It was already mentioned in the introduction that the basic features of the electromagnetic interaction are (i) that it is well known already from classical macroscopic phenomena, and (ii) that it is rather weak due to the smallness of the fine structure constant ($\alpha \approx 1/137$). This smallness constitutes both an advantage and a disadvantage. The advantage is, that already lowest order perturbation treatment is often very reliable, while the disadvantage is that the cross sections for photoreactions and electron scattering are considerably smaller than pure hadronic reactions. I will now discuss the most important one-photon processes in some detail, the formal aspects and what kind of information they provide on hadronic systems.

5.1 Photoabsorption and Electron Scattering

The lowest order processes are the one-photon processes like photon absorption (or emission) and electron scattering which is governed by the exchange of one virtual photon (see Fig. 5.1). The cross section for the absorption of a real photon with four momentum q is given by

$$d\sigma = \frac{(2\pi)^2\alpha}{\omega}\sum_{\lambda\lambda'}\widetilde{W}_{\lambda'\lambda}(\rho_i,\rho_f)\sigma_{\lambda\lambda'}\,, \tag{5.1}$$

with $\sigma_{\lambda\lambda'}$ describing the photon density matrix and

$$\widetilde{W}_{\lambda'\lambda}(\rho_i,\rho_f) = (2\pi)^6\sum_{i,f}\int d^3P_f\delta^{(4)}(P_f-P_i-q)\langle P_f|j_\lambda(0)|P_i\rangle\rho_i\langle P_i|j_{\lambda'}(0)|P_f\rangle\rho_f\,, \tag{5.2}$$

where the initial and final hadronic states are denoted by $|P_{i/f}\rangle$, respectively, without specifying any additional quantum numbers which might be necessary for characterizing the states, and the transverse components of the current operator by j_λ with

$\lambda = \pm 1$ for a spherical basis for the photon polarization vectors. The target is specified by a density matrix ρ_i in order to allow for polarization. In case of an unpolarized target it would just give the average over the intial states. Similarly, the density matrix ρ_f contains the information on the final states which are selected by the experimental arrangement. For example, for the inclusive reaction, f runs over all possible final states with equal weight.

For real photons the general form of the photon density matrix is given by

$$\sigma_{\lambda\lambda'} = \frac{1}{2}(1 + \mathbf{P}^\gamma \cdot \boldsymbol{\sigma})_{\lambda\lambda'}\,, \tag{5.3}$$

where $\lambda = \pm 1$ refers to right or left circularly polarized photons. The vector $\mathbf{P}^\gamma$ describes the photon polarization of the initial photon beam: P_z^γ describes the degree of right or left circular polarization ($P_z > 0$ or < 0, respectively), if the z-axis is chosen in the direction of the photon momentum, and $P_l^\gamma = ((P_x^\gamma)^2 + (P_y^\gamma)^2)^{1/2}$ the degree of linear polarization of the incoming beam. By a proper rotation around the z-axis one can reach $P_y^\gamma = 0$ and then one has linear polarization along the x- or y-axis if $P_x^\gamma < 0$ or $P_x^\gamma > 0$, respectively.

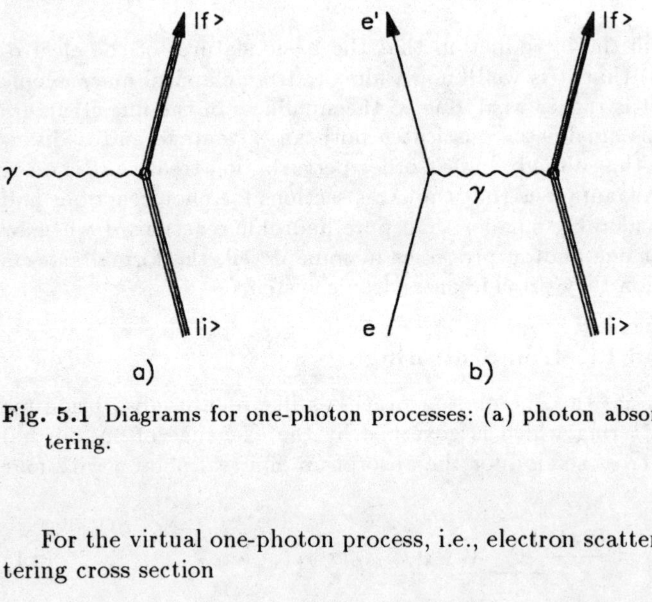

Fig. 5.1 Diagrams for one-photon processes: (a) photon absorption and (b) electron scattering.

For the virtual one-photon process, i.e., electron scattering, one finds for the scattering cross section

$$\frac{d^2\sigma}{dk_2^{lab}d\Omega_e^{lab}} = \frac{2\alpha^2}{(q_\mu^2)^2}\frac{k_2^{lab}}{k_1^{lab}}\sum_{\mu\nu}\eta^{\mu\nu}\widetilde{W}_{\mu\nu}(\rho_i,\rho_f)\,, \tag{5.4}$$

where k_1^{lab} and k_2^{lab} denote the lab frame momenta of the initial and the scattered electrons, respectively, while q_μ^2 is the four momentum transfer squared ($q = k_1 - k_2$). The lepton tensor $\eta^{\mu\nu}$ is determined by the electron kinematics, and the hadron tensor $\widetilde{W}_{\mu\nu}$ is a natural extension of (5.2) to all four covariant indices.

In view of current conservation, it is customary to eliminate the longitudinal current component (parallel to $\mathbf{q}$) by the charge component. Then one finds

$$\sum_{\mu\nu} \eta^{\mu\nu} \widetilde{W}_{\mu\nu} = \sum_{\lambda\lambda'} \tilde{\rho}_{\lambda\lambda'} \widetilde{W}_{\lambda'\lambda} \,, \tag{5.5}$$

where the indices λ, λ' now run over 0, the charge component, and ±1, the two transverse current components. Allowing for longitudinal electron polarization one finds

$$\tilde{\rho}_{\lambda\lambda'} = \rho_{\lambda\lambda'} + h\rho'_{\lambda\lambda'} \,, \tag{5.6}$$

where $|h|$ denotes the degree of longitudinal polarization with $h > 0$ referring to right and $h < 0$ to left helicity. The quantities $\rho_{\lambda\mu}$ and $\rho'_{\lambda\mu}$ ($(\lambda\mu) = (00)$, (11), (01), and (-11) corresponding to L, T, LT, and TT, respectively), are given by

$$\rho_L = \beta^2 Q_\nu^2 \frac{\xi^2}{2\eta} \,, \quad \rho_{LT} = \beta Q_\nu^2 \frac{\xi}{\eta} \sqrt{\frac{\xi+\eta}{8}} \,, \tag{5.7}$$

$$\rho_T = \frac{1}{2} Q_\nu^2 (1 + \frac{\xi}{2\eta}) \,, \quad \rho_{TT} = -Q_\nu^2 \frac{\xi}{4\eta} \,, \tag{5.8}$$

$$\rho'_{LT} = \frac{1}{2} \beta Q_\nu^2 \frac{\xi}{\sqrt{2\eta}} \,, \quad \rho'_T = \frac{1}{2} Q_\nu^2 \sqrt{\frac{\xi+\eta}{\eta}} \,, \tag{5.9}$$

with

$$\beta = \frac{|\mathbf{q}_{lab}|}{|\mathbf{q}_{c.m.}|} \,, \quad \xi = \frac{Q_\nu^2}{\mathbf{q}_{lab}^2} \,, \quad \eta = \tan^2(\frac{\theta_e^{lab}}{2}) \,, \quad Q_\nu^2 = -q_\nu^2 \,, \tag{5.10}$$

where β expresses the boost from the lab to the c.m.-system in which the nuclear current elements are evaluated. Note that $Q_\nu^2 > 0$ according to our metric.

Thus finally, one obtains the well known expression

$$\frac{d^2\sigma}{dk_2^{lab} d\Omega_e^{lab}} = \frac{8\pi\alpha^2}{(q_\mu^2)^2} \frac{k_2^{lab}}{k_1^{lab}} \Big(\rho_L \tilde{f}_L + \rho_T \tilde{f}_T + \rho_{LT} \tilde{f}_{LT} + \rho_{TT} \tilde{f}_{TT} + h(\rho'_T \tilde{f}'_T + \rho'_{LT} \tilde{f}'_{LT}) \Big) \,, \tag{5.11}$$

with four structure functions for unpolarized electrons

$$\tilde{f}_L = \frac{1}{4\pi} \widetilde{W}_{00} \,, \quad \tilde{f}_T = \frac{1}{2\pi} \Re(\widetilde{W}_{11}) \,, \tag{5.12}$$

$$\tilde{f}_{LT} = \frac{1}{2\pi} \Re(\widetilde{W}_{01} + \widetilde{W}_{10}) \,, \quad \tilde{f}_{TT} = \frac{1}{2\pi} \Re(\widetilde{W}_{-11}) \,, \tag{5.13}$$

and two for longitudinally polarized electrons

$$\tilde{f}'_T = \frac{1}{2\pi} \Im(\widetilde{W}_{11}) \,, \quad \tilde{f}'_{LT} = \frac{1}{2\pi} \Im(\widetilde{W}_{01} + \widetilde{W}_{10}) \,, \tag{5.14}$$

where I have used $(\widetilde{W}_{\lambda'\lambda})^* = (-)^{\lambda'+\lambda} \widetilde{W}_{-\lambda-\lambda'}$.

Comparing (5.1) with (5.11) one sees that in electron scattering the virtual photon has in addition to the transverse polarization also a longitudinal one and its density matrix is given by

$$\sigma_{\lambda\lambda'} = \frac{\alpha\omega}{2\pi^2(q_\mu^2)^2}\frac{k_2^{lab}}{k_1^{lab}}\rho_{\lambda\lambda'}dk_2^{lab}d\Omega_2^{lab}\,. \tag{5.15}$$

The explicit form of the structure functions depends on the way how the initial state is prepared and how the final state is analyzed in the specific experimental arrangement. They are in general hermitean forms of the charge and current matrix elements, depend on the variables necessary to specify the final state and contain all available information on the hadronic system. How detailed the information is depends on the specific experimental conditions.

The least detailed information is obtained in the inclusive process, where no analysis of the final hadronic states is done. In this case one finds the following expressions for the cross sections in photoabsorption

$$\sigma_{tot} = \frac{2\pi^2\alpha}{\omega}F_T\,, \tag{5.16}$$

and electron scattering

$$\frac{d^2\sigma}{dk_2^{lab}d\Omega_e^{lab}} = \frac{8\pi\alpha^2}{(q_\mu^2)^2}\frac{k_2^{lab}}{k_1^{lab}}(\rho_L F_L + \rho_T F_T)\,, \tag{5.17}$$

where the only two appearing form factors, the longitudinal and transverse ones, are defined by

$$F_L = \frac{1}{4\pi}\overline{W}_{00}\,, \tag{5.18}$$

$$F_T = \frac{1}{4\pi}\sum_{\lambda=\pm1}\overline{W}_{\lambda\lambda}\,. \tag{5.19}$$

Here the bar over the hadronic tensor indicates that the complete average over all initial and the sum over all possible final states has to be taken in (5.2). These form factors are functions of energy and momentum transfer. It is a characteristic feature of this inclusive process that no interference terms between the different charge and current components appear. Such interference terms, however, will occur for an exclusive reaction where specific initial and/or final states are selected. There are different ways of how to realize an exclusive process. For example, if the final state is a scattering state one may select one or several outgoing particles in a coincidence experiment. Or one may go exclusive by using a polarized target and/or analyzing the polarization of outgoing particles. In all such cases one will be able to get more detailed information on the basic charge and current matrix elements, in particular, on their relative phases. An example will be discussed below.

The various structure functions are completely determined by the charge and transverse current matrix elements according to (5.2). Separating the center-of-mass motion of the hadronic wave functions, e.g.,

$$|P_i\rangle = \frac{1}{(2\pi)^{3/2}}e^{i\mathbf{P}_i\cdot\mathbf{R}}U(P_i)|i\rangle\,, \tag{5.20}$$

and a corresponding expression for the final state $|P_f\rangle$, where $U(P_i)$ denotes the boost operator for boosting the intrinsic rest frame wave function $|i\rangle$ to a total momentum P_i, one can relate the matrix element of the electromagnetic current operator to a matrix element of an intrinsic current operator between intrinsic rest frame wave functions

$$\langle P_f|j_\lambda(0)|P_i\rangle = \frac{1}{(2\pi)^3}\langle f|J_\lambda(\mathbf{q},\mathbf{P})|i\rangle\,. \tag{5.21}$$

Here, the intrinsic current operator in (5.21) is defined by

$$J_\lambda(\mathbf{q},\mathbf{P}) = \int d^3R\, U^\dagger(\mathbf{P}_f)j_\lambda(0)U(\mathbf{P}_i)e^{-i\mathbf{q}\cdot\mathbf{R}}\,, \tag{5.22}$$

with

$$q = P_f - P_i\,, \quad P = P_f + P_i\,. \tag{5.23}$$

Note that the boost operator $U(\mathbf{P})$ is the identity operator for a non-relativistic description. If one wants to include relativistic effects, the boost operator has the general form $U(\mathbf{P}) = \exp(-i\chi(\mathbf{P}))$, where $\chi(\mathbf{P})$ is a hermitean operator depending on the internal dynamics and acting on the intrinsic variables. For more details I refer to [Fri 75b].

For a non-relativistic system the operators are then given by the Fourier transforms

$$J_0(\mathbf{q},\mathbf{P}) = \tilde{\rho}(\mathbf{q})\,, \tag{5.24}$$

$$\mathbf{J}(\mathbf{q},\mathbf{P}) = \mathbf{J}_{int}(\mathbf{q}) + \frac{\mathbf{P}}{2M_A}\tilde{\rho}(\mathbf{q})\,, \tag{5.25}$$

where the operators refer to intrinsic variables relative to the c.m.-system and M_A denotes the total mass of the hadron. They are then evaluated using the multipole decomposition of (2.102).

The main differences between photoabsorption and electron scattering or real and virtual photon processes, respectively, are:

(i) For real photons one has a fixed relation between energy and momentum transfer ($\mathbf{q}^2 = \omega^2$) while for the virtual photons of electron scattering the four momentum is space-like ($\mathbf{q}^2 \geq \omega^2$) allowing an independent variation of energy and momentum transfer within the spacelike region.

(ii) Real photons have transverse polarization so that only the transverse current density contributes while virtual photons have both transverse and longitudinal polarization so that also the longitudinal current density or via charge conservation the charge density contributes.

Qualitatively one may distinguish different regions of the nuclear response to photoabsorption. At low energies below particle emission threshold one finds sharp resonances corresponding to the excitation of bound excited nuclear states. Above particle emission threshold for photon energies of 10 to 30 MeV the dominant feature is the giant dipole resonance. It is a collective nuclear mode which exhausts almost one classical TRK sum rule (see Section 5.3). It can be described in a phenomenological two-fluid model as an oscillation of the proton against the neutron fluid or in a microscopic description as a coherent one-particle one-hole excitation. Increasing the energy, one enters the region of short-range correlations or the so-called quasideuteron region, where the dominant process is the absorption of the photon by a correlated two-nucleon pair (n-p pair) emitted mainly back to back which is dominated by contributions from meson exchange currents. It extends roughly up to pion production threshold. Further up in energy, the dominant mode of absorption is the isobar excitation of a single nucleon to a Δ or to higher nucleon resonances.

In electron scattering one can explore the same dynamical regions but in addition one can exploit the dependence on the momentum transfer. It allows, for example, in the elastic process "cum grano salis" the determination of ground state charge and magnetization densities. Furthermore, a new mechanism appears, the so-called quasi-free process. It corresponds to the situation where the electron is scattered off a bound nucleon transferring energy and momentum in such a manner that the struck nucleon is on-shell in one instant and leaves the nucleus without any further relevant interaction with the residual nucleons which thus merely act as spectators. For this reason, the quasi-free process is essentially determined by the momentum distribution of the struck nucleon and one can use this process, again "cum grano salis", for the determination of the single nucleon wave function.

5.2 Deuteron Break-Up

Now I will turn to the discussion of subnuclear degrees of freedom in the two-body system, i.e., in deuteron disintegration by photons and electrons. The special role of the electromagnetic deuteron break-up is a consequence of (i) the simple structure of the deuteron which allows exact solutions at least in the non-relativistic domain, and (ii) the specific features of the electromagnetic probe as discussed in the preceding sections. The continued interest in this process has persisted over by now almost sixty years because the two-body system is in fact a unique laboratory, in particular for the study of subnuclear d.o.f., for which one has clear evidence. I will limit myself to the non-relativistic approach with a complete treatment of final state interaction and, in some cases, with inclusion of lowest order relativistic corrections. The intrinsic two-body wave functions of intial and final states are obtained by solving the non-relativistic Schrödinger equation with a realistic NN potential, fitted to the experimental NN scattering data.

5.2.1 Photodisintegration

Let me first consider photodisintegration of the deuteron. I will select from the large number of observables like total cross section, angular distribution and polarization observables only a few which clearly illustrate the role of subnuclear d.o.f. in this process. For more details, in particular for a thorough comparison of the theory with experiments I refer to the review of [ArS 91].

In order to distinguish the various ingredients, which enter into the theoretical description, I will use the following nomenclature:

(i) "one-body" for pure one-body current contributions, i.e., no use of Siegert operators,

(ii) "normal (N)" for one-body current plus meson exchange currents as incorporated in the Siegert operators (Siegert MEC),

(iii) "N+MEC" for normal plus explicit exchange currents beyond the Siegert operators,

(iv) "N+MEC+IC" for additional isobar configurations and MEC contributions involving isobar configurations,

(v) "N+MEC+IC+RC" for additional relativistic corrections to the one- and two-body currents.

I will present results obtained with the Bonn r-space potential [MaH 87,Mac 89] as a prototype of a realistic NN-potential, since the different realistic potentials give very similar results. They are based on a calculation in which the different OBE approximations (r-space, q-space, time-ordered) of the Bonn full potential model have been studied in detail [ScA 89b]. But in some cases I will also present results obtained with other realistic potentials like the Paris potential [LaL 80].

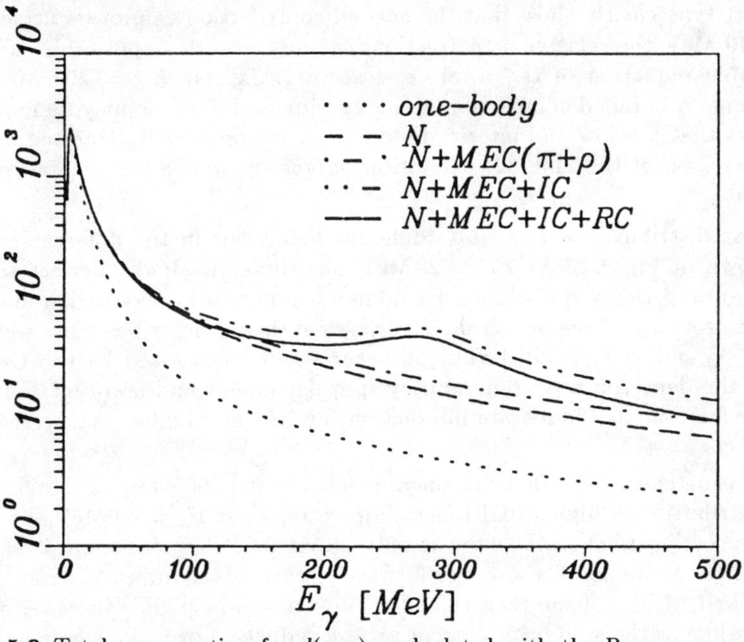

Fig. 5.2 Total cross section for $d(\gamma,p)n$ calculated with the Bonn r-space potential with various contributions from MEC, IC and RC.

Fig. 5.2 shows the theoretical total cross section for two-body disintegration of the deuteron between threshold and 500 MeV photon energy. Beyond the maximum at twice the deuteron binding energy $E_\gamma = 2\epsilon_B = 4.45$ MeV, the cross section drops rapidly with increasing energy, up to π-threshold by a factor of about 50. Then a broad shoulder around $E_\gamma = 260$ MeV indicates the excitation of the $\Delta(1232)$ resonance followed by a further fall-off.

Also shown are the different contributions from MEC, IC and RC. One readily notes the overwhelming importance of MEC above about $E_\gamma = 20$ MeV compared to the pure one-body current. But as already mentioned, most of it is included in the Siegert operators of "N" which constitutes the classical theory. Above $E_\gamma = 80$ MeV the one-body part gives a constant 30 percent contribution. For complex nuclei, it is this energy region where the so-called quasi-deuteron process is the dominant absorption mode. Hence, also in complex nuclei MEC give the dominant contribution in this region. Additional explicit MEC beyond the Siegert operators from π- and ρ-exchange show an almost linear relative increase with energy, e.g., at $E_\gamma = 200$ MeV of about 20 percent.

Isobar configurations which have only a small effect at low energy become of increasing importance above $E_\gamma = 100$ MeV. In particular, the $N\Delta$ configuration in the final state dominates the total cross section for $E_\gamma = 200 - 300$ MeV. In this energy region, where a coupled channel approach is more appropriate, IC contribute by about 60 - 65 percent to the total cross section, mostly through the $M1$-$N\Delta$ excitation. On the other hand, the remaining 35 - 40 percent of "N" being essentially MEC of Siegert type clearly show that the normal contribution cannot be neglected. Above $E_\gamma = 40$ MeV the relativistic corrections, mainly from the spin-orbit current, lead to a sizable reduction of the total cross section, e.g., at $E_\gamma = 200$ MeV by about 15 percent. A detailed comparison with experimental data, as may be found in [ArS 91], shows that the low and intermediate energy region is well described by the theory. In the region of the $\Delta(1232)$ excitation, larger theoretical uncertainties still exist at present.

The angular distributions show quite different behaviour in the different energy regions as shown in Fig. 5.3. At $E_\gamma = 20\,\text{MeV}$ one finds an almost perfect $\sin^2\theta_p$ behaviour because of the overwhelming E1-dominance in this energy region and the small optical anisotropy of the deuteron. The classical theory describes quite well the data. Only at 0^0 and 180^0 small deviations occur which are caused by the D-wave component in the deuteron wave function, by spin dependent and non-central forces like the tensor force in the final state interaction, by $M1$ and higher multipoles and relativistic effects.

With increasing energy the shape of the angular distribution changes significantly due to the contribution of higher multipoles. For example, at $E_\gamma = 60$ MeV the ratio of the 0^0-cross section to the maximum is only about 0.5. The maximum is shifted from 90^0 to 70^0 and one notes a stronger forward-backward asymmetry. Around the maximum explicit MEC beyond Siegert and IC give a non-negligible increase of the cross section, whereas the spin-orbit current gives a reduction over the whole angular range, but most effective around 0^0 and 180^0. The pure one-body current contributes less than 50 percent.

Increasing the energy further, this tendency becomes more and more pronounced. The maximum appears at more forward angles and the whole angular distribution becomes flatter as is seen in Fig. 5.3 for $E_\gamma = 140$ MeV. The contribution of the one-body current decreases to about 25 percent. Explicit MEC beyond Siegert increase the cross section sizably between 40^0 and 140^0 and IC become even larger than explicit MEC, while relativistic effects, mainly the spin-orbit current, give a significant reduction. At this energy one finds already a larger potential model dependence and also the experimental data scatter considerably. On the average the theory appears systematically higher than experimental data [ArS 91].

Furthermore, I would like to consider one example of polarization observables which contain a wealth of additional information on the dynamical properties of the system beyond the one in the differential cross section. While the latter one is the sum of absolute squares of the t-matrix elements, the polarization observables contain

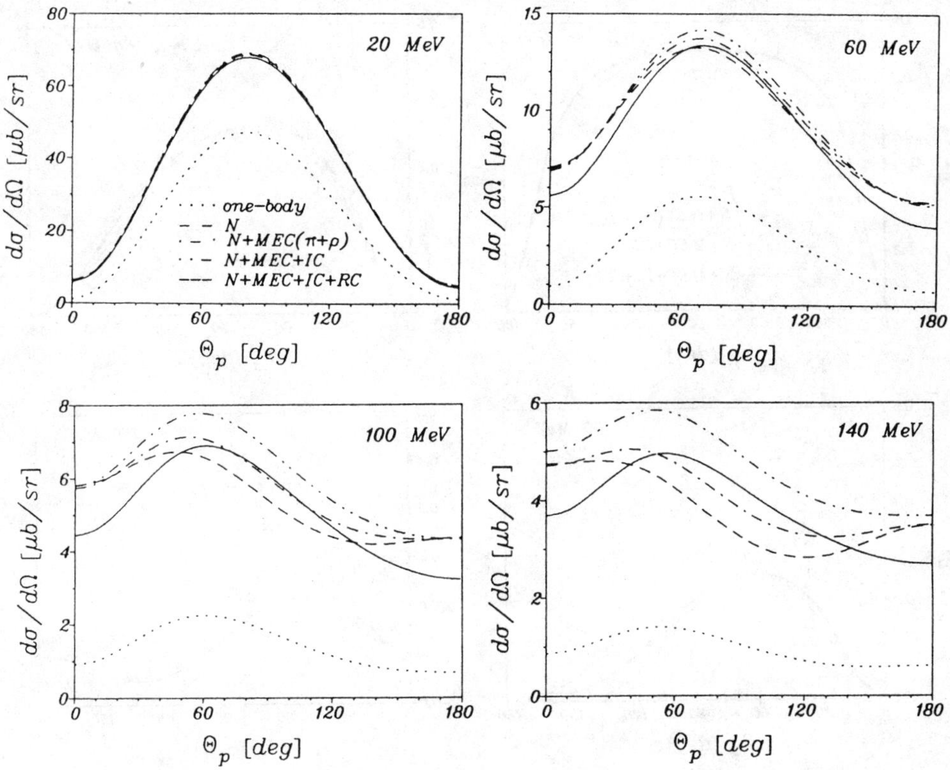

Fig. 5.3 Differential cross section at various energies with different contributions from MEC, IC and RC for the Bonn r-space potential.

interference terms of the t-matrix elements in different combinations. Thus, certain interesting dynamical effects may leave little signature on the differential cross section affecting only small amplitudes but they can be amplified considerably through interference effects in some polarization observables. A systematic study has recently been completed and can be found in [Are 88,ArS 91,ScA 91a,ScA 91b].

Fig. 5.4 shows the asymmetry $\Sigma^l(\theta)$ for linearly polarized photons. For pure $E1$, the asymmetry would be equal to unity, but the increasing importance of other multipoles ($M1$, $E2$,...) reduce Σ^l significantly with increasing energy. Furthermore, one notes a relatively stronger sensitivity to explicit MEC and IC compared to the angular distributions. As is reviewed in [ArS 91], the agreement with experimental data would clearly fail for the classical theory without consideration of subnuclear d.o.f. Only if the latter are included a satisfactory description of the experimental data is achieved.

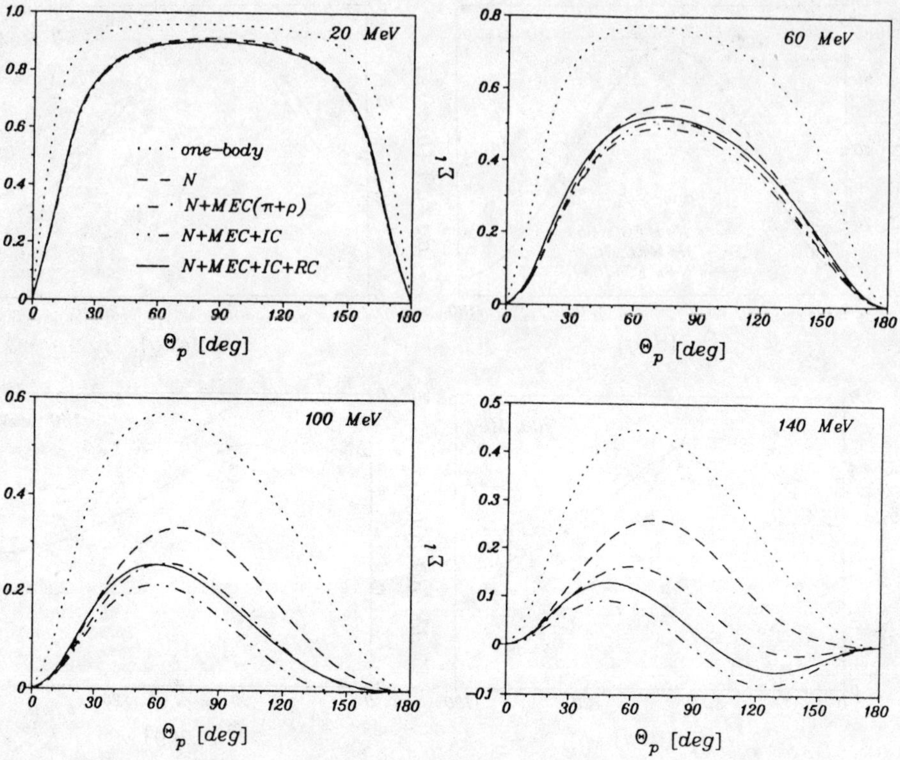

Fig. 5.4 Photon asymmetry Σ^l at various energies with different contributions from MEC, IC and RC for the Bonn r-space potential.

5.2.2 Electrodisintegration

Turning now to electrodisintegration, I will first consider the inclusive process $d(e,e')pn$, where no analysis is made of the hadronic final state, and I will concentrate on the threshold region which is dominated by the excitation of the antibound $^1S_0$-resonance in NN scattering at very low energies. It is the inverse process to thermal n-p radiative capture, which proceeds via $M1$ transition and where MEC and IC give about a 10 percent enhancement [RiB 72,Are 81b]. The advantage of electron scattering in having the momentum transfer at one's disposition becomes now apparent since the relatively small effect of subnuclear d.o.f. in the real photon process can be amplified considerably. The reason for this lies in the fact that with increasing momentum transfer the one-body contribution drops rapidly due to a destructive interference of S- and D-wave contributions and thus the distribution of the momentum transfer onto both nucleons via the two-body operators becomes more favourable. This is shown in Fig. 5.5 where the transverse form factor is shown as obtained from the Rosenbluth separation [SiB 79]. While the longitudinal form factor F_L is well described by the classical theory supporting the Siegert hypothesis, that the charge density is little affected by exchange effects, one finds a large discrepancy in F_T which is only resolved if MEC and IC are added.

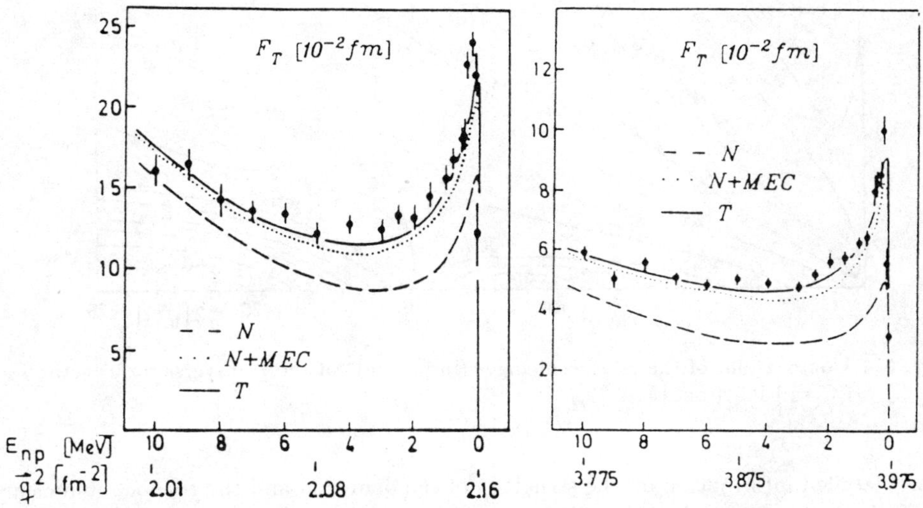

Fig. 5.5 The transverse deuteron form factor in two regions of momentum transfer compared to calculations with the Hamada-Johnston potential for normal (N) theory and additional MEC and IC (T) (from [SiB 79]).

The situation for higher momentum transfers is less clear due to larger uncertainties which arise from electromagnetic and hadronic form factors, potential model dependence and relativistic corrections. Above about $\mathbf{q}^2 = 15$ fm^{-2} the normal and MEC contributions interfere destructively and thus small changes in different calculations can result in large differences in the cross sections. Therefore, one finds considerable variation of the results depending on the potential model and whether one uses G_E^V or F_1^V as e.m. form factor for MEC and also on the form factor parametrization [SiL 88]. Any non-relativistic calculation in the region of high momentum transfer cannot be considered satisfactory as long as the role of relativistic effects and consistent treatment of e.m. form factors has not been clarified.

For a general survey of the role of subnuclear d.o.f. in F_T I show in Fig. 5.6 a contour plot of the relative contributions of MEC and IC to F_T in the $\mathbf{q}^2$–E_{np} plane, which I have taken from [FaA 79]. One readily notes the small influence of subnuclear d.o.f. along the quasi-free ridge as is expected. This kinematic region provides also a good testing ground for the conventional framework. Leaving this region, one finds increasing importance of subnuclear effects. They are particularly strong near the threshold region as discussed above and at higher E_{np} approaching the $\Delta(1232)$ excitation but for not too high momentum transfers.

Now, I will consider one example of a coincidence experiment which offers much

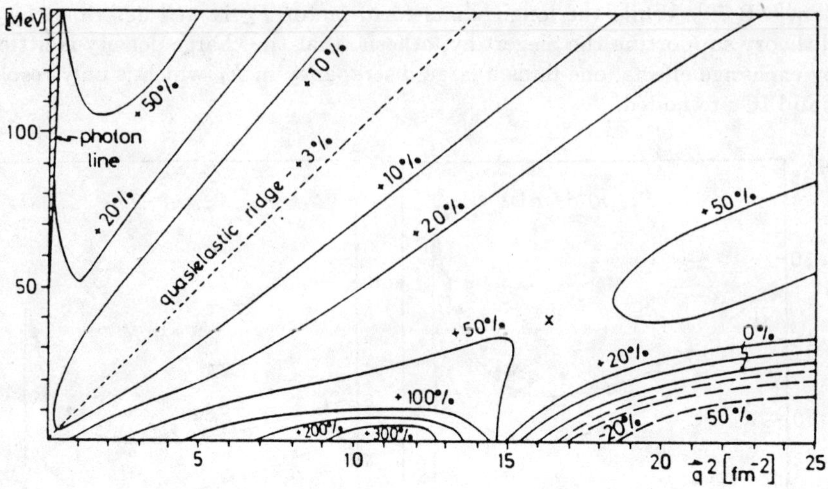

Fig. 5.6 Contour plot of the relative changes (in percent) of the transverse form factor F_T by MEC and IC (from [FaA 79]).

more detailed information on the structure of the deuteron and the role of subnuclear d.o.f. similar to the study of angular distributions in photodisintegration. As we have seen in Section 5.1, the coincidence cross section for unpolarized beam and target is governed by four structure functions $f_\alpha(\theta)$ ($\alpha = L, T, LT, TT$) (see 5.11)

$$\frac{d\sigma}{dk_2^{lab} d\Omega_e^{lab} d\Omega_{np}^{c.m.}} = \frac{\alpha}{6\pi^2 q_\mu^2} \frac{k_2^{lab}}{k_1^{lab}} (\rho_L f_L + \rho_T f_T + \rho_{LT} f_{LT} \cos\phi + \rho_{TT} f_{TT} \cos 2\phi) , \tag{5.26}$$

where the angles θ and ϕ describe the direction of the outgoing proton in the final cm-system with z-axis along $\mathbf{q}$. For details see [ArL 92]. Besides the diagonal and transverse structure functions, one has two additional longitudinal-transverse and transverse-transverse interference terms which each contains different information. This is illustrated for one specific energy and momentum transfer in Fig. 5.7.

The longitudinal structure function f_L shows a general forward peaking which originates from the charge interaction with the proton. f_L is only little affected by IC contributions. The transverse structure function f_T is much more influenced by MEC and IC leading to a drastic increase, in particular in the minimum around 90^0. Among the two interference functions, f_{LT} shows some sensitivity to subnuclear d.o.f., most pronounced in the forward minimum while f_{TT} is very strongly affected by contributions from MEC and IC leading to even a sign change. Even though they are considerably smaller than f_L and f_T, a separation of them will be rewarding for testing our theoretical understanding of subnuclear effects in greater detail.

Finally, I would like to briefly discuss the other aspect of deuteron electrodisintegration experiments, namely to use the deuteron as a neutron target in order to measure the electromagnetic form factors G_{En} and G_{Mn} of the neutron. According to what I have said above, the quasi-free electron scattering off the neutron within

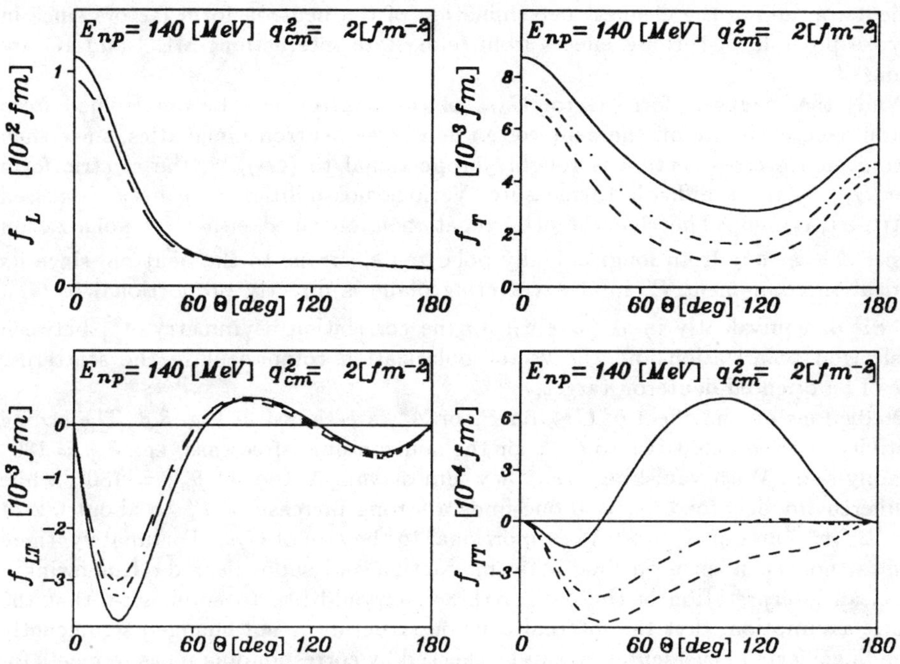

Fig. 5.7 The four structure functions at E_{np} = 140 MeV, q^2 = 2 fm^{-2} for normal theory (dashed) and with additional MEC (dash-dot) and MEC plus IC (full curves) calculated with the Paris potential [LaL 80].

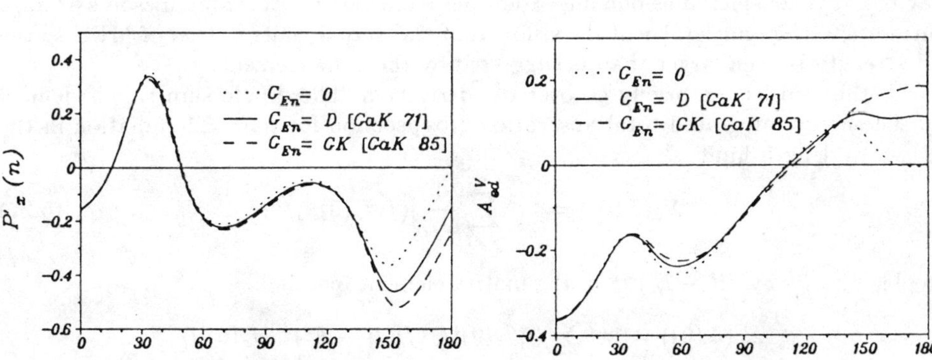

Fig. 5.8 Neutron polarization $P'_x(n)$ (left) and beam-target asymmetry $A^V_{ed}(\pi/2, 0)$ (right) for E_{np} = 120 MeV and q^2 = 12 fm^{-2} with vanishing and non-vanishing neutron electric form factor G_{En}.

the deuteron allows the cleanest determination of the neutron form factors since in this case possible disturbing effects from final state interaction, MEC and IC are minimized.

While the magnetic form factor G_{Mn} of the neutron can be determined from electron backscattering off the deuteron in quasi-free neutron kinematics, since then the coincidence cross section is roughly proportional to $(G_{Mn})^2$, the electric form factor G_{En} is more difficult to measure. Various possibilities have been discussed in [ArC 81,ArL 88]. The cleanest determination is obtained either by polarization transfer $d(\vec{e}, e'\, \vec{n})p$ from longitudinally polarized electrons to the neutron, since its polarization component P'_x in the scattering plane is directly proportional to G_{En} [ArC 81] or equivalently in $\vec{d}\,(\vec{e}, e'n)p$ to the correlation asymmetry A^V_{ed} between the electron polarization and the vector polarization component in the scattering plane of an oriented deuteron target.

Predictions on the effect of G_{En} on P'_x or A^V_{ed} are shown in Fig. 5.8. The strong sensitivity of both quantities to G_{En} for the neutron quasi-free case, i.e., $\theta_{np} = 180^0$ is readily seen. With vanishing G_{En} they almost vanish, too, at $\theta_{np} = 180^0$, while for different models for $G_{En} \neq 0$ one finds a strong increase of P'_x to about 0.2 at $\mathbf{q}^2 = 12\, fm^{-2}$ and of A^V_{ed} to 0.15 proportional to the size of G_{En}. Fortunately, there is almost no dependence on final state interaction and subnuclear d.o.f. allowing a very clean interpretation of the data [ArL 88]. I would like to emphasize, that the implicit assumption, that the internal neutron structure is not changed significantly by binding effects in the deuteron, can be checked by corresponding measurements for the outgoing proton in quasi-free kinematics and comparison with electron scattering on free protons.

5.3 Exchange Effects in the TRK Sum Rule

An early recognition of exchange effects in photonuclear reactions is found in the enhancement of the well-known Thomas-Reiche-Kuhn (TRK) dipole sum rule [Fee 36, Lev 60]. It is considered as one important piece of evidence of strong meson exchange current effects in nuclei. I will show now that the actual contribution of MEC to the $E1$ strength is even larger than is suggested by the enhancement.

To this end let me briefly go over the derivation. The dipole sum rule is defined by the energy integrated total absorption cross section for pure $E1$ radiation in the long-wave-length limit

$$\Sigma_0(\mathbf{J}\,(0)) = 4\pi^2 \sum_{f\neq i} \frac{1}{\omega_{fi}} |\langle f|J_z(0)|i\rangle|^2 \,. \tag{5.27}$$

Replacing ω_{fi}^{-1} by $(H - E_i)^{-1}$ in the matrix element one finds

$$\begin{aligned} \Sigma_0(\mathbf{J}\,(0)) &= 4\pi^2 \sum_f \langle i|J_z(0)|f\rangle\langle f|(H - E_i)^{-1} J_z(0)|i\rangle \\ &= 4\pi^2 \langle i|J_z(0)(H - E_i)^{-1} J_z(0)|i\rangle \,. \end{aligned} \tag{5.28}$$

In the last step I have used closure for the final states which is possible since the initial state does not contribute. Introducing now the Siegert operator from (2.112)

$$\mathbf{J}\,(0) = i[H, \mathbf{D}]\,, \tag{5.29}$$

one finds the well-known TRK sum rule

$$\Sigma_0(\mathbf{J}(0)) = 2\pi^2\langle i|[D_z,[H,D_z]]|i\rangle = \Sigma^{class}(1+\kappa)\,, \tag{5.30}$$

where the classical model-independent dipole sum rule is given by

$$\Sigma^{class} = 2\pi^2\langle i|[D_z,[T,D_z]]|i\rangle \tag{5.31}$$

$$= 2\pi^2\frac{e^2}{M}\frac{NZ}{A}\,. \tag{5.32}$$

The enhancement κ as defined by

$$\kappa = 2\pi^2\langle i|[D_z,[V,D_z]]|i\rangle/\Sigma^{class} \tag{5.33}$$

is usually considered as a measure of exchange current contributions. This interpretation of κ stems from the tacit identification of the classical sum rule Σ^{class} with the one-body contribution. But this is not exactly true because κ results from a combination of two opposite effects, namely from wave function changes due to the exchange character of the potential and from genuine MEC contributions [Are 83].

To this end, I will separate the total current into the one- and two-body contributions and take their expressions in the long-wave-length limit as given in (2.113,2.114). Then one finds for the one-body contribution [Are 83]

$$\begin{aligned}\Sigma_0(\mathbf{J}_{[1]}(0)) &= 4\pi^2\langle i|J_{[1]z}(0)(H-E_i)^{-1}J_{[1]z}(0)|i\rangle \\ &= 4\pi^2\langle i|[D_z,T](H-E_i)^{-1}[T,D_z]|i\rangle\,,\end{aligned} \tag{5.34}$$

while for the classical sum rule one gets

$$\begin{aligned}\Sigma^{class} &= 2\pi^2\langle i|[D_z,H](H-E_i)^{-1}[T,D_z]+h.c.|i\rangle \\ &= 2\pi^2\langle i|J_z(0)(H-E_i)^{-1}J_{[1]z}(0)+h.c.|i\rangle \\ &= \Sigma_0(\mathbf{J}_{[1]}(0))+\mathcal{J}\,,\end{aligned} \tag{5.35}$$

where

$$\begin{aligned}\mathcal{J} &= 2\pi^2\langle i|[D_z,V](H-E_i)^{-1}[T,D_z]+h.c.|i\rangle \\ &= 2\pi^2\langle i|J_{[2]z}(0)(H-E_i)^{-1}J_{[1]z}(0)+h.c.|i\rangle\end{aligned} \tag{5.36}$$

is an interference term between the contributions from one- and two-body currents. Hence we can describe Σ^{class} either as an interference between the total current and the one-body current or as the one-body sum rule $\Sigma_0(\mathbf{J}_{[1]}(0))$ plus the interference term $\mathcal{J}$ which turns out to be positive since exchange currents lead in general to an increase of the $E1$-matrix elements. In other words, the sum rule for the one-body current is smaller than the classical sum rule. This difference must be due to a wave function change by the exchange interaction since the current operators are the same.

On the other hand, one obtains for the difference of the total TRK sum rule and the one-body one

$$\Sigma_0(\mathbf{J}(0)) - \Sigma_0(\mathbf{J}_{[1]}(0)) = \kappa\Sigma^{class} + \mathcal{J}\,, \tag{5.37}$$

which means that the increase of the sum rule by exchange current contributions is bigger than $\kappa\Sigma^{class}$, according to (5.37) again by this interference term $\mathcal{J}$.

These two effects can be easily understood in a simple model of an S-wave deuteron bound in a pure exchange potential. The potential which has to be attractive in the $^3S_1$ state ($T = 0$) becomes repulsive in the ^3P_j states ($T = 1$) which are reached by $E1$ transitions whereas it would remain attractive for a non-exchange type potential. Thus, the P-wave gets repelled and correspondingly the momentum lowered compared to the attractive case. Consequently, the one-body convection current matrix elements being proportional to the momentum are decreased. On the other hand, the transition dipole moment, weighting $\mathbf{r}$, is increased for the repulsive case since the wave function peaks at larger distances and, thus, the sum rule is enhanced.

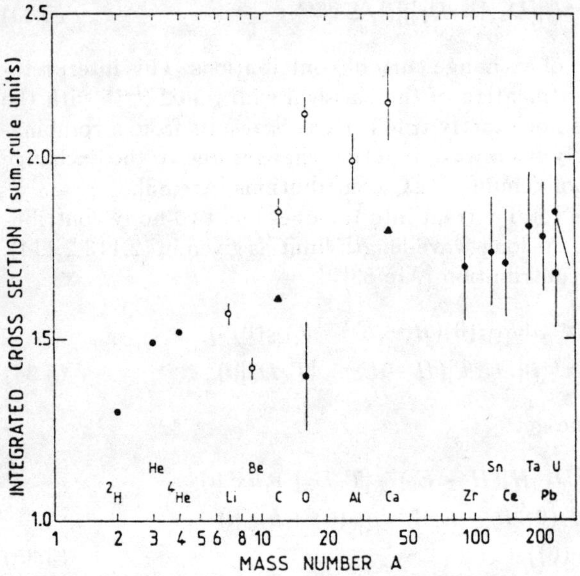

Fig. 5.9 Integrated total photonuclear cross section up to 140 MeV from [Car 86]. The solid circles represent values obtained from the sum of the measured partial photoreaction cross section; the open circles represent those obtained from photon-absorption measurements. The solid triangles represent the results deduced from (γ, γ') experiments.

Theoretical evaluations of the TRK sum rule give an enhancement $\kappa \approx 0.5$ for the deuteron, $\kappa \approx 0.75$ for three-body nuclei and $\kappa \approx 1$ for heavier nuclei ($A \geq 4$) provided one has incorporated a realistic model for short-range correlations including tensor correlations. Unfortunately, this sum rule is not an experimental observable since its definition relies on the unretarded pure $E1$-cross section integrated up to infinity while the experimental cross section includes retardation effects which tend to reduce the $E1$ transitions at higher energies and additional contributions from higher multipoles as well. Furthermore, at higher energies the isobar excitations contribute strongly to the total absorption cross section. For this reason, one usually integrates the experimental total cross section up to the pion-threshold and compares this quantity to the TRK sum rule. Obviously, there is no clear-cut relation between the theoretical TRK sum rule and the integrated experimental cross section and any comparison should be taken with caution. In other words, one can consider the TRK sum rule only as a qualitative estimate for the up to m_π integrated experimental cross section. In this spirit, I will present in Fig. 5.9 experimental results on κ taken from [Car 86]. It appears, that for complex nuclei one finds an enhancement of $\kappa \approx 0.6 - 0.8$, slightly lower than theoretical TRK evaluations. For a survey on the theory see also [Are 79b].

5.4 Photon Scattering

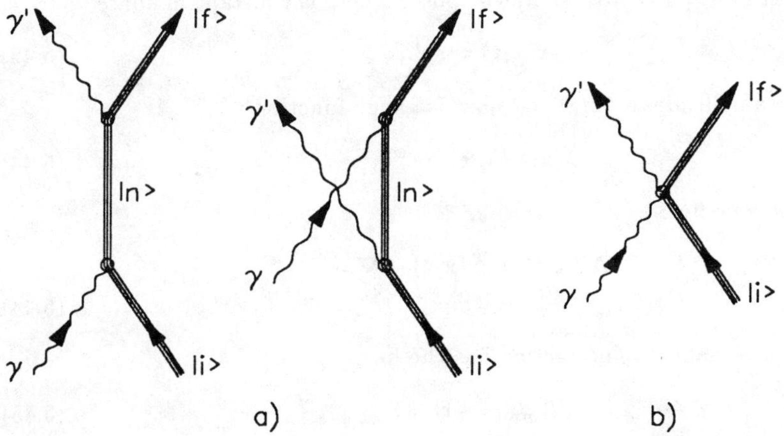

Fig. 5.10 Diagrams for photon scattering: (a) resonance amplitude and (b) two-photon amplitude.

Two-photon processes appear in a variety of different reactions like, for example, photon scattering, bremsstrahlung, Coulomb and dispersion corrections in electron scattering. Here I will concentrate on the photon scattering process [Are 86]. The cross section is given by

$$\frac{d\sigma}{d\Omega} = \frac{k'}{k} |T^{fi}_{\lambda'\lambda}(\mathbf{k}', \mathbf{k})|^2 , \tag{5.38}$$

where the scattering T-Matrix is defined by

$$T^{fi}_{\lambda'\lambda}(\mathbf{k}', \mathbf{k}) = \epsilon^*_\mu(\mathbf{k}', \lambda') \left\{ i \int d^4x e^{ik'\cdot x} \langle P_f | T(j^\mu(x), j^\nu(0)) | P_i \rangle \right. \\ \left. - \int d^4x e^{ik'\cdot x} \langle P_f | B^{\mu\nu}(x, 0) | P_i \rangle \right\} \epsilon_\nu(\mathbf{k}, \lambda) . \tag{5.39}$$

Here, k_μ, $\epsilon_\mu(\mathbf{k}, \lambda)$ and k'_μ, $\epsilon_\mu(\mathbf{k}', \lambda')$ denote incoming and scattered photon momenta and polarization vectors, respectively, P_i and P_f total momenta of initial and final hadron states and T denotes the time-ordering operator.

According to these two terms one may divide the T-matrix into a resonance amplitude R and a two-photon or "seagull" amplitude B

$$T^{fi}_{\lambda'\lambda}(\mathbf{k}',\mathbf{k}) = R^{fi}_{\lambda'\lambda}(\mathbf{k}',\mathbf{k}) + B^{fi}_{\lambda'\lambda}(\mathbf{k}',\mathbf{k}), \tag{5.40}$$

where

$$B^{fi}_{\lambda'\lambda}(\mathbf{k}',\mathbf{k}) = -\langle f|\epsilon^{\prime *}_{\lambda',l'}\tilde{B}_{l'l}(-\mathbf{k}',\mathbf{k})\epsilon_{\lambda,l}|i\rangle, \tag{5.41}$$

and $R^{fi}_{\lambda'\lambda}$ is given below. The corresponding diagrams are shown in Fig. 5.10 representing one-step or sea-gull (B) and two-step or resonance (R) processes.

I would like to point out that the separation is gauge dependent and, therefore, the two terms are not separately observable. This can be easily illustrated in the long-wave-length limit [GoM 31] where the photon field is constant in space

$$\mathbf{A}(\mathbf{x},t) = \mathbf{a}e^{i\omega t} \tag{5.42}$$

with a constant amplitude $\mathbf{a}$. Choosing now as gauge function

$$\lambda(\mathbf{x},t) = \mathbf{x}\cdot\mathbf{a}e^{i\omega t}, \tag{5.43}$$

one finds for the new field

$$\begin{aligned}\mathbf{A}'(\mathbf{x},t) &= \mathbf{A}(\mathbf{x},t) - \boldsymbol{\nabla}\lambda \\ &\equiv 0,\end{aligned} \tag{5.44}$$

and the new electromagnetic interaction has the form

$$H_{e.m.} = -\mathbf{E}\cdot\mathbf{D}, \tag{5.45}$$

where $\mathbf{E}$ denotes the electric field and $\mathbf{D}$ the dipole operator. In this case, since the electromagnetic interaction is now linear in the field variables, the seagull amplitude vanishes and one has only the resonance amplitude.

In general, the detailed form of resonance and seagull amplitudes will depend on the theoretical framework, i.e., which degrees of freedom are treated explicitly. This is because the two-photon amplitude will contain contributions from hidden degrees of freedom in terms of effective operators which otherwise would occur explicitly in the resonance amplitude. For example, the e.m. interaction of a Dirac particle is linear in A and, therefore, one has only a resonance amplitude for Compton scattering. If, however, one projects out the negative energy contributions one obtains an effective seagull term which reduces to the classical Thomson amplitude in the non-relativistic limit.

In the preceding sections we have already discussed the structure and properties of the two-photon operator, in particular, its low energy expansion. The resonance amplitude is obtained by evaluating the time ordering operator in (5.39) and one finds

$$R^{fi}_{\lambda'\lambda}(k',k) = \sum_n \left\{ \int d^3P_n \delta(\mathbf{P}_n - \mathbf{P}_i - \mathbf{k}) N_{ni}(k_0) \langle P_f|j^*_{\lambda'}(0)|P_n\rangle\langle P_n|j_\lambda(0)|P_i\rangle + \begin{pmatrix} k_\mu \leftrightarrow -k'_\mu \\ j_\lambda(0) \leftrightarrow j^*_{\lambda'}(0)\end{pmatrix} \right\}. \tag{5.46}$$

Here I have introduced for convenience using transverse gauge

$$j_\lambda = \epsilon_\lambda(k) \cdot \mathbf{j}\,, \tag{5.47}$$

$$j^*_{\lambda'} = \epsilon^*_{\lambda'}(k') \cdot \mathbf{j}\,, \tag{5.48}$$

$$N_{ni}(k_0) = (P_{n,0} - P_{i,0} - k_0 - i\epsilon)^{-1}\,, \tag{5.49}$$

where P_n denotes the total momentum of the intermediate hadron state. The second term in (5.46) corresponds to the crossed diagram in Fig. 5.10 (a).

The intermediate propagation of the scattering system between absorption and emission leads to a nonlocality of this process preventing the description of the scattering cross section by a simple form factor like in electron scattering. Only in the limiting situation when one can neglect the intermediate propagation of the scattering system, the process can be described by a simple form factor like in x-ray diffraction by crystals. On the other hand, it is just this intermediate propagation of the system, where the whole dynamical spectrum participates, which makes this process so interesting providing different kind of information in terms of generalized polarizabilities compared to the absorptive process.

The concept of generalized polarizabilities has been introduced by Fano in the discussion of photon scattering off a nucleus for the case of pure $E1$-radiation [Fan 60]. The homogeneous electric field $\mathbf{E}$ induces in this case a dipole moment in the nucleus proportional to the applied field

$$P_i = \sum_j \alpha_{ij} E_j\,, \tag{5.50}$$

where α_{ij} denote the components of the electric polarizability tensor. It can be decomposed into a scalar, a vector and a tensor term

$$\begin{aligned}\alpha_{ij} = {} & \frac{1}{3}\delta_{ij} tr\,\alpha + \frac{1}{2}(\alpha_{ij} - \alpha_{ji}) \\ & + \frac{1}{2}(\alpha_{ij} + \alpha_{ji} - \frac{2}{3}\delta_{ij}\mathrm{tr}\,\alpha)\,,\end{aligned} \tag{5.51}$$

or in a more compact notation

$$\alpha = \alpha^{[0]} + \alpha^{[1]} + \alpha^{[2]}\,. \tag{5.52}$$

This decomposition corresponds to the different possible angular momentum transfers ($J = 0$, 1 and 2) onto the target in the scattering of an $E1$-photon. Correspondingly, the scattering amplitude is proportional to this polarizability

$$T_{\lambda'\lambda} \sim \alpha^{[0]}(\epsilon'^*_{\lambda'} \cdot \epsilon_\lambda) + \boldsymbol{\alpha}^{[1]} \cdot (\epsilon'^*_{\lambda'} \times \epsilon_\lambda) + \epsilon'^*_{\lambda'} \cdot \overleftrightarrow{\alpha}^{[2]} \cdot \epsilon_\lambda\,. \tag{5.53}$$

Due to the angular momentum selection rules, only the scalar polarizability will be nonvanishing in elastic scattering off a spin-zero nucleus. But for higher spin nuclei also the vector and tensor polarizabilities will contribute. This concept has been generalized to arbitrary electric and magnetic multipole radiation [ArG 69].

An important feature is the low energy theorem for the elastic scattering amplitude which is a consequence of the low energy theorems which I have already discussed in Section 2 for the current and two-photon operators. To this end one has to derive a low-energy expansion for the resonance amplitude which I will briefly sketch for the terms up to first order in photon energy. Most convenient is the Breit frame where $\mathbf{P}_i = (\mathbf{k}' - \mathbf{k})/2$ and $\mathbf{P}_f = -\mathbf{P}_i$. Using (5.25) and (2.53) one finds for the current matrix elements up to first order

$$\langle P_f | j^*_{\lambda'}(0) | P_n \rangle = \langle i | [H_0^{int}, iD'_{\lambda'} + \frac{1}{2} Q'_{\lambda'}] + iM'_{\lambda'} | n \rangle + \delta_{in} \frac{Ze}{2M_A} \boldsymbol{\epsilon}'^*_{\lambda'} \cdot \mathbf{k}\,, \quad (5.54)$$

$$\langle P_n | j_{\lambda}(0) | P_i \rangle = \langle n | [H_0^{int}, iD_{\lambda} - \frac{1}{2} Q_{\lambda}] - iM_{\lambda} | i \rangle + \delta_{ni} \frac{Ze}{2M_A} \boldsymbol{\epsilon}_{\lambda} \cdot \mathbf{k}'\,, \quad (5.55)$$

where I have introduced as a shorthand notation

$$D'_{\lambda'} = \boldsymbol{\epsilon}'^*_{\lambda'} \cdot \mathbf{D}\,, \quad D_{\lambda} = \boldsymbol{\epsilon}_{\lambda} \cdot \mathbf{D}\,, \quad (5.56)$$

$$M'_{\lambda'} = \boldsymbol{\epsilon}'^*_{\lambda'} \cdot (\mathbf{k}' \times \mathbf{M})\,, \quad M_{\lambda} = \boldsymbol{\epsilon}_{\lambda} \cdot (\mathbf{k} \times \mathbf{M})\,, \quad (5.57)$$

$$Q'_{\lambda'} = \epsilon'^*_{\lambda',l'} Q_{l'l} k'_l\,, \quad Q_{\lambda} = \epsilon_{\lambda,l'} Q_{l'l} k_l\,. \quad (5.58)$$

Furthermore one has

$$N_{ni}(k_0) = \frac{1}{\varepsilon_{ni}} + \frac{k}{\varepsilon^2_{ni}} \quad \text{for } n \neq i \text{ and} \quad N_{ii} = -\frac{1}{k}\,, \quad (5.59)$$

where ε_{ni} denotes the intrinsic excitation energies. Corresponding expressions are obtained for the crossed term. Then one finds for the low energy expansion of the resonance amplitude up to the first order terms

$$\begin{aligned} R^{ii}_{\lambda'\lambda}(\mathbf{k}', \mathbf{k}) = & \langle i | [D'_{\lambda'}, [H_0^{int}, D_{\lambda}]] | i \rangle \\ & + k \langle i | [D'_{\lambda'}, D_{\lambda}] | i \rangle - \frac{1}{k} \langle i | M'_{\lambda'} P_{gs} M_{\lambda} - M_{\lambda} P_{gs} M'_{\lambda'} | i \rangle \\ & - i \frac{Ze}{kM_A} \langle i | (\boldsymbol{\epsilon}_{\lambda} \cdot \mathbf{k}' M'_{\lambda'} - \boldsymbol{\epsilon}'^*_{\lambda'} \cdot \mathbf{k} M_{\lambda}) | i \rangle \\ & - \langle i | [D'_{\lambda'}, M_{\lambda} - \frac{i}{2} [H_0^{int}, Q_{\lambda}]] - [D_{\lambda}, M'_{\lambda'} - \frac{i}{2} [H_0^{int}, Q'_{\lambda'}]] | i \rangle\,, \end{aligned} \quad (5.60)$$

where P_{gs} projects onto the ground state $|i\rangle$. The commutator $[D'_{\lambda'}, D_{\lambda}]$ in the second line of (5.60) vanishes in a non-relativistic approach, but if one considers relativistic corrections it turns out to be of the same size than the other terms linear in k [LaC 60, Fri 77].

Combining the resonance amplitude with the corresponding term of the two-photon amplitude from (2.59)

$$\begin{aligned} B^{ii}_{\lambda'\lambda}(\mathbf{k}', \mathbf{k}) = & -\frac{Z^2 e^2}{M_A} \boldsymbol{\epsilon}'^*_{\lambda'} \cdot \boldsymbol{\epsilon}_{\lambda} - \langle i | [D'_{\lambda'}, [H_0^{int}, D_{\lambda}]] | i \rangle \\ & + \langle i | [D'_{\lambda'}, M_{\lambda} - \frac{i}{2} [H_0^{int}, Q_{\lambda}]] - [D_{\lambda}, M'_{\lambda'} - \frac{i}{2} [H_0^{int}, Q'_{\lambda'}]] \rangle\,, \end{aligned} \quad (5.61)$$

one finds a cancellation for the lowest order term of the scattering amplitude of the contribution from the intrinsic Hamiltonian H_0^{int} to the double commutator so that the remaining part

$$T^{ii}_{\lambda'\lambda}(0,0) = -\frac{Z^2 e^2}{M_A} \boldsymbol{\epsilon}'^*_{\lambda'} \cdot \boldsymbol{\epsilon}_{\lambda} \quad (5.62)$$

is just the classical Thomson amplitude. Similar cancellations between resonance and seagull amplitudes appear for the higher order terms as one can see explicitly for the first order terms in the last lines of (5.60) and (5.61), which, however, vanish in case parity conservation holds as is assumed throughout. This allows to express the scattering amplitude up to second order in the photon momenta in terms of the total charge, the magnetic moment and a few further ground state expectation values representing dynamic properties of the system, like electric polarizability and magnetic susceptibility [Fri 75a,Are 86,ArW 86].

Up to second order one has only contributions from scalar, vector and tensor polarizabilities, i.e.,

$$T^{ii}_{\lambda'\lambda} = \sum_{j=0}^{2} T^{ii\,[j]}_{\lambda'\lambda} \,. \tag{5.63}$$

The scalar part is essentially given by the non-relativistic expressions up to some minor relativistic corrections

$$T^{ii\,[0]}_{\lambda'\lambda} = (-\frac{Z^2e^2}{M_A} + \overline{\alpha}_E k^2)\boldsymbol{\epsilon}'^{*}_{\lambda'}\cdot\boldsymbol{\epsilon}_\lambda + \overline{\chi}_M(\boldsymbol{\epsilon}'^{*}_{\lambda'}\times\mathbf{k}')\cdot(\boldsymbol{\epsilon}_\lambda\times\mathbf{k})\,, \tag{5.64}$$

with

$$\overline{\alpha}_E = \frac{Ze^2}{3M_A}\langle r^2\rangle + \alpha_E^{(0)}\,, \tag{5.65}$$

$$\overline{\chi}_M = \chi_P^{(0)} + \chi_D^{(0)} - \frac{\langle \mathbf{D}^{\,2}\rangle}{2M_A}\,, \tag{5.66}$$

where Z denotes the total charge and M_A the total mass of the hadron. Note that the Breit frame is assumed where $k = k'$.

As additional structure constants one has the static electric polarizability $\alpha_E^{(0)}$, the static para- and diamagnetic susceptibilities $\chi_P^{(0)}$ and $\chi_D^{(0)}$, respectively, the mean-square charge radius $\langle r^2\rangle$ and the mean-square dipole moment $\langle \mathbf{D}^{\,2}\rangle$. They are a measure of how the system responds dynamically to an applied static electric or magnetic field and, hence, characterize the internal constitution of the system. In other words, they contain interesting information on the internal dynamics of the system. Here I will give only the expressions for the electric polarizability

$$\alpha_E^{(0)} = \frac{2}{3}\langle i|\mathbf{D}\cdot\frac{Q_{gs}}{H_0-\epsilon_i}\mathbf{D}|i\rangle\,, \tag{5.67}$$

and the paramagnetic susceptibility

$$\chi_P^{(0)} = \frac{2}{3}\langle i|\mathbf{M}\cdot\frac{Q_{gs}}{H_0-\epsilon_i}\mathbf{M}|i\rangle\,, \tag{5.68}$$

where Q_{gs} projects off the ground state. Both quantities contain substantial contributions from MEC and IC. For further details I refer to [Are 86,ArW 86].

The vector part which has only a contribution of order k is given by

$$\begin{aligned} T^{ii\,[1]}_{\lambda'\lambda} = \frac{i}{k}\Big\{ &\frac{Ze}{M_A}\left(2\mu - \frac{Ze}{M_A}\right)(\boldsymbol{\epsilon}'^{*}_{\lambda'}\times\boldsymbol{\epsilon}_\lambda)k^2 - \mu^2(\boldsymbol{\epsilon}'^{*}_{\lambda'}\times\mathbf{k}')\times(\boldsymbol{\epsilon}_\lambda\times\mathbf{k}) \\ &- \frac{\mu Ze}{M_A}((\boldsymbol{\epsilon}'^{*}_{\lambda'}\times\mathbf{k}')(\boldsymbol{\epsilon}_\lambda\cdot\mathbf{k}') - (\boldsymbol{\epsilon}_\lambda\times\mathbf{k})(\boldsymbol{\epsilon}'^{*}_{\lambda'}\cdot\mathbf{k}))\Big\}\cdot\langle\mathbf{S}\rangle\,, \end{aligned} \tag{5.69}$$

where the magnetic moment of the hadron is given by $\mu\mathbf{S}$. Here the first term is the one of relativistic order and it is important to include it as has already been emphasized above.

In the forward direction, the low energy expression reduces to the simple form

$$T^{ii}_{\lambda\lambda}(\mathbf{k},\mathbf{k}) = -\frac{Z^2e^2}{M_A} + \lambda\left(\mu - \frac{Ze}{M_A}\right)^2 \langle \mathbf{S}\cdot\mathbf{k}\rangle + (\overline{\alpha}_E + \overline{\chi}_M)k^2 . \qquad (5.70)$$

It is interesting to note, that for the structure constants in (5.70) one can obtain sum rules from dispersion relations.

In general, dispersion relations are a consequence of causality and crossing symmetry which lead to certain analytic properties of the elastic forward scattering amplitude $T^{ii}_{\lambda\lambda}(k)$ as function of the complex photon energy. They allow to calculate the forward elastic scattering amplitude in conjunction with the optical theorem from the total absorption cross section. For example, assuming the existence of a once-subtracted dispersion relation for the polarization- and spin-averaged forward elastic scattering amplitude $\overline{T}^{ii}(k)$ one has

$$\begin{aligned}\Re\left(\overline{T}^{ii}(k) - \overline{T}^{ii}(0)\right) &= \frac{2k^2}{\pi}P\int_0^\infty dk'\frac{\Im\left(\overline{T}^{ii}(k')\right)}{k'(k'^2-k^2)}\\ &= \frac{k^2}{2\pi^2}P\int_0^\infty dk'\frac{\overline{\sigma}_{tot}(k')}{k'^2-k^2}, \end{aligned} \qquad (5.71)$$

where in the last line I have used the optical theorem for the averaged total absorption cross section

$$\overline{\sigma}_{tot}(k) = \frac{4\pi}{k}\Im\left(\overline{T}^{ii}(k)\right). \qquad (5.72)$$

Taking the second derivative of (5.71) at $k = 0$ one obtains the well known energy-weighted sum rule for the "average polarizability"

$$\overline{\alpha}_E + \overline{\chi}_M = \frac{1}{2\pi^2}\int_0^\infty dk'\frac{\overline{\sigma}_{tot}(k')}{k'^2}, \qquad (5.73)$$

which can be determined easily from experiment since it converges rapidly enough due to the energy weighting.

In a similar manner one can obtain for the constant of the linear term in (5.70) the Gerasimov-Drell-Hearn (GDH) sum rule [Ger 65,DrH 66]. To this end one considers the difference of elastic forward scattering amplitudes for circularly polarized photons from a completely polarized target with spin parallel and antiparallel to the photon spin, denoted by S_P and S_A, respectively, and assumes an unsubtracted dispersion relation for it

$$\begin{aligned}\Re\left(T^{ii}_{\lambda\lambda}(k,S_P) - T^{ii}_{\lambda\lambda}(k,S_A)\right) &= \frac{2k}{\pi}P\int_0^\infty dk'\frac{\Im\left(T^{ii}_{\lambda\lambda}(k',S_P) - T^{ii}_{\lambda\lambda}(k',S_A)\right)}{k'^2-k^2}\\ &= \frac{k}{2\pi^2}P\int_0^\infty dk'k'\frac{(\sigma^P_{tot}(k',\lambda) - \sigma^A_{tot}(k',\lambda))}{k'^2-k^2}. \end{aligned} \qquad (5.74)$$

Taking now the first derivative at $k = 0$ and using (5.69) one finds the GDH sum rule

$$\langle S_z \rangle \left(\mu - \frac{Ze}{M_A} \right)^2 = \frac{1}{4\pi^2} \int_0^\infty \frac{dk'}{k'} \left(\sigma^P_{tot}(k') - \sigma^A_{tot}(k') \right) , \qquad (5.75)$$

where $\langle S_z \rangle$ denotes the expectation value of the ground state spin component S_z in the state of maximal alignment and $\sigma^{P/A}_{tot}(k')$ the total absorption cross section for circularly polarized photons on a target with spin parallel or antiparallel to the photon spin, respectively. This is also a very interesting relation provided the rhs of (5.75) converges, which has not yet been proven.

For example, in the case of a nucleon it relates the static anomalous magnetic moment to the integrated, energy-weighted difference of the total photoabsorption cross sections of a nucleon with spin parallel and antiparallel to the photon spin. Thus it connects a static ground state property, the anomalous magnetic moment, to the internal excitation spectrum above meson production threshold. Furthermore, because the lhs is positive, it tells us that the integrated, energy-weighted total absorption strength of a circularly polarized photon on a spin-parallel nucleon is bigger than the one on a spin-antiparallel target. Though this sum rule is known over 25 years it is still awaiting its experimental confirmation.

6 Summary and Outlook

In these lectures I have outlined the general structure of the electromagnetic interaction in nuclei with a detailed discussion of the gauge conditions for the electromagnetic current and two-photon operators and the low energy theorems which follow from them. Then I have described the construction of effective operators and the model of nuclear isobar configurations. Finally, after having sketched what kind of information on nuclear structure one can obtain in the various regions of energy and momentum transfers, I have considered several examples of electromagnetic processes on nuclei. In particular, I have considered deuteron disintegration which provides ample and convincing evidence for the presence of subnuclear d.o.f. in nuclei in the sense that the classical nuclear physics description fails to describe a large body of experimental data of various observables while inclusion of subnuclear d.o.f., either as effective operators (MEC) or explicitly as wave function components (IC) leads in general to a satisfactory agreement between theory and experiment. Thus important and clean tests of the hadronic structure are provided in electromagnetic reactions through direct interactions of the electromagnetic probe with the subnuclear d.o.f.

The framework in which these effects have been discussed is built from a model space where mesons, nucleons and isobars are present. It is very important that the corresponding electromagnetic current operators are included in a manner consistent with the interaction model in order to fulfill current conservation as a gauge constraint. Modern models for the NN interaction - realistic potentials - give a reliable overall description of the hadronic structure of the deuteron in the long and intermediate range region, whereas a larger uncertainty remains in the short range part.

On firm grounds is the lowest order isovector π-MEC which is related to the longest range static one-pion exchange mechanism of the NN interaction. It affects most strongly $E1$ transitions already at rather low energies. But this contribution is usually masked by the use of the Siegert operator. Also in $M1$ transitions near threshold one finds clear evidence for the presence of MEC and IC, particularly strong at higher momentum transfers in electrodisintegration, where good agreement with experimental data is achieved up to $\mathbf{q}^2 \approx 6\, fm^{-2}$.

The exclusive reaction $d(e,e'N)N$ allows in principle a much more detailed study of subnuclear d.o.f. but only few experimental data are available till now. Hopefully, the new high-duty cycle machines will change this and will provide us in the near future with a wealth of new data. In particular, the separation of the interesting although smaller interference structure functions will be very rewarding.

I had no time, with the exception of one example, to consider in detail polarization observables in deuteron photodisintegration whose study constitutes a promising field of research and will provide us with a very sensitive tool for testing present models of subnuclear d.o.f. For details I refer to the recent literature, e.g. [Are 92b,Are 93, Don 91]. Also in electrodisintegration only one important aspect in this field has been briefly touched upon, namely the determination of the electric form factor of the neutron in quasi-free electrodisintegration with a polarized beam and either a vector polarized deuteron target or a polarization analysis of the outcoming neutron.

The role of exchange effects in the photonuclear TRK sum rule has been analyzed in detail and it has been shown, that the enhancement of the classical sum rule is a combined effect of exchange current contributions to $E1$-transitions and a wave function change due to the exchange part of the NN interaction acting in opposite direction. But the MEC influence dominates. Finally, elastic photon scattering has been discussed with special emphasis on the manifestation of subnuclear d.o.f. in the low energy domain.

Acknowledgment: I would like to thank the organizers for the invitation and for the warm hospitality during my stay. Furthermore, I am very thankful to M. Kirchbach for many constructive remarks on the manuscript.

References

[AdT 89] Adam, Jr., J., Truhlik, E. and Adamova, D., Nucl. Phys. **A492**, 556 (1989)
[AhB 72] Ahrens, J. et al., Proc. Int. Conf. on Nuclear Structure Studies, Shoda, K. and Ui, H., eds., Res. Rep. Lab. Nucl. Sci., Vol.**5**, 213 (Tohoku University, Sendai 1972)
[AhB 75] Ahrens, J. et al., Nucl. Phys. **A251**, 479 (1975)
[Ahr 85] Ahrens, J., Nucl. Phys. **A446**, 229c (1985)
[ArC 81] Arnold, R.G., Carlson, C.E. and Gross, F., Phys. Rev. **C23**, 363 (1981)
[ArD 71] Arenhövel, H., Danos, M. and Williams, H.T., Nucl. Phys. **A162**, 12 (1971)
[Are 79a] Arenhövel, H., Nukleonika **24**, 273 (1979)
[Are 79b] Arenhövel, H., Lecture Notes in Physics, Vol. **108**, 159 (1979)
[Are 80] Arenhövel, H., Z. Phys. **A297**, 129 (1980)
[Are 81a] Arenhövel, H., Z. Phys. **A302**, 25 (1981)
[Are 81b] Arenhövel, H., Lecture Notes in Physics, Vol. **137**, 136 (1981)
[Are 83] Arenhövel, H., Nuovo Cim. **76A**, 256 (1983)
[Are 86] Arenhövel, H., in New Vistas in Electro-Nuclear Physics, Tomusiak, E.L., Caplan, H.S., and Dressler, E.T., eds., (Plenum Press, New York 1986) p. 251

[Are 87] Arenhövel, H., Prog. Theor. Phys. Suppl. **91**, 1 (1987)
[Are 88] Arenhövel, H., Few-Body Systems **4**, 55 (1988)
[Are 92a] Arenhövel, H., Chin. J. Phys. (Taipei) **30**, 17 (1992)
[Are 92b] Arenhövel, H., Few-Body Syst., Suppl. **5**, 71 (1992)
[Are 93] Arenhövel, H., Proc. Workshop on Electronuclear Physics with Internal Targets and the BLAST Detector (Tempe, Arizona, USA 1992), eds. Alarcon, R. and Butler, M. (World Scientific, Singapore 1993)
[ArG 69] Arenhövel, H. and Greiner, W., Prog. Nucl. Phys. **10**, 167 (1969)
[ArL 88] Arenhövel, H., Leidemann, W. and Tomusiak, E.L., Z. Phys. **A331**, 123 (1988), **A334**, 363 (1989) (Erratum)
[ArL 92] Arenhövel, H., Leidemann, W. and Tomusiak, E.L., Phys. Rev. **C46**, 455 (1992)
[ArS 91] Arenhövel, H. and Sanzone, M., Few-Body Syst., Suppl. **3**, 1 (1991)
[ArW 72] Arenhövel, H. and Weber, H.J., Springer Tracts in Modern Physics **65**, 58 (1972)
[ArW 86] Arenhövel, H. and Weyrauch, M., Nucl. Phys. **A457**, 573 (1986)
[BaK 47] Baldwin, G.C. and Klaiber, G.S., Phys. Rev. **71**, 669 (1947)
[BlH 58] Bloch, C. and Horowitz, J., Nucl. Phys. **8**, 91 (1958)
[Bol 93] Boltzmann, L., "Vorlesungen über Maxwells Theorie der Elektricität und des Lichtes, II. Teil" (J.A. Barth, Leipzig 1893) in "Ludwig Boltzmann, Gesamtausgabe Bd. 2", ed. R.U. Sexl (Akademische Druck- und Verlagsanstalt, Graz 1982)
[BuL 85] Buchmann, A., Leidemann, W. and Arenhövel, H., Nucl. Phys. **A443**, 726 (1985)
[Car 86] Carlos, P., Proc. Vth Int. School of Intermediate Energy Nuclear Physics, Bergere, R., Costa, S., and Schaerf, C., eds., (World Scientific, Singapore 1986) p. 1
[CdA 80] Ciofi degli Atti, C., Prog. Part. Nucl. Phys. **3**, 163 (1980)
[Che 79] Chemtob, M., in "Mesons in Nuclei", Rho, M. and Wilkinson, D.H., eds. (North Holland, Amsterdam, 1979)
[ChG 34] Chadwick, J. and Goldhaber, M., Nature **134**, 237 (1934)
[DFW 66] De Forest, Jr., T. and Walecka, J.D., Adv. Phys. **15**, 1 (1966)
[Don 91] Donnelly, T. W., in "Modern Topics in Electron Scattering", Frois, B. and Sick, I., eds. (World Scientific, Singapore 1991) p. 566
[DrH 66] Drell, S.D. and Hearn, A.C., Phys. Rev. Lett. **16**, 908 (1966)
[FaA 76] Fabian, W. and Arenhövel, H., Nucl. Phys. **A258**, 461 (1976)
[FaA 79] Fabian, W. and Arenhövel, H., Nucl. Phys. **A314**, 253 (1979)
[Fan 60] Fano, U., NBS Technical Note 83 (1960), reprinted in "Photonuclear Reactions", Fuller, E.G. and Hayward, E., eds. (Dowden, Hutchinson & Ross, 1976) p. 338
[Fee 36] Feenberg, E., Phys. Rev. **49**, 328 (1936)
[Fes 62] Feshbach, H., Ann. Phys. (N.Y.) **5**, 357 (1958); **19**, 287 (1962)
[FoW 50] Foldy, L.L. and Wouthuysen, S.A., Phys. Rev. **78**, 29 (1950)
[Fri 75a] Friar, J.L., Ann. Phys. (N.Y.) **95**, 170 (1975)
[Fri 75b] Friar, J.L., Phys. Rev. **C12**, 695 (1975)
[FrF 76] Friar, J.L. and Fallieros, S., Phys. Rev. **C13**, 2571 (1976)
[Fri 76] Friar, J.L., Phys.Rev. Lett. **36**, 510 (1976)
[Fri 77] Friar, J.L., Phys. Rev. **C16**, 1504 (1977)
[Fri 91] Friedman, J.I., Rev. Mod. Phys. **63**, 615 (1991)
[FuS 54] Fukuda, N., Sawada, K. and Taketani, M., Prog. Theor. Phys. **12**, 156 (1954)
[GaH 76] Gari, M. and Hyuga, H., Z. Phys. **A227**, 291 (1976)

[GaH 81] Gari, M. and Hebach, H., Phys. Rep. **72**, 1 (1981)
[GeG54] Gell-Mann, M. and Goldberger, M.L., Phys. Rev. **96**, 1433 (1954)
[GeM 09] Geiger, H. and Marsden, E., Proc. Roy. Soc. London **82**, 495 (1909)
[Ger 65] Gerasimov, S.B., Yad. Fiz. **2**, 598 (1965) (Sov. J. Nucl. Phys. **2**, 430 (1966))
[GoM 31] Goeppert-Mayer, M., Ann. Physik **9**, 273 (1931)
[Gre 76] Green, A.M., Rep. Prog. Phys. **39**, 1109 (1976)
[GrR 87] Gross, F. and Riska, R., Phys. Rev. **C36**, 1928 (1987)
[GrS 79] Green, A.M. and Sainio, M., J. Phys. **G5**, 503 (1979)
[HoM 55] Hofstadter, R. and McAllister, R.W., Phys. Rev. **98**, 217 (1955)
[HoR 73] Hockert, J., Riska, D.O., Gari, M. and Huffmann, A., Nucl. Phys. **A217**, 14 (1973)
[Ken 91] Kendall, H.W., Rev. Mod. Phys. **63**, 597 (1991)
[LaC 60] Lapidus, L.I. and Chou Kuang-Chao, Zh. Eksp. Teor. Fiz. **39**, 1286 (1960) (Sov. JETP **12**, 898 (1961))
[LaL 80] Lacombe, M. et al., Phys. Rev. **C21**, 861 (1980)
[LeA 87] Leidemann, W. and Arenhövel, H., Nucl. Phys. **A465**, 573 (1987)
[Lev 60] Levinger, J.S.,"Nuclear Photodisintegration" (Oxford University Press 1960)
[Low 54] Low, F.E., Phys. Rev. **96**, 1428 (1954)
[LyH 51] Lyman, E.M., Hanson, A.O., and Scott, M.B., Phys. Rev. **84**, 626 (1951)
[Mac 89] Machleidt, R., Adv. Nucl. Phys. Vol. **19**, p. 189 (Plenum, New York, 1989)
[MaH 87] Machleidt, R., Holinde, K. and Elster, Ch., Phys. Rep. **149**, 1 (1987)
[Mat 89] Mathiot, J., Phys. Rep. **173**, 63 (1989)
[Oht 89] Ohta, K., Nucl. Phys. **A495**, 564 (1989)
[RiB 72] Riska, D.O. and Brown, G.E., Phys. Lett. **38B**, 193 (1972)
[Ris 85] Riska, D.O., Physica Scripta **31**, 471 (1985)
[Ris 89] Riska, D.O., Phys. Rep. **181**, 207 (1989)
[Ros 57] Rose, M.E., "Elementary Theory of Angular Momentum" (Wiley, New York, 1957)
[Rut 11] Rutherford, E., Phil. Mag. (6) **21**, 669 (1911)
[SaA 51] Sachs, R.G. and Austern, N., Phys. Rev. **81**, 705 (1951)
[ScA 89a] Schmitt, K.-M. and Arenhövel, H., Few-Body Syst. **6**, 117 (1989)
[ScA 89b] Schmitt, K.-M. and Arenhövel, H., Few-Body Syst. **7**, 95 (1989)
[ScA 91a] Schmitt, K.-M. and Arenhövel, H., Few-Body Syst., **11** (1991)
[ScA 91b] Schmitt, K.-M. and Arenhövel, H., Few-Body Syst., Suppl. **4** (1991)
[Sch 59] Schwinger, J., Phys. Rev. Lett. **3**, 296 (1959)
[ScW 90] Schmitt, K.-M., Wilhelm, P., Arenhövel, H., Cambi, A., Mosconi, B., and Ricci, R., Phys. Rev. **C41**, 841 (1990) (1976)
[SiB 79] Simon, G.G. et al., Nucl. Phys. **A324**, 277 (1979)
[Sie 37] Siegert, A.J.F., Phys. Rev. **52**, 787 (1937)
[SiL 88] Singh, S.K., Leidemann, W. and Arenhövel, H., Z. Phys. **A331**, 509 (1988)
[Tay 91] Taylor, R.E., Rev. Mod. Phys. **63**, 573 (1991)
[WeA 78] Weber, H.J. and Arenhövel, H., Phys. Rep. **36**, 279 (1978)
[WiS 84] Wiringa, R.B., Smith, R.A. and Ainsworth, T.L., Phys. Rev. **C29**, 1207 (1984)
[Yuk 35] Yukawa, H., Proc. Phys.-Math. Soc. Jap. **17**, 48 (1935)

Electro-Excitation of Nucleon Resonances and Meson Production

R. M. Davidson

Institut für Kernphysik, Johannes Gutenberg-Universität, J.-J.-Becher-Weg 45, D-55099 Mainz, Germany

1 Introduction

The photo- and electroproduction of mesons has long been recognized as a powerful tool to study the strong interaction. It was photopion production experiments from protons [1,2] that lead to the discovery of the first nucleon resonance, the Delta(1232). In subsequent years, partial wave analyses of pion-nucleon elastic scattering revealed numerous resonances, and quark models were developed, in part, to explain the occurrence of the new "particles". The resonances were also seen in photoproduction experiments, and besides supplying a consistency check on the masses and total widths of the resonances, analyses of these experiments also supplied the electromagnetic transition amplitudes between the nucleon and the resonance (excited baryon).

The electromagnetic transition amplitudes were quickly realized to be a powerful test of the emerging quark models. One classic example is the electric quadrupole ($E2$) transition between nucleon (N) and Delta (Δ). In the most naive non-relativistic quark model (NRQM) [3], N and Δ have degenerate masses, and $E2$ is identically zero since the N and Δ wave functions contain only l=0 components and the $E2$ operator is proportional to the second-rank spherical tensor, Y_2. The introduction of one-gluon exchange leads to a color-hyperfine interaction between the quarks which splits the N and Δ masses and gives rise to $l \neq 0$ components in the wave functions, and hence a non-zero $E2$. Fitting the strength of the color-hyperfine interaction to reproduce the physical $N - \Delta$ mass difference results in an $E2$ that is in qualitative agreement with experiment; it is small and negative. The ratio of $E2$ to the dominant magnetic dipole transition, $M1$, is about -0.4% [4], compared to the experimental value of $\approx$ -1.1±0.4% [5].

The same analyses that yield the electromagnetic $N - N^*$ (N^* an excited baryon) transition amplitudes also give, to some degree of accuracy, the T-matrix, $t_{\gamma\phi}$, for photo- or electroproduction of mesons from single nucleons. With $t_{\gamma\phi}$ known, photo- and electroproduction of mesons from nuclei may be calculated. The nuclear process, obviously much more complicated than the single nucleon process, offers some exciting prospects, of which we name a few. First, meson production from nuclei serves as a complementary tool to study nuclear structure. For example, electron scattering excites both ΔT = 0 and ΔT = 1 (T is isospin) transitions, while charged pion excitation is ΔT = 1, enabling one in principle to disentangle the two transitions. Second, the photon can probe the interior of the nucleus and produce, for example, a pion deep in the nucleus. In contrast, in pion-nucleus scattering it is difficult to get the pion very far into the nucleus since it is a strongly interacting particle. Thus, pion photoproduction is a tool to study the pion wave function deep in the nucleus. Third, the photoproduction of kaons can be used to produce hypernuclei, i.e. a nucleus in which a nucleon is replaced with a strange baryon such as Λ or Σ. Information then may be obtained on both the kaon-N and $\Lambda - N$ interactions. Fourth, the nucleus can act as a filter for $t_{\gamma\phi}$, changing the relative importance of the pieces that comprise $t_{\gamma\phi}$. Kamalov et al. [6] have shown that due to D-states in ^{3}He, certain polarization observables are quite sensitive to the $E2$ amplitude which was mentioned above.

To date most work, both theoretical and experimental, has been done on photopion production up to a center of momentum energy of about 2.5 GeV. Scant data exist for kaon and eta photoproduction, and the electroproduction data are concentrated around the $\Delta(1232)$ region at a momentum transfer squared roughly less than 3 GeV^2. The situation will change drastically in the next few years. Recent experiments have been done on eta photoproduction at Bates, Mainz, and Bonn. Electroproduction experiments should soon be underway at these institutes, and also at NIKHEF. In a few years, the N^* program should be underway at CEBAF.

Given the on-going interest in the photo- and electroproduction of mesons in the resonance region from both nucleons and nuclei, it is useful to review the theoretical techniques that are used to extract $t_{\gamma\phi}$ from the data. In the next Section, the general formalism is presented for photo- and electroproduction of mesons. In Section 3, we compare the various theories that are used to analyze the data; effective Lagrangians, dispersion relations, isobar models, and dynamical models. Sections 4 and 5 are devoted to some of the interesting results that have so far been obtained.

2 General Formalism

In this section, we present the general formalism for the photo- and electroproduction of pseudoscalar mesons from nucleons:

$$\gamma(K) + N(P_i) \rightarrow B(P_f) + \Phi(Q) \ , \qquad (1)$$

where γ is a photon, either real or virtual, N is a nucleon, B is a member of the baryon octet (N, Λ, Σ) and Φ is a member of the pseudoscalar meson octet (π, K, η). The four-momenta of the particles are shown in parantheses, and the combination of B and Φ in the final state is such that all conservation laws are satisfied, e.g. charge conservation, strangeness conservation, etc. Finally, it is implicitly assumed for electroproduction that the one-photon exchange approximation is valid. Most of the material in this section has been reviewed many times [7], and is included for completeness.

2.1 Kinematics

The Lorentz invariant matrix element for photo- and electroproduction can be written as a function of the Mandelstam variables

$$s = (P_i + K)^2 = (P_f + Q)^2$$

$$t = (Q - K)^2 = (P_i - P_f)^2$$
$$u = (P_i - Q)^2 = (P_f - K)^2 \ , \tag{2}$$

and the four-momenta squared of the incoming and outgoing particles. Since we will take the baryons and mesons to be on-shell, the only additional variable is the K^2 of the photon. We thus have $t_{\gamma\phi} = t_{\gamma\phi}(s, t, u, K^2)$. For a real photon, $K^2 = 0$, while for electroproduction it is understood that $K = K_2 - K_1$, where $K_{2(1)}$ is the four-momentum of the final (initial) electron. Therefore, $K^2 \leq 0$ for the processes we are considering.

The Mandelstam variables satisfy

$$s + t + u = M^2 + M_f^2 + M_\phi^2 + K^2 \ , \tag{3}$$

where M is the nucleon mass, M_f is the final baryon mass, and M_ϕ is the meson mass, and thus only two of the three Mandelstam variables are independent.

It is also useful to know the energies and momenta of the particles in the center of momentum (cm) frame. The total cm energy is $W = \sqrt{s}$, from which one obtains

$$k_0 = \frac{W^2 - M^2 + K^2}{2W} \tag{4}$$

$$\omega = \frac{W^2 - M_f^2 + M_\phi^2}{2W} \ , \tag{5}$$

where k_0 (ω) is the photon (meson) energy in the cm frame. The other energies and momenta in the cm frame may easily be found by using $W = E_i + k_0 = E_f + \omega$, where $E_{i(f)}$ is the energy of the initial (final) baryon in the cm frame, and $\mathbf{k}^2 = \sqrt{E_i^2 - M^2} = \sqrt{k_0^2 - K^2}$, $\mathbf{q}^2 = \sqrt{E_f^2 - M_f^2} = \sqrt{\omega^2 - M_\phi^2}$, where $\mathbf{k}$ and $\mathbf{q}$ are the three momentum in the cm frame of the photon and meson, respectively.

2.2 Invariant Amplitudes

For photoproduction, the S-matrix may be written as

$$S_{\gamma\phi} = \frac{i}{(2\pi)^2}\sqrt{\frac{MM_f}{4\omega k_0 E_i E_f}}\delta^4(P_i + K - P_f - Q)M_{\gamma\phi} \tag{6}$$

and for electroproduction

$$S_{\gamma\phi} = \frac{i}{(2\pi)^{7/2}}\frac{M_e\sqrt{MM_f}}{\sqrt{2\omega E_i E_f K_{10} K_{20}}}\delta^4(P_i + K - P_f - Q)M_{\gamma\phi} \ , \tag{7}$$

where M_e is the mass of the electron, and K_{10}, K_{20} are the energies of the initial and final electrons, respectively. Using Lorentz invariance, gauge invariance, and parity conservation, $M_{\gamma\phi}$ may be written in the form

$$M_{\gamma\phi} = \bar{U}_f \sum_{j=1}^{6} A_j(s, t, u, K^2) M_j U_i \ . \tag{8}$$

The operators M_j contain the Dirac structure of the amplitude, while the invariant functions A_j contain the dependence of the amplitude on the Mandelstam variables. These operators are sandwiched between the initial four-component Dirac spinor, U_i, and the final Dirac spinor, $\bar{U}_f$. The explicit forms of the M_j are [8]

$$\begin{aligned}
M_1 &= -\frac{1}{2}\gamma_5\gamma_\mu\gamma_\nu F^{\mu\nu} \\
M_2 &= \frac{1}{2}\gamma_5\left(P_i + P_f\right)_\mu (2Q - K)_\nu F^{\mu\nu} \\
M_3 &= -\gamma_5\gamma_\mu Q_\nu F^{\mu\nu} \\
M_4 &= -\gamma_5\gamma_\mu\left(P_i + P_f\right)_\nu F^{\mu\nu} - 2\mathrm{M}M_1 \\
M_5 &= \gamma_5 K_\mu Q_\nu F^{\mu\nu} \\
M_6 &= \gamma_5\gamma_\mu K_\nu F^{\mu\nu} \ .
\end{aligned} \tag{9}$$

In these equations, γ_5 and γ_μ are the usual Dirac matrices [9] and $F_{\mu\nu} = \epsilon_\mu K_\nu - \epsilon_\nu K_\mu$. For real photons, ϵ_μ is the polarization vector of the photon, while in electroproduction

$$\epsilon_\mu = \frac{e\bar{U}_2\gamma_\mu U_1}{K^2} \ , \tag{10}$$

where e is the charge of the electron and $\bar{U}_2$, U_1 are the Dirac spinors of the outgoing and incoming electrons, respectively. Each of the M_j is manifestly gauge invariant since $F_{\mu\nu} \rightarrow 0$ if $\epsilon_\mu \rightarrow K_\mu$. For photoproduction, $K_\mu F^{\mu\nu} = 0$, and thus only the amplitudes $A_{1,2,3,4}$ contribute to the physical amplitude for real photons. Finally, the γ_5 matrix appears due to the outgoing pseudoscalar meson.

2.3 Two-Component Form

It is useful to express the amplitude in terms of Pauli matrices and two-component spinors. First, this facilitates the multipole decomposition. Second, it is the most useful form for nuclear physics applications. This is usually carried out in the cm frame with the spinors quantized in the z-direction, which is defined as the direction of the incoming photon. Using the explicit form of the Dirac matrices in terms of the Pauli matrices, and also for example

$$\bar{U}_f = \sqrt{\frac{E_f + M_f}{2M_f}}\chi_f^\dagger\left(1, -\frac{\boldsymbol{\sigma}\cdot\mathbf{q}}{E_f + M_f}\right) \ , \tag{11}$$

it is a matter of algebra to show that

$$iM_{\gamma\phi} = \frac{4\pi W}{\sqrt{MM_f}}\chi_f^\dagger\mathcal{F}\chi_i \ , \tag{12}$$

where the explicit form of $\mathcal{F}$ is

$$\begin{aligned}
\mathcal{F} = {} & i\boldsymbol{\sigma}\cdot\boldsymbol{\epsilon}\mathcal{F}_1 + \boldsymbol{\sigma}\cdot\hat{\mathbf{q}}\boldsymbol{\sigma}\cdot(\hat{\mathbf{k}}\times\boldsymbol{\epsilon})\mathcal{F}_2 + i\boldsymbol{\sigma}\cdot\hat{\mathbf{k}}\hat{\mathbf{q}}\cdot\boldsymbol{\epsilon}\mathcal{F}_3 + i\boldsymbol{\sigma}\cdot\hat{\mathbf{q}}\hat{\mathbf{q}}\cdot\boldsymbol{\epsilon}\mathcal{F}_4 \\
& + i\boldsymbol{\sigma}\cdot\hat{\mathbf{k}}\hat{\mathbf{k}}\cdot\boldsymbol{\epsilon}\mathcal{F}_5 + i\boldsymbol{\sigma}\cdot\hat{\mathbf{q}}\hat{\mathbf{k}}\cdot\boldsymbol{\epsilon}\mathcal{F}_6 - i\boldsymbol{\sigma}\cdot\hat{\mathbf{q}}\epsilon_0\mathcal{F}_7 - i\boldsymbol{\sigma}\cdot\hat{\mathbf{k}}\epsilon_0\mathcal{F}_8 \ .
\end{aligned} \tag{13}$$

Imposing gauge invariance by demanding $\mathcal{F} \to 0$ when $\epsilon_\mu \to K_\mu$ results in two constraints on the $\mathcal{F}_i$'s (one demands that the coefficients of $\boldsymbol{\sigma} \cdot \hat{\mathbf{k}}$ and $\boldsymbol{\sigma} \cdot \hat{\mathbf{q}}$ are separately zero). From these constraints, one may eliminate either $\mathcal{F}_5$ and $\mathcal{F}_6$, or $\mathcal{F}_7$ and $\mathcal{F}_8$. Eliminating $\mathcal{F}_7$ and $\mathcal{F}_8$ results in the above form for $\mathcal{F}$, but with ϵ_μ replaced by

$$a_\mu = \epsilon_\mu - \frac{\epsilon_0}{k_0} K_\mu \ . \tag{14}$$

Note that the time component of a_μ is zero, and thus we have photons with transverse and longitudinal components. On the other hand, eliminating $\mathcal{F}_5$ and $\mathcal{F}_6$ results in

$$\begin{aligned}\mathcal{F} = & \, i\boldsymbol{\sigma} \cdot \mathbf{b}\mathcal{F}_1 + \boldsymbol{\sigma} \cdot \hat{\mathbf{q}}\boldsymbol{\sigma} \cdot (\hat{\mathbf{k}} \times \mathbf{b})\mathcal{F}_2 + i\boldsymbol{\sigma} \cdot \hat{\mathbf{k}}\hat{\mathbf{q}} \cdot \mathbf{b}\mathcal{F}_3 + i\boldsymbol{\sigma} \cdot \hat{\mathbf{q}}\hat{\mathbf{q}} \cdot \mathbf{b}\mathcal{F}_4 \\ & - i\boldsymbol{\sigma} \cdot \hat{\mathbf{q}} b_0 \mathcal{F}_7 - i\boldsymbol{\sigma} \cdot \hat{\mathbf{k}} b_0 \mathcal{F}_8 \ ,\end{aligned} \tag{15}$$

where

$$b_\mu = \epsilon_\mu - \frac{\boldsymbol{\epsilon} \cdot \hat{\mathbf{k}}}{|\mathbf{k}|} K_\mu \ . \tag{16}$$

In this case, $b_0 \neq 0$, but $\mathbf{b} \cdot \mathbf{k} = 0$, and we thus have photons with transverse and Coulomb (also called scalar) components.

With this latter choice, the $\mathcal{F}$'s may be expressed in terms of the A's as follows:

$$\begin{aligned}
\mathcal{F}_1 &= \frac{S_1 S_2 (W-M)}{8\pi W}\left[A_1 + (W-M)A_4 + \frac{Q\cdot K}{(W-M)}(A_3 - A_4) - \frac{K^2}{(W-M)}A_6\right] \\
\mathcal{F}_2 &= \frac{qk(W+M)}{8\pi W S_1 S_2}\left[-A_1 + (W+M)A_4 + \frac{Q\cdot K}{(W+M)}(A_3 - A_4) - \frac{K^2}{(W+M)}A_6\right] \\
\mathcal{F}_3 &= \frac{qkS_2}{8\pi W S_1}\left[\left(W^2 - M^2 - \frac{1}{2}K^2\right)A_2 + (W+M)(A_3 - A_4) + K^2 A_5\right] \\
\mathcal{F}_4 &= \frac{q^2 S_1}{8\pi W S_2}\left[-\left(W^2 - M^2 - \frac{1}{2}K^2\right)A_2 + (W-M)(A_3 - A_4) - K^2 A_5\right] \\
\mathcal{F}_7 &= -\frac{qS_1}{8\pi W S_2}\left\{(E_i - M)A_1 - \left[\mathbf{q}\cdot\mathbf{k}\left(\frac{3}{2}k_0 - 2W\right) + \mathbf{k}^2\left(W - \frac{3}{2}\omega\right)\right]A_2\right. \\
&\quad - (W+M)(E_i - M)A_4 + [Q\cdot K - \omega(W-M)](A_3 - A_4) \\
&\quad \left. - (k_0 Q\cdot K - \omega K^2)A_5 + (W+M)(E_i - M)A_6\right\} \\
\mathcal{F}_8 &= \frac{kS_2}{8\pi W S_1}\left\{(E_i + M)A_1 - \left[\mathbf{q}\cdot\mathbf{k}\left(\frac{3}{2}k_0 - 2W\right) + \mathbf{k}^2\left(W - \frac{3}{2}\omega\right)\right]A_2\right. \\
&\quad + (W-M)(E_i + M)A_4 - [Q\cdot K - \omega(W+M)](A_3 - A_4) \\
&\quad \left. - (k_0 Q\cdot K - \omega K^2)A_5 - (W-M)(E_i + M)A_6\right\} \ ,
\end{aligned} \tag{17}$$

where $S_1 = \sqrt{E_i^2 + M^2}$, $S_2 = \sqrt{E_j^2 + M_j^2}$, $q = |\mathbf{q}|$, and $k = |\mathbf{k}|$. These equations may be inverted to obtain the A's in terms of the $\mathcal{F}$'s.

2.4 Multipole Decomposition

As the resonances have definite parity and angular momentum, it is extremely useful when analyzing the data to work with eigenamplitudes of parity and angular momentum, i.e. to work with the multipoles. The allowed multipoles are (transverse) electric, $E_{l\pm}$, (transverse) magnetic, $M_{l\pm}$, and Coulomb, $C_{l\pm}$, where l is the meson-baryon orbital angular momentum, and "$\pm$" indicates the total angular momentum of $J = l \pm 1/2$. Defining L to be the photon's angular momentum, the electric and Coulomb multipoles have parity $(-1)^L$ with $L = l \pm 1$ and the magnetic multipoles have parity $(-1)^{L+1}$ with $L = l$. To give an example, the $\Delta - N$ transition is $3/2^+ \rightarrow 1/2^+$, and thus the photon may have $L = 1, 2$ with positive parity. For $L = 1$, only magnetic dipole ($M1$) is allowed, while for $L = 2$, the transition can be either electric quadrupole ($E2$) or Coulomb quadrupole ($C2$). We then expect to see resonant structure in the M_{1+} (due to $M1$), E_{1+} (due to $E2$), and C_{1+} (due to $C2$) multipoles.

The expansion of the $\mathcal{F}$'s in multipoles is:

$$\mathcal{F}_1 = \sum_{l=0}^{\infty} [P'_{l+1}(x)E_{l+} + P'_{l-1}(x)E_{l-} + lP'_{l+1}(x)M_{l+} + (l+1)P'_{l-1}(x)M_{l-}]$$

$$\mathcal{F}_2 = \sum_{l=0}^{\infty} [(l+1)P'_l(x)M_{l+} + lP'_l(x)M_{l-}]$$

$$\mathcal{F}_3 = \sum_{l=0}^{\infty} [P''_{l+1}(x)E_{l+} + P''_{l-1}(x)E_{l-} - P''_{l+1}(x)M_{l+} + P''_{l-1}(x)M_{l-}]$$

$$\mathcal{F}_4 = \sum_{l=0}^{\infty} [-P''_l(x)E_{l+} - P''_l(x)E_{l-} + P''_l(x)M_{l+} - P''_l(x)M_{l-}]$$

$$\mathcal{F}_7 = \sum_{l=0}^{\infty} [-(l+1)P'_l(x)C_{l+} + lP'_l(x)C_{l-}]$$

$$\mathcal{F}_8 = \sum_{l=0}^{\infty} [(l+1)P'_{l+1}(x)C_{l+} - lP'_{l-1}(x)C_{l-}] \quad , \qquad (18)$$

where P_l are Legendre polynomials, $x = \cos\theta$ with θ the cm scattering angle, and a prime denotes differentiation with respect to x.

These equations can be inverted to obtain the multipoles in terms of the $\mathcal{F}$'s:

$$E_{l+} = \int_{-1}^{1} \frac{dx}{2(l+1)} \left[P_l(x)\mathcal{F}_1 - P_{l+1}(x)\mathcal{F}_2 + \frac{l}{2l+1}(P_{l-1}(x) - P_{l+1}(x))\mathcal{F}_3 + \frac{l+1}{2l+3}(P_l(x) - P_{l+2}(x))\mathcal{F}_4\right]$$

$$E_{l-} = \int_{-1}^{1} \frac{dx}{2l} \left[P_l(x)\mathcal{F}_1 - P_{l-1}(x)\mathcal{F}_2 - \frac{l+1}{2l+1}(P_{l-1}(x) - P_{l+1}(x))\mathcal{F}_3 + \frac{l}{2l-1}(P_l(x) - P_{l-2}(x))\mathcal{F}_4\right]$$

$$M_{l+} = \int_{-1}^{1} \frac{dx}{2(l+1)} \left[P_l(x)\mathcal{F}_1 - P_{l+1}(x)\mathcal{F}_2 - \frac{1}{2l+1}(P_{l-1}(x) - P_{l+1}(x))\mathcal{F}_3\right]$$

$$M_{l-} = \int_{-1}^{1} \frac{dx}{2l} \left[-P_l(x)\mathcal{F}_1 + P_{l-1}(x)\mathcal{F}_2 + \frac{1}{2l+1}(P_{l-1}(x) - P_{l+1}(x))\mathcal{F}_3 \right]$$

$$C_{l+} = \int_{-1}^{1} \frac{dx}{2(l+1)} \left[P_{l+1}(x)\mathcal{F}_7 + P_l(x)\mathcal{F}_8\right]$$

$$C_{l-} = \int_{-1}^{1} \frac{dx}{2l} \left[P_{l-1}(x)\mathcal{F}_7 + P_l(x)\mathcal{F}_8\right] \ . \tag{19}$$

Only the transverse multipoles contribute to photoproduction. The longitudinal multipoles are related to the Coulomb multipoles by $L_{l\pm} = C_{l\pm}k_0/k$. The helicity elements $A_{l\pm}$ and $B_{l\pm}$ are also sometimes used, and are related to the multipoles by

$$(l+1)E_{l+} = A_{l+} + \frac{l}{2}B_{l+}$$

$$(l+1)M_{l+} = A_{l+} - \frac{l+2}{2}B_{l+}$$

$$(l+1)E_{(l+1)-} = A_{(l+1)-} - \frac{l+2}{2}B_{(l+1)-}$$

$$(l+1)M_{(l+1)-} = A_{(l+1)-} + \frac{l}{2}B_{(l+1)-} \ . \tag{20}$$

The $A_{l\pm}$ have helicity 1/2 while the $B_{l\pm}$ have helicity 3/2; the final state has helicity 1/2 since the meson carries no spin. One of the predictions of perturbative quantum chromodynamics (pQCD) is that helicity is conserved, and thus at some large $-K^2$ the $A_{l\pm}$ should dominate over the $B_{l\pm}$. The outstanding question, which must be answered experimentally, is how large $-K^2$ has to be before the perturbative domain of QCD is reached.

2.5 Isospin

First recall that the nucleons form an isodoublet, the pions an isotriplet, the kaons an isodoublet, and the sigmas an isotriplet. Both the η and Λ are isosinglets. Assuming that the strong interaction conserves isospin and the electromagnetic current transforms as an isoscalar plus the third component of an isovector,

$$j_\mu = j_\mu^S + j_\mu^{V_3} \ , \tag{21}$$

it is easy to find the isospin structure of the amplitude. The initial state has T=1/2, and in the cases of $\gamma N \rightarrow N\pi$ and $\gamma N \rightarrow \Sigma K$, the final isospin may be 1/2 or 3/2. An isoscalar photon obviously results in a $T = 1/2$ final state, and one normally defines

$$\langle 1/2|j^S|1/2\rangle \equiv A^0 \ , \tag{22}$$

where A is understood to be a generic symbol representing A_i, $\mathcal{F}_i$, or one of the multipoles. For an isovector photon, the final state may be $T = 1/2$ or 3/2:

$$\langle 1/2|j^{V_3}|1/2\rangle \equiv A^1$$

$$\langle 3/2|j^{V_3}|1/2\rangle \equiv A^3 \ . \tag{23}$$

Defining for later use $A^+ = (A^1+2A^3)/3$ and $A^- = (A^1-A^3)/3$, the physical channels for pion production can be expressed as

$$\begin{aligned}\langle n\pi^+|T|\gamma p\rangle &= \sqrt{2}(A^0 + A^-)\\ \langle p\pi^0|T|\gamma p\rangle &= (A^0 + A^+)\\ \langle p\pi^-|T|\gamma n\rangle &= \sqrt{2}(A^0 - A^-)\\ \langle n\pi^0|T|\gamma n\rangle &= -(A^0 + A^-)\ .\end{aligned} \tag{24}$$

Therefore, only three of the four reactions are independent. For $K\Sigma$ production, the above form holds with the replacements on the left-hand-side (LHS) of (24) $(p,n) \to (K^+, K^0)$ and $(\pi^+,\pi^0,\pi^-) \to (\Sigma^+,\Sigma^0,\Sigma^-)$. For the reactions $\gamma N \to K\Lambda$ and $\gamma N \to N\eta$, the final state has $T = 1/2$, and thus

$$\begin{aligned}\langle p\eta|T|\gamma p\rangle &= A^0 + A^1\\ \langle n\eta|T|\gamma n\rangle &= A^0 - A^1\ ,\end{aligned} \tag{25}$$

with similar equations for $K\Lambda$ production replacing on the LHS of (25) η with Λ and (p,n) with (K^+, K^0).

At times there has been interest in a possible isotensor component in the photon, in which case all four of the $\gamma N \to N\pi$ reactions would be independent. Unfortunately, there is not enough high-quality data for $\gamma n \to n\pi^0$ to put a tight bound on the isotensor component. Another place to look for the isotensor component is in the photo- and electroexcitation of $T = 3/2$ resonances. If there is no isotensor component, then

$$\langle N^*(3/2,1/2)|T|\gamma p\rangle = \langle N^*(3/2,-1/2)|T|\gamma n\rangle\ , \tag{26}$$

and thus a difference between the two photo-couplings is a signal for an isotensor component in the photon.

2.6 Crossing Symmetry and C-Invariance

If we apply crossing symmetry, we may relate the amplitude for the process

$$\gamma B(P_i) \to B'(P_f)\Phi(Q) \tag{27}$$

to the amplitude for the process

$$\gamma \bar{B}'(-P_i) \to \bar{B}(-P_f)\Phi(Q)\ . \tag{28}$$

This latter process may be at an unphysical point. Now applying charge conjugation invariance, we may relate (28) to

$$\gamma B'(-P_i) \to B(-P_f)\bar{\Phi}(Q)\ , \tag{29}$$

where "bars" denote anti-particles. In the case of π and η production from nucleons, we are back to the original amplitude, but with s and u interchanged. Taking into account the charge conjugation properties of the particles and the Dirac matrices appearing in (9), one may easily determine how the A_i behave under the interchange of s and u for pion production:

$$A_i^{+,0}(s,t,u) = \chi A_i^{+,0}(u,t,s)$$
$$A_i^-(s,t,u) = -\chi A_i^-(u,t,s) \ , \qquad (30)$$

where χ = 1 for i = 1,2,4 and χ = -1 for i = 3,5,6. The crossing properties for η production are the same as those for the +, 0 amplitudes. For kaon production, the best one can do is relate, for example, $A_i(s,t,u)$ for $\gamma N \rightarrow \Lambda K$ to $A_i(u,t,s)$ for $\gamma\Lambda \rightarrow N\bar{K}$.

2.7 Unitarity and T-Invariance

Time-reversal invariance can be used to relate radiative meson capture to photomeson production. Of particular interest is the reaction

$$\pi^- p \rightarrow \gamma n \ , \qquad (31)$$

which is obviously related to

$$\gamma n \rightarrow p\pi^- \ . \qquad (32)$$

Thus, with proton targets we are able to experimentally determine three of the four photopion reactions, and then use isospin invariance to get the remaining reaction $\gamma n \rightarrow n\pi^0$.

Time-reversal invariance also implies that the S-matrix is symmetric, which is one of the ingredients that goes into deriving Watson's theorem [10], which we now discuss. The unitarity of the S-matrix

$$S^\dagger S = 1 \ , \qquad (33)$$

obviously constrains the number of independent S-matrix elements. Let us consider a simple problem with two channels. Assuming T-invariance, the most general S-matrix takes the form

$$S = \begin{pmatrix} \eta e^{2i\delta_1} & \pm i\sqrt{1-\eta^2}e^{i(\delta_1+\delta_2)} \\ \pm i\sqrt{1-\eta^2}e^{i(\delta_1+\delta_2)} & \eta e^{2i\delta_2} \end{pmatrix} \ . \qquad (34)$$

Thus the S-matrix is defined in terms of one elasticity parameter, η, and two phases δ_1 and δ_2. In terms of the T-matrix defined as

$$T = \frac{(S-1)}{2i} \ , \qquad (35)$$

we have

$$T_{jj} = \frac{i}{2}\left(1 - \eta e^{2i\delta_j}\right)$$
$$T_{12} = \frac{\sqrt{1-\eta^2}}{2}e^{i(\delta_1+\delta_2+n\pi)} \ . \qquad (36)$$

It should be obvious from the form of (34) that we are in the partial wave representation of the S-matrix, i.e. the S-matrix has been block diagonalized into states of good parity and total angular momentum.

Now let us take channel 1 to be the πN channel and channel 2 to be the γN channel. Ignoring the electromagnetic interaction compared to the strong interaction ($\eta \approx 1$), one obtains

$$T_{11} = \sin\delta_1 e^{i\delta_1}$$
$$T_{12} = |T_{12}|e^{i(\delta_1 + n\pi)} \ , \tag{37}$$

where we have ignored the Compton phase shift, δ_2, compared to the πN elastic scattering phase shift, δ_1. Therefore, the phase of a photo- or electroproduction multipole is given by the πN scattering phase shift in that channel, which is Watson's theorem.

Actually we have cheated in the above derivation of Watson's theorem because we do not really have a two channel problem. First, we have assumed that isospin is good so that all the possible πN elastic scattering states can be reduced to a combination of $T = 3/2$ and $1/2$. Second, there is really more than one γN channel. For example, both $M1$ and $E2$ photons can lead to a final πN state with $J = T = 3/2$. Fortunately, Fermi [11] has shown that as long as the off-diagonal components of the S-matrix are small, one obtains

$$T_{ij} = |T_{ij}|e^{i(\delta_i + \delta_j + n\pi)} \ . \tag{38}$$

Thus, ignoring the Compton phase shift, one finds, for example, that both the E^3_{1+} and M^3_{1+} multipoles have the same phase δ_{33}, where δ_{33} is the πN elastic scattering phase shift in the $J = T = 3/2$ channel.

The applicability of Watson's theorem breaks down above the two pion threshold, or to be more definite, when η is no longer approximately 1. Above the two pion threshold, the constraint imposed by unitarity is much more complicated [12]. One must in principle consider an infinite-dimensional S-matrix since there is a three-body final state, and all possible combinations of angular momentum and spin which add together to give the total J^P of the initial γN state must be considered. In practice, it is found that two of the outgoing particles are often correlated so that only a few of the angular momentum combinations are important. For example, up to a photon energy of about 700 MeV, two pion photoproduction is described well by the mechanism [13]

$$\gamma N \to \pi\Delta \to \pi(\pi N) \ . \tag{39}$$

3 Construction of the Amplitude

Before turning to a discussion of the specific models of photo- and electromeson production, it is useful to summarize the common features of these models. The first ingredients are the Born terms calculated with either pseudoscalar (PS) or pseudovector (PV) ΦNB coupling (see Fig. 1). In principle, these terms have no free parameters, but in some cases the ΦNB coupling is not well determined. The Born terms are generally assumed to dominate the higher partial waves in which resonances are not present. We add that the tree level amplitudes obtained from PS and PV couplings differ by a term proportional to the anomalous magnetic moment and this "equivalence breaking" term appears only in the E_{0+} and M_{1-} multipoles. In the case of charged meson production, the higher partial waves are important due to meson-pole term (Fig. 1C), which has a very slow convergence in partial waves. In Fig. 2, we show the differential cross-section for $\gamma p \rightarrow n\pi^+$ when only s and p waves are kept (dashed line) compared to the full calculation (solid line). The difference between these two curves is primarily due to the t-channel pion exchange (Fig. 1C). It is amusing to note that in the usual gauge this term is zero in the forward direction. However, the individual multipoles are not zero, and hence one needs to keep many multipoles in order to achieve cancellation in the forward direction.

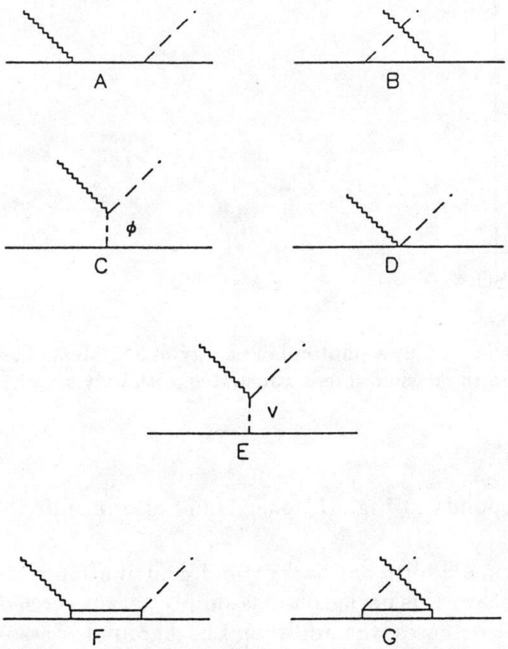

Fig. 1. Diagrammatic representation of the photomeson production amplitude. Wavy lines are photons, dashed lines are mesons, and solid lines are baryons. A-C represent the PS Born terms while D also appears for the PV Born terms. It should be understood that the Feynman rules are different for PS and PV. E represents t-channel vector meson exchanges, while F and G represent the s and u channel resonance exchanges, respectively.

The second piece of the amplitude is the resonant contributions which are normally written in a relativistic Breit-Wigner form

$$A_R = \frac{M_R\sqrt{\Gamma_\gamma \Gamma_\phi}}{s - M_R^2 + iM_R\Gamma_T} , \qquad (40)$$

where M_R is the "mass" of the resonance, Γ_γ is the γN decay width, Γ_ϕ is the ΦB decay width and Γ_T is the total decay width of the resonance. In most cases M_R, Γ_ϕ and Γ_T are borrowed from πN scattering in order to reduce the number of parameters. It should be emphasized that one needs not only the strengths of the decay widths, but also the energy dependence, which can introduce additional parameters. Apart

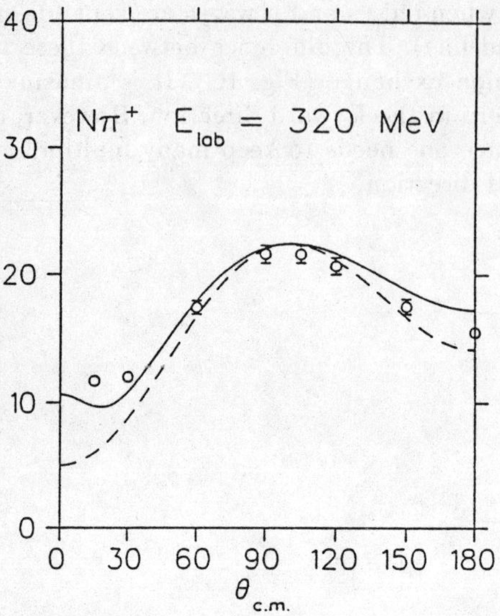

Fig. 2. The differential cross section for $\gamma p \to n\pi^+$ at a photon lab energy of 320 MeV. The solid line is the full calculation of [14], while the dashed line is calculated with only s and p waves. The data are from [15].

from the above amplitude, which corresponds to Fig. 1F, one should also include its crossing partner, Fig. 1G.

The last piece of the amplitude is some additional background contribution containing additional parameters. In many cases, this background is simply parameterized as some smooth function of energy. In other cases, the additional background is associated with single particle exchanges such as t-channel vector meson exchanges, Fig. 1E, whose couplings are known to some degree from other experiments, or can be estimated theoretically.

The final problem is to add all the pieces together such that unitarity is satisfied, at least approximately. For this reason, and others, the theoretical analysis of the data is on much better footing in the $\Delta(1232)$ region than in the higher resonance region. In the $\Delta(1232)$ region, the only important strong channel is the πN channel, and Watson's theorem may be used. This results in many simplifications and reduces the amount of phenomenology one needs to put into the amplitude. It is fair to say that if we do not understand the amplitude in the $\Delta(1232)$ region, we have little chance of understanding it at higher energies.

After the amplitude has been constructed, one fits the parameters of the model to the data and extracts some number for Γ_γ. In practice, many difficulties arise in the fitting procedure. First, different data sets do not always agree and one has to decide which data sets to renormalize and by how much. Second, important polarization data may be lacking which makes it difficult to separate overlapping resonances even if they have different quantum numbers. Third, one is trying to minimize a function with many parameters, often 20 or more, and there may be many local minima with similar χ^2. One then has to find some other criteria to prefer one solution over the others.

3.1 Isobar Models

Isobar models, also referred to as energy dependent multipole analyses, have been widely used to analyze the photo- and electromeson production data. To give a short list of references, photopion production has been considered in [16,17,18], photokaon production in [19,20], photoeta production in [21], and η and π electroproduction in [22]. Since most of these analyses are minor variations of Walker's work [16], it is sufficient here to review this work. The model consists of the electric Born terms, i.e. the Born terms without the anomalous magnetic moments, some additional background which is constrained to be a smooth function of energy, and Breit-Wigner resonances parameterized as

$$A(W) = A(W_0)\left(\frac{k_0 q_0}{kq}\right)^{1/2} \frac{W_0\sqrt{\Gamma\Gamma_\gamma}}{s_0 - s - iW_0\Gamma} , \tag{41}$$

where the subscript "0" indicates the value of the quantity evaluated at the resonance energy W_0. The explicit form for the widths is

$$\Gamma = \Gamma_0\left(\frac{q}{q_0}\right)^{2l+1}\left(\frac{q_0^2+X^2}{q^2+X^2}\right)^{l}$$
$$\Gamma_\gamma = \Gamma_0\left(\frac{k}{k_0}\right)^{2j_\gamma}\left(\frac{k_0^2+X^2}{k^2+X^2}\right)^{j_\gamma} . \tag{42}$$

For the $\Delta(1232)$ the damping factor, X, was taken to be 0.16 GeV, and for all other resonances X was taken to be 0.35 GeV.

It was originally hoped that the resonances would give totally the imaginary part of the amplitude. However, it became evident that the additional background needed to be complex in order to obtain a good fit to the data. In the early works, the additional background was simply fitted at a few energies and interpolated at other energies [17]. In other works [18,22,21], some explicit functional form for the additional background was assumed. In particular, in Ref. [18], the multipoles were parameterized as

$$A = A_I(1 + iT_\pi) + A_R T_\pi , \tag{43}$$

where A_I and A_R are real and T_π represents the πN partial waves. The resonant structure of the photoproduction amplitude is implicitly contained in the resonant structure of the πN partial waves. The resonant term, A_R, was parameterized as

$$A_R = \frac{M_\pi}{q}\left(\frac{k}{q}\right)^{l}\sum_{n=0}^{N} p_n\left(\frac{\omega}{M_\pi}\right)^{n} , \tag{44}$$

and the background piece, A_I, consists of two parts: the full PS Born terms, plus an additional background piece, A_Q, parameterized as

$$A_Q = \sum_{n=0}^{M} \bar{p}_n Z Q_{l+n}(Z) , \tag{45}$$

where $Q_l(Z)$ are Legendre functions of the second kind and

$$Z = \frac{\sqrt{q^2 + 4M_\pi^2}}{q} \ . \tag{46}$$

A_Q is supposed to represent the two pion cut.

Unlike most previous parameterizations which did not explicitly attempt to satisfy unitarity in the higher resonance region, the above form is motivated by assuming that there are two dominant strong channels contributing in a given multipole, i.e. the πN channel and one other channel such as $\pi\Delta$ or ηN. For many resonances, there are two dominant decay channels, but there are exceptions. For example, the $S_{31}(1620)$ decays strongly into πN, $\pi\Delta$ and ρN. Also, it is not clear if the background mechanism can be broken down into a few dominant channels. One must also worry if A_I and A_R have poles on the real axis as they are functions of the K-matrix elements [18] which do in general have poles on the real axis. A_R causes no problem since it is a ratio of K-matrix elements and the pole drops out. The structure of A_I is more complicated and one must assume that the K-matrix residues "factorize" in order to cancel the poles. Finally, we should also mention here that there is no reason to assume that the total width should have the same energy dependence as the πN width when the resonance decays into more than one channel, as is the case with (41).

3.2 Fixed-t Dispersion Relations

Due to a large number of parameters, the isobar models are able to give a good fit to the data, and have been extremely successful in extracting the resonant couplings from the data. Despite these successes, theoretical objections have been raised about this approach. These objections may be found in several places [7], but may be summarized as follows: are the amplitudes obtained in isobar models, which are not manifestly crossing symmetric, consistent with fixed-t dispersion relations?

Assuming analyticity and using crossing symmetry, the invariant amplitudes for pion photo- and electroproduction satisfy [7,8]

$$\begin{aligned} ReA_i^{\pm,0}(s,t,K^2) = &\left[\frac{1}{s-M^2} + \xi\frac{1}{u-M^2}\right] \Gamma_i^{\pm,0}(t,K^2) + \frac{\Gamma_{ti}^{\pm,0}}{(t-M_\pi^2)} \\ &+ \frac{\mathcal{P}}{\pi}\int_{(M+M_\pi)^2}^{\infty} ds' \left[\frac{1}{(s'-s)} + \xi\frac{1}{(s'-u)}\right] ImA_i^{\pm,0}(s',t,K^2) \end{aligned} \tag{47}$$

where ξ is $\pm$ 1 depending on the crossing property of the amplitude (30), and Γ_i and Γ_{ti} are the residues of the one-nucleon and one-pion exchange diagrams shown in Fig. 1. Note that the dispersion relations give a definite relationship between the real and imaginary parts of the amplitude. The Γ_i are given by

$$\begin{aligned} \Gamma_1^{\pm,0} &= \frac{eg}{2} F_1^{v,s}(K^2) \\ \Gamma_2^{\pm,0} &= 2\Gamma_5^{\pm,0} = \frac{-egF_1^{v,s}(K^2)}{t-M_\pi^2} \\ \Gamma_3^{\pm,0} &= \Gamma_4^{\pm,0} = \frac{-eg}{2} F_2^{v,s}(K^2) \\ \Gamma_6^{\pm,0} &= 0 \ . \end{aligned} \tag{48}$$

For Γ_{ti}, only the fifth element of the (-) amplitude is non-zero

$$\Gamma_{t5}^{-} = \frac{2eg}{K^2}\left[F_\pi(K^2) - F_1^v(K^2)\right] \ , \tag{49}$$

where $F_\pi(K^2)$ is the pion electromagnetic form factor, $F_1^{v,s}$ are the nucleon electric isovector and isoscalar form factors and $F_2^{v,s}$ are the anomalous magnetic moment isovector and isoscalar form factors. In particular,

$$F_{1,2}^v(K^2) = F_{1,2}^p(K^2) - F_{1,2}^n(K^2)$$
$$F_{1,2}^s(K^2) = F_{1,2}^p(K^2) + F_{1,2}^n(K^2) \ , \tag{50}$$

with the normalizations at $K^2 = 0$ of $F_\pi(0) = F_1^p(0) = 1$, $F_1^n(0) = 0$ and $F_2^{p,n} = \kappa_{p,n}/(2M)$, where $\kappa_{p,n}$ are the anomalous magnetic moments of the proton and neutron, respectively.

Using (17), (19) and (47), one may obtain a set of coupled singular integral equations for the multipoles [8,23]. As t is fixed, it is evident when integrating over s' that x' is not fixed, and for $s' < s$ there is the possibility that $|x'| > 1$. Since it is not clear if the multipole expansion converges outside the physical range, the main predictive power of dispersion relations is in the $\Delta(1232)$ region, where it can be shown that up to photon lab energies of about 450 MeV the expansion does converge for any s'. This region has been studied by many authors [8,23,24] and excellent agreement with the data is obtained, essentially parameter free.

In the higher energy region, the dispersion integral is normally broken down into two [25] or three [26] regions. For $W' < 2$ GeV, the imaginary part of the multipoles is parameterized as either Breit-Wigners [26] as in [16], or in terms of the K-matrix [25]. In some multipoles, an additional smooth background was included in the dispersion integral. For $W' > 2.5$ GeV, the invariant amplitudes are assumed to have a Regge form

$$\mathrm{Im}A_i^{\pm,0} = b_i^{\pm,0}(t)\left(\frac{s}{s_0}\right)^{\alpha(t)-1} + \bar{b}_i^{\pm,0}\left(\frac{s}{s_0}\right)^{\bar{\alpha}(t)-1} \ , \tag{51}$$

with $s_0 \approx 2.5$ GeV. In the region $2 < W' < 2.5$ GeV, Crawford [26] used the form

$$\mathrm{Im}A = \ \mathrm{Im}A(\mathrm{resonance})\frac{s_2 - s}{s_2 - s_1} + \ \mathrm{Im}A(\mathrm{Regge})\frac{s - s_1}{s_2 - s_1} \ , \tag{52}$$

with $W_1 = 2$ GeV and $W_2 = 2.5$ GeV. In order to avoid problems with convergence of the multipole expansion, only data with $-t < 1.5$ GeV^2 were fitted. It should be emphasized that in contrast to isobar models, only the imaginary part of the amplitude needs to be parameterized. The dispersion integrals provide the real part.

In the case of electroproduction, complications arise since F_1^n and F_π are not well known. Adler [27] was able to obtain a good fit to the electroproduction data in the $\Delta(1232)$ region assuming $F_\pi(K^2) = F_1^p(K^2)$ and $G_E^n = 0$. Mistretta et al. [28] fitted F_π to the π^+ electroproduction data with $G_E^n = 0$ and found results consistent with a monopole form factor

$$F_\pi(K^2) = \left(1 + \frac{K^2}{\Lambda^2}\right)^{-1} \ , \tag{53}$$

where $\Lambda = 0.56 \pm 0.08$ GeV. Devenish and Lyth [29] considered electroproduction in the higher resonance region in an attempt to extract the $N^* - N$ transition form factors. F_π was taken from ρ dominance.

3.3 Effective Lagrangians

The use of isobar models and fixed-t dispersion relations are the two main methods used to extract the photo-couplings. However, from a nuclear physics point of view, these two methods have a drawback. In these approaches, the amplitude is obtained explicitly in the πN rest frame and thus must be transformed to a different frame when considering the nuclear process. Due to Fermi motion of the nucleons in the nucleus, it is desirable to have an amplitude that is explicitly frame independent. This is one of the main motivations of the effective Lagrangian approach. The other advantage is that compared to the isobar model, the number of parameters can be reduced, and one has a physical interpretation of the background.

The effective Lagrangian (EL) approach arose as a simple method to obtain the soft pion results based on current algebra and the hypothesis of partially conserved axial vector current (PCAC) [30]. One simply writes down an EL that is Lorentz invariant, gauge invariant and chirally symmetric. Evaluation of the amplitude in the tree approximation yields the soft pion results, and provides a smooth extrapolation away from the soft pion point. However, in the light of the chiral perturbation theory (CHPT) calculations [31], it is not clear if this extrapolation is really smooth. To illustrate the EL approach, we consider the case of pion photoproduction in the $\Delta(1232)$ region. One starts with the PV πNN Lagrangian, applies minimal substitution, and inserts the anomalous magnetic moment terms by hand to arrive at [32]

$$\begin{aligned}\mathcal{L}_{PV} = & \frac{f}{M_\pi}\bar{N}\gamma_\mu\gamma_5\boldsymbol{\tau}N\cdot\partial^\mu\boldsymbol{\pi} - e\bar{N}\gamma_\mu Q_N N A^\mu \\ & + \frac{e}{4M}\bar{N}(\kappa^0+\kappa^+\tau_3)\sigma_{\mu\nu}NF^{\mu\nu} - ie\pi_-\overleftrightarrow{\partial}^\mu\pi_+A_\mu \\ & + i\frac{ef}{M_\pi}\bar{N}\gamma_\mu\gamma_5\left[Q_N,\boldsymbol{\tau}\right]N\cdot\boldsymbol{\pi}A^\mu \ , \end{aligned} \tag{54}$$

where $\kappa^0 = (\kappa^p+\kappa^n)/2$, $\kappa^+ = (\kappa^p-\kappa^n)/2$, and $f^2/(4\pi)\approx 0.08$ is the PV πNN coupling constant. Also, $Q_N = 1+\tau_3$ where $\boldsymbol{\tau}$ are the usual Pauli isospin matrices, A^μ is the photon field, and $F^{\mu\nu} = \partial^\nu A^\mu - \partial^\mu A^\nu$. It should be emphasized that the EL approach described here is a model and not the same as CHPT. For example, it is inconsistent from the point of view of CHPT to add in the anomalous magnetic moment terms without considering other contributions, such as pion loops, which appear at the same order.

Evaluation of this EL in the tree approximation leads to E_{0+} multipoles that are in good agreement with experiment [33,34] at threshold [35], and reproduce to order M_π/M the low energy theorems [31,36,37] for this process. This success prompts one to believe that at higher energies one can still expect the EL to give a reasonable description of the non-resonant part of the amplitude. To obtain agreement with the data in the $\Delta(1232)$ region, one must construct an EL describing the $\pi N\Delta$ and $\gamma N\Delta$ vertices [32]. These interactions are normally taken to be [38,14]

$$\begin{aligned}\mathcal{L}_{\pi N\Delta} &= \frac{g_{\pi N\Delta}}{M_\pi}\partial^\mu\boldsymbol{\pi}\bar{N}\cdot\mathbf{T}O_{\mu\nu}(Z)\Delta^\nu + H.c. \\ \mathcal{L}^1_{\gamma N\Delta} &= \frac{ieg_{1\Delta}}{2M}\bar{\Delta}^\mu O_{\mu\lambda}(Y)\gamma_\nu\gamma_5 T_3^\dagger NF^{\nu\lambda} + H.c. \\ \mathcal{L}^2_{\gamma N\Delta} &= \frac{-eg_{2\Delta}}{4M^2}\bar{\Delta}^\mu O_{\mu\nu}(X)\gamma_5 T_3^\dagger(\partial_\lambda N)F^{\nu\lambda} + H.c. \ , \end{aligned} \tag{55}$$

where Δ^μ is the Δ field, $\mathbf{T}$ are the $1/2 \leftrightarrow 3/2$ isospin transition matrices and

$$O_{\mu\nu}(Z) = g_{\mu\nu} + [(1+4Z)A/2 + Z]\,\gamma_\mu\gamma_\nu \ . \tag{56}$$

As expected, there are two independent $\gamma N\Delta$ couplings whose strengths are given by $g_{1\Delta}$ and $g_{2\Delta}$, and may be related to $E2$ and $M1$ (see below). Furthermore, $g_{\pi N\Delta}$ is the $\pi N\Delta$ coupling constant related to the width of the Δ. For virtual photons, there is one additional $\gamma N\Delta$ coupling which is most easily constructed by shifting the derivative in $\mathcal{L}^2_{\gamma N\Delta}$ from the nucleon field to $F^{\nu\lambda}$. Note the presence of the additional parameters A, X, Y, Z which are related to problems involving relativistic treatments of spin 3/2 fields [12]. The above vertices have been chosen such that the parameter A, which also appears in the Δ propagator, will not appear in any physical matrix element. The remaining parameters X, Y, Z are usually fitted to the data.

The last piece of the amplitude comes from t-channel vector meson exchanges, in particular ω and ρ. The EL is [38]

$$\mathcal{L}_V = \frac{e\lambda_V}{4M_\pi}\epsilon_{\mu\nu\lambda\sigma}F^{\mu\nu}V^{\lambda\sigma}\pi - g_{V1}\bar{N}\gamma_\mu N V^\mu + \frac{g_{V2}}{4M}\bar{N}\sigma_{\mu\nu}NV^{\mu\nu} \ , \tag{57}$$

where V^μ is either the ω or ρ field, $V^{\mu\nu} = \partial^\nu V^\mu - \partial^\mu V^\nu$, $g_{V1,2}$ are the two VNN coupling constants, and λ_V describes the $V \rightarrow \gamma\pi$ decay. The VNN couplings are not well known, but the vector meson contributions are small in the Δ region, and uncertainties in these couplings can essentially be absorbed into uncertainties in X, Y, Z which are fitted to the data.

One now evaluates the amplitude in the tree approximation, and obtains a result that is gauge invariant, Lorentz invariant and crossing symmetric. However, the amplitude is real and has a pole on the real axis at $W = M_\Delta$. Unitarization of the amplitude, i.e. removal of the pole and consistency with Watson's theorem, has been considered in [39,40] and more recently in [14]. As the unitarization is carried out in the cm frame, the explicit frame independence of the amplitude is destroyed. In the original work of Blomqvist and Laget (BL) [40], the Born terms were kept real and the Δ pole was moved to the complex plane by adding a width to the propagator

$$s - M_\Delta^2 \rightarrow s - M_\Delta^2 + iM_\Delta\Gamma \ . \tag{58}$$

The resulting amplitude was frame independent and gauge invariant, but not unitary. The single nucleon data were reasonably well produced, but the M^3_{1+} multipole was not well reproduced due to the violation of Watson's theorem. A unitary version of the amplitude was also supplied by BL, but it was the non-unitary version that was commonly used in nuclear calculations. Since then, the problem of unitarization and frame transformations has been solved by Wittman [41] and Laget [42].

In the case of η and K production, the EL approach becomes quite similar to the isobar model in the sense that there are numerous resonances and vector meson exchanges that must be considered, and the couplings are generally not well known. Furthermore, the EL approach is not theoretically well founded in this energy region, and the constraint of unitarity is usually ignored. Despite the numerous resonances and vector meson exchanges, the EL approach in this energy region still contains fewer parameters than the isobar model. The kaon production amplitude [43,44] has been constructed mainly to be used as a tool in hypernuclear studies, although it can be used in principle to study the $N^*K\Lambda$ decay widths. In the case of η production, the EL approach has recently been used to extract the $S_{11}(1535) \to \gamma p$ transition amplitude [45].

The electroproduction of pions in the Δ region has been considered by Olsson [46]. The electromagnetic form factors must be inserted by hand, which can lead to problems with gauge invariance since a priori $F_\pi(K^2) \neq F_1^p(K^2)$. The Born terms with form factors are in general not gauge invariant, and hence cannot be projected into the form (9) without a trick. Replacement of ϵ by K results in, for example, a term

$$[F_1^p(K^2) - F_\pi(K^2)]\, K^2 \; . \tag{59}$$

To make the amplitude gauge invariant, one simply adds an additional term that is formally zero, but upon replacement of ϵ with K cancels the above term. In this case the additional term would be

$$-\,[F_1^p(K^2) - F_\pi(K^2)]\, K \cdot \epsilon \; . \tag{60}$$

This brings us to an important question regarding the validity of the EL approach in the Δ region and at higher energies. The form factors are not only functions of K^2, but of all the four momenta that enter the vertex, so that, for example, $F_2 = F_2(K^2, P_i^2, P_f^2)$. In photoproduction, it is the case that either P_i^2 or P_f^2 is not equal to M^2, and thus one must worry about form factors even with real photons [47], not to mention the form factors at the strong vertex. Stated differently, one may ask about the importance of the loop corrections. The unitarization of the amplitude considered in [14,39,40] amounts to keeping the imaginary part of the loops, which generates the unitary corrections, but ignoring the real part of the loops. In CHPT, the one-loop corrections at threshold are of order $1 \times 10^{-3}\ M_\pi^{-1}$ [31], where in the same units the $p\pi^0$ amplitude is about -2, and the charged pion amplitude is about $\pm$ 30. The parameters X, Y, Z may partially contain the real part of the loop corrections since they are fitted to the data. Also, it is found [14,38] that pion photoproduction favors a relatively large $g_{\omega 2}$ coupling, perhaps a consequence of ignoring the loop corrections. An important by-product of this is that the threshold Δ contributions found in [35] could be entirely different if loops are explicitly included.

3.4 Dynamical Models

To overcome some of the problems of the EL approach, the simple unitarization procedure and the disregard of the real part of higher order contributions, so-called dynamical models have arisen in the past few years. The original works [48,49] were confined to the 33 channel both in πN scattering and photoproduction, but now have been extended to all important multipoles in the Δ region, both for photoproduction [50] and electroproduction [51]. The work starts with the construction of the πN T-matrix either in a separable form [50] or based on a meson exchange model [52,53]. For example, the separable form results in the Lippmann-Schwinger equation [50]

$$T_{\pi\pi}(\mathbf{k}_f, \mathbf{k}_i, E) = V_{\pi\pi}(\mathbf{k}_f, \mathbf{k}_i, E) + \int dk\mathbf{k}^2 V_{\pi\pi}(\mathbf{k}_f, \mathbf{k}, E) G_{\pi N}(\mathbf{k}, E) T_{\pi\pi}(\mathbf{k}, \mathbf{k}_i, E) \; , \tag{61}$$

where

$$G_{\pi N} = [E - E_\pi(\mathbf{k}) - E_n(\mathbf{k}) + i\epsilon]^{-1} \ . \tag{62}$$

The potential was taken to have pole parts and non-pole parts:

$$V_{\pi\pi}(k_f, k_i, E) = f_0(k_f) g_B^0(E) f_0(k_i) \pm h_0(k_f) h_0(k_i) \ , \tag{63}$$

where $g_B^0 = (E - M_{0B})^{-1}$, and B denotes either N or Δ. Thus one starts with bare nucleons and deltas that get dressed by pions.

Once $T_{\pi\pi}$ is known, one solves the Lippman-Schwinger equation for $T_{\pi\gamma}$

$$T_{\pi\gamma} = V_{\pi\gamma} + T_{\pi\pi} G_{\pi N} V_{\gamma\pi} \ , \tag{64}$$

where $V_{\pi\gamma}$ is constructed out of a particular three-dimensional reduction of the tree amplitude discussed in Sect. 3.3. The reduction was chosen so that the amplitude is gauge invariant, but it is not known if this reduction is unique. Furthermore, all the Born terms are multiplied by the same form factor, which is not expected on physical grounds.

Apart from the parameters describing $T_{\pi\pi}$, there are three parameters entering the pion photoproduction amplitude; a cutoff Λ and the bare strengths of the $E2$ and $M1$ denoted by G_E^0 and G_M^0 (see below). The numbers quoted are the bare parameters, and hence the difference with previous determinations of these quantities. The model used in [50] gave a good fit to the M_{1+}^3 and E_{1+}^3 multipoles, but the fit to the data was not as good as obtained with the previously mentioned techniques. Use of the meson exchange model [53] drastically improved the fit to the photoproduction data, and also gave a good fit to the electroproduction data. The amplitude constructed in this manner is unitary, non-perturbative, and believed to be highly suitable for nuclear calculations. However, due to the introduction of form factors and the three dimensional reduction of the amplitude, it is not known if the LETs, which are one of the driving forces behind the EL approach, are satisfied in this model. Further comparison of this model with the EL approach is given in the next section.

4 Single-Nucleon Results

The isobar model and fixed-t dispersion relations, despite some early disagreements, have now converged to common photo-couplings. For the most part, the differences between results obtained using the different methods is no larger than the differences obtained by different groups using the same method. The current results [54] obtained for some of the 4-star resonances are shown in Table 1 compared with two quark model predictions [55,56]. The large decay amplitudes are well known, but the smaller amplitudes need improvement. The quark model predictions are in qualitative agreement with these results, but there are significant disagreements. The photo-couplings are expressed in terms of the helicity amplitudes $A_{1/2}$ and $A_{3/2}$, and are in units of 10^{-3} GeV$^{-1/2}$. In terms of $A_{l\pm}$ and $B_{l\pm}$, $A_{1/2}$ and $A_{3/2}$ are normally defined to be proportional to the imaginary part of the resonance contribution at $W = M_R$. In particular,

$$\begin{aligned} \mathrm{Im}\,(A_{l\pm}^R) &= \pm f C_{\pi N}^I A_{1/2} \\ \mathrm{Im}\,(B_{l\pm}^R) &= \mp f \left[\frac{16}{(2j-1)(2j+3)}\right]^{1/2} C_{\pi N}^I A_{3/2} \ , \end{aligned} \tag{65}$$

where $C^I_{\pi N}$ is an appropriate isospin Clebsch-Gordon coefficient and

$$f = \left[\frac{kM\Gamma_\pi}{(2j+1)qM_R\Gamma^2}\right] \quad . \tag{66}$$

All kinematical variables are evaluated at $W = M_R$. The total photon decay width is then given by

$$\Gamma_\gamma = \frac{k_R^2 M}{\pi M_R}\frac{2}{(2J+1)}\left[|A_{3/2}|^2 + |A_{1/2}|^2\right] \quad . \tag{67}$$

It should be emphasized that the photon decay amplitudes calculated in a quark model are not necessarily the same physical quantities that are extracted from the data.

Table 1. The $N^* \to N\gamma$ helicity amplitudes from the recent PDG listings [54] compared with two quark model predictions [55,56]. The helicity amplitudes are in units of 10^{-3} GeV$^{-1/2}$.

Resonance	$A_{1/2}$[54]	$A_{3/2}$[54]	$A_{1/2}$[55]	$A_{3/2}$[55]	$A_{1/2}$[56]	$A_{3/2}$[56]
P11(1440)	-68 ± 5		+27		-5	
S11(1535)	+74 ± 11		+156		+97	
S11(1650)	+48 ± 16		0		0	
P13(1720)	+52 ± 39	-35 ± 24	+100	-30	+42	-54
P33(1232)	-141 ± 5	-258 ± 12	-108	-187	-101	-176
S31(1620)	+30 ± 10		+47		+86	
D35(1930)	-30 ± 40	-10 ± 35				

4.1 Energy Independent Multipole Analyses

Another method used to determine the multipoles is the energy independent multipole analysis, which we discuss briefly since some results from these analyses are presented below. On the surface the idea is fairly simple. The multipoles are functions of W and K^2, and ideally at each W and K^2 there are data for the differential cross section and polarization observables over a wide angular range. Since the observables are bilinears of the multipoles, one may simply fit the multipoles for $l \leq l_{max}$ to the data while taking the Born terms for $l > l_{max}$.

This procedure is generally regarded as the least biased way to determine the multipoles, but its success depends on the availability of a large amount of data. As the multipoles are complex quantities, this approach is most successful in the $\Delta(1232)$ region where Watson's theorem may be used to half the number of parameters fitted to the data. As with isobar models, one is performing a many parameter fit to the data and there are often many local minima. The problem is compounded in this case since the fit is done independently at each W and K^2, and multipoles are often obtained which do not have a smooth energy dependence. Plentiful and accurate data, particularly for the polarization observables, alleviates this problem to a large extent. Once the multipoles have been obtained in this manner, one is still faced with the problem of extracting the $N - N^*$ electromagnetic transition amplitudes from the multipoles.

4.2 $E2$ and $M1$ Transition Amplitudes

There has been on-going interest in the $E2$ $N-\Delta$ transition, and thus it is useful to present some of the recent results on this topic. Taking results listed in the most recent particle data review (PDG) [54], one finds ratios of $E2/M1$ ranging from -0.9% to -2.4% with an average of -1.5 ± 0.5% , while other analyses report values ranging from +3.7% [48] to -3.0% [50]. This range of values illustrates the uncertainty in the data and model dependence in the extraction of $E2$ and $M1$, which we discuss below.

Status of M^3_{1+} and E^3_{1+}. In the $\Delta(1232)$ region, energy independent multipole analyses have been used to obtain the l = 0,1 multipoles. This is extremely useful since the resonant part of the amplitude is isolated in two multipoles, M^3_{1+} and E^3_{1+}, and one can test how well the background in these multipoles is known by comparing how well the model agrees with the other, nonresonant, multipoles. In Fig. 3 we show the real (circles) and imaginary (triangles) parts of the M^3_{1+} multipole obtained from [59] compared with the fit of Ref. [14]. The fit looks good, but due to the small error bars on the M^3_{1+}, the χ^2_{DF} is about 5. However, it seems that the error bars are generally underestimated, as different analyses do not agree within the quoted errors, see Table 2. To eliminate some of the difference due to different phase shifts, the results are presented for N_M, where

$$M^3_{1+} = N_M e^{i\delta_{33}} \ . \tag{68}$$

Table 2. Values of N_M, Eq. (68), in units of $10^{-3}M_\pi^{-1}$ obtained from different energy independent multipole analyses as a function of the photon lab energy, K_L (in MeV).

K_L(MeV)	BD[59]	PS[62]	MIR[60]	GRU[63]	GET[61]
300	38.7±0.2	38.4±0.2	38.39±0.09	38.9±1.81	38.68±0.08
320	40.1±0.2	40.2±0.2	39.79±0.07	40.53±0.94	41.47±0.06
350	35.7±0.2	36.5±0.2	33.92±0.06	36.36±0.22	33.41±0.09
360	33.4±0.2	34.4±0.1	34.06±0.07		32.35±0.12
380	28.7±0.2	30.2±0.2	28.90±0.07	30.56±0.61	29.24±0.09
400	24.8±0.2	26.0±0.2	25.24±0.08	26.94±0.68	25.90±0.09

Let us now look at the E^3_{1+} multipole which is shown in Fig. 4A,B. It does not seem to have much of a resonant structure. The two curves shown are with $E2/M1$ = -0.4% and $E2/M1$ = 0. Thus, there are two important remarks to be made. First, near resonance the E^3_{1+} is very small (compare to the M^3_{1+}) and one may reasonably doubt that the E^3_{1+} is as well known as is claimed. Second, the resonant contribution to this multipole is very small adding to the difficulty of extracting $E2$ from the data. In Table 3, we present the quantity N_E, defined similarly to N_M, obtained in the different multipole analyses.

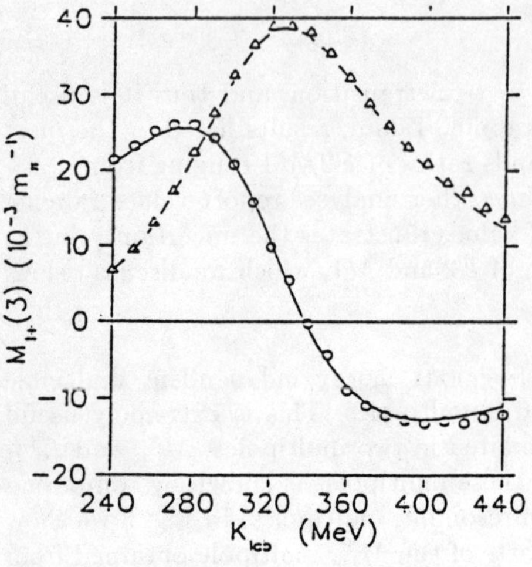

Fig. 3. The real (solid line and circles) and imaginary (dashed line and triangles) parts of the M^3_{1+} multipole. The calculation is from [14] and the data are from [59].

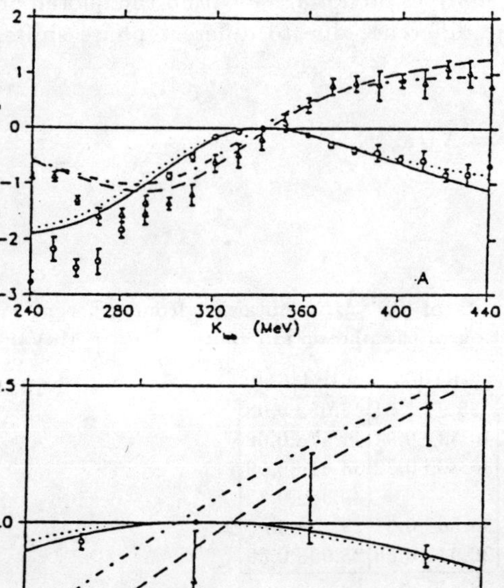

Fig. 4. The real (circles) and imaginary (triangles) parts of the E^3_{1+} multipole as determined in [59] . The solid (real part) and dashed (imaginary part) lines are the full calculations of [14], while the dashed and dashed-dotted lines are with $E2 = 0$. Fig. 4B is a blow-up of the resonance region.

Table 3. Values of N_E in units of $10^{-3} M_\pi^{-1}$ obtained from different energy independent multipole analyses as a function of the photon lab energy, K_L.

K_L(MeV)	BD	PS	MIR	GRU	GET
300	-1.62±0.13	-1.74±0.25	-1.98±0.07	-1.71±0.11	-1.83±0.06
320	-0.67±0.18	-0.69±0.24	-1.17±0.08	-1.23±0.12	-0.82±0.08
350	0.09±0.17	-0.10±0.10	-0.25±0.06	-0.43±0.09	-0.15±0.07
360	0.44±0.15	0.00±0.21	-0.02±0.06		-0.14±0.07
380	0.90±0.17	0.62±0.36	0.38±0.08	0.10±0.08	0.49±0.10
400	1.04±0.15	-0.21±0.45	0.79±0.09	0.49±0.07	0.70±0.11

Model Dependence of *E*2 and *M*1. Apart from the question about the accuracy of the multipoles, there is the important question of model dependence in the extraction of $E2$ and $M1$. As mentioned above, dynamical models have been used to try to extract the bare couplings, while most other approaches extract essentially Breit-Wigner parameters. The model dependence in the extraction of $E2$ and $M1$ has been investigated in Ref. [14] using the EL approach, and can most easily be seen by considering πN scattering in the 33 channel. Starting from the effective Lagrangian for the $\pi N \Delta$ vertex (55), one obtains in the tree approximation a contribution to the partial wave f_{1+}^3

$$q f_{1+}^{\Delta} = \frac{M_\Delta \Gamma_\pi(W)}{M_\Delta^2 - W^2} \equiv \frac{1}{\epsilon} \equiv \tan\delta_R \ , \tag{69}$$

with the pion decay width Γ_π given by

$$\Gamma_\pi(W) = \frac{g_{\pi N\Delta}^2 (E_f + M) q^3 (W + M_\Delta)}{24\pi M_\pi^2 W M_\Delta} \ . \tag{70}$$

Note that the effective Lagrangian gives a prediction for the energy dependence of the pion decay width. Taking into account a background contribution at the tree level of the form $q f_{1+}^B \equiv \tan\delta_B$, one now has to combine these amplitudes together to obtain the total f_{1+}^3. Various prescriptions have been given by Olsson [39] and Noelle [57], and we will also consider the K-matrix method. The three different unitarization methods give for δ_{33}

$$\tan\delta_{33} = \frac{1 + \epsilon \tan\delta_B}{\epsilon + \eta \tan\delta_B}, \tag{71}$$

where η =+1, -1, 0 for the Olsson, Noelle, and K-matrix methods, respectively. An interesting feature is that value(s) of $\tan\delta_{33}$ can be found in all three methods at which $g_{\pi N\Delta}$ and M_Δ can be found independently of $\tan\delta_B$. In particular, $g_{\pi N\Delta}$ and M_Δ can be determined by the energies at which $\tan\delta_{33} = \pm 1, \infty, -i$ for the Olsson, K-matrix, and Noelle methods, respectively. Thus, $\tan\delta_R$ is known and δ_B can be determined from (71) using the experimental δ_{33}. This method, however, does not tell us how reasonable the energy dependence of the width is, and therefore it is useful to assume some form for $\tan\delta_B$ and fit the parameters to δ_{33}. Taking $\tan\delta_B = a(q/\mu)^3 + b(q/\mu)^5$, one finds $(g_{\pi N\Delta}, M_\Delta)$ = (1.94, 1217), (2.46, 1250), and (2.16, 1232) for the Olsson, Noelle and K-matrix methods, respectively (M_Δ in MeV). These values agree within a few percent with those obtained from the special values of $\tan\delta_{33}$. In each case, assuming an error of 1 degree for δ_{33} at every energy, the χ_{DF}^2 is 0.45 and $a \approx 0.1$,

$b \approx$ -0.01. Note the model dependence of $g_{\pi N\Delta}$ and M_Δ, which will carry over into the determination of $E2$ and $M1$. This model dependence should come as no surprise since $g_{\pi N\Delta}$ and M_Δ are defined at different energies in the different models.

Having determined $\Gamma_\pi(W)$, which gives a good description of δ_{33}, we now proceed to the (γ, π) channel. For the resonant multipoles, the different methods give

$$A_{Ol} = e^{i\delta_{33}} \left[A_B \cos(\delta_{33} - \delta_B) + \sqrt{1 - \frac{A_B^2}{N^2}} N \sin(\delta_{33} - \delta_B) \right] , \tag{72}$$

$$A_N = e^{i\delta_{33}} \left[A_B \cos\delta_{33} + N \sin(\delta_{33} - \delta_B) \right] , \tag{73}$$

$$A_K = e^{i\delta_{33}} \left[A_B \cos\delta_{33} + \frac{N}{\epsilon} \cos\delta_{33} \right] , \tag{74}$$

where A refers to either the E_{1+}^3 or the M_{1+}^3 multipole, and the subscript denotes the unitarization method (Ol=Olsson, N=Noelle, K=K-matrix). Note that Watson's theorem is manifestly satisfied, and the functional form of the resonant multipoles is different in the various unitarization methods. A_B is the projection of the background (all terms excluding the s-channel Δ) into the resonant multipole. Also,

$$N_M = C \left[g_{1\Delta}(3W + M) - g_{2\Delta} \frac{W(W - M)}{2M} \right] , \tag{75}$$

$$N_E = -C(W - M) \left[g_{1\Delta} - g_{2\Delta} \frac{W}{2M} \right] \quad ; \; C = \frac{e\sqrt{E_i + M} k \mu}{4 g_{\pi N\Delta} \sqrt{E_f + M} q^2 M (W + M)} , \tag{76}$$

The gauge couplings, $g_{1\Delta}$ and $g_{2\Delta}$, can be fitted to the data and determine $E2$ and $M1$. Specifically,

$$M1 = \frac{e}{12M} \sqrt{\frac{k_\Delta}{M_\Delta M}} \left[g_{1\Delta}(3M_\Delta + M) - g_{2\Delta} \frac{M_\Delta (M_\Delta - M)}{2M} \right] \tag{77}$$

$$E2 = \frac{-e}{6M} \frac{k_\Delta}{(M_\Delta + M)} \sqrt{\frac{k_\Delta M_\Delta}{M}} \left[g_{1\Delta} - g_{2\Delta} \frac{M_\Delta}{2M} \right] , \tag{78}$$

where k_Δ is the photon energy evaluated at W = M_Δ. Sometimes $E2$ and $M1$ are expressed in terms of G_M and G_E:

$$E2 = -eG_E \sqrt{\frac{M_\Delta^2 - M^2}{8M^3}}$$

$$M1 = eG_M \sqrt{\frac{M_\Delta^2 - M^2}{8M^3}} , \tag{79}$$

so that $E2/M1$ = -G_E/G_M. The helicity amplitudes $A_{1/2}$ and $A_{3/2}$ are also often quoted, and are related to $E2$ and $M1$ by

$$A_{1/2} = -\frac{1}{2}(M1 + 3E2)$$

$$A_{3/2} = -\frac{\sqrt{3}}{2}(M1 - E2) . \tag{80}$$

Fitting the gauge couplings to various extant multipole sets [59]-[63] using the three different unitarization methods, one finds that the extracted $E2$ and $M1$ are quite sensitive to the method of unitarization. This is not surprising considering that the different methods give quite different values for $g_{\pi N\Delta}$, and $N_{E,M} \propto 1/g_{\pi N\Delta}$ (see (75,76)). Considering all the multipole sets, it was found: $M1$ = 250±16, $E2$ = -4.05±0.91, and $E2/M1$ = -1.63±0.37% for Olsson's method; $M1$ = 333±16, $E2$ = -8.34±3.45, and $E2/M1$ = -2.50±0.98% for Noelle's method; $M1$ = 283±10, $E2$ = -4.72±0.96, and $E2/M1$ = -1.68±0.32% for the K-matrix method, where $E2$ and $M1$ are in units of 10^{-3} GeV$^{-1/2}$. The errors here reflect the spread of values obtained from fitting the different data sets, and thus the model dependence induces greater uncertainty in $E2$ and $M1$ than the uncertainty in the multipole data base. Considering all the fits, including those with different ω couplings, the final results [14] are $M1$ = 285±37, $E2$ = -4.60±2.58, and $E2/M1$ = -1.57±0.72%.

If one chooses to identify the resonance parameters $E2$ and $M1$ with the imaginary parts of the multipoles at the energy at which the phase shift is 90^0 [54,64], then one may obviously extract these model independently from the multipoles. This is equivalent, in the case of the $\Delta(1232)$, to associating the $E2$ and $M1$ with the K-matrix residues [5]. In Ref. [5], this analysis was carried out on the existing energy independent multipole data base. Defining

$$M1 = Im M^3_{1+} \left[\frac{8M_\Delta q\pi\Gamma}{3Mk}\right]^{1/2} \tag{81}$$

$$E2 = Im E^3_{1+} \left[\frac{8M_\Delta q\pi\Gamma}{3Mk}\right]^{1/2}, \tag{82}$$

one finds the results given in Table 4. The errors on the multipoles have been increased [5] such that the different multipole sets are in statistical agreement in the sense used by the PDG [54]. The errors on $E2$ and $M1$ would be significantly smaller if this increase in the multipole error was not taken into account [14].

Table 4. The values of $E2$ and $M1$ in units of 10^{-3} GeV$^{-1/2}$ extracted [5] from the imaginary part of the E^3_{1+} and M^3_{1+} multipoles at resonance.

Data	$E2$	$M1$	$E2/M1$(%)
BD[59]	-1.7 ± 2.9	289 ± 26	-0.59 ± 1.01
MIR[60]	-2.9 ± 1.6	280 ± 30	-1.04 ± 0.58
GET[61]	-2.3 ± 2.8	286 ± 30	-0.80 ± 0.98
PS[62]	-2.2 ± 3.0	278 ± 38	-0.79 ± 1.08
GRU[63]	-5.9 ± 2.7	313 ± 30	-1.88 ± 0.88

The dynamical models avoid the problem of how to unitarize the amplitude since the Lippmann-Schwinger equation satisfies unitarity. This approach does not, however, eliminate the model dependence in extracting $E2$ and $M1$, as is evidenced by the various values of $E2/M1$ that have been extracted using this approach; +3.7% [48], -3.0% [50], -2.2% [53]. The bare couplings evidently are very sensitive to $T_{\pi\pi}$, and one may ask why should one bother with the bare couplings. The idea is that the quark model has no dynamical pions, and thus the pion cloud could give some important dressing to the $\gamma N \Delta$ couplings. Thus, in the dynamical models one starts with bare couplings which are dressed by pions, and one may extract a bare coupling, which, it is claimed, should be compared with the NRQM. Indeed the results obtained in [50] of $A_{1/2}$ = -97 and $A_{3/2}$ = -191 agree quite well with the NRQM [4]. However, in [53], the bare couplings are $A_{1/2}$ = -134 and $A_{3/2}$ = -255, in good agreement with the PDG listings. Furthermore, in Ref. [58], the pion cloud was considered in the NRQM. The pion cloud obviously also effects the properties of the nucleon, and after adjusting the parameters so that the agreement with $\langle r_p^2 \rangle$, κ_p and κ_n was not destroyed, it was found that there was essentially no change in $A_{1/2}$ and $A_{3/2}$.

In the near future, experiments will be performed that are highly sensitive to the E_{1+}^3 multipole. Already high quality data has been taken at Brookhaven for the quantity $d\sigma_{\parallel}/d\sigma_{\perp}$ [65]. The results are shown in Fig. 5 compared with calculations of [14] and [50], including the full calculations and calculations with $E2 = 0$. One sees that the data is good enough to distinguish between the full calculations and the calculations with $E2 = 0$, but these models do not agree with the data. It might be thought that the difference between [14] and [50] is due to the different treatments of the background-resonance separation in the E_{1+}^3 channel, but this is not the case. The discrepancy arises primarily due to the differences in the non-resonant multipoles, particularly the M_{1-}^1 multipole (see Fig. 6). This multipole receives a large contribution from the u-channel Δ which was included in [14], but omitted in [50]. Turning off the u-channel Delta in [14] readily reproduces the calculation in [50] of this observable, Fig. 7. Once this is taken into account, one sees that the two calculations with and without the $E2$ are very similar. We also learn that this observable is very sensitive to multipoles other than the E_{1+}^3, and hence can only be a good constraint on $E2$ once the other multipoles are known to high precision.

In light of this, it is difficult to say what exactly this data implies for the E_{1+}^3. It is interesting that this data is difficult to fit simultaneously with other data, indicating some inconsistencies in the data base. If one forces a fit to this data, it is found that $E2/M1$ increases by a factor of 2 [64]. Finally, we warn the reader that it is not always clear what is meant to set $E2 = 0$ in a calculation. For example, in the EL approach, this could mean that one takes just the background, B, in this multipole without unitarization, or one may take a unitarized background such as $B \cos\delta e^{i\delta}$. Finally, one could take N_E, from Eq. (76) to vanish at $W = M_\Delta$, and hence at other values of W one does have a "resonant" contribution. These different possibilities are shown in Fig. 8, compared with the full calculation of [14].

4.3 The S11(1535) → γp Decay Amplitude

In pion production, both T=1/2 and T=3/2 resonances may contribute, whereas in η production only T=1/2 resonances can contribute thereby providing a nice tool to separate the T=1/2 resonances from the T=3/2 resonances. The S11(1535) resonance, which peaks slightly above the ηp threshold, is known to decay roughly 50% into ηN and 50% into πN. Thus by analyzing either η or π photoproduction one may extract the S11 → γp decay amplitude, $A_{1/2}$. Past determinations of $A_{1/2}$ for the S11(1535)

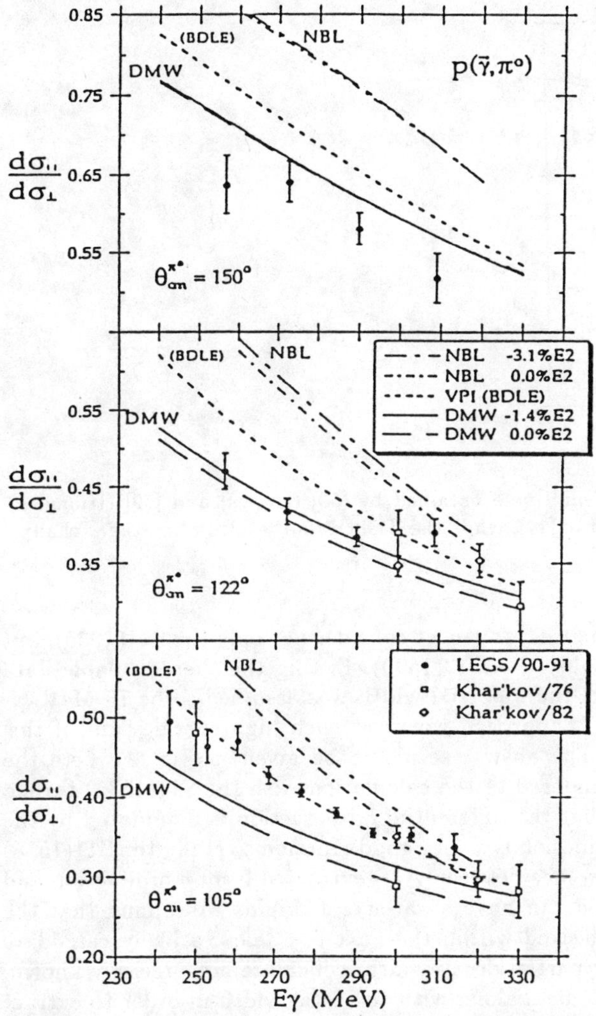

Fig. 5. The Brookhaven data [65] for $d\sigma_{\parallel}/d\sigma_{\perp}$ compared with the calculations of [14] (solid line, full calculation; dashed line, $E2 = 0$), the calculations of [50] (long dashed-dotted line, full; short dashed-dotted line $E2 = 0$), and the direct multipole input of [59]. Figure taken from [65].

were based on photopion production and resulted in $A_{1/2} = 73 \pm 14$ in the usual units of 10^{-3} GeV$^{-1/2}$. This number is smaller than that predicted by most quark models, see Table 1.

Recently Benmerrouche and Mukhopadhyay analyzed the η photoproduction data using the EL approach [45]. The background contribution was constructed from the nucleon Born terms and t-channel ω and ρ exchange as in the case of pion photoproduction, but both PV and PS ηNN couplings were considered. As the energy is high

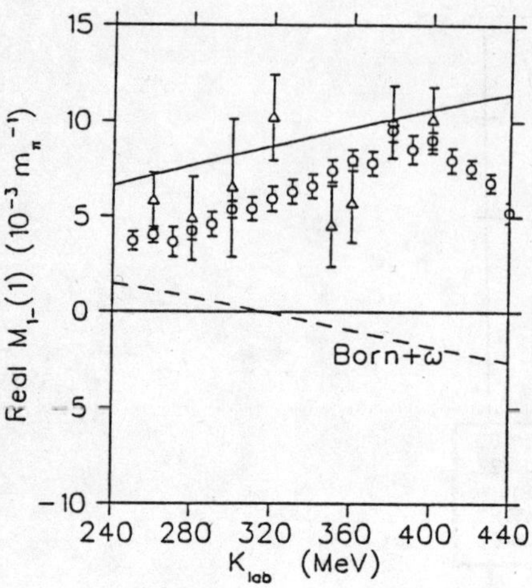

Fig. 6. The real part of the M_{1-}^{1} multipole obtained by [59] (circles) and [62] (triangles). The solid line is the full calculation of [14], while the dashed line results when the u-channel Δ contribution is turned off.

enough to excite numerous resonances, one must consider not only the S11(1535), but also other resonances, most notably the D13(1520). Fitting the then available data resulted in $A_{1/2} = 95 \pm 11$ where the η partial width was assumed to be 75 MeV.

One may wonder why the EL approach works at such high energies and if this biased the extraction of $A_{1/2}$. The answer seems to be given in Fig. 9 where the full calculation (solid line) is compared to the calculation with the S11(1535) turned off (dashed line). It is obvious that the differential cross section is dominated by the S11(1535), and thus η photoproduction is a very good reaction to study the S11(1535) $\rightarrow \gamma p$ decay amplitude. The difference between $A_{1/2}$ extracted from η production and that obtained from π production can have a variety of origins. Assuming that the data is correct, which will be checked within the next few years, a likely candidate for the discrepancy is the η or π partial decay width, which are not precisely known. It is necessary to know these numbers along with the total width in order to extract $A_{1/2}$ from the data, see (66).

4.4 Perturbative QCD at 3 GeV²?

As stated above, pQCD predicts that $B_{l\pm}/A_{l\pm} \rightarrow 0$ as $Q^2 = -K^2$ becomes large and the question is how large Q^2 needs to be. For the $N - N^*$ transition form factors, this implies that $A_{3/2}/A_{1/2} \rightarrow 0$ as $Q^2 \rightarrow \infty$, and some resonances, such as the D13(1520) and F15(1688), exhibit this behavior already at $Q^2 = 3$ GeV^2. However, this behavior is explained by NRQMs and it is necessary to consider resonances which do not exhibit this feature in a NRQM, such as the $\Delta(1232)$. In fact, until recently it was believed that the $\Delta(1232)$ does not exhibit this behavior at 3 GeV^2. In terms of $E2$ and $M1$, the pQCD prediction is $E2/M1 \rightarrow 1$ as $Q^2 \rightarrow \infty$, i.e. a change of sign and an increase in two orders of magnitude compared to the real photon ratio.

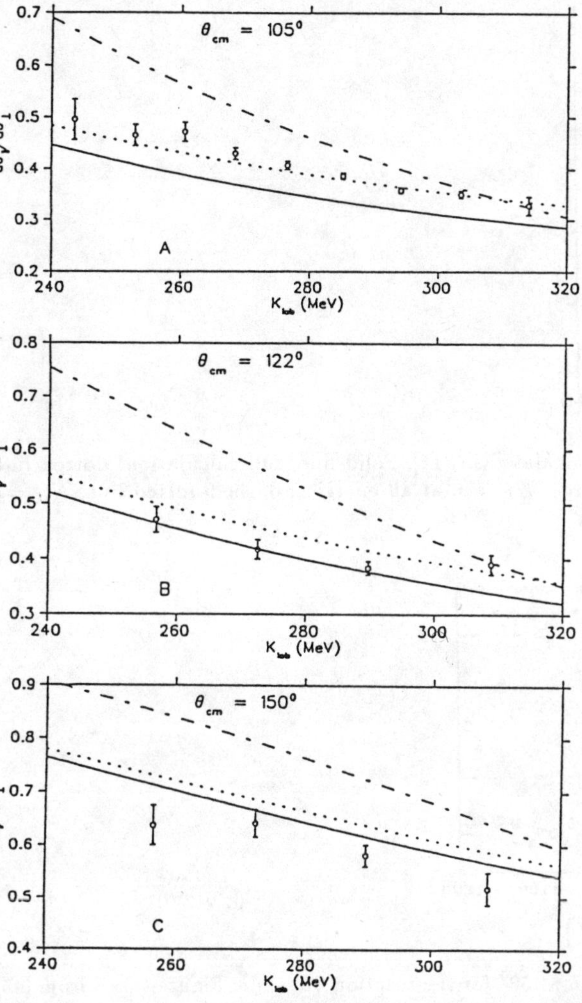

Fig. 7. The Brookhaven data [65] compared with three variations of the model in Ref. [14]. Solid, full calculation; dashed-dotted, no u-channel Δ; dotted, $E_{1+}^{\pi^0} \to 2.1 E_{1+}^{\pi^0}$.

As electroproduction is not as well experimentally studied as photoproduction, one examines the electroproduction data for $p\pi^0$ production, since the background is assumed to be small, and extracts from the data the ratio $Re(E_{1+}^* M_{1+})/|M_{1+}|^2$. This ratio should approach 1 in the pQCD regime, but previous analyses of the data at 3 GeV^2 found the ratio to be 0.05±0.05 [66]. However, Warren and Carlson [67] have made the important observation that the data is analyzed assuming M_{1+} dominance. To be more specific, one assumes that only s and p waves contribute to the amplitude, and using the differential cross section one may solve for $|M_{1+}|^2$ and the interference terms between M_{1+} and the smaller multipoles. The interference terms

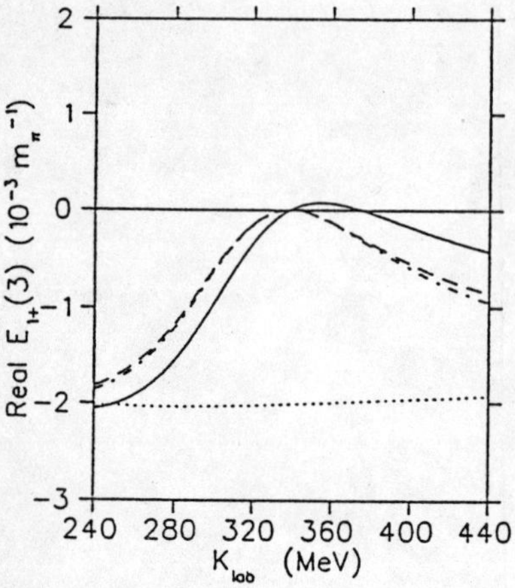

Fig. 8. The real part of E_{1+}^3 as calculated in [14]. Solid line, full calculation; dotted line, background contribution; dashed line, $N_E = 0$ at all energies; dashed-dotted line, $N_E = 0$ only at $W = M_\Delta$.

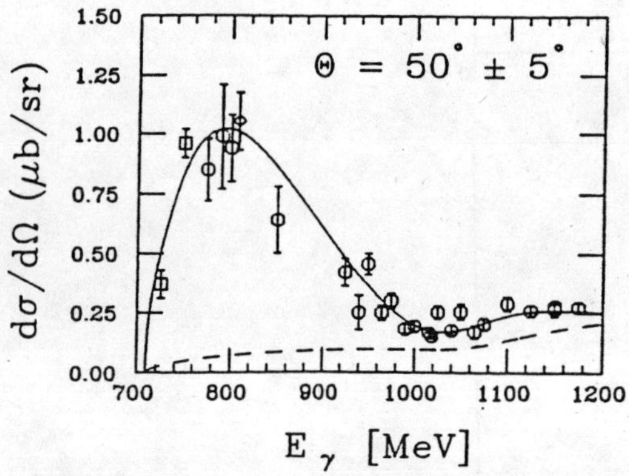

Fig. 9. The differential cross section at 50^0 for the reaction $\gamma p \rightarrow \eta p$. Figure taken from [45].

amongst the smaller multipoles are assumed negligible, and the approach is assumed to be successful if the extracted interference terms are much smaller than $|M_{1+}|^2$. The important result of Ref. [67] is that this is not the only possible scenario at $Q^2 =$ 3 GeV^2. If one instead assumes A_{1+} dominance (see (20)) and solves for $|A_{1+}|^2$ and the interference terms with A_{1+}, then the interference terms are again small resulting in $E_{1+}/M_{1+} = 0.70 \pm 0.09$. Furthermore, the data is inconsistent with either E_{1+} or B_{1+} dominance. Evidently this is a place where more data is needed to resolve the question of which scenario is correct. If it turns out that the amplitude is A_{1+} dominated, one should not jump to the conclusion that pQCD sets in at 3 GeV^2. A comprehensive experimental study of the entire resonance region needs to be made before it can be decided where pQCD sets in.

5 Results from Nuclei

Given the single nucleon amplitude, the nuclear transition amplitude is usually constructed in the impulse approximation. Denoting the free on-shell transition amplitude acting on the ith nucleon as t_i, we have for the nuclear transition amplitude, T,

$$T = \sum_i^A t_i = \sum_{\alpha,\beta} \langle \alpha | t | \beta \rangle a_\alpha^\dagger a_\beta \ . \tag{83}$$

Thus, in 2nd quantized notation, t is a one-body operator destroying a particle in state β and creating a particle in state α. We evidently obtain

$$\langle F, \pi | T | I, \gamma \rangle = \sum_{\alpha,\beta} \langle F | a_\alpha^\dagger a_\beta | I \rangle \langle \alpha, \pi | t | \beta, \gamma \rangle \ . \tag{84}$$

The operator $a_\alpha^\dagger a_\beta$ may be written in terms of irreducible tensors in spin-isospin space such that

$$\langle F | a_\alpha^\dagger a_\beta | I \rangle = \sum \text{ Clebsches } \langle F || \left[a_\alpha^\dagger \otimes a_\beta \right]_{J,T} || I \rangle \ . \tag{85}$$

The reduced matrix element, $\langle F || \left[a_\alpha^\dagger \otimes a_\beta \right]_{J,T} || I \rangle$, contains the nuclear structure information which is common to other reactions such as elastic and inelastic electron scattering, β decay, etc.

The part $\langle \alpha, \pi | t | \beta, \gamma \rangle$ is obviously different for different reactions. For pion production,

$$\langle \alpha, \pi | t | \beta, \gamma \rangle = \langle \alpha | \phi^*(\mathbf{q}, \mathbf{r}) t e^{i\mathbf{k}\cdot\mathbf{r}} | \beta \rangle \ , \tag{86}$$

where $|\alpha\rangle$ and $|\beta\rangle$ are the single nucleon wave functions and t has spin dependent and and spin independent parts

$$t = \boldsymbol{\sigma} \cdot \mathbf{K} + L \ . \tag{87}$$

In the plane wave impulse approximation (PWIA), the pion wave function $\phi^*(\mathbf{q}, \mathbf{r})$ is taken to be a plane wave

$$\phi^*(\mathbf{q}, \mathbf{r}) = e^{-i\mathbf{q}\cdot\mathbf{r}} \ . \tag{88}$$

In the distorted wave impulse approximation (DWIA), ϕ^* is taken from analyses of pion-nucleus scattering using either optical potentials or the Δ-hole model [68].

The formalism as presented here is obviously in configuration space whereas the single nucleon operator is normally derived in momentum space and has non-local terms like

$$\left[(P_i + K)^2 - M^2 \right]^{-1} \ . \tag{89}$$

The momenta appearing in (89) are obviously gradient operators in configuration space and an expansion of these terms in the momentum is necessary. To avoid this problem, the nuclear amplitude may be constructed in momentum space. The price that one pays is the six-dimensional integral

$$\langle \alpha, \pi | t | \beta, \gamma \rangle = \int d^3p d^3q \psi_\alpha^\dagger(\mathbf{p}') \phi^*(\mathbf{q}', \mathbf{q}) t(\mathbf{p}, \mathbf{p}', \mathbf{k}\mathbf{q}') \psi_\beta(\mathbf{p}) \ , \tag{90}$$

where $\mathbf{k} + \mathbf{p} = \mathbf{q}' + \mathbf{p}'$. To handle Coulomb corrections in momentum space, an additional integration is needed [41]. Fortunately, the Coulomb corrections are important only at small energies and can be ignored in the Δ region.

5.1 The $^{14}N(\gamma,\pi^+)^{14}C_{g.s.}$ Reaction

One sees that the γ, π reaction on nuclei is sensitive to the nuclear structure, the pion wave function, the single nucleon wave functions, and of course the elementary amplitude. All these features are nicely appearing in the reaction $^{14}N(\gamma,\pi^+)^{14}C_{g.s.}$. It is commonly assumed that only the 1p shell nucleons contribute to this reaction, and thus we need four numbers specifying the strengths of $p_{3/2}, p_{1/2} \rightarrow p_{3/2}, p_{1/2}$. These numbers as well as the single nucleon wave functions are constrained by electron scattering, μ capture, β decay, etc. In Fig. 10 we show the reaction for three different sets of wave functions [69,70,71]. The difference between the results using the Cohen-Kurath (CK) [69] and the other wave functions is quite large. The interesting difference between the wave functions of Ennslin et al. (E) [70] and Huffman et al. (H1) [71] is in the treatment of the β decay matrix element. The Gamow-Teller operator, $\boldsymbol{\sigma} \cdot \mathbf{j}_l$ ($\mathbf{j}_l$ the leptonic current) is responsible for β decay, and the dominant piece of the π^+ amplitude at low energies is proportional to the operator $\boldsymbol{\sigma} \cdot \boldsymbol{\epsilon}$. One way to obtain the long lifetime of ^{14}C is to use the nuclear structure to suppress the Gamow-Teller matrix element at the one-body level, thereby suppressing the differential cross section for $^{14}N(\gamma,\pi^+)^{14}C_{g.s.}$ at small energies and forward angles. This approach was taken by Ensslin et al. The other idea is that the one-body Gamow-Teller matrix element is not suppressed, but is accidentally canceled by higher order processes such as meson exchange. This approach was taken by Huffman et al. Evidently, measurement of π^+ production at forward angles and small energies could distinguish these two approaches. However, if meson exchange effects are large for β decay, one must also worry how they effect photopion production.

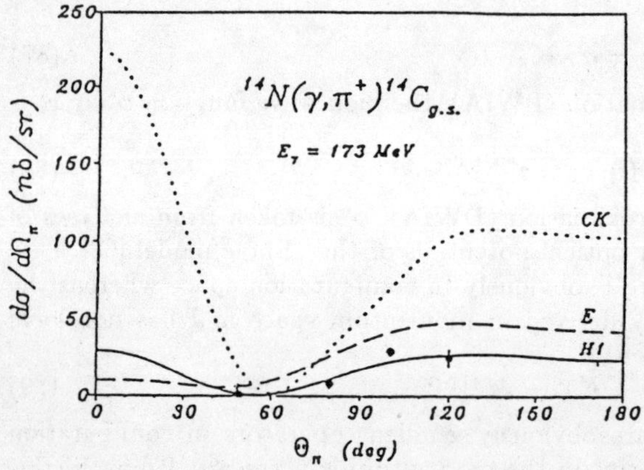

Fig. 10. The differential cross section for $^{14}N(\gamma,\pi^+)^{14}C_{g.s.}$ at a photon lab energy of 173 MeV calculated with three different sets of wave functions; Cohen-Kurath (CK) [69], Ennslin et al. (E) [70], and Huffman et al. (H1) [71]. Data and figure are from [73].

The dependence of the nuclear transition amplitude on the elementary amplitude and the pion wave function is clearly seen in Fig. 11, where we show the calculation of [41] compared to the preliminary data of Stoler and Seneviratne at a photon lab energy of 260 MeV. The dependence on the pion wave function shows up when comparing the solid and dashed curves. The solid curve uses a fully unitarized elementary amplitude with pion wave functions taken from the Δ-hole model, while the dashed curve is with pion wave functions obtained from an optical potential [72], again using the full operator. Replacing the unitarized elementary operator with the non-unitary version of the BL operator [40], but with pion wave functions from the optical potential, results in the dashed-dotted line. Thus, a comparison of the dashed and dashed-dotted curves shows the importance of the elementary amplitude.

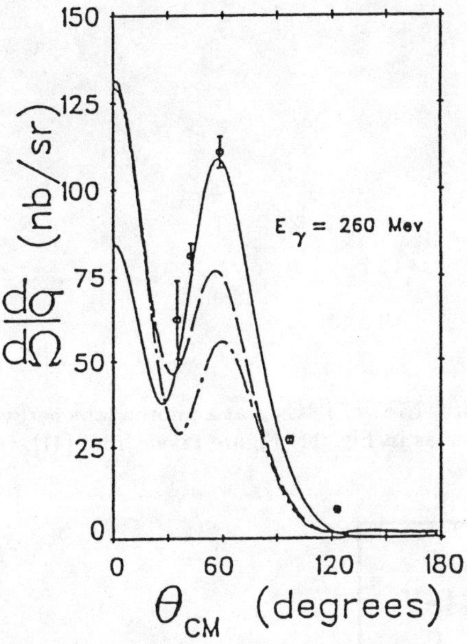

Fig. 11. Calculated [41] differential cross section for $^{14}N(\gamma,\pi^+)^{14}C_{g.s.}$ at a photon lab energy of 260 MeV compared with the preliminary data of Stoler and Seneviratne. Solid line, unitarized operator with pion wave functions taken from Δ-hole model; dashed line, unitarized operator with pion wave functions taken from optical potential; dashed-dotted line, non-unitary operator of BL with pion wave functions taken from optical potential. Figure taken from [41].

At higher energies, the sensitivity to the pion wave functions remains, but the sensitivity to the elementary operator is reduced, as is shown in Fig. 12. The calculations are from [41] and the data at 320 MeV are from [74]. The discrepancy between theory and experiment has now been largely removed, although disagreement still persists. The calculation of Ref. [41] was not a Δ-hole calculation, only the pion wave functions were taken from the Δ-hole model. The full Δ-hole calculation was carried out by Tiator et al. [75] and is compared with the most recent data at this energy [76], see Fig. 13. The main difference between the results of [41] and [75] is the lowering of the first minimum and an increase of the maximum around 45^0. The data have also come down significantly compared to the earliest results.

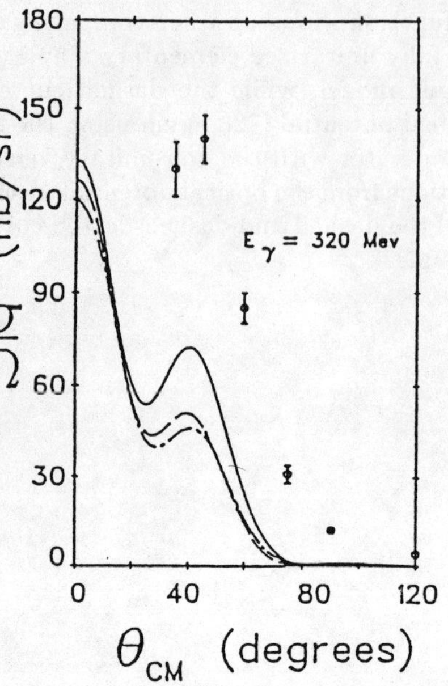

Fig. 12. Calculated [41] differential cross section for $^{14}N(\gamma, \pi^+)^{14}C_{g.s.}$ at a photon lab energy of 320 MeV compared with the data of [74]. Curves as in Fig. 11. Figure taken from [41].

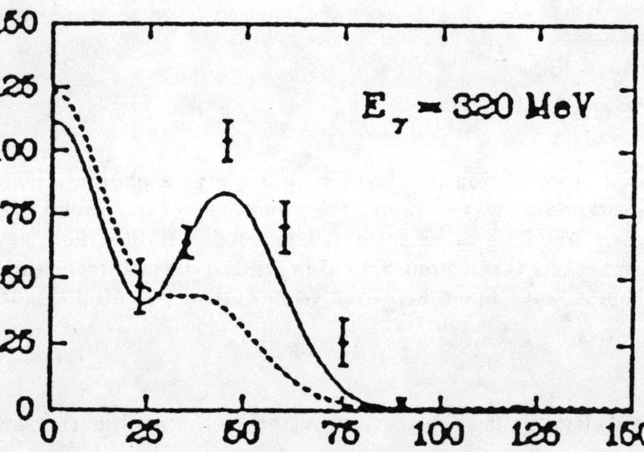

Fig. 13. Calculated [75] differential cross section for $^{14}N(\gamma, \pi^+)^{14}C_{g.s.}$ at a photon lab energy of 320 MeV compared with the data of [76]. Solid line, Δ-hole calculation; dashed line, DWIA calculation. Data and figure are from [76].

5.2 The Nucleus as a Filter

Kamalov et al. [6] recently considered polarization observables for ${}^3He(\gamma,\pi^+){}^3H$ using a coupled channels formalism. First consider the reaction $\gamma p \to n\pi^+$ and the sensitivity of the photon asymmetry, Σ, to the $E2$. This result is shown in Fig. 14, and one sees that Σ is not too sensitive to $E2$. We add that $E2 = 0$ in this calculation corresponds to the dotted line in Fig. 8, and thus it is perhaps more accurately stated that Σ does not show much sensitivity to E^3_{1+}.

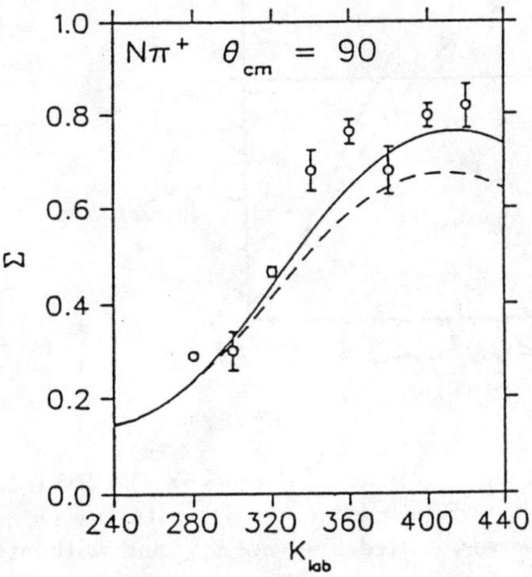

Fig. 14. Photon asymmetry, Σ, for $\gamma p \to n\pi^+$ at 90 degrees. Solid line, full calculation of [14]; dashed line, no E^{Δ}_{1+}, i.e. only background terms in the E^3_{1+} multipole.

Let us now consider ${}^3He(\gamma,\pi^+){}^3H$, where the wave functions are obtained by solving Faddeev equations with realistic $N-N$ interactions, and are thus believed to be under control. The resulting wave functions are about 90% S-states and 8% D-states. Keeping only the S-states and considering only the multipoles $E_{l\pm}$, $M_{l\pm}$ with $l = 0,1$ for the elementary amplitude, one obtains [6]

$$\Sigma_S \propto \frac{1}{2}|M_{1+}|^2 - \frac{3}{2}|E_{1+}|^2 + \text{Re}\left[E^*_{1+}(M_{1+} - M_{1-}) + M^*_{1+}M_{1-}\right] \ , \qquad (91)$$

which is the same combination of multipoles as in the single nucleon case and hence is not expected to show much sensitivity to $E2$. Now taking into account D-states, one finds an additional contribution to Σ due to the interference of S and D states:

$$\Sigma_{SD}(90^0) \propto 2qk\text{Re}(E_{0+}f^*) - q^2|E_{0+}|^2 - (4k^2+q^2)|f|^2 \ , \qquad (92)$$

where $f = 3E_{1+} - M_{1+} + M_{1-}$. Since the E_{0+} is large for π^+ production, one sees that the $S-D$ state interference, which is of course absent in the single nucleon case, leads to an enhanced sensitivity to E_{1+}, and hence $E2$. Using the full elementary amplitude, one sees the expected sensitivity to $E2$ in the PWIA, Fig 15., and the sensitivity remains in the full coupled channels formalism, as is shown in Fig. 16.

It is also interesting to note that the target asymmetry at 30^0 is determined almost entirely by the Δ, see Fig. 17. Therefore, this is a good place to look for medium modifications of the Δ.

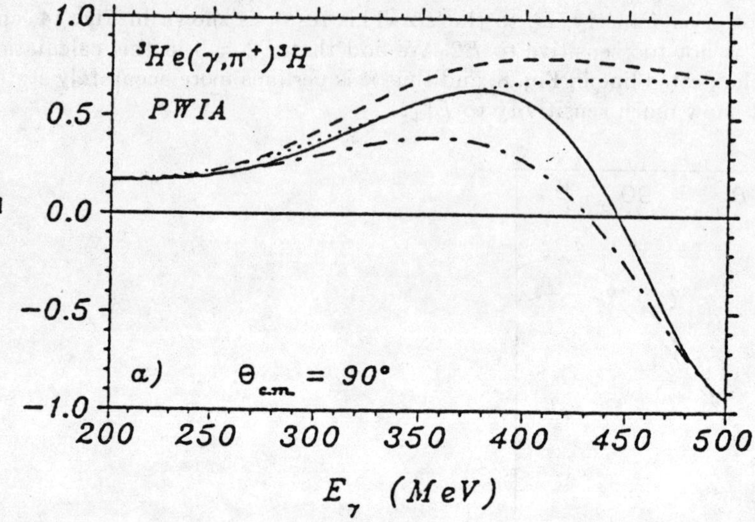

Fig. 15. Calculated [6] photon asymmetry for ^{3}He$(\gamma, \pi^+)^3$H at 90 degrees in the PWIA. Solid line, full production operator with D-waves; dashed-dotted line, no E_{1+}^{Δ} with D-waves; dashed line, full production operator without D-waves; dotted line, no E_{1+}^{Δ} and without D-waves. Figure taken from [6].

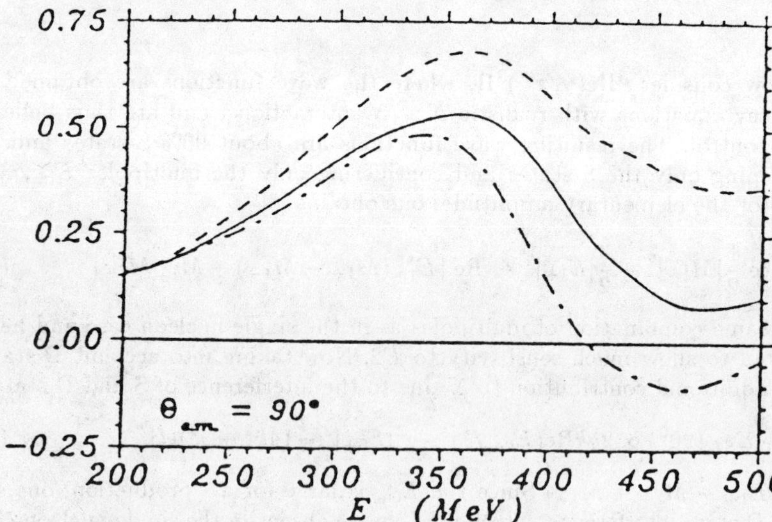

Fig. 16. Full coupled channels results of [6] for the photon asymmetry of ^{3}He$(\gamma, \pi^+)^3$H at 90 degrees. Curves as in Fig. 15. Figure taken from [6].

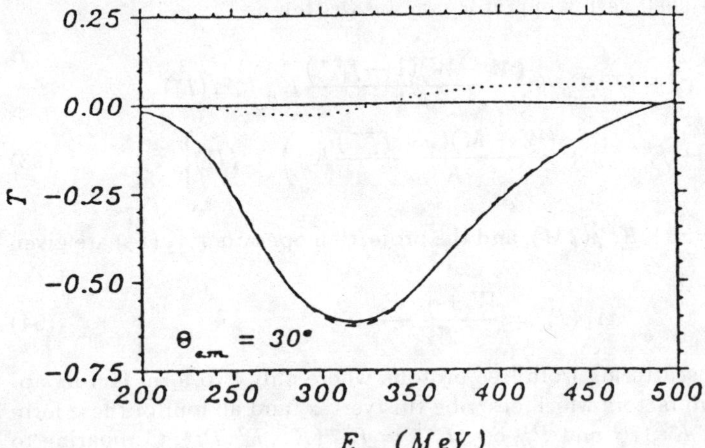

Fig. 17. Full coupled channels results of [6] for the target asymmetry of $^3\mathrm{He}(\gamma,\pi^+)^3\mathrm{H}$ at 30 degrees. Solid line, with D-states; dashed line, without D-states; dotted line, with D-states, but Born terms only in production operator. Figure taken from [6].

5.3 Off-shell Form Factors

Electron scattering from nuclei has long been used to study nuclear structure. As in the case of meson production from nuclei, theoretical calculations of this process are usually based on the impulse approximation with prescriptions for the nucleon currents founded on information obtained from the free nucleon. In a simple physical picture, in electron scattering from nuclei one is probing the electromagnetic structure of bound nucleons, and may naively test if nucleons in nuclei are swollen by 5-10% as was suggested in [77] as a possible explanation of the EMC effect.

This simple physical picture is evidently wrong. The nuclear current, which should obviously be conserved, cannot be written as a sum of one-body currents since this sum is in general not conserved. To achieve current conservation, it is necessary to include exchange currents. While it is known how to treat this problem nonrelativistically [78], a fully relativistic formulation of the problem is lacking.

Apart from the problem of gauge invariance, another problem arises when one describes the nuclear current in terms of nucleon currents. Let us recall the example considered in Ref. [79]. Quasi-free electron scattering from nuclei is perhaps the best place to study the electromagnetic properties of nucleons in nuclei. For quasi-free kinematics, the virtual photon has the right combination of energy and momentum to put the struck proton instantly on-shell without need for further interaction of the proton with the residual nucleus. We thus have the situation that the virtual photon strikes an off-shell proton with four-momentum P_i ($P_i^2 < M^2$) and knocks it out of the nucleus such that the final four-momentum, P_f, satisfies $P_f^2 = M^2$.

Evidently, for this process we need to know the γpp vertex when the initial proton is off-shell. The half off-shell vertex operator may be written as

$$\bar{U}_f \left[\left(f_1^{++}\gamma_\mu + f_2^{++}\Sigma_\mu + \frac{(M-W)(1-f_1^{++})}{K^2} K_\mu \right) \Lambda_+(P_i) + \left(f_1^{+-}\gamma_\mu + f_2^{+-}\Sigma_\mu + \frac{(W+M)(1-f_1^{+-})}{K^2} K_\mu \right) \Lambda_-(P_i) \right] , \quad (93)$$

where $W = \sqrt{P_i^2}$, $\Sigma_\mu = i\sigma_{\mu\nu}K^\nu/(2M)$, and the projection operators $\Lambda_\pm(P_i)$ are given by

$$\Lambda_\pm(P_i) = \frac{W \pm \gamma \cdot P_i}{2W} . \quad (94)$$

In contrast to electron scattering from free protons where only two form factors appear, there are four form factors which describe this vertex, and all four of these form factors are functions of K^2, P_i^2 and P_f^2, e.g. $f_1^{++} = f_1^{++}(K^2, P_i^2, P_f^2)$. Comparing to (50),

$$f_1^{++}(K^2, M^2, M^2) = F_1^p(K^2)$$
$$f_2^{++}(K^2, M^2, M^2) = F_2^p(K^2) . \quad (95)$$

We also have $f_1^{+-}(0, P_i^2, M^2) = 1$. Similar equations describe the γnn vertex.

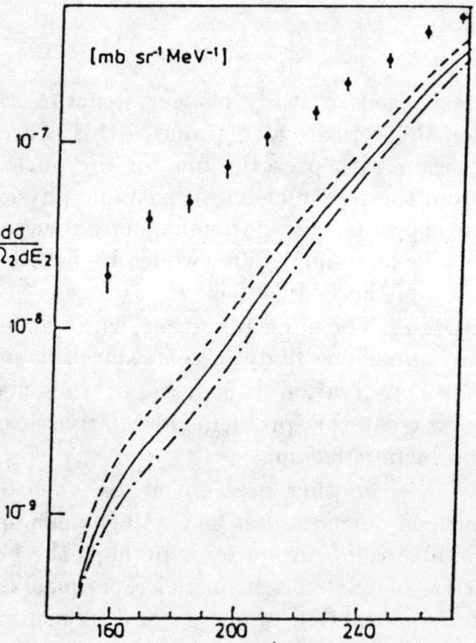

Fig. 18. Cross section for inclusive electron scattering from ^{3}He at 0.8 GeV^2 as a function of the energy loss of the electron. The data are from [81] and the calculations and figure are from [79]. Dashed curve, P_i^2 in form factors included; solid curve, no P_i^2 dependence in form factors; dashed-dotted curve, j_0 expressed in terms of j_z.

To get an idea how the off-shell form factors effect the electron scattering cross sections, a simple one-loop pion model was used to calculate $f_i^{++}(K^2, P_i^2, M^2)$ [79] and they were normalized at $P_i^2 = M^2$ to the experimental form factors

$$f_{1,2}^{++}(K^2, P_i^2, M^2) = F_{1,2}^{exp}(K^2) \left[\frac{f_{1,2}^{++}(K^2, P_i^2, M^2)}{f_{1,2}^{++}(K^2, M^2, M^2)} \right]_{model} . \tag{96}$$

Ignoring the f_i^{+-} terms and using a standard prescription to construct the nuclear current from the nucleon current [80], Naus et al. [79] calculated in the PWIA inclusive electron scattering from ^{3}He. Their results at $Q^2 = 0.8$ GeV^2 are shown in Fig. 18 compared to the data of [81]. The solid curve results when the P_i^2 dependence of the form factors is ignored, while taking this dependence into account results in the dashed curve. Evidently, the extra P_i^2 dependence can be quite important under these kinematical conditions. The dashed-dotted curve is obtained when j_0 is expressed in terms of j_z, hence showing that this prescription to construct the nuclear current is not gauge invariant. Although these results are based on a simple model, it should be clear that small deviations from "expected" results should not be taken as evidence for a change of nucleon properties in the nucleus.

6 Conclusions

The field of photo- and electroproduction of mesons is nearly 40 years old, but interest in this reaction continues. In these lecture notes, we have tried to highlight the interesting physics in this energy region by presenting a few topics of current interest. The topics presented here, and the list of references, are by no means exhaustive, but we hope it is sufficient to give the reader an introduction into this field. On the experimental side, research is moving forward. Within the next few years, high quality data will emerge for both the single-nucleon and nuclear processes, posing a challenge for theorists to keep pace.

Acknowledgments

I would like to thank Prof. W. Plessas for the invitation to this school and the organizing committee for their efforts in arranging what turned out to be a superb school. This work was supported by the Deutsche Forschungsgemeinschaft (SFB 201).

References

1. J. Steinberger and A.S. Bishop, Phys. Rev. **78**, 494(L) (1950).
2. H.L. Anderson, E. Fermi, E.A. Long and D.E. Nagle, Phys. Rev. **85**, 936 (1952).
3. C. Becchi and G. Morpurgo, Phys. Rev. **146**, 352 (1965).
4. N. Isgur, G. Karl and R. Koniuk, Phys. Rev. **D25**, 2394 (1982).
5. R.M. Davidson and N.C. Mukhopadhyay, Phys. Rev. **D42**, 20 (1990).
6. S.S. Kamalov, L. Tiator and C. Bennhold, Nucl. Phys. **A457**, 599 (1992).
7. A. Donnachie, "Photo- and Electroproduction Processes" in *High Energy Physics*, vol. 5 (E.H.S. Burhop, ed., Academic Press, New York, 1972), pp. 1-185.

8. G.F. Chew, M.L. Goldberger, F.E. Low and Y. Nambu, Phys. Rev. **106**, 1345 (1957).
9. J.D. Bjorken and S.D. Drell, *Relativistic Quantum Mechanics* (McGraw-Hill, New York, 1964).
10. K.M. Watson, Phys. Rev. **95**, 228 (1954).
11. E. Fermi, Supp. Nuovo Cimento **10**, 529 (1955).
12. M. Benmerrouche, R.M. Davidson and N.C. Mukhopadhyay, Phys. Rev. **C39**, 2339 (1989).
13. J.M. Laget, Phys. Rep. **69**, 1 (1981).
14. R.M. Davidson, N.C. Mukhopadhyay and R.S. Wittman, Phys. Rev. **D43**, 71 (1991).
15. D. Menze, W. Pfeil and R. Wilcke, Bonn Compilation of Pion Photoproduction Data (unpublished).
16. R.L. Walker, Phys. Rev. **182**, 1729 (1969).
17. W.J. Metcalf and R.L. Walker, Nucl. Phys. **B76**, 253 (1974).
18. R.A. Arndt, R.L. Workman, Z. Li and L.D. Roper, Phys. Rev. **C42**, 1853 (1990); ibid 1864 (1990).
19. H. Thom, Phys. Rev. **151**, 1322 (1966).
20. H. Tanabe, M. Kohno and C. Bennhold, Phys. Rev. **C39**, 741 (1989).
21. H.R. Hicks, S.R. Deans, D.T. Jacobs, D.W. Lyons and D.L. Montgomery, Phys. Rev. **D7**, 2614 (1973).
22. H. Breuker et al., Z. Phys. **C13**, 113 (1982). H. Breuker et al., Z. Phys. **C17**, 121 (1983).
23. F.A. Berends, A. Donnachie and D.L. Weaver, Nucl. Phys. **B4**, 1 (1967).
24. A. Jurewicz, Phys. Rev. **D28**, 1604 (1983).
25. I. Arai, in *Baryon 1980*, Proceedings of the IVth International Conference on Baryon Resonances, N. Isgur, ed. (University of Toronto, Toronto, 1980).
26. R.L. Crawford, in *Baryon 1980*, Proceedings of the IVth International Conference on Baryon Resonances, N. Isgur, ed. (University of Toronto, Toronto, 1980).
27. S.L. Adler, Ann. Phys. (NY) **50**, 189 (1968).
28. C. Mistretta et al., Phys. Rev. Lett. **20**, 1523 (1968).
29. R.C.E. Devenish and D.H. Lyth, Nucl. Phys. **B93**, 109 (1975).
30. S. Weinberg, Phys. Rev. Lett. **18**, 188 (1967).
31. V. Bernard, N. Kaiser, J. Gasser and U.-G. Meißner, Phys. Lett. **268B**, 291 (1991). V. Bernard, N. Kaiser, and U.-G. Meißner, Nucl. Phys. **B383**, 442 (1992). V. Bernard, N. Kaiser, and U.-G. Meißner, πN Newsletter No. 7, 62 (1992). See also Meißner's lectures at this school.
32. R.D. Peccei, Phys. Rev. **181**, 1902 (1969).
33. E. Mazzucato et al., Phys. Rev. Lett. **57**, 3144 (1986).
34. R. Beck et al., Phys. Rev. Lett. **65**, 1841 (1990).
35. R.M. Davidson and N.C. Mukhopadhyay, Phys. Rev. Lett. **60**, 748 (1988).
36. P. de Baenst, Nucl. Phys. **B24**, 633 (1970).
37. H.W.L. Naus, J.H. Koch and J.L. Friar, Phys. Rev. **D41**, 2852 (1990).
38. M.G. Olsson and E.T. Osypowski, Nucl. Phys. **B87**, 399 (1975); Phys. Rev. **D17**, 174 (1978).
39. M.G. Olsson, Nucl. Phys. **B78**, 55 (1974).
40. K.I. Blomqvist and J.M. Laget, Nucl. Phys. **A280**, 405 (1977).
41. R. Wittman and N.C. Mukhopadhyay, Phys. Rev. Lett. **57**, 1113 (1986).
42. J.M. Laget, Nucl. Phys. **A481**, 765 (1988).
43. R.A. Adelseck, C. Bennhold and L.E. Wright, Phys. Rev. **C32**, 1681 (1985).
44. R.A. Williams, C.-R. Ji and S. Cotanch, Phys. Rev. **C46**, 1617 (1992).
45. M. Benmerrouche and N.C. Mukhopadhyay, Phys. Rev. Lett. **67**, 1070 (1991).
46. M.G. Olsson, E.T. Osypowski and E.H. Monsay, Phys. Rev. **D17**, 2938 (1978).
47. H.W.L. Naus and J.H. Koch, Phys. Rev. **C36**, 2459 (1987).

48. H. Tanabe and K. Ohta, Phys. Rev. **C31**, 1876 (1985).
49. S.N. Yang, J. Phys. **G11**, L205 (1985).
50. S. Nozawa, B. Blankleider and T.-S. H. Lee, Nucl. Phys. **A513**, 459 (1990).
51. S. Nozawa and T.-S. H. Lee, Nucl. Phys. **A513**, 511 (1990); ibid 543 (1990).
52. B. Pearce and B. Jenning, Nucl. Phys. **A528**, 655 (1991).
53. T.-S. H. Lee and B. Pearce, Nucl. Phys. **A530**, 532 (1991).
54. Review of Particle Properties, Particle Data Group, Phys. Rev. **D45** (1992).
55. R.P. Feynman, M. Kislinger and F. Ravndal, Phys. Rev. **D3**, 2706 (1971).
56. T. Kubota and K. Ohta, Phys. Lett. **65B**, 374 (1976).
57. P. Noelle, Prog. Theor. Phys. **60**, 778 (1978).
58. I. Guiasu and R. Koniuk, Phys. Rev. **D36**, 2757 (1987).
59. F.A. Berends and A. Donnachie, Nucl. Phys. **B84**, 342, (1975).
60. I.I. Miroshnichenko et al., Yad. Fiz. **32**, 659 (1980) [Sov. J. Nucl. Phys. **32**, 339 (1980).]
61. V.A. Get'man et al., Yad. Fiz. **38**, 385 (1983) [Sov. J. Nucl. Phys. **38**, 230 (1983).]
62. W. Pfeil and D. Schwela, Nucl. Phys. **B45**, 379 (1971).
63. V.F. Grushin et al., Yad. Fiz. **38**, 1448 (1983) [Sov. J. Nucl. Phys. **38**, 881 (1983).]
64. R.L. Workman, R.A. Arndt and Z. Li, Phys. Rev. **C46**, 1546 (1992).
65. G. Blanpied et al., Phys. Rev. Lett. **69**, 1880 (1992).
66. Particle Data Group, M. Roos et al., Phys. Lett. **111B**, 182ff (1982).
67. G.A. Warren and C.E. Carlson, Phys. Rev. **D42**, 3020 (1990).
68. H. Hirata, F. Lenz and K. Yazaki, Ann. Phys. **108**, 116 (1977).
69. S. Cohen and D. Kurath, Nucl. Phys. **73**, 1 (1965).
70. N. Ensslin et al., Phys. Lett. **153B**, 203 (1985).
71. R.L. Huffman et al., Phys. Lett. **139B**, 203 (1984).
72. K. Stricker, H. McManus and J.A. Carr, Phys. Rev. **C19**, 929 (1979).
73. K. Rohrich et al., Phys. Lett. **153B**, 203 (1985).
74. P.K. Teng et al., Phys. Lett. **177B**, 25 (1986).
75. L. Tiator et al., Nucl. Phys. **A485**, 565 (1988).
76. L. Ghedira et al., Phys. Rev. **C41**, 653 (1990).
77. J.V. Noble, Phys. Rev. Lett. **46**, 412 (1981).
78. See H. Arenhövel's lectures at this school.
79. H.W.L. Naus, S.J. Pollock, J.H. Koch, and U. Oelfke, Nucl. Phys. **A509**, 717 (1990).
80. T. de Forest, Jr., Nucl. Phys. **A392**, 232 (1983).
81. S. Rock et al., Phys. Rev. **C26**, 1592 (1982).

Present Electron and Photon Scattering Experiments up to GeV Energies

Berthold H. Schoch

Physikalisches Institut, Friedrich-Wilhelms-Universität, D-53090 Bonn, Germany

1 Introduction

In the past it has been demonstrated that the electromagnetic probe is very well suited to investigate static and dynamic properties of nuclei and nucleons. High precision data taken in complete kinematics provide the experimental information in order to check the results of microscopic calculations. Fig. 1 shows the differential cross section as a function of the momentum transfer **q** for the elastic scattering of electrons on ^{208}Pb. Over twelve orders of magnitude the cross section has been measured [1].

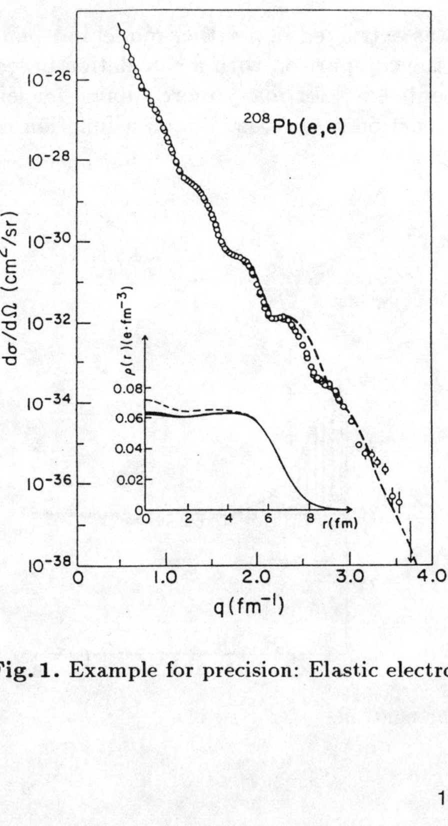

Fig. 1. Example for precision: Elastic electron on lead.

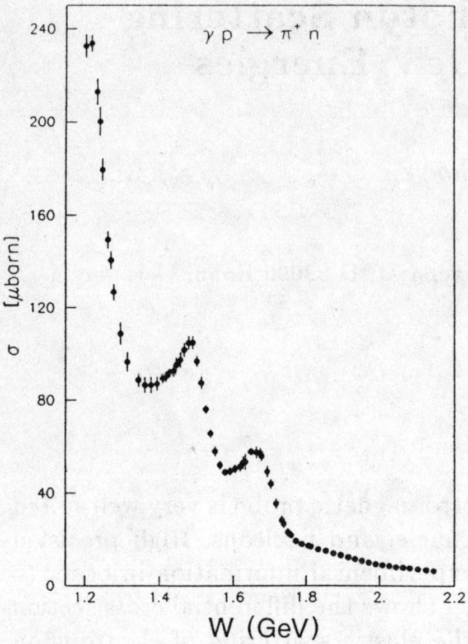

Fig. 2. Example for precision: Total cross section for single pion production on the proton.

From these data the charge distribution was extracted in a rather model independent way, see inset. The dashed curve shows the comparison with a calculation based on a mean field approach. Similar measurements exist for many more spin=0 nuclei. Fig. 2 shows the total cross section for the reaction γ+p $\Rightarrow$ π^++n as a function of the invariant mass of the system [2].

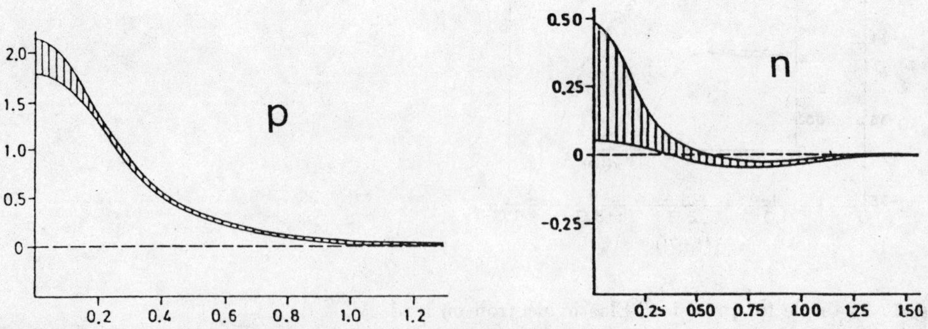

Fig. 3. Charge distribution of the proton and the neutron.

Clearly visible are the excitations of three resonances. A bremsstrahlung beam has been used for these experiments. No microscopic models of the nucleon exist, so far, which are refined enough to describe the measurements in the energy region shown in Fig. 2. These two examples demonstrate the potential of the electromagnetic probe to determine static properties like the charge distribution and dynamic properties like the excitation of inner degrees of freedom. However, the cases shown are special and well selected examples. Fig. 3 shows the charge distribution of the proton and of the neutron [3]. In both cases but especially for the neutron the charge distribution is not known well enough in order to confront the data with the results of calculations. The total cross section for photoproduction of positive pions on the proton shown in Fig. 2 is only part of the total cross section on the proton as shown in Fig. 4 as the shaded area. Above about $k_\gamma = 8$ GeV the major part of the cross section is due to multiple pion production. The information for these processes is rather poor. In order to understand the photon-nucleon process at these high energies precise data for these channels are a precondition. Therefore, large acceptance detectors in combination with a tagged photon beam have to be operational. In order to see what kind of equipment is required to make the best use of the properties of the electromagnetic probe a more general discussion of the electromagnetic interaction seems to be in order.

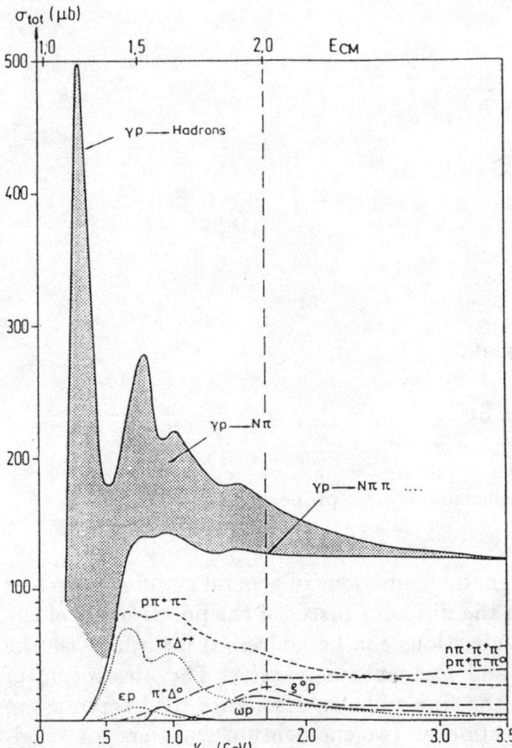

Fig. 4. Total photon absorption cross section on the proton.

2 The Electromagnetic Interaction

The interaction takes place by the exchange of a virtual photon (see Fig. 5) described by the vector potential $\mathbf{A}_\mu$. Because of the polarisation degrees of freedom of the virtual photon three different pieces of the interaction amplitude can be separated experimentally. By selection of a spin = 0 nucleus in the first example of the introduction it was possible to explore the Coulomb piece of the interaction. In general, for an experimental separation of the three pieces - Coulomb, transverse electric, magnetic - the polarisation of the virtual photon has to be explored. By performing a coincidence experiment of the type A(e,e'x) the partial linear polarisation of the virtual photon in the scattering plane (e-e'-plane) can be explored by detecting the product particle of the reaction in a cone around the direction of the virtual photon (see Fig. 6).

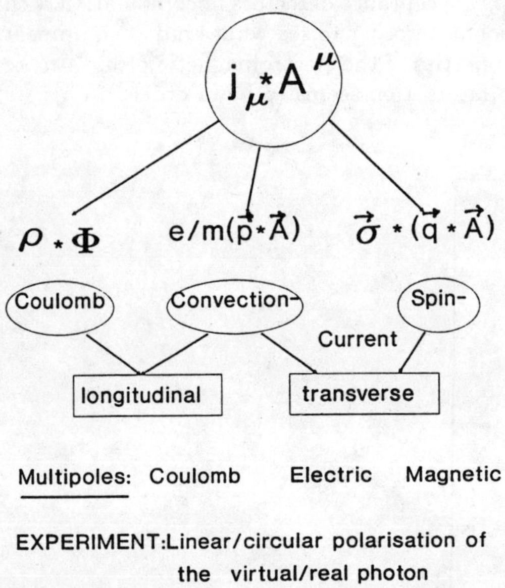

Fig. 5. The different projections upon the electromagnetic probe.

In order to study spin currents or magnetic transitions in general circular polarised virtual/real photons must be used. With the different pieces of the polarisation of the virtual photon at hand different physics questions can be addressed for otherwise the same kinematic conditions like energy and momentum transfer. The absorption of virtual photons in the energy region of the first nucleon resonance may serve as an example. Fig. 7 shows the total cross section at two momentum transfers for longitudinal/scalar photons as well as transverse photons. The experiment was performed in Bonn [4] by using the classical Rosenbluth method in order to separate the scalar and transverse pieces. The transverse cross section is dominated by the (3,3)-nucleon resonance like in the case of real photons shown in Fig. 4. The absorption cross section for scalar photons, however, exhibits no resonance shape at all. The shape of the cross section is rather flat starting at the production threshold for pions and is mainly due to a process schematically shown in Fig. 8, the so-called pion pole term.

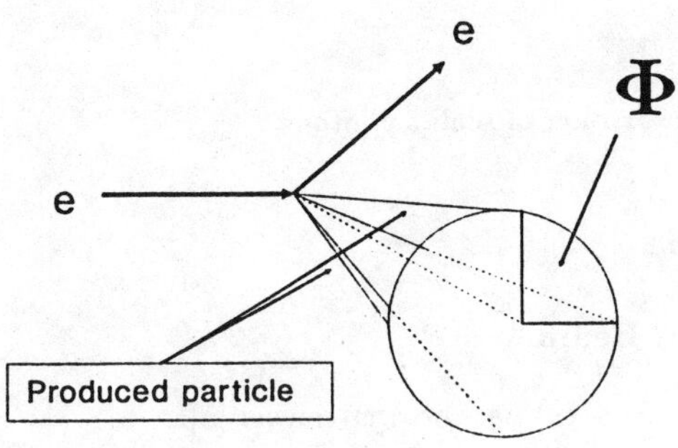

Fig. 6. Out of plane detection of a produced particle in order to explore the partial linear polarisation of the virtual photon.

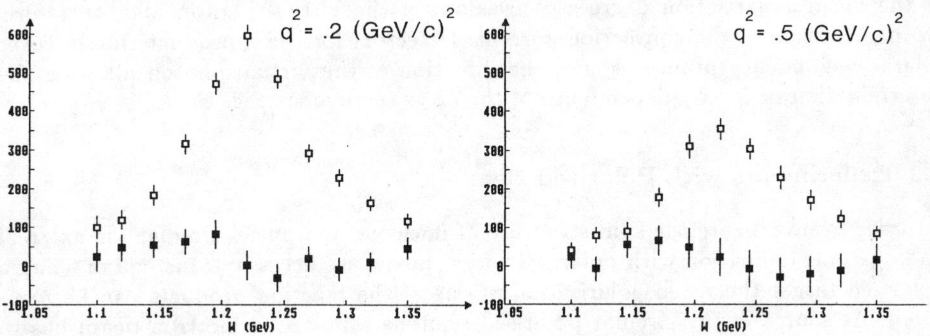

Fig. 7. The total cross section in the Δ-region for transverse (squares) and scalar (full squares) photons.

This different absorption behaviour for scalar and transverse photons, respectively, holds for higher excitation energies, too: Transverse virtual/real photons as a probe are well suited to excite the nucleon resonances, scalar/longitudinal virtual photons can be used to knock out the charged constituents e.g. in nuclei or nucleons. A further discrimination of the transverse pieces of the cross section due to convection and spin currents becomes possible by exploring the linear and circular polarisation of the virtual/real photons.

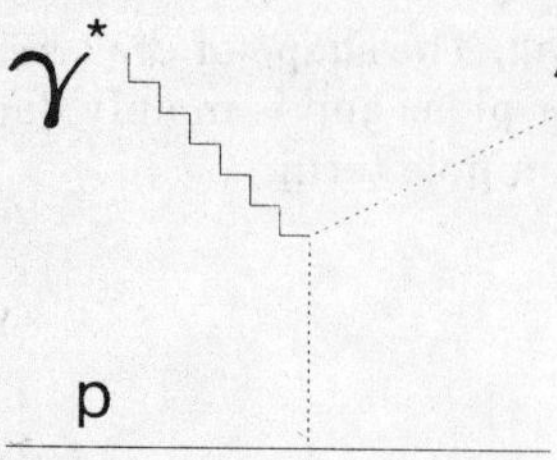

Fig. 8. The pion pole term dominating the absorption of scalar photons.

3 Consequences for Experiments

3.1 Experiments with an Unpolarised Beam

The absorption cross section for a reaction A(γ^*,x) can be written as [5]

$$\sigma = \sigma_T + \epsilon \cdot \sigma_L + \sqrt{\epsilon(\epsilon+1)} \cdot \sigma_{TL} \cdot \cos \Phi_x + \epsilon \cdot \sigma_{TT} \cdot \cos 2\Phi_x \ . \tag{1}$$

Thereby, T stands for transverse and L for longitudinal, ϵ, depending on the electron kinematics only, describes the degree of γ^*-polarisation. The classical solution to separate the longitudinal/transverse pieces of the cross section due to a Rosenbluth plot asks for accurate absolute cross section measurements in different kinematical regions. Coincidence measurements out of the (e,e')-reaction plane allow an exploration of the linear polarisation degrees of freedom of the virtual photon and, therefore, an investigation of the convection current. Detectors for the x-product should cover a large angular acceptance around the direction of the virtual photon allowing the determination of the Φ_x-dependence of the cross section(see Fig. 5).

3.2 Experiments with Polarised Beams

In order to investigate the spin structure of nucleons and nuclei circular virtual/real photons in combination with polarised target nuclei are necessary. Instead of using a polarised target the recoil polarisation of one of the reaction products can be measured. As source of the circular polarised photons a polarised electron beam has to be prepared.

3.3 Status

Due to the development of high duty cycle accelerators covering a wide energy range, and providing polarised beams, the preconditions for exclusive precision experiments are given. Combined with sophisticated detectors and targets new classes of experiments will yield the information to uncover the subtle structure of nucleons and nuclei. The high duty cycle electron accelerators in the energy range from 100 MeV up to 4 GeV in operation or under construction are located in:
Lund (Sweden), Darmstadt (Germany), Saskatoon (Canada), Amsterdam (Netherlands), Mainz (Germany), Boston (U.S.), Bonn (Germany), Newport News (U.S.).

4 Selected Experiments

In the following chapters results of experiments using the high duty cycle electron beams of ELSA(Bonn) and MAMI A (Mainz) will be presented.

4.1 The Investigation of the N-Δ Transition

Tensor forces in the N-N interaction lead to a d-state contribution in the ground state wave function of the deuteron and, therefore, a deformation. Tensor forces between quarks in the nucleon can lead to a deformation, too. Observable could be such a contribution in the transition between the nucleon to its first $\Delta(3,3)$-resonance via the measurement of the quadrupole amplitude. The N-Δ transition is known to be determined by magnetic dipole absorption. In order to observe the small quadrupole amplitude it is mandatory to exploit the interference between these two amplitudes. It can be shown [6] that the reaction p(e,e'π^0)p allows to extract information about this interference. Thereby, it is not necessary to measure absolute cross sections. The interference appears in the σ_{TL}-piece of the cross section given in equation (1). In the actual experiment the process p(e,e'γ) has been measured. By selecting the high energy tail of the photon spectrum the direction of the π^0 could be fixed [7], thereby, the decay characteristics of the π^0 into two photons were exploited (see Fig. 9). Contributions of bremsstrahlung and the photon decay of the Δ-resonance have been calculated and found to be of the few percent level.

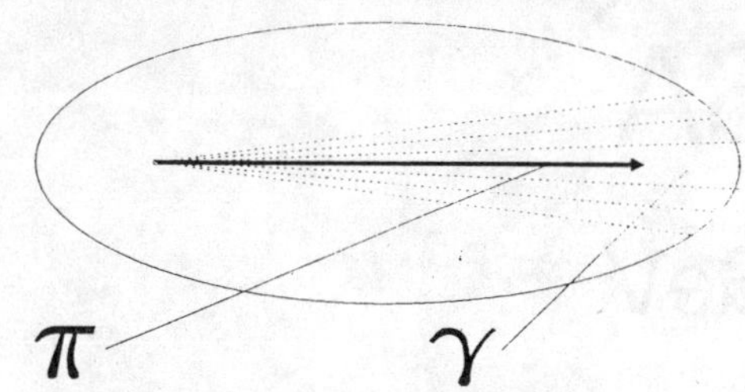

Fig. 9. The direction of the pion momentum can be reconstructed by observing the high energetic photon of the two photon decay of the pion.

The experimental set-up is shown in Fig. 10. The electrons with a momentum of 1.2 GeV/c extracted out of ELSA (Fig. 11) hit a hydrogen target.

The inelastically scattered electrons were detected and momentum analysed in a magnetic spectrometer with a large momentum acceptance ($\Delta p/p = \pm 0.12$). Therefore, a large part of the resonance could be covered in one setting. The photons, fixing the direction of the π^0, were detected in an array of 176 lead glass detectors. The extracted Φ_π-distribution is shown in Fig. 12.

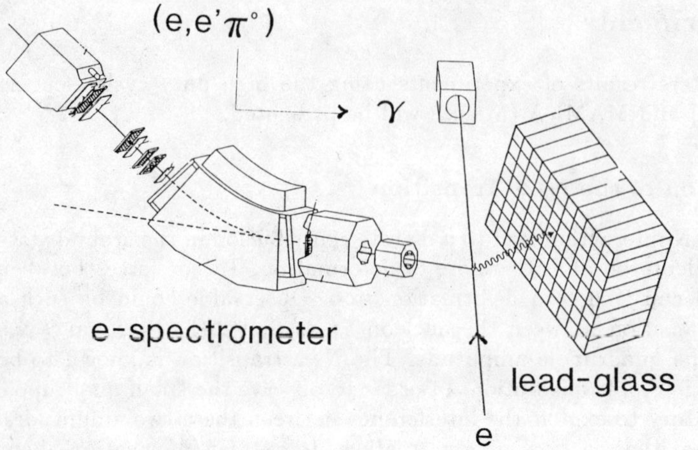

Fig. 10. Set-up for the electro-production of pions.

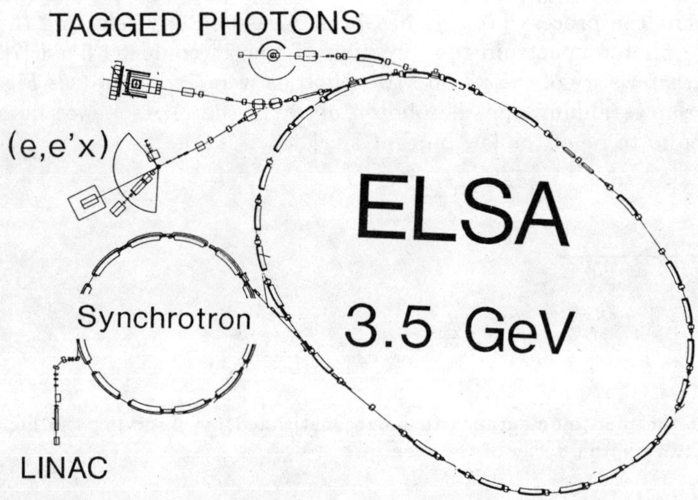

Fig. 11. The ELSA facility.

With a few assumptions already used in previous investigations [8] the ratio S_{1+}/M_{1+} was extracted as shown in Fig. 13. Thereby, S_{1+} stands for the scalar quadrupole amplitude and M_{1+} for the magnetic dipole transition.

For the first time a value at low momentum transfer was extracted by using the same procedure as the measurements performed at higher q^2. An increase in the absolute value is observed. In [6] such a behaviour has been reported as a result of calculations and in [9] as a result of a multipole analysis. More precise data in this low q^2-region will yield a data base which will allow to extract transition densities as a function of r.

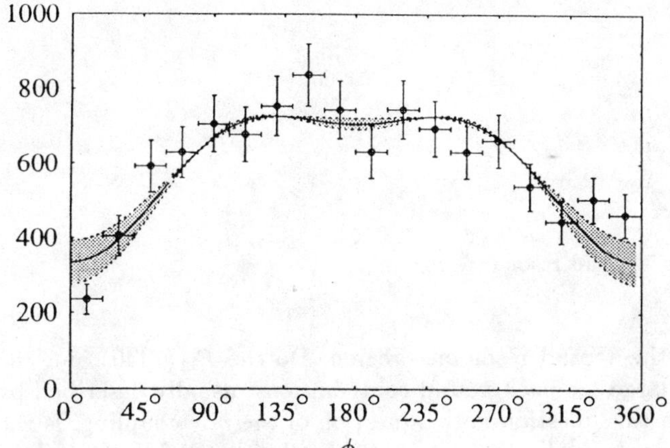

Fig. 12. The azimuthal distribution.

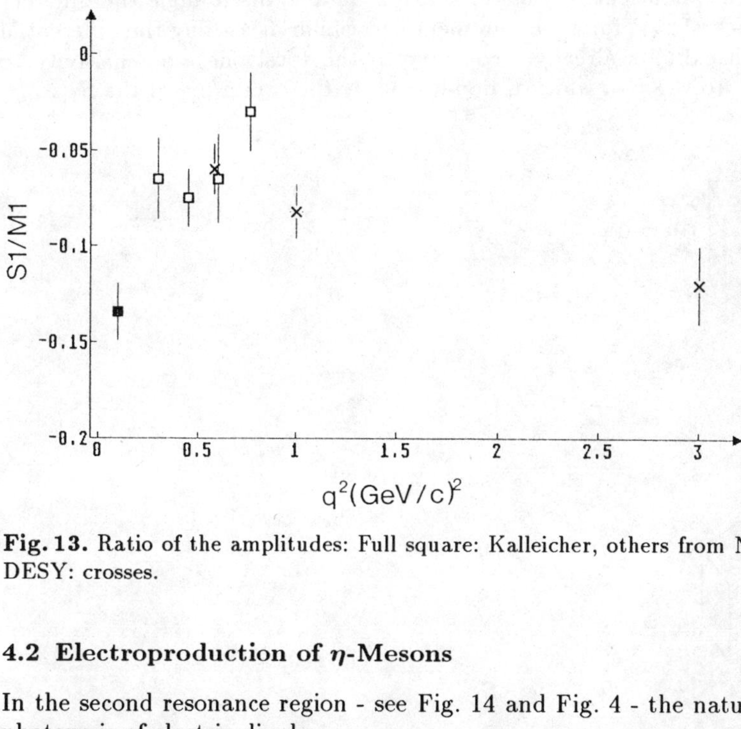

Fig. 13. Ratio of the amplitudes: Full square: Kalleicher, others from NINA: open squares, DESY: crosses.

4.2 Electroproduction of η-Mesons

In the second resonance region - see Fig. 14 and Fig. 4 - the nature of the absorbed photons is of electric dipole.

Thereby, the excitation of the $D_{13}(1520)$-resonance dominates. This resonance decays predominantly into the the π-N channel. The decay of the $S_{11}(1535)$ has a large branching ratio into the η-channel. Therefore, the total $\gamma + p \rightarrow \eta + p$ absorption cross section reflects the excitation of the S_{11}-resonance. This strong coupling of the S_{11} to the η-N channel cannot be explained by calculations using microscopic models of the nucleon. Besides the contribution of the S_{11} there should be other contributions

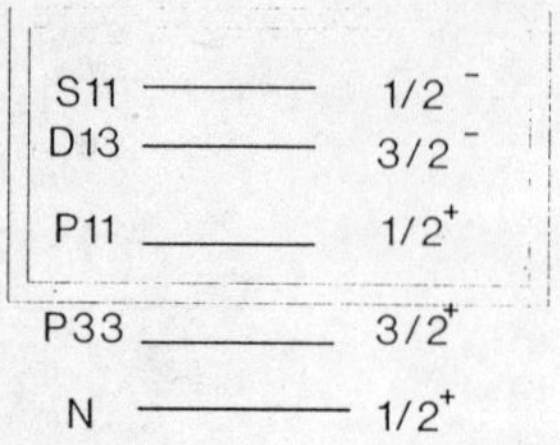

Fig. 14. The resonances of the "second resonance region".

to the p(e,e'η)p channel in the second resonance region. Do the $P_{11}(1440)$ and the $D_{13}(1520)$ contribute? How large are background contributions, usually described by Born terms? Connected with this question is the question of the η-N coupling. Is the pseudoscalar or pseudovector scheme better suited to describe the coupling and how large is the η-N coupling constant? Given the bunch of questions it is clear that one experiment cannot be sufficient in yielding enough data to disentangle the different pieces. The obvious first step for a measurement consists in measuring the differential cross section which exhibits already - contrary to the total one - a sensitivity to multipoles different to the E_{0+} which is dominated by the excitation of the S_{11}.

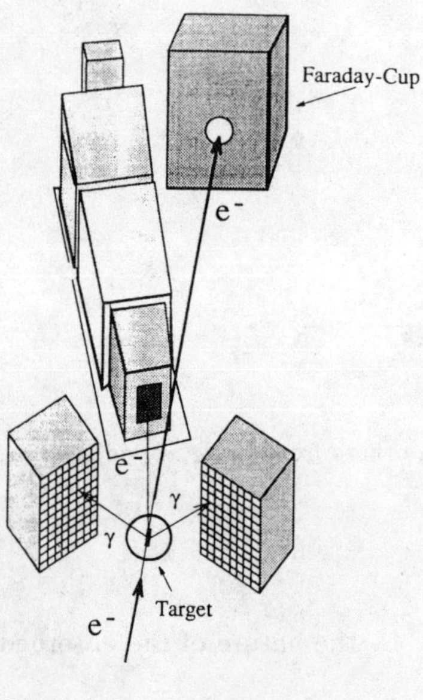

Fig. 15. Set-up for η-production.

Fig. 15 shows, schematically, the set-up used for the experiment p(e,e'η)p. The electron detection corresponds to the already described p(e,e'π^0) experiment. The $\eta \rightarrow 2\gamma$-decay was used to identify the η-kinematics. In one spectrometer setting an energy of 150 MeV was covered. In one given geometry of the lead glass blocks the whole $0 \leq \theta_\eta \leq 180^0$ range could be explored and - in addition - the Φ_η-dependence (see equation (1)) of the cross section be extracted. Fig. 16 shows the total cross section as a function of the invariant mass W.

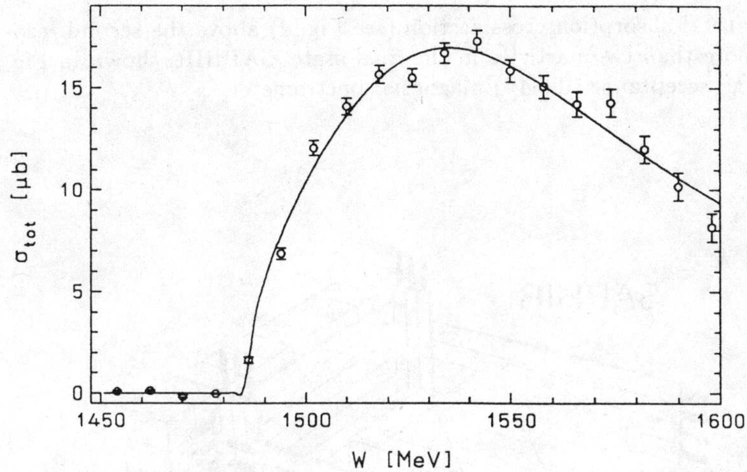

Fig. 16. Total cross section.

Fig. 17 shows the differential cross section as well as the Φ_η-asymmetry on top of the S_{11}-resonance. All data are very preliminary and should serve here as a demonstration of the kind of accuracy one can expect [17].

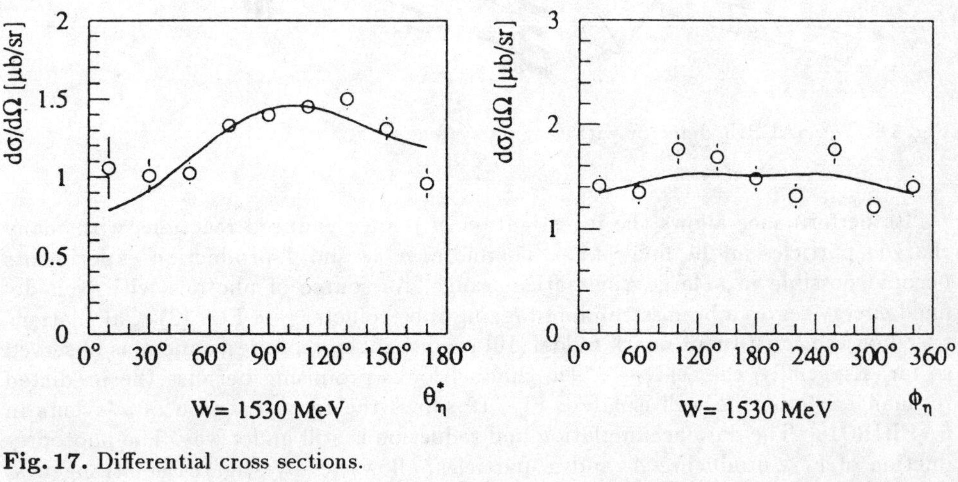

Fig. 17. Differential cross sections.

In conclusion, looking at the total set of data clear signatures of multipoles besides the E_{0+} can be seen.

4.3 Photoproduction of K^+

A large part of the total absorption cross section (see Fig. 4) above the second resonance region has more than two particles in the final state. SAPHIR, shown in Fig. 18, stands for a large acceptance ($0.6 \cdot 4\pi$) magnetic spectrometer.

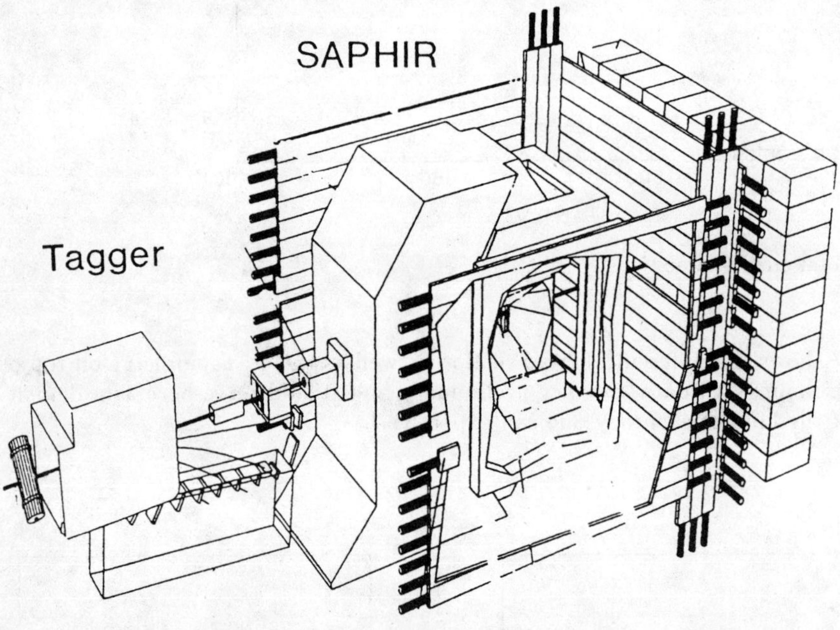

Fig. 18. The SAPHIR detector with tagging system.

Its performance allows the investigation of photon induced reactions with many charged particles in the final state. Therefore, ρ, ω and Φ-production experiments become possible in a large kinematical range. As source of photons with well defined energy serves a bremsstrahlung tagging spectrometer (see Fig. 18). Calculations based on the constituent quark model [10] predict many more resonances as observed so far. Especially, the $\gamma+\mathrm{N}\rightarrow N^{*}+\omega$ channel looks promising because the predicted resonances should be well isolated. Fig. 19 shows the reconstruction of ω-events in SAPHIR [18]. The data accumulation and reduction is still under way. The photoproduction of K^+, producing Λ- and Σ-particles, allows to investigate the strangeness

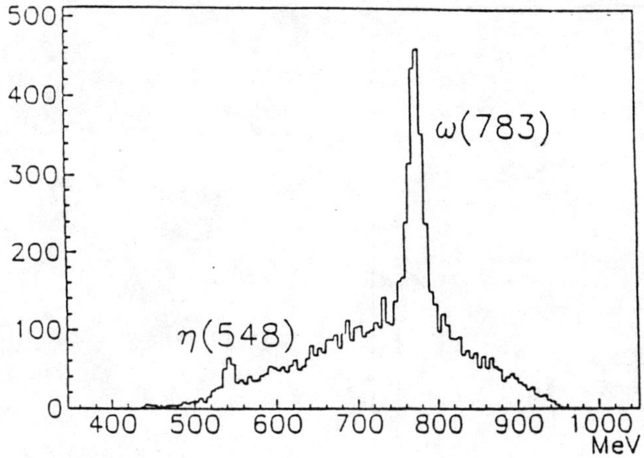

Fig. 19. Reconstruction of the ω-meson mass with SAPHIR.

degrees of freedom. Thereby, the polarisation of the Λ and Σ-particles can be reconstructed as shown in Fig. 20 for the Σ.

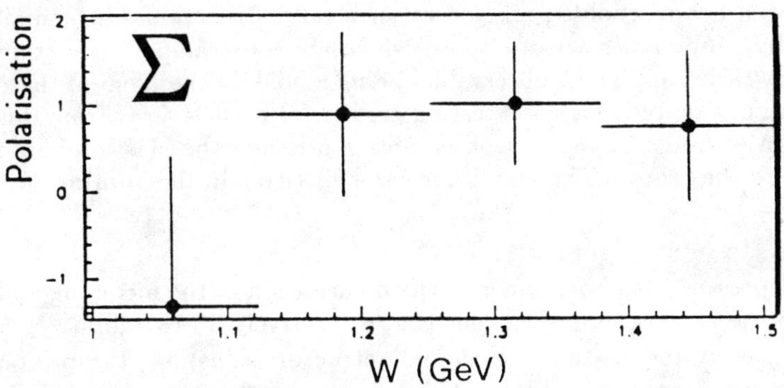

Fig. 20. Polarisation of the Σ.

This is the first measurement of the Σ-polarisation in a photoinduced experiment in this energy region. Very interesting will be the investigation of the t-dependence of the photoproduction of the above mentioned vector mesons with polarised photons. Thereby, by the kinematics of the decay products, the polarisation of these mesons can be reconstructed. Fig. 21 shows old data from DESY [11] of the $\gamma - \pi$ reaction. The π^-/π^+ ratio approaches already at rather low t-values the value of .25 which can be expected by assuming that the photon knocks off a d- and a u-quark from the neutron and the proton, respectively.

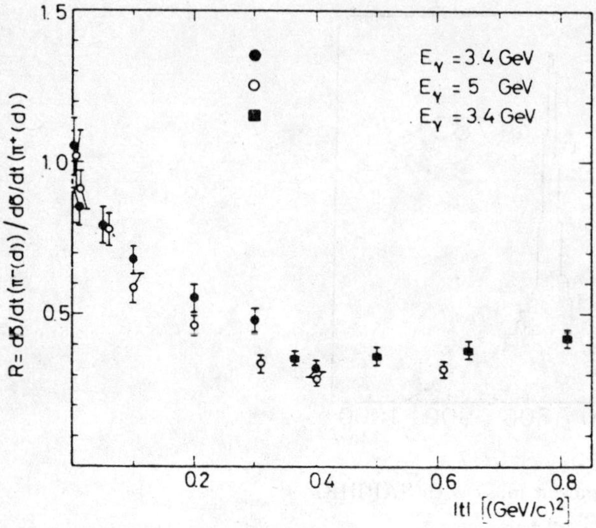

Fig. 21. The ratio of single pion production on the nucleon.

4.4 Experimental Tests of Low Energy Theorems

High flux tagged photon beams enable precision experiments of reactions with small cross sections. Thereby, observables are of interest which are constrained - at certain kinematical points - by theorems. These observables being model independent at these special kinematical regions represent a good testing ground for models away from this special kinematics. An example of such an observable represents the elastic photon scattering amplitude on the proton. The amplitude can be written in the limiting case $\omega \rightarrow 0$ as:

$$T(\omega) = f_1 + i\omega \cdot f_2(\epsilon \times \epsilon' \cdot \sigma) \tag{2}$$

where ω, ϵ, ϵ' and σ represent the photon energy, the polarisation of the incoming and scattered photon and the polarisation of the nucleon, respectively. In this limit, $f_1 = -e/m$ and f_2=-$\frac{2\alpha\cdot\kappa^2}{m^2}$. α, m and κ stand for the fine structure constant, the proton mass and the anomalous magnetic moment. The low energy theorem expresses in this case nothing else than the fact that in the long wave length limit the static quantities like charge, magnetic moment and mass determine the amplitude. Away from these low energy points quadratic terms in the amplitude become important. The electric and magnetic polarisabilities of the nucleon can be extracted from this energy dependence of the amplitudes [12,13]. These, however, are very sensitive and, therefore, discriminating for the various models of the nucleon.

A similar constrained amplitude can be found in the photoproduction of pions on the nucleon at the production threshold. Fig. 22 shows the total cross sections in the threshold region for photon scattering, neutral and charged pion production. Very close to the production threshold the s-wave dominates the cross section. Following the reasoning of Gell-Mann [14] stating that in such a case the electric dipole amplitude - there are no other s-wave amplitudes - must be proportional to the electric dipole

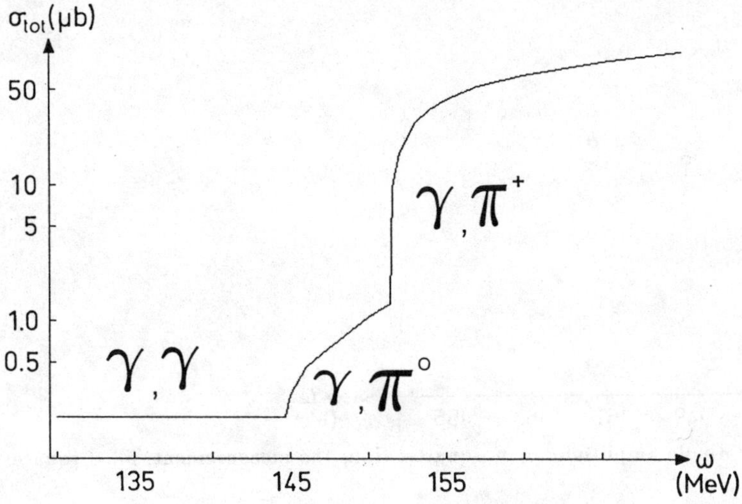

Fig. 22. Total photoabsorption cross section close to the photoproduction thresholds on the proton.

moment in the final state, the relative strength of the π^0 and π^+-cross sections can be explained, at least qualitatively. In the case of the neutron the π^0-cross section should be zero. The different theoretical approaches to this problem are explained in U. Meißner's lectures.

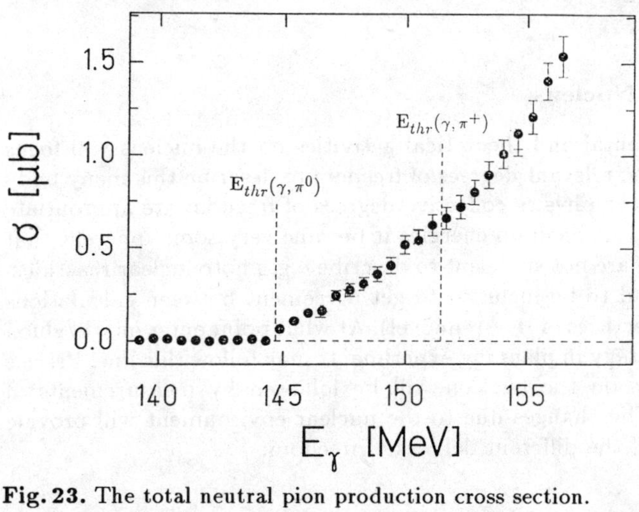

Fig. 23. The total neutral pion production cross section.

Fig. 23 shows the result of a (γ, π^0) measurement [15] with MAMi A demonstrating the power of the new tagged photon sources. From the angular distribution s- and p-wave amplitudes can be extracted. The s-wave amplitude is shown in Fig. 24. Close

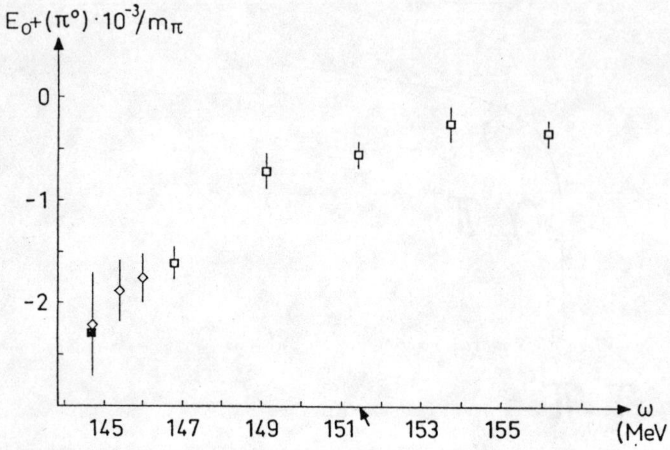

Fig. 24. The electric dipole amplitude: open squares: from the measurement, filled square: low energy theorem.

to threshold, only the total cross section values have been taken. At threshold, the prediction of the low energy theorem agrees with the extracted amplitude. However, see to this point also the discussion in Meißner's lectures. The large variation as a function of energy cannot be explained so far. Also the p-wave, below the π^+-threshold shows an unexpected energy dependence. The results underline the statements made at the beginning of this chapter: The low energy theorems provide fixed points for observables. Points close by get sensitive for models.

4.5 Experiments on the Nucleus

A large part of the experimental and theoretical activities on the nucleus will focus on the question: What are the relevant degrees of freedom to describe this many body problem. In low energy nuclear physics collective degrees of freedom are appropriate to describe deformations etc. At medium energies it became very soon apparent that nucleonic degrees of freedom are not sufficient to describe e.g. photonuclear reactions. Meson exchange currents had to be included to get agreement between calculations and measurements (see the lectures of H. Arenhövel). At what point enter quark/gluon degrees of freedom? The strategy in planning experiments may follow this line: Precise and exclusive measurements on the nucleon will be followed by measurements of the very same observable. The changes due to the nuclear environment will provide signatures of the interplay of the different degrees of freedom.

Coherent Pion-Production As a last example along these lines results from an experiment will be presented which was performed with the tagged photon beam of MAMi A: The coherent production of neutral pions. The amplitude M_{1+} - due to the excitation of the (3,3)-resonance - dominates already for photon energies larger than 5 MeV above threshold. The $\gamma-\pi^0$ mode allows the investigation of a coherent process.

Taken the - good - assumption that the elementary amplitudes for π^0-production on the neutron and proton are equal, a A^2-dependence of the cross section can be expected. It would be interesting to observe the energy dependence of the cross section at the same t over the whole width of the resonance. Do width and position change? By changing t - that means changing the momentum transfer to the nuclear system keeping the photon energy constant - the ability of the nuclear system to take over momentum will be checked like e.g. in elastic electron scattering. A comparison should shed light on the reaction mechanism. Fig. 25 shows the results of the measurement for the differential cross section for C^{12} and C^{13} [16]. The cross section is the same for $\theta_\pi \geq 65^0$. This indicates that the last loosely bound neutron in C^{13} does not contribute to the coherent sum for larger t-values. At lower t-values the full A^2-dependence shows up. So far no calculation exists which describes these experimental findings.

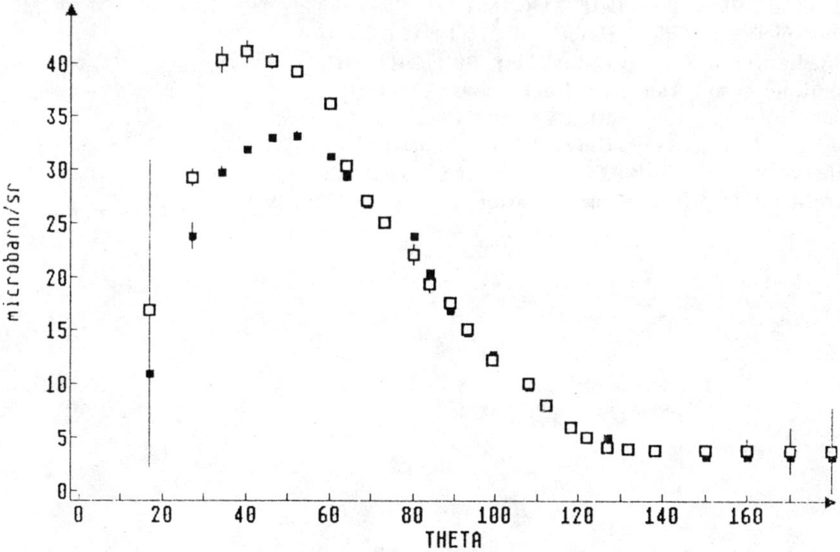

Fig. 25. Differential cross sections on the carbon isotopes.

Acknowledgment: It is my pleasure to thank all the people who worked at various times in a collaborative effort to gather the reported results: The groups from Glasgow (R.O. Owens et al.), Edinburgh (D. Branford et al.), Giessen (V. Metag et al.), Bonn (T. Reichelt et al.) as well as the accelerator groups in Mainz (K.H. Kaiser et al.) and Bonn (D. Husmann et al.) and last not least the PhD-students referred to in the text and the others who are still hard working.

References

1. B. Frois et al., Phys. Rev. Lett. **38** (1977) 152
2. W. Heise, PhD-thesis 1988, Univ. of Bonn, unpublished
3. G.G. Simon et al., Z. Naturforsch. **35a** (1979) 1
4. K. Baetzner, PhD-thesis 1972, Univ. of Bonn
5. D.H. Lyth, in Electromagnetic Interactions with Hadrons I, Edited by A. Donnachie and G. Shaw, Plenum Press, New York, 1978
6. P. Christillin and G. Dillon, J. Phys. **G16** (1990) 805
7. F. Kalleicher, PhD-thesis 1993, Univ. of Mainz, unpublished
8. R.S. Siddle et al., Nucl. Phys. **B35** 93(1971) 93
9. R.L. Crawford, Nucl.Phys. **B28** (1971) 573
10. N. Isgur and G. Karl, Phys. Rev. **D19** (1979) 2653
11. P. Heide et al., Phys. Rev. Lett. **21** (1968) 248
12. F.J. Federspiel et al., Phys. Rev. Lett. **67** (1991) 67
13. J. Schmiedmayer et al., Phys. Rev. Lett. **66** (1991) 1015
14. M. Gell-Mann et al., Ann. Rev. Nucl. Sci. **4** (1954) 219
15. R. Beck et al., Phys. Rev. Lett. **65** (1990) 1841
16. R. Gothe, PhD-thesis 1988, Univ. of Mainz, unpublished
17. M. Wilhelm, PhD-thesis 1993, Univ. of Bonn, unpublished
18. W. Schwille et al., private communication

Partons and QCD: Hadron Structure and Basics of the Standard Model

W.J. Stirling

Departments of Mathematical Sciences and Physics, University of Durham, Durham DH1 3LE, United Kingdom

1 Introduction

Quantum Chromodynamics (QCD), the gauge field theory which describes the interactions of coloured quarks and gluons, is one of the components of the $SU(3) \times SU(2) \times U(1)$ Standard Model. At short distances, equivalently high energies, the effective coupling is small and the theory can be studied using perturbative techniques. Some of the most direct information on QCD comes from studying the short-distance structure of hadrons using electroweak gauge boson probes – deep inelastic scattering. Here we see the asymptotic property directly, as the gauge bosons scatter incoherently off the weakly interacting quarks and gluons. From such experiments we learn how the partons (i.e. the quarks and gluons) share the momentum and quantum numbers of the hadron. By studying how the 'structure functions' vary with the momentum transferred by the probe, precision measurements of the short-distance coupling can be made.

In these lectures we will discuss the parton structure of hadrons and its relation to perturbative Quantum Chromodynamics. In the next section, some of the fundamental properties of the theory are reviewed, in particular how the 'running' coupling α_s is defined. In Section 3 we introduce the parton model of short-distance hadron structure and discuss how the basic 'scaling property' is modified by perturbative QCD corrections. In Section 4 the parton distributions are studied in detail, and in Section 5 some applications to hadron-hadron collisions are reviewed. Lack of time precludes an in-depth treatment of most of these issues, but further details can of course be found in the literature (see for example [1]).

2 Basics of Perturbative QCD

2.1 The QCD Lagrangian

The QCD Lagrangian is, up to gauge-fixing terms,

$$\begin{aligned}
\mathcal{L}_{QCD} &= -\frac{1}{4} F^{(a)}_{\mu\nu} F^{(a)\mu\nu} + \sum_q \bar{\psi}^q_i \left(i\gamma^\mu (D_\mu)_{ij} - m_q \delta_{ij} \right) \psi^q_j \\
F^{(a)}_{\mu\nu} &= \partial_\mu A^a_\nu - \partial_\nu A^a_\mu + g_s f_{abc} A^b_\mu A^c_\nu \\
(D_\mu)_{ij} &= \delta_{ij} \partial_\mu - i g_s T^a_{ij} A^a_\mu
\end{aligned} \qquad (1)$$

where g_s is the QCD coupling constant, T^a_{ij} and f_{abc} are the $SU(3)$ colour matrices and structure constants, respectively, the $\psi^q_i(x)$ are the 4-component Dirac spinors associated with each quark field of colour i and flavour q, and the $A^a_\mu(x)$ are the eight Yang-Mills gluon fields. From this Lagrangian, the Feynman rules can be derived in the usual way – the rules for a covariant gauge are summarized in Table 1.

Explicit forms for the $SU(3)$ colour matrices and structure constants can be found, for example, in the *Review of Particle Properties* [2]. The following are some useful identities:

$$\begin{aligned}
[T^a, T^b] &= i f^{abc} T^c \\
\{T^a, T^b\} &= d^{abc} T^c + \frac{1}{3}\delta^{ab} \\
f^{acd} f^{bcd} &= C_A \delta^{ab} \\
(T^a T^a)_{ij} = T^a_{ik} T^a_{kj} &= C_F \delta_{ij} \\
\mathrm{Tr}(T^a T^b) = T^a_{ij} T^b_{ji} &= T_F \delta^{ab} \\
C_A = N_c &= 3 \\
C_F = \frac{N_c^2 - 1}{2N_c} &= \frac{4}{3} \\
T_F &= \frac{1}{2} \\
\mathrm{Tr}(T^a T^b T^c) &= \frac{i}{4} f^{abc} + \frac{1}{4} d^{abc} \\
f^{abc} f^{abc} &= 24 \\
d^{abc} d^{abc} &= \frac{40}{3}
\end{aligned} \tag{2}$$

where summation over repeated indices is understood.

2.2 The QCD Coupling Constant

Quantum Chromodynamics is an asymptotically free gauge field theory, that is, the strength of the interaction between the quarks and gluons becomes weaker in the short-distance limit. In QCD the renormalized coupling can be defined in a variety of ways, for example from the 'dressed' qqg or ggg vertices. Renormalization of the coupling necessitates the introduction of a scale μ – effectively the scale at which the ultra-violet loop divergences are subtracted off. A dimensionless physical quantity R which depends on some energy scale Q will depend also on μ both *explicitly* and *implicitly* through the renormalized coupling, i.e. $R = R(\mu/Q, \alpha_s(\mu))$. The fact that such a quantity should not depend on the arbitrary scale μ (when calculated to all orders in perturbation theory) leads to an equation for the μ dependence of the renormalized coupling:

$$\frac{\mu^2}{\alpha_s(\mu^2)} \frac{\partial \alpha_s(\mu^2)}{\partial \mu^2} = -\frac{\alpha_s(\mu^2)}{4\pi}\beta_0 - \left(\frac{\alpha_s(\mu^2)}{4\pi}\right)^2 \beta_1 - \left(\frac{\alpha_s(\mu^2)}{4\pi}\right)^3 \beta_2 + \ldots$$

$$\beta_0 = 11 - \frac{2}{3} n_f$$

Table 1. Feynman rules for QCD in a covariant gauge.

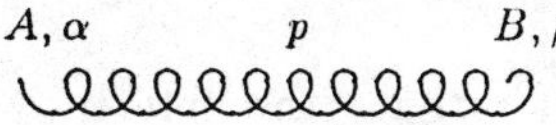	$\delta^{AB}\left[-g^{\alpha\beta}+(1-\lambda)\frac{p^{\alpha}p^{\beta}}{p^{2}+i\varepsilon}\right]\frac{i}{p^{2}+i\varepsilon}$
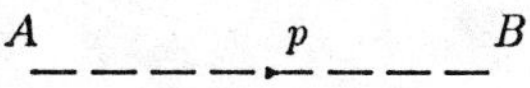	$\delta^{AB}\frac{i}{p^{2}+i\varepsilon}$
	$\delta^{ab}\frac{i}{(\hat{p}-m+i\varepsilon)_{ji}}$
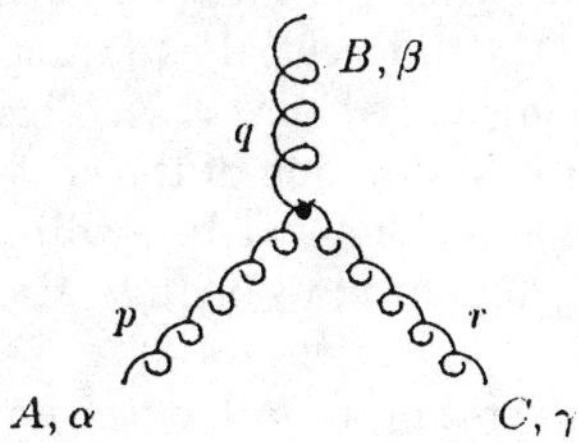	$-gf^{ABC}\left[g^{\alpha\beta}(p-q)^{\gamma}+g^{\beta\gamma}(q-r)^{\alpha}+g^{\gamma\alpha}(r-p)^{\beta}\right]$ (all momenta incoming)
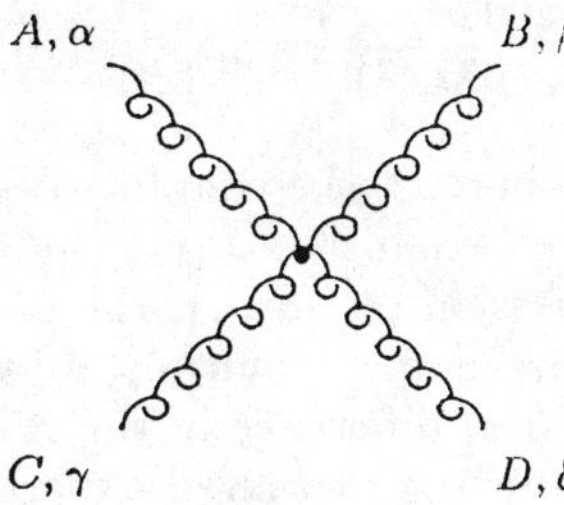	$-ig^{2}f^{XAC}f^{XBD}(g_{\alpha\beta}g_{\gamma\delta}-g_{\alpha\delta}g_{\beta\gamma})$ $-ig^{2}f^{XAD}f^{XBC}(g_{\alpha\beta}g_{\gamma\delta}-g_{\alpha\gamma}g_{\beta\delta})$ $-ig^{2}f^{XAB}f^{XCD}(g_{\alpha\gamma}g_{\beta\delta}-g_{\alpha\delta}g_{\beta\gamma})$
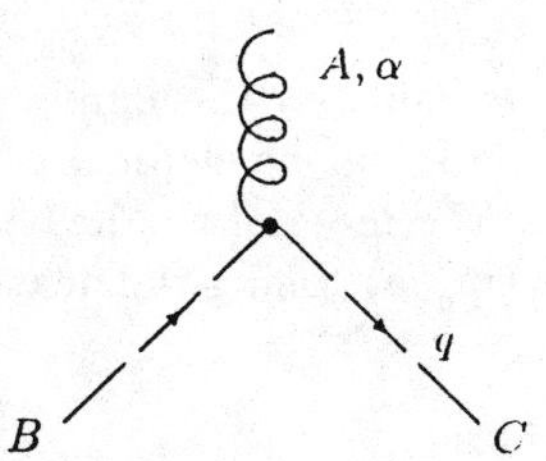	$gf^{ABC}q^{\alpha}$
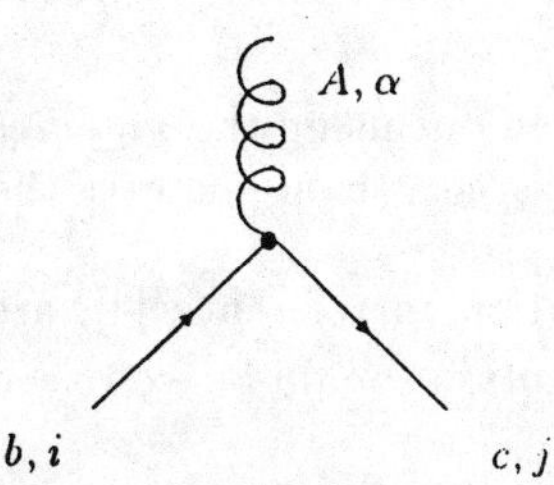	$-ig\left(t^{A}\right)_{cb}(\gamma^{\alpha})_{ji}$

$$\beta_1 = 102 - \frac{38}{3} n_f$$
$$\beta_2(\overline{\rm MS}) = \frac{2857}{2} - \frac{5033}{18} n_f + \frac{325}{54} n_f^2 . \tag{3}$$

Retaining only the first two terms on the right-hand side and solving the differential equation for $\alpha_s(\mu^2)$ gives

$$\frac{1}{\alpha_s} + b_1 \log\left(\frac{b_1 \alpha_s}{1 + b_1 \alpha_s}\right) = b_0 \log\frac{\mu^2}{\Lambda^2}, \tag{4}$$

with

$$b_0 = \frac{\beta_0}{4\pi}, \quad b_1 = \frac{\beta_1}{4\pi\beta_0}. \tag{5}$$

Note that a constant of integration in the form of a dimensionful parameter Λ has been introduced – replacing Λ by $c\Lambda$ also gives a solution to the differential equation. The convention chosen here is such that the left-hand side vanishes when $\mu = \Lambda$. This is the standard definition of the 'two-loop' coupling constant as a function of the scale μ and the fundamental QCD scale parameter Λ. It is adequate for 'next-to-leading order' phenomenology. The above expression for α_s can be generalized to include also the β_2 term [3].

An explicit form for α_s can be obtained by expanding in inverse powers of $\log(\mu^2/\Lambda^2)$:

$$\alpha_s(\mu^2) = \frac{12\pi}{(33 - 2n_f)\log(\mu^2/\Lambda^2)}\left[1 - \frac{6(153 - 19n_f)}{(33 - 2n_f)^2}\frac{\log\log(\mu^2/\Lambda^2)}{\log(\mu^2/\Lambda^2)} + \ldots\right], \tag{6}$$

which illustrates the characteristic 'asymptotic freedom' property – the coupling decreases monotonically as μ^2 increases. Note, however, that this expansion corresponds to a slightly different definition of Λ from the implicit expression (4) for α_s, the expansion of which would contain a term $\sim$ const./$\log^2$. The freedom to multiply Λ by a constant can be used to remove this term. There is a $\mathcal{O}(15\%)$ difference in the Λ's defined by (4) and (6). Since in practice it is usually α_s which is measured experimentally, it is important when comparing Λ values to check that the same equation has been used to determine Λ from the coupling constant.

A second difficulty with the above definitions is that Λ depends on the number of active flavours. Values of Λ for different numbers of flavours can be defined by imposing the continuity of α_s at the scale $\mu = m$, where m is the mass of the heavy quark. For example, for the b-quark threshold: $\alpha_s(m_b^2, 4) = \alpha_s(m_b^2, 5)$. Using the next-to-leading order form (4) for $\alpha_s(Q^2)$ one can show that

$$\Lambda(4) \approx \Lambda(5)\left(\frac{m_b}{\Lambda(5)}\right)^{\frac{2}{25}}\left[\ln\left(\frac{m_b^2}{\Lambda(5)^2}\right)\right]^{\frac{963}{14375}} . \tag{7}$$

In practice, most higher order QCD corrections are carried out using the *modified minimal subtraction* ($\overline{\rm MS}$) regularization scheme. To be consistent, then, one uses the above results for $\alpha_s(\mu^2)$ with $\Lambda \equiv \Lambda_{\overline{\rm MS}}$.

Some recent α_s measurements, taken from the compilation in reference [4], are shown in Fig.1. The lines indicate different values of $\Lambda^{(5)}_{\overline{\rm MS}}$. Caution should be exercised when comparing the precision of the various measurements, as the errors have different meanings in different processes. A central value of $\Lambda^{(5)}_{\overline{\rm MS}} \simeq 150$ MeV is indicated.

Historically, the dimensionful parameter Λ (in some particular renormalization scheme, e.g. $\overline{\rm MS}$) has been taken as *the* fundamental parameter of QCD, to be measured by experiment. However, it is just as natural to take instead the value of the renormalized coupling at some reference mass scale. The reason this was not done was that there was no obvious 'standard' scale, high enough so that the coupling there was perturbatively small. The situation has changed with the advent of high precision α_s measurements at LEP. It now seems sensible to adopt $\alpha_s^{\overline{\rm MS}}(M_Z^2)$ as the fundamental parameter, rather than $\Lambda_{\overline{\rm MS}}$. (The small uncertainty in M_Z is completely negligible compared to the uncertainty in the coupling itself). Of course using the above equations it is straightforward to convert between the two. Table 2 gives the conversion between $\Lambda^{(5)}_{\overline{\rm MS}}$ and $\alpha_s(M_Z^2)$ using the definition given in (4).

Table 2. $\alpha_s(M_Z^2)$ for various $\Lambda^{(5)}_{\overline{\rm MS}}$.

$\Lambda^{(5)}_{\overline{\rm MS}}$ (MeV)	$\alpha_s(M_Z^2)$
50	0.0970
100	0.1060
150	0.1122
200	0.1170
250	0.1210
300	0.1245
350	0.1277
400	0.1305
450	0.1332
500	0.1356
550	0.1379
600	0.1401

3 Deep Inelastic Scattering

The original, and still the most powerful, test of perturbative QCD is the breaking of Bjorken scaling in deep inelastic lepton-hadron scattering. Nowadays, structure function analyses not only provide some of the most precise tests of the theory but also determine the momentum distributions of partons in hadrons for use as input in predicting cross sections in high energy hadron-hadron collisions. In this section we first describe the basic features of the parton model in deep inelastic scattering and then discuss how the picture is modified by perturbative corrections.

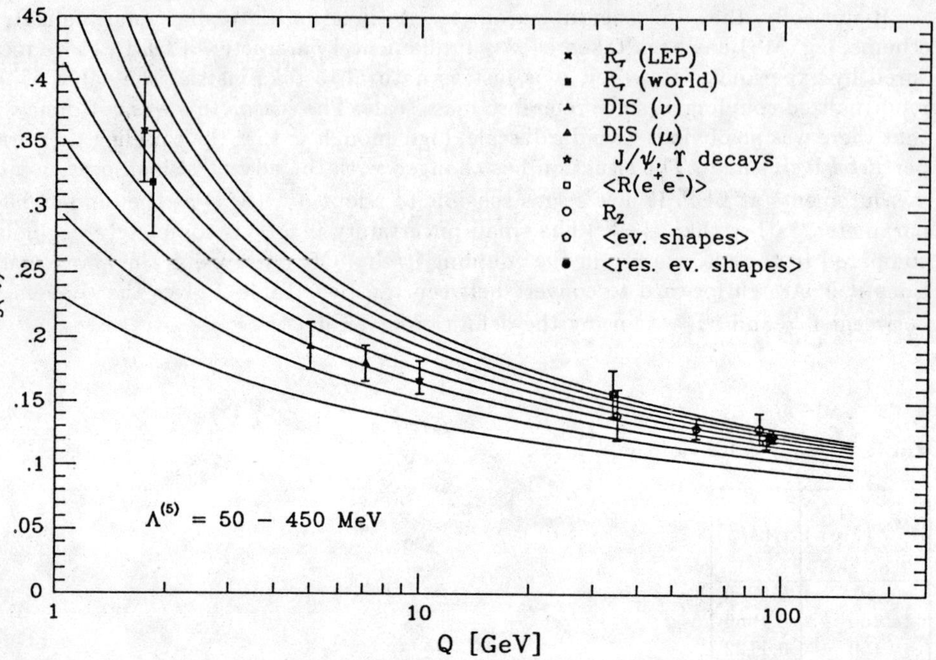

Fig. 1. A compilation [4] of α_s measurements from different processes. The lines correspond to 50 MeV intervals of $\Lambda^{(5)}_{\overline{\rm MS}}$.

3.1 The Parton Model

Consider the deep inelastic process $lp \to lX$. Label the incoming and outgoing lepton four-momenta by k^μ and k'^μ, respectively, the incoming proton momentum by p^μ ($p^2 = M^2$) and the momentum transfer by $q^\mu = k^\mu - k'^\mu$. The standard deep inelastic variables are defined by:

$$
\begin{aligned}
Q^2 &= -q^2 \qquad p^2 = M^2 \\
x &= \frac{Q^2}{2p\cdot q} = \frac{Q^2}{2M(E-E')} \\
y &= \frac{q\cdot p}{k\cdot p} = 1 - E'/E \\
s &= (k+p)^2 = M^2 + \frac{Q^2}{xy},
\end{aligned} \tag{8}
$$

where the energies are defined in the rest frame of the target. The structure functions $F_i(x, Q^2)$ are then defined in terms of the lepton scattering cross sections. For charged lepton scattering, $lp \to lX$,

$$\frac{d^2\sigma^{em}}{dxdy} = \frac{4\pi\alpha^2(s-M^2)}{Q^4}\left[\left(\frac{1+(1-y)^2}{2}\right)2xF_1^{em} + (1-y)(F_2^{em}-2xF_1^{em}) - \frac{M^2}{s-M^2}xyF_2^{em}\right], \tag{9}$$

and for neutrino (antineutrino) scattering, $\nu(\bar{\nu})p \to lX$,

$$\frac{d^2\sigma^{\nu(\bar{\nu})}}{dxdy} = \frac{G_F^2(s-M^2)}{2\pi}\left[\left(1-y-\frac{M^2}{s-M^2}xy\right)F_2^{\nu(\bar{\nu})} + y^2xF_1^{\nu(\bar{\nu})} + (-)y(1-y/2)xF_3^{\nu(\bar{\nu})}\right]. \tag{10}$$

In the quark-parton model, these structure functions are related to the quark 'distribution functions' or 'densities' $q(x,Q^2)$, where $q(x,Q^2)dx$ is the probability that a parton carries a momentum fraction x of the target nucleon's momentum when probed (by a gauge boson γ^*, W or Z) at energy scale Q. Thus

$$\begin{aligned} F_2^{\nu} &= 2x[d+s+\bar{u}+\bar{c}] \\ xF_3^{\nu} &= 2x[d+s-\bar{u}-\bar{c}] \\ F_2^{\bar{\nu}} &= 2x[u+c+\bar{d}+\bar{s}] \\ xF_3^{\bar{\nu}} &= 2x[u+c-\bar{d}-\bar{s}] \\ F_2^{em} &= x[\frac{4}{9}(u+u+c+\bar{c})+\frac{1}{9}(d+\bar{d}+s+\bar{s})] \\ 2xF_1 &= F_2. \end{aligned} \tag{11}$$

Note that when the nature of the target is unambiguous the notation $q(x,Q^2)$ and $g(x,Q^2)$ for the quark and gluon densities can be used, otherwise a general notation is $f_{a/A}(x,Q^2)$, where a = u, d, ... g and A = p, n, Fe, Cu, etc. In the 'naive' parton model the structure functions *scale*, i.e. $F(x,Q^2) \to F(x)$ in the asymptotic (Bjorken) limit: $Q^2 \to \infty$, x fixed. In fact, it was the observation of scaling in the original SLAC experiments that provided the first evidence of parton structure in the hadron. To a first approximation, therefore, one can take the parton distributions to be functions of x only: $q(x,Q^2) \to q(x)$. We shall see below how perturbative QCD induces logarithmic deviations from scaling, exactly in line with more recent high-precision experimental measurements.

The picture which emerges is of a proton consisting of three valence quarks (uud) and a 'sea' of $q\bar{q}$ pairs and gluons. In the most simple version of this parton model the sea would be (three) flavour symmetric and hence the net quark distributions would be given by $u = u_V + S$, $d = d_V + S$ and $s = \bar{s} = \bar{u} = \bar{d} = S$, with the sum rules

$$\int_0^1 u_V(x)dx = 2\int_0^1 d_V(x)dx = 2 \tag{12}$$

$$\int_0^1 x[u_V(x)+d_V(x)+6S(x)]dx = 1\,. \tag{13}$$

These represent conservation of proton quantum numbers and momentum fraction, respectively.

Fig.2 shows some representative modern quark and gluon distributions (the MRS-D_0' distributions of reference [5]) extracted from deep inelastic scattering and other processes. The scale is $Q^2 = 20$ GeV^2, a typical value for fixed-target deep inelastic experiments. At this scale the charm quark distribution is very small and is not shown.

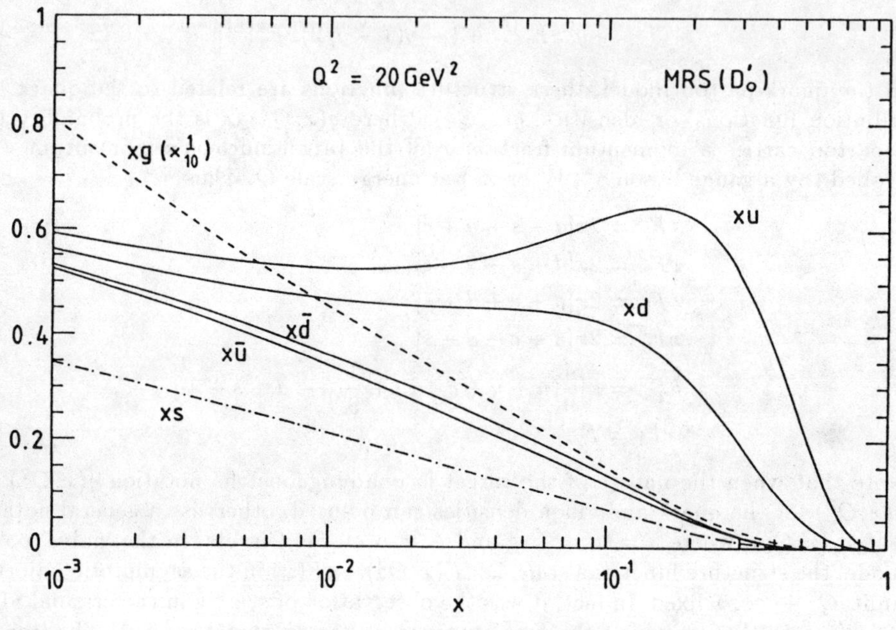

Fig. 2. Quark and gluon distributions at $Q^2 = 20$ GeV^2 from reference [5].

3.2 Scaling Violations – the Altarelli-Parisi Equations

In QCD, Bjorken scaling is broken by logarithms of Q^2. Physically, a quark in the proton can emit a gluon, with probability determined by α_s, and lose momentum as a result. Since the higher the Q^2 the more phase space is available for gluon emission, the expectation is that parton distributions should shrink to small x as Q^2 increases, with the rate of shrinkage being controlled by α_s.

In describing quantitatively the way in which scaling is violated it is convenient to define singlet and non-singlet quark distributions:

$$F^{NS} = q_i - q_j, \qquad F^S = \sum_i (q_i + \bar{q}_i). \tag{14}$$

The non-singlet structure functions have non-zero values of flavour quantum numbers such as isospin or baryon number. The variation with Q^2 of these functions is described by the so-called Altarelli-Parisi equations [6]:

$$\begin{aligned} Q^2\frac{\partial F^{NS}}{\partial Q^2} &= \frac{\alpha_s(Q^2)}{2\pi} P^{qq} * F^{NS} \\ Q^2\frac{\partial F^{S}}{\partial Q^2} &= \frac{\alpha_s(Q^2)}{2\pi} \left(P^{qq} * F^{S} + 2n_f P^{qg} * g\right) \\ Q^2\frac{\partial g}{\partial Q^2} &= \frac{\alpha_s(Q^2)}{2\pi} \left(P^{gq} * F^{S} + P^{gg} * g\right), \end{aligned} \tag{15}$$

where $*$ denotes a convolution integral:

$$f * g = \int_x^1 \frac{dy}{y} f(y) g\left(\frac{x}{y}\right). \tag{16}$$

In leading order the Altarelli-Parisi splitting functions are

$$\begin{aligned} P^{qq} &= \frac{4}{3}\left(\frac{1+x^2}{1-x}\right)_+ \\ P^{qg} &= \frac{1}{2}(x^2 + (1-x)^2) \\ P^{gq} &= \frac{4}{3}\left(\frac{1+(1-x)^2}{x}\right) \\ P^{gg} &= 6\left(\frac{1-x}{x} + x(1-x) + \left(\frac{x}{1-x}\right)_+\right) \\ &\quad - \left(\frac{1}{2} + \frac{n_f}{3}\right)\delta(1-x). \end{aligned} \tag{17}$$

Note the 'plus prescription' for those functions which are singular as $x \rightarrow 1$:

$$\int_0^1 dx f(x)\, (g(x))_+ = \int_0^1 dx (f(x) - f(1)) g(x). \tag{18}$$

The Altarelli-Parisi equations can be solved analytically by defining *moments* (formally, the Mellin transforms) of the structure functions, $M_n^{NS} = \langle F^{NS}\rangle_n \equiv \int_0^1 dx x^{n-1} F^{NS}$ etc. The convolution integral then becomes a simple product. Introducing the leading order expression for the QCD coupling constant (see Section 2.2),

$$\alpha_s(Q^2) = \frac{4\pi}{\beta_0 \log(Q^2/\Lambda^2)}, \tag{19}$$

one obtains, for the non-singlet solution,

$$M_n^{NS}(Q^2) = M_n^{NS}(Q_0^2)\left(\frac{\alpha_s(Q^2)}{\alpha_s(Q_0^2)}\right)^{-d_n}, \tag{20}$$

where $d_n = 2\langle P^{qq}\rangle_n/\beta_0$. Note that $d_1 = 0$ and that $d_n < 0$ for $n \geq 2$, which implies that the parton distributions decrease and increase with increasing Q^2 at large and small x, respectively. Solutions for the singlet and gluon moments can be found in a similar way, by first diagonalizing the coupled equations.

The precision of contemporary deep inelastic data demands that the QCD predictions are calculated to next-to-leading order. This amounts to the replacements (shown schematically):

$$P(x) \to P^{(0)}(x) + \frac{\alpha_s}{2\pi} P^{(1)}(x)$$
$$F = \sum q \to F^{(1)} = \sum C * q, \qquad C = \delta(1-x) + O(\alpha_s). \tag{21}$$

Beyond leading order, the definition of parton distributions (like α_s) becomes scheme dependent. It is conventional nowadays to work in the $\overline{\rm MS}$ scheme, where for example

$$\begin{aligned} F_2^{\mu p}(x,Q^2) &= \sum_q e_q^2 x q^{\overline{\rm MS}}(x,Q^2) + \alpha_s^{\overline{\rm MS}}(Q^2) \sum_q x \int_x^1 \frac{dz}{z} c_q(z) q^{\overline{\rm MS}}\left(\frac{x}{z},Q^2\right) \\ &\quad + x \int_x^1 \frac{dz}{z} c_g(z) g^{\overline{\rm MS}}\left(\frac{x}{z},Q^2\right) . \end{aligned} \tag{22}$$

An example of a next-to-leading order QCD fit [5] to recent high-precision data on $F_2^{\mu p}$ from the BCDMS [7] and NMC [8] collaborations is shown in Fig.3.

A more complete discussion of how structure function measurements are made and how detailed QCD fits are performed can be found in the lectures of Milsztajn [10]. In summary, the observed scaling violations are exactly in line with perturbative QCD expectations and lead to some of the most precise measurements of the strong coupling (see Fig.1).

4 Parton Distributions in Hadrons

4.1 Introduction

As we have seen in the previous sections, distributions of the quarks (and gluons) in the proton are determined by values of the structure functions $F_i(x,Q^2)$ measured in muon (or electron)-nucleon (or nucleus) and neutrino-nucleus deep inelastic scattering experiments. It is relevant to ask why we should devote so much effort to the study of the distributions of partons in the proton. There are two main reasons – one experimental and one theoretical. First, a detailed knowledge of parton distributions is an essential ingredient in all 'hard' interactions involving protons, and so they are needed to estimate the production rates of the various 'hard' processes that may occur at current and forthcoming colliders (see Section 5). However, the parton structure is interesting in its own right. In particular, novel perturbative QCD effects are expected to become apparent at small x (i.e. $x \lesssim 10^{-3}$). The reason is that at small x the sum over soft gluon emission off the incoming parton leads to a power series in $\alpha_s \log(1/x)$, which on resummation, via the Lipatov (or BFKL) equation [11], suggests that the gluon behaves as

$$xg \sim x^{-\lambda} \tag{23}$$

as $x \to 0$, with λ predicted, typically, to be about 0.5. Such an increase in $xg(x,Q^2)$, as x decreases, cannot go on indefinitely. If the density of gluons becomes too large

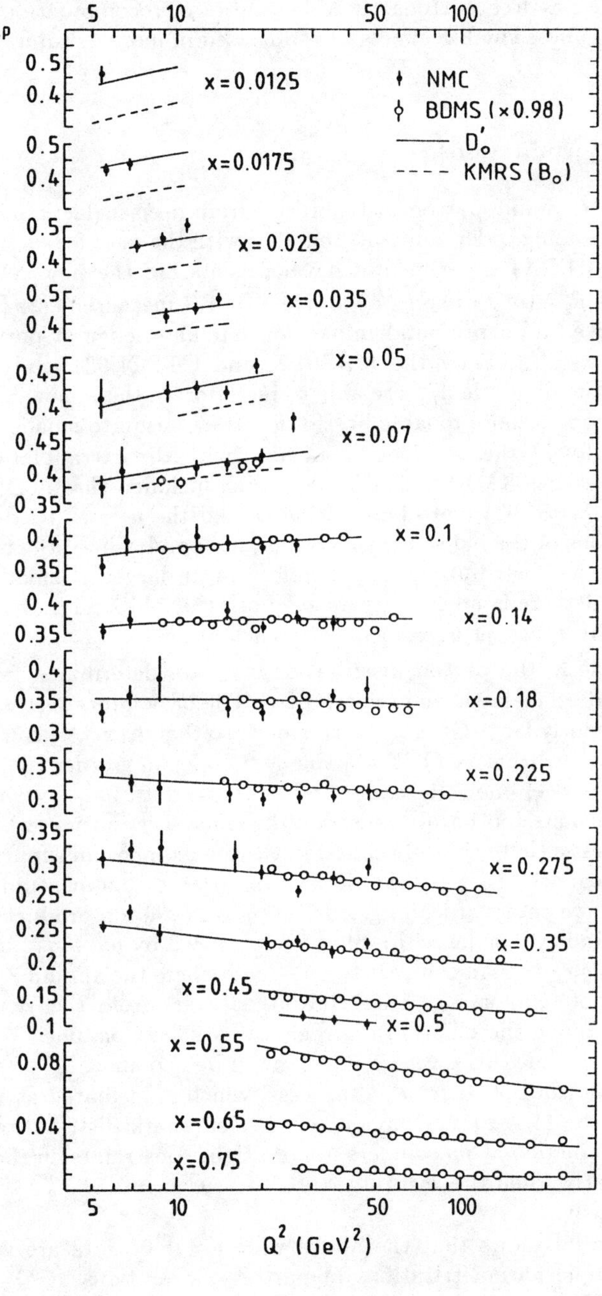

Fig. 3. The BCDMS [7] and NMC [8] measurements of $F_2^{\mu p}$, together with the fits by KMRS [9] and MRS [5]. The NMC data were not available for KMRS.

they can no longer be treated as free partons, and the effects of recombination or shadowing must be included. These small x effects are addressed in more detail in the lectures of Kwiecinski [12].

4.2 Survey of Parton Determinations

There is a long history of determining parton distributions from deep-inelastic data. Table 3 lists just the next-to-leading order analyses, together with the data fitted. The most significant recent (pre-HERA) experimental developments are the new NMC measurements [8] of $F_2^{\mu p}$ and $F_2^{\mu n}$ (or rather $F_2^{\mu D}$) and the CCFR measurements [13] of $F_2^{\nu N}$ and $xF_3^{\nu N}$. They have had a profound impact on our knowledge of parton distributions, especially for $x \lesssim 0.1$. Only the MRS('92) and CTEQ('93) analyses (that is, the three below the line in Table 3) were able to incorporate these new data, and so should be able to provide a more reliable basis for extrapolation to smaller x.

Figure 3 [5] serves to highlight the problems associated with the extrapolation. It shows (dashed curves) an earlier KMRS global fit [9] which included the BCDMS data, together with the new MRS('92) fit to both BCDMS and the new NMC data sets. We see that extrapolations of the old solutions considerably undershoot the new data at small x. The new quark distributions are therefore much larger at small x. On the other hand for $x \gtrsim 0.1$ there is no change; we see both the 'old' and 'new' μp data and the partons are all in excellent agreement with each other.

As is apparent from Table 3, the parton distributions, f_i, are determined from global fits to a wide range of deep inelastic and related data. The basic procedure is to parametrize the f_i at a sufficiently large Q_0^2, say 4 or 5 GeV^2, so that $f_i(x, Q^2)$ can be calculated at higher Q^2 from perturbative QCD via the next-to-leading order (NLO) Altarelli-Parisi equations. The deep-inelastic muon and neutrino data pin down the valence and sea quark distributions, but hardly constrain the gluon distribution, which only enters directly at NLO; essentially the only constraint is the momentum sum rule which shows that the gluon carries just less than 50% of the proton's momentum at Q_0^2. On the other hand the gluon enters at leading order in prompt photon production in hadron-hadron collisions. Indeed for 'large' p_T photons produced by $pp \to \gamma X$, the dominant QCD subprocess is $gq \to \gamma q$, in contrast to $p\bar{p} \to \gamma X$ where the annihilation process $q\bar{q} \to \gamma g$ is much more important. The relevant data are from the WA70 collaboration [15] which determine the gluon in the region $x \sim 0.35$. Combined with the momentum sum rule constraint, this gives an input gluon behaviour $\propto (1-x)^{5.3}$ at large x. Data for the Drell-Yan $pN \to \mu^+\mu^- X$ process, which is mediated at LO by $q_{\rm val}\bar{q}_{\rm sea} \to \gamma^*$, constrain the $(1-x)^{\eta_s}$ behaviour of the sea quark distributions. Finally data on W and Z production at $\bar{p}p$ colliders impose tight constraints on the u and d distributions, particularly when the accurate NMC measurements of $F_2^{\mu n}/F_2^{\mu p}$ have to be fitted simultaneously.

The recent MRS analysis [5] shows that the above wide range of data are well described by remarkably simple parametrizations of parton distributions at $Q_0^2 =$ 4 GeV^2; in total only about 15 parameters are required. An example of the quality of the fit to the recent high precision deep-inelastic data has already been shown in Fig. 3. In order to provide an indication of the possible range of extrapolations to smaller values of x, two sets of partons are presented which describe the data equally well –

Table 3. NLO determinations of parton distributions [14,9,16,5,17] together with the data used in the various analyses. Data marked † were used in preliminary form.

	μ-DIS	ν-DIS	Prompt γ	D-Yan	W, Z
MRS '88	EMC + ..	CDHSW	AFS($+J/\psi$)	–	–
DFLM '88	(EMC + ..)	CHARM + ..	–	E288 + ..	–
ABFOW '89	BCDMS	–	WA70	–	–
HMRS '90	EMC BCDMS NMC(n/p)	CDHSW	WA70	E605	–
MT '90	EMC BCDMS	CDHSW	–	E288 E605	–
KMRS '90 (sets B_0,B_-)	BCDMS NMC(n/p)	CDHSW	WA70	E605	–
MRS (Apr '92) (sets D_0,D_-)	BCDMS NMC(p,n)†	CDHSW CCFR†	WA70	E605	(UA2, CDF
MRS (Nov '92) (sets D_0',D_-')	BCDMS NMC(p,n)	CCFR	WA70	E605	(UA2, CDF)
CTEQ ('93)	BCDMS NMC(p,n)	CCFR	WA70 E706,UA6	E605	

one set omitting and one set including the Lipatov behaviour, (23), of the gluon and sea quark distributions at small x. Namely

$$\begin{aligned} xg(x,Q_0^2) &\sim \text{constant} \qquad &\text{for the } D_0' \text{ set,} \\ xg(x,Q_0^2) &\sim x^{-\frac{1}{2}} \qquad &\text{for the } D_-' \text{ set,} \end{aligned} \tag{24}$$

as $x \to 0$; and similarly for $x\bar{q}(x,Q_0^2)$ since the sea is driven at small x by $g \to q\bar{q}$.

The NMC results for the Gottfried sum rule (see below) suggest that the non-strange sea is not $u \leftrightarrow d$ symmetric. Indeed from the lack of Regge $\rho - a_2$ exchange degeneracy we would expect

$$\bar{d} - \bar{u} \;=\; Ax^{\beta}(1-x)^{\eta_s}\,, \tag{25}$$

where the behaviour at small x is governed by the intercept of the meson Regge trajectory, so that $\beta \simeq -\alpha_R(0) \simeq -0.5$. In the MRS-$D_0'$ and D_-' distributions $\bar{d} - \bar{u}$ is parametrized in this way [5,16]. The MRS strange sea distribution is taken to be

$$s \;=\; \tfrac{1}{4}(\bar{u} + \bar{d}) \tag{26}$$

at Q_0^2, where the (factor of 2) suppression is motivated by the recent CCFR results [19] for $\nu N \to \mu^+\mu^- X$. We note that in the recent CTEQ('93) analysis [17], where the strange quark distribution is freely and independently parametrized, the strange sea turns out to be of comparable size to the non-strange sea, in disagreement with the dimuon data. Further discussion of this issue can be found in reference [20].

4.3 Flavour Asymmetry in the Non-Strange Sea

It is difficult to determine $\bar{u}$ and $\bar{d}$ separately from the deep inelastic data. In fact, prior to 1992, analyses assumed that the non-strange sea was flavour symmetric, that is $\bar{u} = \bar{d}$. The motivation for allowing $\bar{u} \neq \bar{d}$ came from comparing the NMC measurement [21]

$$\int_{0.004}^{0.8} \frac{dx}{x} \left(F_2^{\mu p} - F_2^{\mu n}\right) = 0.227 \pm 0.007\,(\text{stat.}) \pm 0.014\,(\text{sys.}) \tag{27}$$

at $Q^2 = 4\ \text{GeV}^2$ with the Gottfried-sum-rule expectation

$$\begin{aligned} I_{\text{GSR}} &= \int_0^1 \frac{dx}{x} \left(F_2^{\mu p} - F_2^{\mu n}\right) = \tfrac{1}{3} \int_0^1 \frac{dx}{x} \left(u_V - d_V\right) + \tfrac{2}{3} \int_0^1 \frac{dx}{x} \left(\bar{u} - \bar{d}\right) \\ &= \tfrac{1}{3}, \qquad \text{if } \bar{u} = \bar{d}\,. \end{aligned} \tag{28}$$

A straightforward comparison of (27) and (28) implies $\bar{u} \neq \bar{d}$. In the MRS fits [16,5] $\bar{d} - \bar{u}$ is parametrized according to (25) such that I_{GSR} is in the range $0.24 - 0.26$. It is interesting to note that, even including the NMC data, it is still possible to maintain $\bar{u} = \bar{d}$ and to obtain an equally good global description of the data at the expense of a contrived small x behaviour of the valence distributions, see set S_0 and especially the discussion in [16]. However, as mentioned above, Regge arguments suggest that it is more reasonable to allow $\bar{u} \neq \bar{d}$, as in sets D_0' and D_-'.

The structure of $\bar{d} - \bar{u}$ is not determined by available data. All we really know is that on average $\bar{d} > \bar{u}$ at small x. The situation is summarized in Fig. 4 which shows the ingredients of the Gottfried sum rule for various sets of partons with $\bar{u} \neq \bar{d}$. The central part of the figure is a visual display of the integrand of the sum rule (27). The data are from NMC [23]. The top part of the figure shows the accumulated contribution to the sum rule integrated down from $x = 1$; thus the limiting values of the curves as $x \to 0$ give I_{GSR} of (27). The central curves in the shaded bands correspond to MRS(D_0') partons [5] with $\bar{d} - \bar{u}$ parametrized as in (25); these partons give $I_{\text{GSR}} = 0.256$. The shaded bands correspond to variations about this solution [22] with the parametrization (25) extended to

$$\bar{d} - \bar{u} = A x^{\beta} (1 - x)^{\eta_s} (1 + \gamma x + \delta x^2) \tag{29}$$

with A chosen to maintain $I_{\text{GSR}} = 0.256$. All these extended fits are evidently consistent with the deep inelastic data.

A powerful and independent method for obtaining further information on $\bar{d} - \bar{u}$ [24] is to compare (Drell-Yan) lepton-pair production of pp and pn origin. Fig. 5 shows the

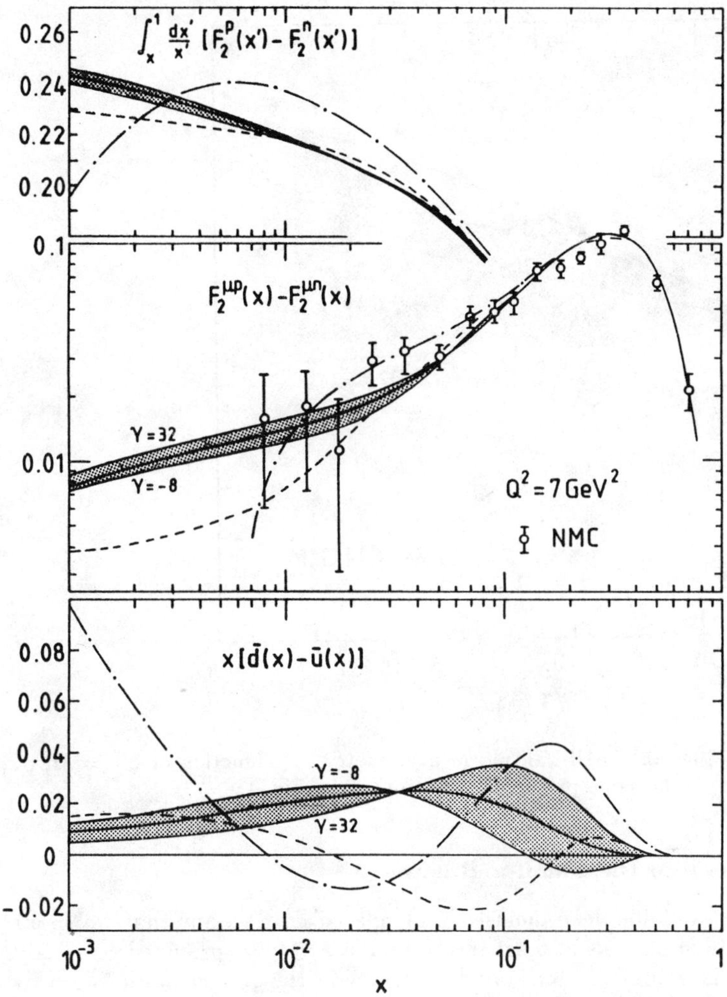

Fig. 4. The ingredients of the Gottfried sum rule for various sets of parton distributions, as described in the text. The figure is taken from reference [22].

predictions for the asymmetry between Drell-Yan production on proton and neutron targets

$$A_{\rm DY} = \frac{\sigma_{pp} - \sigma_{pn}}{\sigma_{pp} + \sigma_{pn}} \tag{30}$$

as a function of $\sqrt{\tau} = M_{l^+l^-}/\sqrt{s}$, at $y = 0$, $p_{\rm beam} = 450$ GeV/c. Because $u > d$ in the proton, the asymmetry is positive for sets with $\bar{d} - \bar{u}$ zero or small at large x. A measurement [25] of the asymmetry to an accuracy of a few per cent will provide a powerful discriminator of the possible behaviours of $\bar{d} - \bar{u}$.

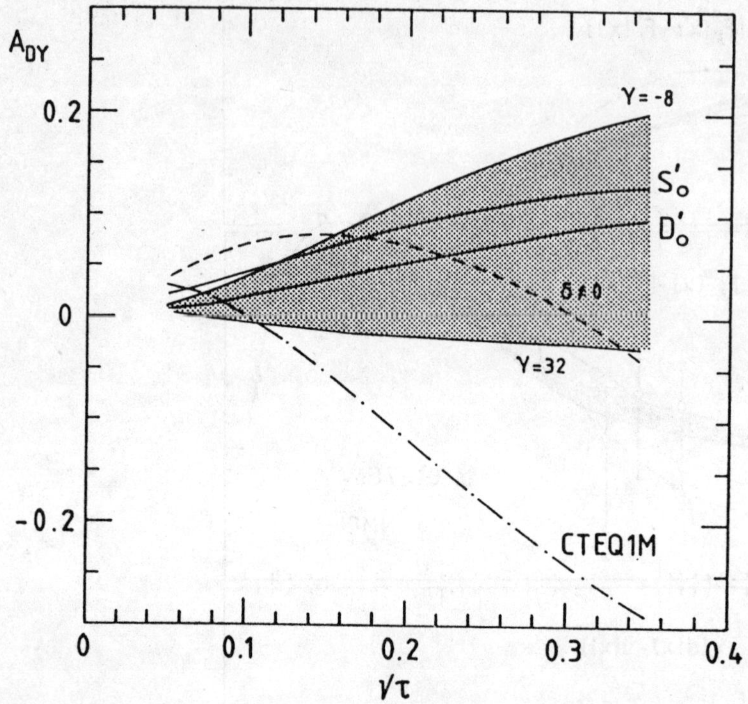

Fig. 5. Predictions [22] for the Drell-Yan pp/pn asymmetry as a function of $\sqrt{\tau} = M/\sqrt{s}$. The curves correspond to the same parton distributions as in Fig. 4.

4.4 Extrapolations into the Small-x Regime

The precision of the available deep-inelastic and related data means that we should now have reliable sets of partons at least in the region $x \gtrsim 0.05$, provided reasonable behaviour of $\bar{d} - \bar{u}$ is assumed. Extrapolations outside the region of the data are notoriously unreliable. All we can do is to extrapolate the various sets to small x and show the general trends.

For example, the parton sets MRS-D_0' and MRS-D_-' [5] are essentially identical for $x \gtrsim 0.02$ but (by construction) are dramatically different at small x. In particular they lead to the very different predictions for F_2^{ep} in the HERA regime, as shown by the continuous curves in Fig. 6.

Also shown are (i) the CTEQ1M and CTEQ1MS predictions [17] and (ii) the predictions from the 'dynamical quark' model of Glück, Reya and Vogt (GRV) [18]. In this model, 'valence-like' distributions at very low Q ($Q_0^2 = 0.3$ GeV^2) are evolved and fitted to MRS valence quark distributions at a higher Q. The growth seen for GRV at small x is thus a result of Altarelli-Parisi Q^2 evolution. By coincidence, it is not unlike that seen for D_-', at least in the HERA regime. It is probably fair to say that much of the attractiveness of the original GRV model was lost when, besides the valence quarks (which could be regarded as of constituent origin), it was

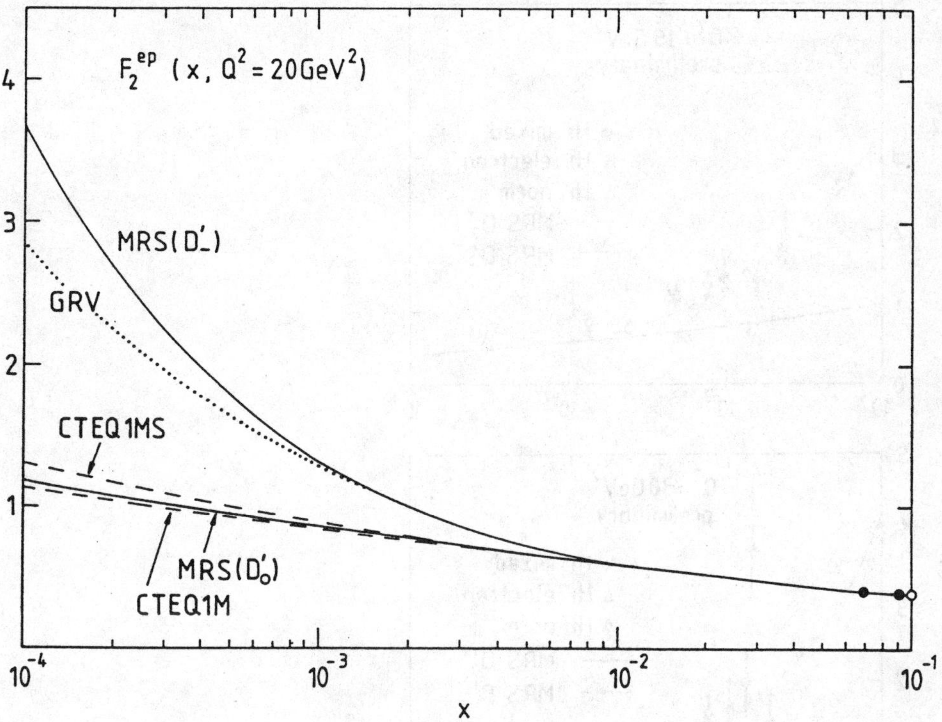

Fig. 6. Predictions for F_2^{ep} at $Q^2 = 20$ GeV^2 obtained from the parton sets from references [5,17,18].

found necessary to also input valence gluon and sea quark contributions in order to respectively account for the WA70 prompt photon data [15] and to give the observed x shape of the $F_2^{\mu p}$ data. Nevertheless, unlike D'_, GRV is a parameter-free prediction at small x. (Subsequent to the School, preliminary H1 data on F_2^{ep} were announced. These are shown in Fig.7, and indicate a strong preference for the singular D'_ and GRV predictions.)

The measurements of F_2^{ep} at HERA probe primarily the sea quark distribution at small x. Methods for directly extracting the gluon distribution from HERA measurements of the longitudinal structure function F_L [27] and $dF_2/d\log Q^2$ have been suggested. These are based on the observation that in the leading-order expressions

$$F_L(x,Q^2) = \frac{\alpha_s(Q^2)}{\pi}\left\{\frac{4}{3}\int_x^1 \frac{dy}{y}\left(\frac{x}{y}\right)^2 F_2(y,Q^2) + 2\sum_q e_q^2 \int_x^1 \frac{dy}{y}\left(\frac{x}{y}\right)^2\left(1-\frac{x}{y}\right) yg(y,Q^2)\right\} \quad (31)$$

$$\frac{dF_2(x,Q^2)}{d\log Q^2} = \frac{\alpha_s(Q^2)}{2\pi}\left\{\int_x^1 \frac{dy}{y}\left(\frac{x}{y}\right) P_{qq}\left(\frac{x}{y}\right) F_2(y,Q^2)\right.$$

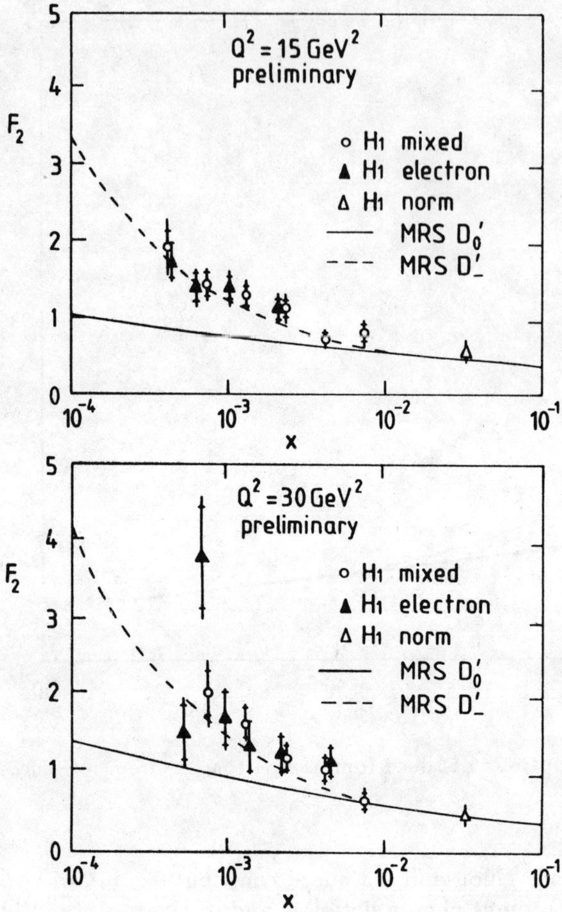

Fig. 7. Preliminary H1 data [26] on the F_2 structure function at HERA together with the extrapolations resulting from MRS parton sets D_0' and D_-' [5]. The data are subject to an additional overall normalization uncertainty of $\pm 12\%$.

$$+2\sum_q e_q^2 \int_x^1 \frac{dy}{y}\left(\frac{x}{y}\right) P_{qg}\left(\frac{x}{y}\right) yg(y,Q^2)\Bigg\} \tag{32}$$

it is the gluon term on the right-hand side that dominates at small x in each case. The clear correlations shown in Fig. 8 lend support to the case for making precise measurements of these two quantities, leading to qualitative estimates of the underlying gluon behaviour.

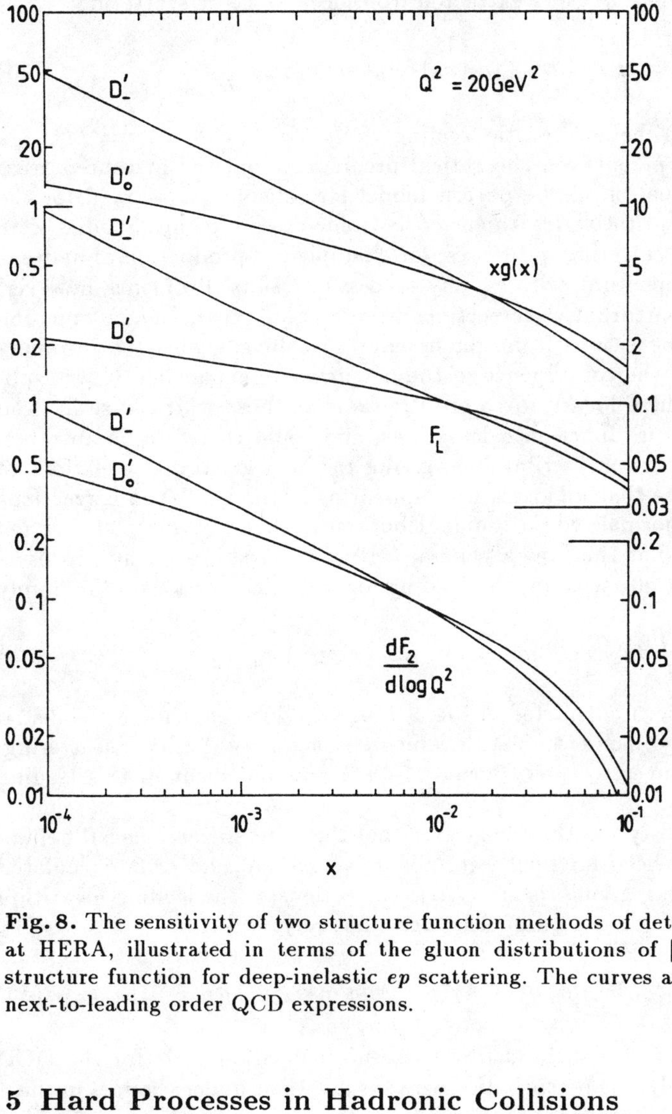

Fig. 8. The sensitivity of two structure function methods of determining the small x gluon at HERA, illustrated in terms of the gluon distributions of [5]. F_L is the longitudinal structure function for deep-inelastic ep scattering. The curves are calculated using the full next-to-leading order QCD expressions.

5 Hard Processes in Hadronic Collisions

5.1 Introduction

It was first pointed out by Drell and Yan [33] that parton model ideas developed for deep inelastic scattering could be extended to certain processes in hadron-hadron collisions. The paradigm process was the production of a massive lepton pair by quark-antiquark annihilation – the Drell-Yan process – and the hadronic cross section σ was to be obtained by weighting the subprocess cross section $\hat{\sigma}$ for $q\bar{q} \to \mu^+\mu^-$ with the parton distribution functions $f_{q/A}(x)$ extracted from deep inelastic scattering:

$$\sigma_{AB} = \int dx_a dx_b \, f_{a/A}(x_a) f_{b/B}(x_b) \, \hat{\sigma}_{ab\to X} \ . \tag{33}$$

The domain of validity is the asymptotic 'scaling' limit $M^2_{l^+l^-}, s \to \infty$, $\tau = M^2_{l^+l^-}/s$ fixed. The good agreement between theoretical predictions and the measured cross sections provided confirmation of the parton model formalism, and allowed for the first time a rigorous, quantitative treatment of hadronic cross sections. Studies were extended to other 'hard scattering' processes, for example the production of hadrons with large transverse momentum, with equally successful results. Problems, however, appeared to arise when perturbative corrections from real and virtual gluon emission were calculated. Large logarithms from gluons emitted collinear with the incoming quarks appeared to spoil the convergence of the perturbative expansion. It was subsequently realised that these logarithms were the same as those which arise in deep inelastic scattering structure function calculations, and could therefore be absorbed in the definition of the parton distributions, giving rise to logarithmic violations of scaling. The key point was that *all* logarithms appearing in the Drell-Yan corrections could be factored into renormalized parton distributions in this way, and *factorization theorems* which showed that this was a general feature of hard scattering processes were derived [34]. Taking into account the leading logarithm corrections, (33) simply becomes:

$$\sigma_{AB} = \int dx_a dx_b \, f_{a/A}(x_a, Q^2) f_{b/B}(x_b, Q^2) \, \hat{\sigma}_{ab\to X} \ , \tag{34}$$

corresponding to the structure depicted in Fig. 9. The Q^2 which appears in the distribution functions is a large momentum scale which characterises the hard scattering, e.g. $M^2_{l^+l^-}$, p_T^2, ... Changes to the Q^2 scale of $\mathcal{O}(1)$ are equivalent in this leading logarithm approximation.

The final step in the story was the recognition that the *finite* corrections left behind after the logarithms had been factored were not universal and had to be calculated separately for each process, giving rise to $\mathcal{O}(\alpha_s)$ corrections to the leading logarithm (or leading order (LO)) cross section of (34). Schematically

$$\sigma_{AB} = \int dx_a dx_b \, f_{a/A}(x_a, M^2) \, f_{b/B}(x_b, M^2) \times [\hat{\sigma}_0 + \alpha_s(\mu^2) \, \hat{\sigma}_1 + ...]_{ab\to X} \ . \tag{35}$$

Here M^2 is the *factorization scale* and μ is the renormalization scale for the QCD running coupling. Formally, the perturbation series is invariant under changes in these parameters, the M and μ dependence of the coefficients, e.g. $\hat{\sigma}_1$, exactly compensating the explicit dependence of the parton distributions and the coupling constant. This compensation becomes more exact as more terms are included in the perturbation series. To avoid unnaturally large logarithms reappearing in the perturbation series it is sensible to choose M and μ values of order the typical momentum scales of the hard scattering process, and $M = \mu$ is also often assumed.

In general, all the important hadronic processes have now been calculated to next-to-leading order (NLO), i.e. up to and including the $\hat{\sigma}_1$ terms. One process – the Drell-Yan process – is even calculated to one order higher (see below). As will be illustrated in the following section, this allows a very high degree of precision in a wide

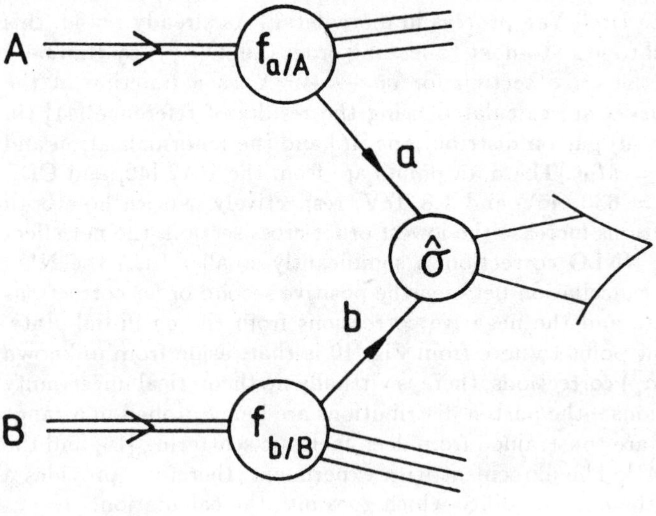

Fig. 9. The basic structure of hard scattering processes in perturbative QCD.

variety of processes. In many cases, the residual renormalization and factorization scale dependence is weak, and the precision of the theoretical prediction is limited only by uncertainties in the knowledge of the parton distributions, in particular the relatively poorly known gluon distribution.

What, then, are the most important applications of this formalism? Nowadays one can, for example, (i) attempt to measure α_s, particularly from those processes involving α_s at leading order, i.e. in $\hat{\sigma}_0$; (ii) obtain information on parton distributions complementary to that from deep inelastic scattering; (iii) attempt to extend the domain of applicability to other kinematic regions, in particular to small x, where additional complications arise [12]; (iv) perform precision analyses of signals and backgrounds for *new physics*, e.g. top, Higgs, supersymmetric particle production etc. There are many examples of situations where the ability to detect a signal for new particle production depends crucially on the accuracy of the Standard Model background estimate.

5.2 Higher-Order Corrections to Inclusive Processes

As discussed in Section 5.1, many hard scattering cross sections in hadronic collisions have now been calculated to next-to-leading order. Some of the most important are:

1. The Drell-Yan process (γ^*, $W^\pm$, Z^0 production) [35].
2. Large p_T prompt photon production [36].
3. Heavy quark pair production [37].
4. Electroweak gauge boson pair production ($Z^0Z^0, W^+W^-, W^\pm\gamma, ...$) [38].
5. Higgs production by gg fusion [39].
6. Inclusive large p_T jet production [40].

As an example of the impact of the higher order corrections on the cross section, it is instructive to consider the Drell-Yan process in more detail. As already noted, this has recently been calculated to next-to-next-to-leading order (i.e. $O(\alpha_s^2)$) by Hamberg et al. [41]. Fig. 10 shows the cross section for $p\bar{p} \to W^{\pm}X$ as a function of the collider energy $\sqrt{s}$. The curves are calculated using the results of reference [41] (in the $\overline{\rm MS}$ scheme) with MRS(S0) parton distributions [16] and the renormalization and factorization scales $M = \mu = M_W$. The data points are from the UA2 [42] and CDF [43] collaborations, at $\sqrt{s}$ = 630 GeV and 1.8 TeV, respectively. Notice how both the NLO and NNLO corrections increase the lowest order cross section, the net effect being of order +30%. The NNLO correction is significantly smaller than the NLO correction, due to a partial cancellation between the positive second order corrections involving the $q\bar{q}$ initial state and the negative corrections from the qg initial state. Perhaps the most important point to note from Fig. 10 is that, aside from unknown (and presumably small) $O(\alpha_s^3)$ corrections, there is virtually no theoretical uncertainty associated with the predictions – the parton distributions are being probed in a range of $x \sim M_W/\sqrt{s}$ where they are constrained from deep inelastic scattering [16] and the scale dependence is weak [41]. The agreement with experiment, therefore, provides a powerful test of the whole theoretical edifice which goes into the calculation.

I am grateful to the organizers for all their help before, during and after the School. I would also like to thank my collaborators Alan Martin and Dick Roberts for their help with Section 4.

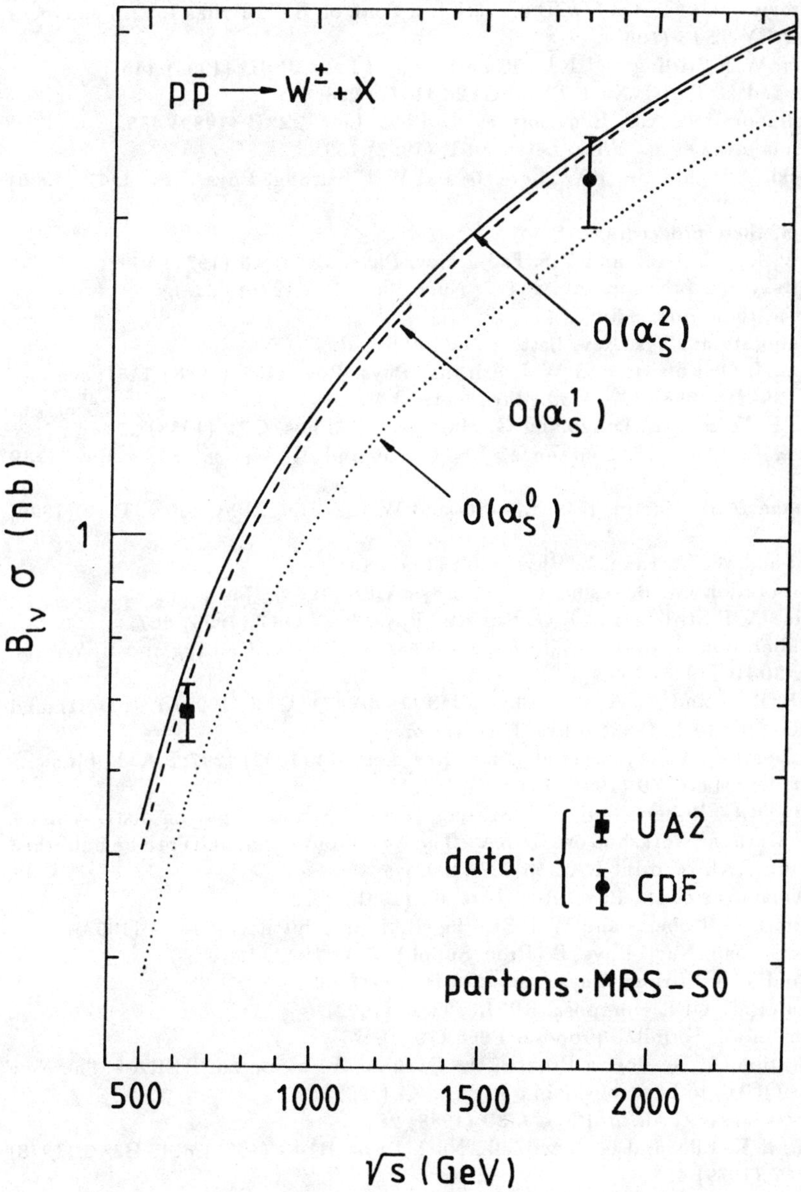

Fig. 10. Theoretical predictions for the $W^{\pm}$ cross section in $p\bar{p}$ collisions, as a function of $\sqrt{s}$, with data from UA2 [42] and CDF [43].

References

1. *The Structure of the Proton*, R.G. Roberts, Cambridge University Press (1990).
2. *Review of Particle Properties*, Phys. Lett. **239B** (1990) 1.
3. W.J. Marciano, Phys. Rev. **D29** (1984) 580.
4. S. Bethke, review talk 'Tests of QCD' at 26th Int. Conf. on HEP, Dallas 1992, Heidelberg preprint HD-PY-92-13 (1992).
5. A.D. Martin, W.J. Stirling and R.G. Roberts, Phys. Lett. **306B** (1993) 145.
6. G. Altarelli and G. Parisi, Nucl. Phys. **B126** (1977) 298.
7. BCDMS collaboration: A.C. Benvenuti et al., Phys. Lett. **223B** (1989) 485.
8. NMC: P. Amaudruz et al., Phys. Lett. **295B** (1992) 159.
9. J. Kwiecinski, A.D. Martin, R.G. Roberts and W.J. Stirling, Phys. Rev. **D42** (1990) 3645.
10. A. Milsztajn, these proceedings.
11. E.A. Kuraev, L.N. Lipatov and V.S. Fadin, Sov. Phys. JETP **45** (1977) 199.
    Ya.Ya. Balitsky and L.N. Lipatov, Sov. J. Nucl. Phys. **28** (1978) 822.
12. J. Kwiecinski, these proceedings.
13. CCFR collaboration: Phys. Rev. Lett., to be published.
14. A.D. Martin, R.G. Roberts and W.J. Stirling, Phys. Rev. **D37** (1988) 1161; see also Phys. Lett. **206B** (1988) 327; Mod. Phys. Lett. **A4** (1989) 1135.
    M. Diemoz, F. Ferroni, E. Longo and G. Martinelli, Z. Phys. **C39** (1988) 21.
    P. Aurenche, R. Baier, M. Fontannaz, J.F. Owens and M. Werlen, Phys. Rev. **D39** (1989) 3275.
    P.N. Harriman, A.D. Martin, R.G. Roberts and W.J. Stirling, Phys. Rev. **D42** (1990) 798.
    J.G. Morfin and W.-K. Tung, Z. Phys. **C52** (1991) 13.
15. WA70 collaboration: M. Bonesini et al., Z. Phys. **C38** (1988) 371.
16. A.D. Martin, W.J. Stirling and R.G. Roberts, Phys. Rev. **D47** (1993) 867.
17. CTEQ collaboration: J. Botts, J. Morfin, J. Owens, J. Qiu, W.-K. Tung and H. Weerts, Phys. Lett. **304B** (1993) 159.
18. M. Glück, E. Reya and A. Vogt, Z. Phys. **C48** (1990) 471, **C53** (1992) 127; Dortmund preprint DO-TH-93-02 (1993), Phys. Lett. (in press).
19. CCFR collaboration: C. Foudas et al., Phys. Rev. Lett. **64** (1990) 1207; S.A. Rabinowitz et al., Phys. Rev. Lett. **70** (1993) 134.
20. A.D. Martin, R.G. Roberts and W.J. Stirling, 'Present status of parton distributions', Proc. of the Durham Workshop on 'HERA – The New Frontier for QCD', to be published in J. Phys. **G**, RAL preprint RAL-93-027 (1993).
21. NMC: P. Amaudruz et al., Phys. Rev. Lett. **66** (1991) 2712.
22. A.D. Martin, R.G. Roberts and W.J. Stirling, RAL preprint RAL-93-014 (1993).
23. NMC: E.M. Kabuß, Nucl. Phys. B (Proc. Suppl.) **29A** (1992) 1.
24. S.D. Ellis and W.J. Stirling, Phys. Lett. **256B** (1991) 258.
25. M.C. Abreu et al., CERN proposal SPSLC 92-15 (1992).
    G.T. Garvey et al., Fermilab proposal P866 (1992).
26. H1 collaboration: A. De Roeck, Proc. of the Durham Workshop on 'HERA – The New Frontier for QCD', to be published in J. Phys. **G** (1993).
27. A.M. Cooper-Sarkar et al., Z. Phys. **C39** (1988) 281.
28. G. Altarelli, R.K. Ellis and G. Martinelli, Nucl. Phys. **B143** (1978) 521; **B146** (1978) 544(e); **B147** (1979) 461.
    J. Kubar-Andre and F.E. Paige, Phys. Rev. **D19** (1979) 221;
    J. Kubar-Andre, M. Le Bellac, J.L. Meunier and G. Plaut, Nucl. Phys. **B175** (1980) 251.

29. R. Hamberg, T. Matsuura and W.L. van Neerven, Nucl. Phys. **359** (1991) 343.
30. P. Nason, S. Dawson and R.K. Ellis, Nucl. Phys. **B303** (1988) 607.
W. Beenakker et al., Phys. Rev. **D40** (1989) 54.
31. A. Djouadi, M. Spira and P.M. Zerwas, Phys. Lett. **264B** (1991) 440.
32. S.D. Ellis, Z. Kunszt and D.E. Soper, Phys. Rev. Lett. **64** (1990) 2121; University of Washington preprint UW-PT-91-13 (1991).
33. S.D. Drell and T.M. Yan, Ann. Phys. (N.Y.) **66** (1971) 578.
34. See for example: J.C. Collins and D.E. Soper, Ann. Rev. Nucl. Part. Sci. **37** (1987) 383, and references therein.
35. G. Altarelli, R.K. Ellis and G. Martinelli, Nucl. Phys. **B143** (1978) 521; **B146** (1978) 544(e); **B147** (1979) 461.
J. Kubar-Andre and F.E. Paige, Phys. Rev. **D19** (1979) 221.
J. Kubar-Andre, M. LeBellac, J.L. Meunier and G. Plaut, Nucl. Phys. **B175** (1980) 251;
36. P. Aurenche, R. Baier, M. Fontannaz and D. Schiff, Nucl. Phys. **B286** (1987) 509; **B286** (1987) 553; **B297** (1988) 661.
37. P. Nason, S. Dawson and R.K. Ellis, Nucl. Phys. **B312** (1989) 607.
W. Beenakker, H. Kuijf, W.L. van Neerven and J. Smith, Phys. Rev. **D40** (1989) 54.
38. J. Smith, D. Thomas and W.L. van Neerven, Z. Phys. **C44** (1989) 267.
B. Mele, P. Nason and G. Ridolfi, Nucl. Phys. **B357** (1991) 409.
J. Ohnemus and J.F. Owens, Phys. Rev. **D43** (1991) 3626.
B. Bailey, J. Ohnemus and J.F. Owens, Phys. Rev. **D46** (1992) 2018.
39. D. Graudenz, M. Spira and P.M. Zerwas, preprint DESY-92-149 (1992).
A. Djouadi, M. Spira and P.M. Zerwas, Phys. Lett. **264B** (1991) 440.
40. S.D. Ellis, Z. Kunszt and D.E. Soper, Phys. Rev. Lett. **64** (1990) 2121; Proc. Lepton-Photon Symp. and Int. Europhysics Conf. on High Energy Physics, Geneva, Switzerland (1991), eds. S. Hegarty, K. Potter and E. Quercigh, World Scientific (1992), Vol.1, p.417.
F. Aversa, M. Greco, P. Chiapetta and J.Ph. Guillet, Phys. Rev. Lett. **65** (1990) 401.
41. R. Hamberg, T. Matsuura and W.L. van Neerven, Nucl. Phys. **B345** (1990) 331; Nucl. Phys. **B359** (1991) 343.
W.L. van Neerven and E.B. Zijlstra, Nucl. Phys. **B382** (1992) 11.
42. UA2 collaboration: J. Alitti et al., Phys. Lett. **276B** (1992) 365.
43. CDF collaboration: F. Abe et al., Phys. Rev. **D44** (1991) 29.

Electroweak Reactions in the Non-Perturbative Regime of QCD

Ulf-G. Meißner

Institute for Theoretical Physics, University of Berne, Sidlerstraße 5,
CH-3012 Berne, Switzerland

1 Introduction

These lectures are concerned with the structure of hadrons at low energies, where the strong coupling constant is large. Most of the hadronic world discussed here will be made up of the light u, d and s quarks since these are the constituents of the low-lying hadrons. The best way to gain information about the strongly interacting particles is thus the use of well-understood probes, such as the photon or the massive weak gauge bosons. At very low energies, the dynamics of the strong interactions is governed by constraints from chiral symmetry. This leads to the use of effective field theory methods which in the present context is called baryon chiral perturbation theory. In the first part of these lectures, I will briefly outline the basic framework of this effective field theory and use photo-nucleon processes to discuss the strengths and limitations of it. The basic degrees of freedom are the pseudoscalar Goldstone bosons chirally coupled to the matter fields like e.g. the nucleons. The very low-energy face of the low-lying baryons is therefore of hadronic nature, essentially point-like Dirac particles surrounded by a cloud of Goldstone bosons. The information about the underlying quark structure is only indirect since one translates e.g. quark masses into pseudoscalar meson masses. Nevertheless, certain aspects of the quark structure of the nucleon (or other baryons) can be extracted by making clever use of the well-known couplings of the quarks to the electroweak gauge bosons. For that, it is important to understand the notion of hadron form factors, electromagnetic or weak ones. These encode certain aspects of the quark structure of the nucleon. In addition, parity-violating electron scattering off nucleons or nuclei or neutrino-nucleon scattering allows one to measure the weak form factors related to the neutral currents which complement the information gained from the electromagnetic and weak charged current form factors. This will be the theme of the second part of these lectures. However, since one is dealing with small effects like e.g. left-right asymmetries in polarized electron scattering off protons at low energies, one also has to worry about the effects of the heavy quarks c, b and t and about radiative corrections, which are nominally suppressed by powers of $(\alpha/4\pi)$ but sometimes enhanced by large logarithms or by some judiciously large factors. I will give a short state of the art summary on these topics at the end of these lectures.

Naturally, I can only cover a small fraction of the many interesting phenomena related to low energy hadron physics. I have chosen to mostly talk about the nucleon since after all it makes up large chunks of the stable matter surrounding us and also is a good intermediary between the nuclear and the high energy physicists present at this school. Most of the methods presented here can easily be applied to other problems, and as it will become obvious at many places, we still have a long way to go to understand all the intriguing features of the nucleon in a systematic and controlled fashion. Whenever possible, I will avoid to talk about models, with the exception of some circumstances where they offer some qualitative insight into certain aspects of hadron structure.

2 Chiral Perturbation Theory with Nucleons

2.1 Chiral Symmetry in QCD

Consider the QCD Lagrangian for the three light flavors u, d and s,

$$\mathcal{L}_{QCD} = \bar{q}(i\gamma_\mu D^\mu - \mathcal{M})q + \mathcal{L}_{\text{gluon}} \tag{2.1}$$

with $q = $ (u, d, s), and I have suppressed all color indices. The quark fields can be chosen such that the quark mass matrix is diagonal,

$$\mathcal{M} = \begin{pmatrix} m_u & & \\ & m_d & \\ & & m_s \end{pmatrix} . \tag{2.2}$$

The entries in (2.2) are the current quark masses. At the typical hadronic scale of $\Lambda \simeq 1$ GeV, these can be considered small. This holds certainly for the u and d quarks, the s quark is somewhat heavier ($\simeq 175$ MeV) which makes it more difficult to deal with. To a good first approximation, it is, however, legitimate to set the current quark masses to zero. In that case, the QCD Lagrangian exhibits a flavor $SU(3)_L \times SU(3)_R \times U(1)_{L+R} \times U(1)_{L-R}$ symmetry. The vectorial $U(1)$ is related to the baryon number and the axial $U(1)$ is afflicted by the anomaly which gives the η' a mass of 1 GeV. The chiral group has therefore sixteen independent generators. This leads to sixteen conserved left- and right-handed currents,

$$J^a_{\mu,A} = \bar{q}_A \gamma_\mu \frac{\lambda^a}{2} q_A \quad (a = 1, \ldots, 8\,; A = L, R) \tag{2.3}$$

or, equivalently, eight conserved vector and eight conserved axial-vector charges. These form an $SU(3)$ algebra. However, this symmetry of the Lagrangian is not present in the ground state or the spectrum of the strongly interacting particles. Something very similar to the ferromagnet below the Curie temperature happens. The magnet Hamiltonian is rotationally invariant but below T_c spontaneous magnetization sets in and all spins align in one direction breaking the symmetry of the Hamiltonian. Similarly in QCD only the vector charges are conserved allowing us to order the hadrons in multiplets of certain isospin and hypercharge quantum numbers. Nevertheless, the information about the axial charges is not lost. Goldstone's theorem [1] tells us that for any broken generator there exists a massless boson, in this case of pseudoscalar nature. This means that we should expect eight massless pseudoscalar mesons in the particle data listings. This is not the case. The fact of the matter is that the Goldstone bosons pick up a small mass related to the small current quark masses, like e.g.

$$M_\pi^2 = (m_u + m_d)B + O(\mathcal{M}^2) \tag{2.4}$$

where $B = - < 0|\bar{q}q|0 > /F_\pi^2$ is the order parameter of the spontaneous symmetry breaking and F_π measures the strength of the non-vanishing transition amplitude

$$< 0|A_\mu^i|\pi^j >= i\delta^{ij} p_\mu F_\pi \ . \tag{2.5}$$

In what follows, I will assume that $B \sim 1$ GeV, i.e. that the quark mass expansion of the pseudoscalar meson masses is indeed governed by the leading term linear in the quark masses. It is interesting to observe that (2.4) gives us a recipe how to translate quark into meson masses. Evidently, at low energies the almost massless Goldstone bosons are the dominant degrees of freedom. This has been the theme of Heiri Leutwyler's lectures [2] at the Schladming school two years ago and I refer the reader for a more detailed discussion of these underlying ideas. Also a recent review [3] is available which can be used as a first reading.

2.2 Chiral Perturbation Theory (Mesons)

The purest reaction of low-energy QCD is the elastic scattering of pions from pions in the threshold region. I will use this as an example to discuss the principles underlying the so-called chiral perturbation theory (CHPT). It is convenient to collect the Goldstone bosons in a matrix-valued field $U(x) = \exp(i\phi(x)/F_\pi)$, in the standard notation. To lowest order in the momenta and quark masses, the interactions of the Goldstone bosons e.g. can be described by the gauged non-linear σ-model,

$$\mathcal{L}^{(2)} = \frac{F^2}{4}\mathrm{Tr}[\nabla_\mu U^\dagger \nabla^\mu U] + \frac{F^2 B}{4}\mathrm{Tr}[\mathcal{M}(U + U^\dagger)] \ , \tag{2.6}$$

with ∇_μ a covariant derivative including the coupling to the external electroweak fields. The second term in (2.6) reflects the explicit symmetry breaking from the quark masses. Both terms are of chiral dimension two since they contain either two derivatives or one quark mass (which is equivalent to two derivatives, see (2.4)). To lowest order, the strong interactions are therefore given in terms of two parameters. These are F, the pion decay constant in the chiral limit, and B related to the quark condensate (remember that I assume B to be substantially larger than F). Calculating tree diagrams from this effective Lagrangian leads to the celebrated current algebra results. In particular, one recovers Weinberg's prediction for the $\pi\pi$ scattering amplitude, $A(s,t,u) = (s - M_\pi^2)/F_\pi^2$ [4]. Clearly, this smells like an expansion in energy since close to threshold, the c.m. energy $\sqrt{s}$ becomes very small and, consequently, the interaction between the Goldstone bosons weak. This is a particular imprint of the spontaneously broken chiral symmetry which allows for a "perturbative" expansion in the non-perturbative regime of QCD. Obviously, the effective Lagrangian (2.6) cannot be the whole story. First, one would like to know how big the corrections from higher order terms (with more derivatives and/or quark mass insertions) are. Second, and more important, tree diagrams are always real. Twenty years ago at this school,

Lehmann [5] demonstrated that unitarity forces one to consider loop diagrams. Based on the counting rules developed by Weinberg [6], which state that n-loop diagrams are suppressed by powers of q^{2n} (with q a genuine small momentum), Gasser and Leutwyler [7] have systematized the whole approach for QCD. Let us consider the corrections which appear at next-to-leading order. First, there are the one-loop diagrams mandated by unitarity. Second, there are higher order contact terms with more derivatives or quark mass insertions. These are accompanied by a-priori unknown coupling constants, the so-called low-energy constants. Their values can be fixed from phenomenology [7] or understood to a high degree of accuracy from the QCD version of the vector-meson dominance principle [8]. It is at this next-to-leading order where the various equivalent lowest order approaches differ. For example, the linear σ-model is a perfect candidate to be used instead of (2.6), but at next-to-leading order it simply gives low-energy constants at variance with the data [7]. Let me return to the example of $\pi\pi$ scattering. The amplitude expanded to one loop accuracy takes the form

$$T_{\pi\pi}(s,t,u) = A(s,t,u) + B(s,t,u) + C(s,t,u) \ , \tag{2.7}$$

where $B(s,t,u)$ includes the loop (unitarity) effects and $C(s,t,u)$ the contribution from the higher dimensional operators. That this gives indeed a better description of the $\pi\pi$ phase shifts than the tree level prediction $A(s,t,u)$ can be seen in ref. [9]. In what follows, I will always confine myself to the one loop approximation. For a discussion about its range of applicability, see e.g. ref. [10].

2.3 Inclusion of Matter Fields

In this section I will be concerned with adding the low-lying spin-1/2 baryons (N, Σ, Λ, Ξ) to the effective field theory. The inclusion of such matter fields is less straightforward since these particles are not related to the symmetry violation. However, their interactions with the Goldstone bosons is dictated by chiral symmetry. Let us denote by B the conventional 3×3 matrix representing the spin-1/2 octet fields,

$$B = \begin{pmatrix} \frac{1}{\sqrt{2}}\Sigma^0 + \frac{1}{\sqrt{6}}\Lambda & \Sigma^+ & p \\ \Sigma^- & -\frac{1}{\sqrt{2}}\Sigma^0 + \frac{1}{\sqrt{6}}\Lambda & n \\ \Xi^- & \Xi^0 & -\frac{2}{\sqrt{6}}\Lambda \end{pmatrix} \tag{2.8}$$

It is most convenient to choose a non-linear realization of the chiral symmetry [11] so that B transforms as

$$B \to K B K^\dagger \tag{2.9}$$

where K is a complicated function that does not only depend on the group elements $g_{L,R}$ of the $SU(3)_{L,R}$ but also on the Goldstone boson fields collected in $U(x)$, i.e. $K(x) = K(g_L, g_R, U(x))$ defines a local transformation. Expanding K in powers of the Goldstone boson fields, one realizes that a chiral transformation is linked to absorption or emission of pions, kaons and etas (which was the theme in the days of "current algebra" techniques). These topics are discussed in more detail in [3]. The lowest order meson-baryon Lagrangian is readily constructed. Let us restrict the discussion to processes with one incoming and one outgoing baryon, such as πN scattering, threshold pion photoproduction or baryon Compton scattering (otherwise, we would have to add contact n-fermion terms with $n \geq 4$). In that case, it takes the form

$$\mathcal{L}_{\rm MB}^{(1)} = {\rm Tr}\left\{ i\bar{B}\gamma^\mu D_\mu B - m_0 \bar{B}B + \frac{1}{2}D\bar{B}\gamma^\mu\gamma_5\{u_\mu, B\} + \frac{1}{2}F\bar{B}\gamma^\mu\gamma_5[u_\mu, B]\right\} \quad (2.10)$$

with m_0 the average octet mass (in the chiral limit) and D_μ the usual covariant derivative. There are two axial-vector type couplings multiplied by the conventional F and D constants. In the case of two flavors, there is only one such term proportional to $g_A = F + D$. Notice that the lowest order effective Lagrangian contains one derivative and therefore is of dimension one as indicated by the superscript '(1)'. In contrast to the meson sector (2.6), odd powers of the small momentum q are allowed (thus, to leading order, no quark mass insertion appears since $\mathcal{M} \sim q^2$). It is instructive to expand (2.10) in powers of the Goldstone and external fields. From the vectorial term, one gets the minimal photon-baryon coupling, the two-Goldstone seagull (Weinberg term) and many others. Expansion of the axial-vectors leads to the pseudovector meson-baryon coupling, the celebrated Kroll-Rudermann term and much more. However, as already stated, it is not sufficient to calculate tree diagrams from the lowest order effective theory. In the presence of baryons, the loop expansion is more complicated than discussed in the previous section. First, since odd powers in q are allowed, a one-loop calculation of order q^3 involves contact terms of dimension two and three, i.e. combinations of zero or one quark mass insertions with zero to three derivatives. These terms are collected in $\mathcal{L}_{\rm MB}^{(2,3)}$ and a complete list of them can be found in Krause's paper [12]. Second, the finiteness of the baryon mass in the chiral limit and that its value is comparable to the chiral symmetry breaking scale $\Lambda \sim M_\rho$ complicates the low energy structure. This has been discussed in detail by Gasser et al. [13]. Let me just give one illustrative example. The one loop contribution to the nucleon mass not only gives the celebrated non-analytic contribution proportional to $M_\pi^3 \sim \mathcal{M}^{3/2}$ but also an infinite shift of m_0 which has to be compensated by a counterterm of dimension zero. It is a general feature that loops produce analytic contributions at orders below what one would naively expect (e.g. below q^3 from one loop diagrams). Therefore, in a CHPT calculation involving baryons one has to worry more about higher order contributions than it is the case in the meson sector. There is one way of curing this problem, namely to go into the extreme non-relativistic limit [14] and consider the baryons as very heavy (static) sources. Then, by a clever definition of velocity-dependent fields, one can eliminate the baryon mass term from the lowest order effective Lagrangian and expand all interaction vertices and baryon propagators in increasing powers of $1/m_0$. This is similar to a Foldy-Wouthuysen transformation you all know from QED. In this limit one recovers a consistent derivative expansion. For example, the one loop contribution of the Goldstone bosons to the baryon self-energy is nothing but the non-analytic M_ϕ^3 terms together with three contact terms from $\mathcal{L}_{\rm MB}^{(2)}$. However, one has to be somewhat careful still. The essence of the heavy mass formalism is that one works with old-fashioned time-ordered perturbation theory. So one has to watch out for the appearance of possible small energy denominators (infrared singularities). This problem has been addressed by Weinberg [15] in his discussion about the nature of the nuclear forces. The dangerous diagrams are the ones where cutting one pion line (this only concerns pions which are not in the asymptotic in- or out-states) separates the diagram into two disconnected pieces (one therefore speaks of reducible diagrams). These diagrams should be inserted in a Schrödinger equation or a relativistic generalization thereof with the irreducible ones entering as a potential. So the full CHPT machinery is applied to the irreducible diagrams. This should be kept in mind. For the purposes I am discussing, we do not need to worry about these complications. Being aware of them, it is then straightforward to apply baryon CHPT to many nuclear and particle physics problems [3,16,17]. I will illustrate this on two particular examples in the next sections. Before doing that, however, I would like to stress that all these calculations are only in their infancy. It is believed that for a good quantitative description one has to perform systematic calculations to order q^4, i.e. beyond next-to-leading order. There are many indications of this coming from the baryon masses, hyperon non-leptonic decays and so on. At present, only some rather unsystematic attempts have been done which incorporate some q^4 effects. In particular, it has been argued that one should include the low-lying spin-3/2 decuplet in the effective theory from the start [16,18]. A critical discussion of this approach can be found in ref. [19]. Clearly, more work is needed to clarify these issues. Let me now proceed and discuss two applications of the framework outlined here. These are calculations in flavor $SU(2)$ were the loop corrections are considerably smaller. As I will show, they shed some light on the non-perturbative structure of the nucleon.

2.4 Threshold Pion Photo- and Electroproduction

Let us consider the reaction $\gamma(k) + N_1(p_1) \to \pi^a(q) + N_2(p_2)$ with $N_{1,2}$ denoting protons and/or neutrons and 'a' refers to the charge of the produced pion. In the case of real photons ($k^2 = 0$) one talks of photoproduction whereas for virtual photons (radiated off an electron beam) the process is called electroproduction. Of particular interest is the threshold region where the photon has just enough energy to produce the pion at rest or with a very small three-momentum. In this kinematical regime, it is advantageous to perform a multipole decomposition since at threshold only the S-waves survive. These multipoles are labelled $E\,(M)_{l\pm}$, with $E\,(M)$ for electric (magnetic), $l = 0, 1, 2, \ldots$ the pion orbital angular momentum and the $\pm$ refers to the total angular momentum of the pion-nucleon system, $j = l \pm 1/2$. They parametrize the structure of the nucleon as probed with low energy photons. Let us first concentrate on the photoproduction case. At threshold, the differential cross section takes the form

$$\frac{|\mathbf{k}|}{|\mathbf{q}|}\frac{d\sigma}{d\Omega}\bigg|_{\mathbf{q}\to 0} = (E_{0+})^2 \tag{2.11}$$

i.e. it is entirely given in terms of the electric dipole amplitude E_{0+} (which at threshold is real). The quark mass expansion of E_{0+} is given by

$$E_{0+} = \frac{-ie}{8\pi m}\frac{\mu(2+\mu)}{(1+\mu)^{3/2}} f(\bar{m}/\Lambda_{\mathrm{QCD}}) \tag{2.12}$$

with $\mu = M_\pi/m \simeq 1/7$ and $\bar{m}$ the renormalization group invariant quark masses. Modulo logarithms, one can expand f in powers of μ,

$$f = \frac{1}{\mu} f_{-1} + f_0 + \mu f_1 + \mathcal{O}(\mu^2) \quad . \tag{2.13}$$

For charged pions f_{-1} is non-vanishing and one is lead to the famous Kroll-Rudermann result [20] which states that E_{0+} for $\gamma p \to \pi^+ n$ and $\gamma n \to \pi^- p$ can be expressed in terms of the strong pion-nucleon coupling constant, the nucleon mass and some kinematical factors. Consequently, in these cases E_{0+} is finite in the chiral limit. Matters are different for the case of neutral pions. Here, f_{-1} vanishes and thus the leading term in the expansion of E_{0+} goes like μ (μ^2) for $\gamma p \to \pi^0 p$ ($\gamma n \to \pi^0 n$). The amplitude is therefore sensitive to the chiral symmetry breaking which makes it a good object to study experimentally and theoretically. If one performs a one loop CHPT calculation, one can get unambiguously the next-to-leading order terms in the quark mass expansion of the electric dipole amplitude. For the production of the proton, the threshold value is [21]

$$E_{0+} = -\frac{eg_{\pi N}}{8\pi m}\mu\left\{1 - \left[\frac{1}{2}(3+\kappa_p) + \left(\frac{m}{4F_\pi}\right)^2\right]\mu + \mathcal{O}(\mu^2)\right\} \quad , \tag{2.14}$$

with $\kappa_p = 3.71$ the anomalous magnetic moment of the proton. The second term in the square brackets is the new one found in the CHPT calculation. It stems from the so-called triangle diagram and its crossed partner. If one expresses E_{0+} in terms of the conventional Lorentz invariant functions $A_{1,2,3,4}$, one finds that these diagrams give a contribution $\delta A_1 = (eg_{\pi N}/32F_\pi^2)\,\mu$, i.e. they are non-analytic in the quark masses since $\mu \sim \sqrt{\hat{m}}$. Remember also that the electric dipole amplitude at threshold is proportional to μA_1 and therefore one gets this novel contribution at next-to-leading order in the quark mass expansion. Clearly, the expansion in μ is slowly converging, the coefficient of the term of order μ^2 is so large that it compensates the leading term proportional to μ. Therefore, for a meaningful prediction one has to go further in the expansion. This has been done in the relativistic formalism which sums up some of the higher order corrections [22]. One finds (at threshold) $E_{0+} = -1.33 \pm 0.09$ using the conventional units of $10^{-3}/M_{\pi^+}$. The error reflects solely the uncertainty in estimating the finite contact term contributing at order q^3. This is somewhat below the generally accepted experimental value of -2.0 ± 0.2 (see e.g. Bernstein and Holstein [23] or Drechsel and Tiator [24] and references therein). However, one should point out that the calculation was performed in the isospin limit with $M_{\pi^\pm} = M_{\pi^0}$ and $m_p = m_n$. It is a very tough problem to include isospin-breaking in a systematic fashion - I invite you to solve it. To get an idea about the effects related to it, one can perform a toy calculation and set the pion and nucleon masses by hand on their physical values in the various diagrams. Then, one finds $E_{0+} = -1.97$, which is an encouraging number but it has to be confirmed by a better calculation. The chief reason for the difference to the isospin-symmetric case is that the contribution of the triangle diagram now vanishes below the $\pi^+ n$ threshold. It is furthermore interesting to observe that the one-loop corrections to the electric dipole amplitudes of charged pion photoproduction move the prediction closer to the data. These chiral corrections are suppressed by powers of μ^2 and $\mu^2 \ln \mu$ with respect to the leading Kroll-Ruderman term ($\sim \mu^0$) and therefore a more accurate experimental determination of these quantities is called for. Let me now consider the electroproduction process. First, the low-energy theorems for the two S-wave multipoles E_{0+} and L_{0+} (for virtual photons, there is also a longitudinal coupling to the nucleon) have been discussed in the light of CHPT in ref. [25]. Let me stress that the LETs derived in refs. [21,25] are the ones implied by QCD and that they can be tested experimentally. Notice that there still seems to be some confusion concerning the meaning and the interpretation of these theorems [26]. However, I will not dwell on this topic here. Instead, I want to direct your attention to some new data. Welch et al. [27] have published the S-wave cross section for the reaction $\gamma^\star p \to \pi^0 p$ very close to the photon point. This measurement is a quantum step compared to previous determinations which mostly date back to the seventies when pion electroproduction was still a hot topic in particle physics. In this experiment, k^2 varied between -0.04 and -0.10 GeV^{-2} and the S-wave cross section could be extracted with an unprecedented accuracy (see fig. 2 in [27]). This is also the kinematical regime where a CHPT calculation might offer some insight. Indeed, in ref. [28] it was shown that the k^2-dependence of this cross section seems to indicate the necessity of loop effects. With conventional models including e.g. form factors and the anomalous magnetic moment coupling the trend of the data cannot be described. However, the corrections from the one loop diagrams to the tree level prediction are substantial. This gives further credit to the previously made statement that a calculation beyond next-to-leading order should be performed. The last topic I want to address in this section concerns the determination of the nucleon axial radius from charged pion electroproduction. Let me briefly explain how the axial form factor comes into play. The basic matrix element to be considered is the time-ordered product of the electromagnetic (vector) current with the interpolating pion field sandwiched between nucleon states. Now one can use the PCAC relation and express the pion field in terms of the divergence of the axial current. Thus, a commutator of the form [V,A] arises. Current algebra tells us that this gives an axial current between the incoming and outgoing nucleon fields and, alas, the axial form factor. The isospin factors combine in a way that they form a totally antisymmetric combination which cannot be probed in neutral pion production. These ideas were formalized in the venerable low-energy theorem (LET) due to Nambu, Lurié and Shrauner [29] for the isospin–odd electric dipole amplitude $E_{0+}^{(-)}$ in the chiral limit,

$$E_{0+}^{(-)}(M_\pi = 0, k^2) = \frac{e g_A}{8\pi F_\pi}\left\{1 + \frac{k^2}{6} r_A^2 + \frac{k^2}{4m^2}(\kappa_V + \frac{1}{2}) + \mathcal{O}(k^3)\right\}. \tag{2.15}$$

Therefore, measuring the reactions $\gamma^\star p \to \pi^+ n$ and $\gamma^\star n \to \pi^- p$ allows to extract $E_{0+}^{(-)}$ and one can determine the axial radius of the nucleon, r_A. This quantity measures the distribution of spin and isospin in the nucleon, i.e. probes the Gamov–Teller operator $\sigma \cdot \tau$. A priori, the axial radius is expected to be different from the typical electromagnetic size, $r_{\rm em} \simeq 0.85$ fm. It is customary to parametrize the axial form factor $G_A(k^2)$ by a dipole form, $G_A(k^2) = (1 - k^2/M_A^2)^{-2}$ which leads to the relation $r_A = \sqrt{12}/M_A$. The axial radius determined from electroproduction data is typically $r_A = 0.59 \pm 0.04$ fm whereas (anti)neutrino-nucleon reactions lead to somewhat larger values, $r_A = 0.65 \pm 0.03$ fm. This discrepancy is usually not taken seriously since the values overlap within the error bars. However, it was shown in ref. [30] that pion loops modify the LET (2.15) at order k^2 for finite pion mass. In the heavy mass formalism, the coefficient of the k^2 term reads

$$\frac{1}{6} r_A^2 + \frac{1}{4m^2}(\kappa_V + \frac{1}{2}) + \frac{1}{128 F_\pi^2}(1 - \frac{12}{\pi^2}) \tag{2.16}$$

where the last term in (2.16) is the new one. This means that previously one had extracted a modified radius, the correction being $3(1 - 12/\pi^2)/64 F_\pi^2 \simeq -0.046$ fm^2.

This closes the gap between the values of r_A extracted from electroproduction and neutrino data. It remains to be seen how the $1/m$ suppressed terms will modify the result (2.16). Such investigations are underway.

2.5 Nucleon Compton Scattering

Consider low-energy (real) photons scattering off a spin-1/2 particle. In forward direction, the scattering amplitude takes the form

$$T_{1/2}(\omega) = f_1(\omega^2)\, \epsilon_f^* \cdot \epsilon_i + i\omega\, f_2(\omega^2)\, \sigma \cdot (\epsilon_f^* \times \epsilon_i) \tag{2.17}$$

where ω is the photon frequency and the $\epsilon_{i,f}$ are the polarization vectors in the initial and final state, respectively. The energy expansion of the spin-dependent amplitude $f_1(\omega^2)$ reads

$$f_1(\omega^2) = -\frac{e^2 Z^2}{4\pi m} + (\alpha+\beta)\,\omega^2 + \mathcal{O}(\omega^4) \tag{2.18}$$

where the first energy-independent term is nothing but the Thomson amplitude mandated by gauge invariance. Therefore, to leading order, the photon only probes some global properties like the mass or electric charge of the spin-1/2 target. At next-to-leading order, the non-perturbative structure is parametrized by two constants, the so-called electric and magnetic polarizabilities (more correctly, one might want to talk about "magnetic susceptibility", however, I will use the common language). A detailed discussion of the CHPT calculation of these fundamental two-photon observables is given in ref. [3] (see also the references quoted there). Here, I just want to add a few educational remarks. For the proton, the Thomson term is non-vanishing and it is interesting to compare its magnitude to the one from the polarizability contribution using $(\alpha+\beta)_p = 14.2 \cdot 10^{-4}$ fm^3. The recent Illinois Compton scattering experiment [31] to determine α_p and β_p was performed at photon energies ranging from 32 to 72 MeV. While at the lowest ω the polarizability term amounts to a 31 per cent correction to the Thomson term contribution ($= -1.22 \cdot 10^{-4}$ fm), at 72 MeV the second term in the energy expansion is 1.5 times as large as the leading term. Clearly, higher order (ω^4) contributions have to be taken into account at such energies. Another remark concerns the behaviour of $\alpha_{p,n}$ and $\beta_{p,n}$ in the chiral limit. These quantities diverge as $1/M_\pi$ as the pion mass tends to zero. This is expected since the two photons probe the long-ranged pion cloud, i.e. there is no more Yukawa suppression as in the case for a finite pion mass. Now a well-known dispersion sum rule relates $(\alpha+\beta)$ to the total nucleon photoabsorption cross section. The latter is, of course, also well-behaved in the chiral limit which at first sight seems to be at variance with the behaviour of the expansion of the scattering amplitude. But be aware that the general form of (2.18) has been derived under the assumption that there is a well defined low-energy limit. Similar observations can also be made concerning the chiral expansion of the $\pi\pi$ scattering amplitude discussed in subsection 2. Let me now focus on the spin-dependent amplitude $f_2(\omega^2)$ which has an expansion analogous to (2.18),

$$f_2(\omega^2) = f_2(0) + \gamma\,\omega^2 + \mathcal{O}(\omega^4). \tag{2.19}$$

By the optical theorem, the imaginary part of f_2 is related to the photoabsorption cross sections for circularly polarized photons on polarized nucleons,

$$\mathrm{Im}\, f_2(\omega^2) = -\frac{1}{8\pi}\Big[\sigma_{3/2}(\omega) - \sigma_{1/2}(\omega)\Big] \tag{2.20}$$

where the indices 3/2 and 1/2 refer to the total γN helicities. We will come back to this relation later on. Let me now turn to the discussion of the Taylor coefficients $f_2(0)$ and γ in eq. (2.19). There exists a celebrated LET for $f_2(0)$ due to Low, Gell-Mann and Goldberger [32]. Using very general principles like gauge invariance, Lorentz invariance and crossing symmetry, they showed that

$$f_2(0) = -\frac{e^2\kappa^2}{8\pi m^2} \,. \tag{2.21}$$

Here, κ denotes the anomalous magnetic moment of the particle the photon scatters off. In CHPT, one can easily derive the LET making use of the heavy baryon formalism as shown in ref. [33]. It stems from some contact terms which are nominally suppressed by powers of 1/m. To one-loop order, one can give $f_2(\omega^2)$ in closed form since only a few loop diagrams contribute (in addition to the contact terms giving the LET),

$$f_2(\omega^2) = -\frac{e^2\kappa^2}{8\pi m^2} + \frac{e^2 g_A^2}{32\pi^3 F^2}\Big[\frac{M^2}{\omega^2}\arcsin^2(\frac{\omega}{M}) - 1\Big] \,. \tag{2.22}$$

From this the slope parameter γ follows immediately. Notice that γ is again the sum of an electric and magnetic piece, but in the absence of any data, I will only discuss the sum. The em decomposition can be supplied upon request. To leading order, γ is identical for the p and the n,

$$\gamma = \frac{e^2 g_A^2}{96\pi^3 F_\pi^2 M_\pi^2} = 4.4\cdot 10^{-4}\ \mathrm{fm}^4 \ . \tag{2.23}$$

In ref. [33] it was shown that including 1/m suppressed effects in the relativistic formalism together with an estimate of the $\Delta(1232)$ resonance contribution leads to $\gamma_p = -1.5$ and $\gamma_n = -0.5$ (in conventional units), in fair agreement with the estimate based on the dispersion sum rule

$$\gamma = -\frac{1}{4\pi^2}\int_{M_\pi}^\infty \frac{d\omega}{\omega^2}\Big[\sigma_{3/2}(\omega) - \sigma_{1/2}(\omega)\Big] \tag{2.24}$$

which follows directly from eq. (2.20). Using the helicity cross sections provided by Workman and Arndt [34] one gets $\gamma_p^{\rm emp} = -0.89$ and $\gamma_n^{\rm emp} = -0.53$. Clearly, a direct experimental determination of these observables is called for. To give an idea about the size of the effect induced by the "spin-polarizability" γ, let us compare the leading order term $f_2(0) = -5.19\cdot 10^{-4}$ fm^2 with the correction from $\gamma\omega^2$ at various energies. For $\omega = 50$ MeV, this term amounts to a 6 per cent reduction of the leading one. At 100 MeV photon energy, the correction has grown to 22 per cent. For comparison, evaluating the full one-loop amplitude (2.22), one finds corrections of 7 and 37 percent for $\omega = 50$ and 100 MeV, respectively. These latter numbers will, of course, be changed by higher loop corrections. We notice that the influence of the next-to-leading order term is much less pronounced than in the case of the spin-independent amplitude and thus an accurate determination of γ and its electric and magnetic components will be more difficult. The last topic I want to address in this section is the so-called Drell-Hearn-Gerasimov [35] sum rule and its extension to virtual photons. For photons with four-momentum $k^2 \leq 0$ the extended DHG sum rule takes the form

$$I(k^2) = \int_{\omega_{\rm thr}}^{\infty} \frac{d\omega}{\omega} \left[\sigma_{1/2}(\omega, k^2) - \sigma_{3/2}(\omega, k^2) \right] . \tag{2.25}$$

For real photons, this reduces to the DHG sum rule, which has not yet been tested experimentally. For the proton, one has $I(0) = -\pi e^2 \kappa_p^2 / 2m^2 = -0.53\ \mathrm{GeV}^{-2}$. At large $k^2 \simeq -10\ \mathrm{GeV}^2$, the EMC [36] data tell us that I is positive and thus has to undergo a sign change. CHPT allows one to calculate the slope of $I(k^2)$ in the vicinity of the photon point [37]. While the most recent phenomenological analysis gives a kink structure at small k^2 [38], in CHPT one finds that $I(k^2)$ increases monotonically. At present, one cannot decide upon this fine detail but has to wait for the experimental determinations which are possible with CW machines and are planned (MAMI, CEBAF, ...). Finally, I wish to stress again that CHPT allows to systematically explore the consequences of the spontaneously broken chiral symmetry of QCD. It is **not** a model as should have become clear from the above discussion but rather an exact non-perturbative method.

3 The Quark Structure of the Nucleon

In this section I will be concerned with the quark structure of the nucleon as revealed from measurements using electroweak probes. In particular, one can address the questions surrounding the possible admixtures of strange quark components into the proton's wave function. Since these are mostly small effects, I will also discuss the effects of heavy quark and radiative corrections to the processes which allow one to determine the electroweak form factors.

3.1 Currents, Sizes and Form Factors of the Nucleon

The starting point of the discussion are the well-known electromagnetic and axial currents based on photon and charged vector boson exchanges. Let us first consider the vector current J_μ stemming from the one-photon exchange. Between nucleon states of momentum p and helicity s (in general, I will suppress helicitiy indices) Lorentz-invariance, parity and charge conjugation allow us to write

$$< N(p', s') | J_\mu | N(p, s) > = e \bar{u}(p', s') \left[F_1(q^2) \gamma_\mu + F_2(q^2) \frac{i \sigma_{\mu\nu} q^\nu}{2m} \right] u(p, s) \tag{3.1}$$

with $q_\mu = (p' - p)_\mu$ the four-momentum transfer, m denotes the nucleon mass and $u(p, s)$ a conventional nucleon spinor. The finite extension (non-perturbative structure) is parametrized in terms of the Dirac and Pauli form factors (ffs), $F_1(q^2)$ and $F_2(q^2)$, respectively. Gauge invariance of the electromagnetic interactions demands $\partial_\mu J^\mu(x) = 0$. The physical interpretation of these ffs is particularly transparent in the Breit (brick wall) frame, in which the photon transfers no energy, $q_\mu = (0, \mathbf{q})$. The time component of J_μ is the charge density, $J_0 = \rho$ and the space components lead to the distribution of magnetism,

$$
\begin{aligned}
< N(\mathbf{q}/2, s')|J_0|N(-\mathbf{q}/2, s) > &= eG_E(q^2)\delta_{s's} \\
&= e[F_1(q^2) - \tau F_2(q^2)]\delta_{s's} \\
< N(\mathbf{q}/2, s')|\mathbf{J}|(-\mathbf{q}/2, s) > &= \frac{e}{2m} G_M(q^2)\chi^\dagger_{s'} i\sigma \times \mathbf{q}\chi_s \\
&= \frac{e}{2m}[F_1(q^2) + F_2(q^2)]\chi^\dagger_{s'} i\sigma \times \mathbf{q}\chi_s
\end{aligned}
\tag{3.2}
$$

with χ_s a two-component spinor and $\tau = -q^2/4m^2 = Q^2/4m^2$ ($Q^2 > 0$ for space-like photons). Fourier transformation of the electric charge density leads to

$$
\begin{aligned}
\int \rho(\mathbf{r}) j_0(qr) d^3r &= \int \rho(\mathbf{r}) d^3r - \frac{1}{6}Q^2 \int \rho(\mathbf{r}) r^2 d^3r + \ldots \\
&= Z - \frac{1}{6}Q^2 < r^2 >_E + \ldots
\end{aligned}
\tag{3.3}
$$

where the first term is obviously the electric charge of the particle (here, the proton or the neutron) and the second one defines the electric charge radius. It is conventional to normalize these charge radii via

$$
< r^2 >= \frac{6}{G(0)} \frac{dG(q^2)}{dq^2}\bigg|_{q^2=0}
\tag{3.4}
$$

with the exception of the neutron charge radius since in that case $G(0) = 0$ so that $< r_E^2 >_n$ is simply six times the slope of the neutron form factor. The electric (E) and magnetic (M) ffs are normalized as follows

$$
G_E^p(0) = 1, \quad G_E^n(0) = 0, \quad G_M^p(0) = \mu_p = 2.793, \quad G_M^n(0) = \mu_n = -1.913 \ . \tag{3.5a}
$$

For the various radii, the presently available data are [39]

$$
\begin{aligned}
&< r_E^2 >_p^{1/2} = 0.86 \pm 0.01\,\text{fm}, \quad < r_M^2 >_p^{1/2} = 0.86 \pm 0.06\,\text{fm}, \\
&< r_E^2 >_n = -0.12 \pm 0.01\,\text{fm}^2, \quad < r_M^2 >_n^{1/2} = 0.88 \pm 0.07\,\text{fm} \ .
\end{aligned}
\tag{3.5b}
$$

A few remarks on these numbers are in order. First, the typical electromagnetic (em) size is $r_{\text{em}} \simeq 0.85$ fm and the negative value of the neutron charge radius squared shows that a surplus of negative charge in the exterior region of the neutron must exist. Most disturbing is the fact that in particular the magnetic charge radius of the neutron, which is a quantity as fundamental as say the magnetic moment, is only poorly known. As pointed out by Arenhövel and Schoch at this school [40] experiments are under way to remedy the situation. At larger momentum transfers, it has become common to use the so-called dipole fits, which proved to be embarrassingly good up to several GeV^2 on typical logarithmic plots when elastic electron-proton scattering experiments were performed in the sixties and early seventies as discussed by Taylor at this school (see e.g. ref. [41]). The dipole fit reads

$$
G_E^p(Q^2) = \frac{G_M^p(Q^2)}{\mu_p} = \frac{G_M^n(Q^2)}{\mu_n} = G_D(Q^2) = \frac{1}{(1 + Q^2/0.71\,\text{GeV}^2)^2} \ . \tag{3.6}
$$

For the electric ff of the neutron, the "trivial" dipole fit is $G_E^n(Q^2) = 0$, however, in nature this ff is small but non-vanishing. It can be approximated by [42]

$$G_E^n(Q^2) = -\frac{\mu_n \tau}{1+Q^2/0.42\,\mathrm{GeV}^2} G_D(Q^2)\,. \tag{3.7}$$

In refs. [43], you can take a look at the quality of these fits on a linear scale for Q^2 up to 4 GeV^2 - it is obvious that the dipole fit is at most a fair approximation. Also, it has never been given a sound theoretical foundation. While the dipole mass $M_D = 0.843$ GeV is suggestive of vector meson exchange (like in the vector meson dominance approach to photon-hadron couplings), such a mechanism would naturally lead to a monopole fall-off. In case of the neutron electric ff, which is determined indirectly from elastic scattering off the deuteron and subtraction of the proton ff, a major uncertainty stems from the use of the underlying model of the nucleon-nucleon force as discussed in ref. [42]. Here, experiments using polarized electrons and polarized helium targets are hoped to improve the theoretical uncertainties. Definitively, we need a better determination of these fundamental nucleon properties. It is also instructive to see what perturbative QCD tells us about the large Q^2 behaviour of these ffs. Based on the quark counting rules [44], which state that to distribute the large photon momentum equally to three quarks in the nucleon, two gluon exchanges are necessary, one finds that $F_1(Q^2) \sim 1/Q^4$ and $F_2(Q^2) \sim 1/Q^6$ since the gluon propagator goes like $1/Q^2$ and in the case of F_2 an extra $1/Q^2$ is needed for the helicity flip. At which value of Q^2 this asymptotic behaviour sets in is not known, certainly way beyond 1 GeV^2 as originally thought. Notice that the dipole fits to the nucleon ffs have the correct large Q^2 behaviour. This is a further mystery surrounding these fits. It might be based on the fact that a simple meson-cloud model of the nucleon also leads to such fall-offs (in the one-boson exchange approximation).

Let me now switch to the axial current (related to the exchange of charged vector bosons). Its nucleon matrix element reads

$$< N(p')|A_i^\mu(0)|N(p) >= \bar{u}(p')\big[G_A(q^2)\gamma^\mu + \frac{G_P(q^2)}{2m}q^\mu\big]\gamma_5 \frac{\tau_i}{2} u(p) \tag{3.8}$$

where $G_A(q^2)$ is the axial ff. The induced pseudoscalar ff $G_P(q^2)$ is essentially pole dominated and only accessible via scattering processes involving heavy leptons (like e.g. muons). I will not consider it in what follows. $G_A(0)$ is nothing but the axial-vector coupling constant measured in neutron β-decay, $g_A = 1.26$. The Q^2-dependence of $G_A(Q^2)$ follows again a dipole form,

$$G_A(Q^2) = \frac{g_A}{(1+Q^2/M_A^2)^2}\ . \tag{3.9}$$

The value of the axial cut-off mass was already discussed in subsection 2.4, it translates into a typical axial size of $r_A \simeq 0.65$ fm. This is considerably smaller than the typical em size. Therefore, the size of the nucleon depends on the probe one uses. In fact, the hierarchy of these various nucleon sizes can nicely be understood in the topological soliton model of the nucleon as spelled out in detail in ref. [45]. In a nutshell, the argument goes as follows. The isoscalar baryon number current, whose space-integral is the baryon number, is given by the pion fields in the soliton with an extension of $r_B = 0.5$ fm. The isoscalar photon, however, sees in addition the virtual ω-meson content of the nucleon. This leads to an isoscalar charge radius of $< r_E^2 >_{I=0} = r_B^2 + 6/M_\omega^2 = (0.8\ \mathrm{fm})^2$ in nice agreement with the data. By a similar argument, one expects a somewhat smaller axial radius since the additional factor now carries the axial vector mass of 1.1 GeV and one is thus naturally led to an axial extension of 0.65 fm.

3.2 A Toy Model: Quark Distributions from Nucleon Form Factors

To understand how the different electroweak properties as parametrized by the various ffs can give insight into the quark structure of the nucleon in the non-perturbative regime, let us consider a simple (too simple) model. The proton and the neutron are made up of valence u and d quarks and a sea of quark-antiquark pairs. We assume that the p and the n are simply related by the interchange of one valence u with one valence d quark. For the sake of simplicity, let us forget the gluons and just imagine some confinement mechanism which keeps the current quarks within the typical hadronic volume of 1 fm^3. Denoting by J^j_μ the em current associated to the quarks of flavor $j = u, d, s, c, b, t$, we can write [46]

$$\begin{aligned} < N(p',s')|J_\mu|N(p,s) > &=< N(p',s')|\sum_j J^j_\mu|N(p,s) > \\ &= e\sum_j \bar{u}(p',s')Q_j(\gamma_\mu F_1^j + \frac{i\sigma_{\mu\nu}q^\nu}{2m}F_2^j)u(p,s) \end{aligned} \tag{3.10}$$

where Q_j denotes the quark charges and the quarks are considered as point-like Dirac particles with no anomalous magnetic moment. In terms of the electric and magnetic ffs, this decomposition takes the form

$$G^{(p,n)}_{E,M} = \sum_j Q_j G^{j\,(p,n)}_{E,M} \; . \tag{3.11}$$

These expressions are exact under the assumptions made as long as the sum extends over all quark flavors. Furthermore, the G^j contain contributions from quarks as well as from antiquarks. Therefore, the $q\bar{q}$ sea does not contribute to the total charge since $Q_j = -\bar{Q}_j$. Since the p and the n as well as the u and the d quark are considered to be essentially the same particles differing only by an isospin rotation, it should be possible to separate the response of the various quark flavors [47]. For doing that, one has to make further assumptions. First, one postulates a local isospin invariance of the type $J^{u,p}_\mu(x) = J^{d,n}_\mu(x)$ and $J^{d,p}_\mu(x) = J^{u,n}_\mu(x)$ (which has never been tested and is doubtful in the light of the recent structure function measurements). Second, we assume that all other quark flavors do not contribute, i.e. $J^{q,p}_\mu(x) = J^{q,n}_\mu(x) = 0$ for $q = s, c, b, t$. This allows to uniquely give the em u and d distributions in terms of appropriate combinations of the em ffs of the proton and the neutron. The final formulae are obtained by normalizing with respect to the total quark charges, e.g. pulling out a factor 4/3 for the u quark distribution in the proton. This leads to [48]

$$\begin{aligned} G^u_{E,M} &= G^p_{E,M} + \frac{1}{2}G^n_{E,M} \; , \\ G^d_{E,M} &= G^p_{E,M} + 2G^n_{E,M} \; . \end{aligned} \tag{3.12}$$

It is instructive to get some numbers from eq. (3.12). First, let us stick in the magnetic moments. We find $G^u_M(0) = 1.836$ and $G^d_M(0) = -1.033$, which is not very different from the $SU(3)$ result for constituent quarks. Notice, however, that here we are considering current quarks. The implications of this result are discussed in more detail in Beck's lecture [48]. Similarly, for the electric charge radii one finds $< r_E^2 >_u = 0.68 \pm 0.02$ fm^2 and $< r_E^2 >_d = 0.51 \pm 0.02$ fm^2 which is a 35 per cent difference. To further separate the valence from the sea quarks, one needs additional assumptions about the radial distributions as detailed in ref. [48]. In any case, this discussion should only be considered illustrative, the model is too crude to attach a deep physical significance to these numbers. However, in a very simple fashion it demonstrates how a clever combination of measurable hadron form factors allows us to make statements about quark distributions. This problem will now be tackled in a more serious fashion.

3.3 Electroweak Currents

In this subsection, I will define the electroweak currents related to the exchanges of the various vector bosons of the minimal standard model. These spin-1 particles are the massless photon and the massive $W^\pm$ and Z^0 bosons. At tree level, the couplings of these to the various quark flavors are specified completely in terms of the $U(1)$ gauge coupling g and the $SU(2)$ gauge coupling g'. Equivalently, one can use the em coupling e and weak mixing angle θ_W since $g = e \sin\theta_W = e\, s$ and $g' = g/4\cos\theta_W = g/4\,c$. Also, the vector boson masses are related at tree level via $M_Z = M_W/c$. For the most recent values, see Altarelli's lectures at this school [49]. Let us consider again the photon current which is purely vectorial and its couplings to the light quarks u, d and s. One has

$$
\begin{aligned}
J_\mu^\gamma &= \sum_{q=u,d,s} Q_q \bar{q}\gamma_\mu q = \frac{2}{3}(\bar{u}\gamma_\mu u) - \frac{1}{3}(\bar{d}\gamma_\mu d + \bar{s}\gamma_\mu s) \\
&= \sum_{q=u,d,s} \bar{q}\gamma_\mu \frac{1}{2}(\lambda^3 + \frac{1}{\sqrt{3}}\lambda^8) q \\
&= \frac{1}{2}(\bar{u}\gamma_\mu u - \bar{d}\gamma_\mu d) + \frac{1}{6}(\bar{u}\gamma_\mu u + \bar{d}\gamma_\mu d - 2\bar{s}\gamma_\mu s) \quad .
\end{aligned}
\tag{3.13}
$$

I have exhibited the two most frequently used conventions. In the second line, the decomposition in terms of Gell-Mann's flavor $SU(3)$ matrices is shown. One notices that the photon is blind to the singlet piece $\sim \lambda^0$. In the third line, the photon current is decomposed into its isovector (IV) and isoscalar (IS) components. This current leads to the em form factors $G_{E,M}^\gamma$ or $F_{1,2}^\gamma$ discussed before (the label 'γ' is needed to differentiate these ffs from the ones related to the weak neutral current as discussed below). The charged weak currents are flavor-changing and lead to the isovector axial ff discussed before. Further information comes from the weak neutral current mediated by the Z^0. Its coupling to the quark has a vectorial and an axial-vector piece. Ignoring the overall coupling strength g' and defining $V_q = \bar{q}\gamma_\mu q$ for a quark of flavor q, the vector part at tree level reads

$$
\begin{aligned}
V_\mu^Z &= (\frac{1}{4}-\frac{2}{3}s^2)V_u + (-\frac{1}{4}+\frac{1}{3}s^2)(V_d+V_s) \\
&= \frac{1}{2}(\frac{1}{2}-s^2)(V_u-V_d) + \frac{1}{6}(\frac{1}{2}-s^2)(V_u+V_d-2V_s) - \frac{1}{12}(V_u+V_d+V_s) \quad (3.14) \\
&= (\frac{1}{2}-s^2)\frac{1}{2}(V_u-V_d) + \frac{1}{6}s^2(V_u+V_d-2V_s) - \frac{1}{4}V_s
\end{aligned}
$$

where again the second line gives the $SU(3)$ components and the third line the decomposition into IS, IV and explicit strange pieces. Similarly, with $A_q = \bar{q}\gamma_\mu\gamma_5 q$ we have for the axial-vector current

$$
\begin{aligned}
A_\mu^Z &= -\frac{1}{4}A_u + \frac{1}{4}(A_d+A_s) \\
&= -\frac{1}{4}(A_u-A_d) - \frac{1}{12}(A_u+A_d-2A_s) + \frac{1}{12}(A_u+A_d+A_s) \quad (3.15) \\
&= -\frac{1}{2}\Big[\frac{1}{2}(A_u-A_d)\Big] + \frac{1}{4}A_s\ .
\end{aligned}
$$

Two remarks are in order. First, the Z^0 couples to the singlet component $\sim \lambda^0 \sim (u+d+s)$ and, second, there is no isoscalar axial coupling at tree level. For the following discussion, let me express the ffs related to the Z^0-exchange in the flavor $SU(3)$ basis following Kaplan and Manohar [50],

$$
\begin{aligned}
&F_{1,2}^Z = \sum_{\alpha=0,3,8} (b_\alpha - a_\alpha s^2)F_{1,2}^\alpha\,; \quad F_{1,2}^\gamma = \sum_{\alpha=0,3,8} a_\alpha F_{1,2}^\alpha\,, \\
&G_1 = -\sum_{\alpha=0,3,8} b_\alpha G_A^\alpha = -\sum_{\alpha=0,3,8} b_\alpha G_1^\alpha\,, \qquad (3.16) \\
&a^0 = 0\,, \quad a^3 = 1\,, \quad a^8 = \frac{1}{\sqrt{3}}\,; \quad b^0 = -\frac{1}{4}\,, \quad b^3 = \frac{1}{2}\,, \quad b^8 = \frac{1}{2\sqrt{3}}\,.
\end{aligned}
$$

The value of the a's and b's are subject to radiative and heavy quark corrections as discussed below. For completeness, I have also given the em ffs in this basis. Using for the $SU(3)$ generators $\mathrm{Tr}(T^aT^b) = \delta^{ab}/2$ and supplying the isosinglet currents with an overall factor 1/3, the normalizations of the various ffs are given as follows. From the proton and the neutron charge, we have

$$
F_1^3(0) = 1/2\,, \quad F_1^8(0) = \sqrt{3}/2\ . \qquad (3.16a)
$$

Similarly, the p and n anomalous magnetic moments lead to

$$
F_2^3(0) = (\kappa_p - \kappa_n)/2\,, \quad F_2^8(0) = \sqrt{3}(\kappa_p + \kappa_n)/2\ . \qquad (3.16b)
$$

$F_1^0(0)$ is equal to unity since it gives the baryon number. However, this does not exclude a finite extension of $s\bar{s}$ pairs leading to a strange electric radius (in complete analogy to the neutron charge radius). The value of the singlet anomalous magnetic moment is not known. If there are no strange quarks in the proton, one has $F_2^0(0) = \kappa_p + \kappa_n = -0.12$. So a measurement of $F_2^0(0)$ will give information about the possible strange quark contribution to the anomalous magnetic moment, $F_2^0(0) = \kappa_p + \kappa_n + \mu_s$. The triplet and octet normalizations of the axial ff are given by

$$
G_1^3(0) = g_A = F + D = 1.26\,, \quad G_1^8(0) = (3F-D)/\sqrt{3} = 0.32\,. \qquad (3.16c)
$$

The isosinglet charge $G_1^0(0)$ is again sensitive to the strange quark content and can e.g. be determined in polarized deep inelastic lepton scattering as discussed by Reya at this school [51] (see also ref. [36]). Alternatively, neutrino scattering off nucleons or nuclei might be used to get a handle on $G_1^0(0)$. This will be discussed below. At present, no experimental information is available about the Q^2-dependence of the singlet and octet axial ffs, so one either has to resort to some assumptions or models (see section 3.5). Clearly, the singlet axial charge and the singlet anomalous magnetic moment are the best objects to study the strange quark content of the nucleon using electroweak probes. The other obvious candidate is the so-called pion-nucleon σ-term which measures the strength of the strange matrix element $m_s < p|\bar{s}s|p >$. Combining πN data, dispersion theory and constraints from chiral symmetry, Gasser et al. [52] have given the most precise determination of this quantity and its t-dependence. From this analysis one deduces that $m_s < p|\bar{s}s|p > \simeq 130$ MeV. This is a sizeable though not dramatic strange quark effect. After defining the electroweak form factors, I will now discuss some experiments which allow to extract them and the related strange quark matrix elements.

3.4 Parity-Violating Electron Scattering

Consider the scattering of polarized electrons on a nucleon or a nucleus. Due to the $\gamma - Z^0$ interference, parity violation (pv) occurs and this allows to determine combinations of the electromagnetic and weak neutral ffs. The basic Feynman diagrams are simply the one-photon exchange proportional to $F_{1,2}^\gamma$ and the Z^0-exchange proportional to $F_{1,2}^Z$ and G_1. To get an idea about the magnitude of this interference effect, let us perform some dimensional analysis. The electromagnetic amplitude reads

$$\mathcal{A}_{\rm em} = < f|J^\mu_{\rm em}\frac{1}{Q^2}j_{\mu,{\rm em}}|i > \sim \frac{\alpha}{Q^2} , \tag{3.17}$$

where $j_{\mu,{\rm em}}$ is the well-known em lepton current, $1/Q^2$ the photon propagator and $J^\mu_{\rm em}$ the hadronic vector current parametrized by the em ffs. The parity-violating weak amplitude takes the form

$$\mathcal{A}_{\rm w}^{\rm pv} = < f|V^\mu a_\mu + A^\mu v_\mu|i > \sim \frac{G_F}{\sqrt{2}} , \tag{3.18}$$

which is a product of the lepton (v_μ, a_μ) and hadron (V_μ, A_μ) vector and axial-vector currents, the latter containing the weak form factors. $G_F = 1.16 \cdot 10^{-5}$ GeV^{-2} is the Fermi constant. The total cross section can approximately be written as

$$\sigma = |\mathcal{A}_{\rm em} + \mathcal{A}_{\rm w}^{\rm pv}|^2 \simeq |\mathcal{A}_{\rm em}|^2 + 2|\mathcal{A}^*_{\rm em}\mathcal{A}_{\rm w}^{\rm pv}| . \tag{3.19}$$

The observable of interest is the left-right asymmetry A,

$$A = \frac{\sigma_R - \sigma_L}{\sigma_R + \sigma_L} \simeq \frac{|\mathcal{A}^*_{\rm em}\mathcal{A}_{\rm w}^{\rm pv}|}{|\mathcal{A}_{\rm em}|^2} \sim \frac{G_F m^2}{4\pi\alpha}\frac{Q^2}{m^2} = 1.1 \cdot 10^{-4}\frac{Q^2}{{\rm GeV}^2} . \tag{3.20}$$

For a typical four-momentum of $Q^2 = 0.1$ GeV^2 in the non-perturbative regime we can expect an asymmetry of 10^{-5}. This is a small number but within the reach of present day technology. Historically, the first measurement of this asymmetry using a deuterium target at SLAC was performed at Q^2 around 20 GeV^2 [53]. To be specific, consider now as the target the proton. An elementary calculation gives

$$\begin{aligned} A_{ep} &= -\frac{G_F Q^2}{\sqrt{2}\pi\alpha} \frac{\left[\epsilon G_E^\gamma G_E^Z + \tau G_M^\gamma G_M^Z - \frac{1}{2}(1-4s^2)\sqrt{1-\epsilon^2}\sqrt{\tau(1+\tau)}G_M^\gamma G_1\right]}{\epsilon (G_E^\gamma)^2 + \tau (G_M^\gamma)^2} \\ &= -\frac{G_F Q^2}{\sqrt{2}\pi\alpha}[A_E + A_M + A_A] \end{aligned} \tag{3.21}$$

with $1/\epsilon = [1 + 2(1+\tau)\tan^2(\theta/2)] = 0\ldots 1$ given by the scattering angle of the electron. In the second line, I have split the asymmetry into an electric, magnetic and axial piece. It is instructive to consider various kinematical regimes. At forward angles ($\theta \to 0$, $\epsilon \to 1$), the electric part dominates at low Q^2 and the asymmetry is thus sensitive to the strange electric ff. At higher Q^2, due to the τ prefactor, the magnetic part takes over and one can access $G_M^s(Q^2)$. Notice that in forward direction the axial contribution is not only suppressed by $1-4s^2 \simeq 0.08$ but also by $\sqrt{1-\epsilon^2}$. In backward direction, matters are different since as $\theta \to 0$, ϵ also tends to zero. So at low Q^2 one can get $G_M^s(Q^2)$ with some contamination from the axial part. This is the kinematical regime where the SAMPLE experiment [54] hopes to determine the strange anomalous magnetic moment. This warrants a closer look. In a simplified analysis, we set $s^2 = 1/4$, $\theta = \pi$ and work at some low Q^2, say 0.2 GeV^2. Then, $A(Q^2)$ takes the simple form [55]

$$A(Q^2) \simeq -\frac{G_F Q^2}{\sqrt{2}\pi\alpha}\frac{G_M^Z}{G_M^\gamma} \simeq -\frac{G_F Q^2}{4\sqrt{2}\pi\alpha}\frac{\kappa_p - F_2^0(0)}{\kappa_p + 1} = -\frac{G_F Q^2}{4\sqrt{2}\pi\alpha} R \tag{3.22}$$

assuming that the Q^2-dependence is similar for G_M^Z and G_M^γ. The value of R ranges from $R = 1$ to $R = 0.28$ for $F_2^0(0) = -1\ldots 1$. For no strangeness contribution, we have $F_2^0(0) = -0.12$ and thus $R = 0.64$. To give an idea about what one can expect, let me consider some models. The $SU(3)$ Skyrme model with a proper treatment of the symmetry breaking leads to $F_2^0(0) = -0.17\ldots -0.25$ [56], the NJL model to $F_2^0(0) = -0.16\ldots -0.26$ [57] and vector meson dominance supplemented with the ϕ-meson (i.e. explicit strange quarks) to $F_2^0(0) = -0.43 \pm 0.09$ [58]. Similarly, one can also discuss the strange electric radius, i.e. the radius related to the operator $\bar{s}\gamma_\mu s$. A simple model based on a kaon cloud surrounding the nucleon leads to processes like $p \to \Lambda K^+, \Sigma^0 K^+, \ldots \to p$ and lets us expect that $< r_E^2 >_s$ is positive since the positive charge sits in the cloud of the kaons. This is, however, a very crude estimate. In fact, most models tend to give a small and negative $< r_E^2 >_s$, like e.g. VMD plus the ϕ-meson -0.14 ± 0.04 fm^2 [58] or the $SU(3)$ Skyrmion -0.10 ± 0.05 fm^2 [56]. These are fairly small numbers and it appears doubtful that the strange electric radius can be determined accurately from a proton target in the near future (assuming, of course, that its value is as small as indicated by the models). Here, nuclei might come in handy. For isoscalar, spin-0 nuclei such as ^{4}He or ^{12}C, one finds for the asymmetry [59]

$$A = \frac{G_F Q^2}{4\sqrt{2}\pi\alpha}\left[s^2 + \frac{G_E^s}{2(G_E^p + G_E^n)}\right] , \tag{3.23}$$

which at very low Q^2 allows to extract the weak mixing angle and at somewhat higher momenta the strange electric ff. Of course, one has to take into account nuclear structure issues like pv level mixing and so on. A lucid discussion of these topics can be found in refs. [59,60]. Clearly, pv electron scattering is an interesting field and many more aspects of it not covered here can e.g. be found in the proceedings of the CALTECH workshop [61].

3.5 Neutrino and Antineutrino Scattering off Nucleons and Nuclei

Neutrino scattering off nucleons or nuclei offers another possibility of exploring the strange quark content of the nucleon. The neutrino couples via the Z^0 to the quarks and thus probes isosinglet, triplet and octet components. Let us first consider the differential cross section for elastic neutrino/antineutrino-proton scattering,

$$\frac{d\sigma}{dQ^2} = \frac{G_F^2 m^2}{8\pi E_\nu^2}\left[A \pm B\,W + C\,W^2\right] , \tag{3.24}$$

with E_ν the neutrino energy, $W = (4E_\nu m - Q^2)/m^2$ and the '$\pm$' refers to the case of neutrinos (antineutrinos). The functions A, B and C depend on the ffs G_1 and $F_{1,2}^Z$ (I omit the superscript 'Z' on F_1 and F_2),

$$\begin{aligned} A &= 4\tau\left[G_1^2(1+\tau) - 4(1-\tau)(F_1^2 - \tau F_2^2) + 16\tau F_1 F_2\right] , \\ B &= -8\tau G_1(F_1 + F_2) , \\ C &= G_1^2/4 + F_1^2 + \tau F_2^2 . \end{aligned} \tag{3.24a}$$

To demonstrate this in more detail, let me simplify the analysis by assuming a Q^2 value of about 0.5 GeV^2 and retaining only the terms linear in the strange ffs. This allows to recast (3.24) in the form [62]

$$\frac{d\sigma}{dQ^2} = \sigma_{\text{non-strange}}\left[1 - 0.72F_1^s(Q^2) - 0.78F_2^s(Q^2) - 1.23G_1^s(Q^2)\right] . \tag{3.24b}$$

Obviously, to extract the strange matrix element $< p|\bar{s}\gamma_\mu\gamma_5 s|p > = G_1^s(0)$, one has to know something about $F_{1,2}^s(Q^2)$ or make some assumptions. Ahrens et al. [63] and Kaplan and Manohar [50] have shown that setting $F_1^s = F_2^s = 0$ and assuming that all three axial ffs have the same dipole mass ($M_A = 1.032$ GeV), the extracted value of $G_1^s(0)$ is in agreement with the one obtained from polarized deep inelastic μp scattering [36]. However, as already stressed in ref. [63], this result is very sensitive to the actual value of the dipole mass M_A. This issue was further addressed by Bernard et al. [64]. They used the topological chiral soliton model of the nucleon to calculate the isosinglet axial ff since in the framework of this model the known triplet ff is well described. It was found that the cut-off mass in the singlet channel is actually 20 per cent larger than the one for the triplet and octet ffs. Redoing the analysis with these ffs, one finds that the value of $G_1^s(0)$ is reduced by a factor of three. A similar analysis has recently been performed by Garvey et al. [65]. In conclusion, the often claimed agreement concerning the value of $G_1^s(0)$ from the neutrino and EMC data can only be considered accidental. To avoid the problems of extrapolating from the typical Q^2 of 0.5 to 1 GeV^2 in elastic νp or $\bar{\nu}p$ scattering, it was recently proposed that (anti)neutrino-induced quasi-free nucleon knock-out off nuclei might give a better handle on the strange matrix elements [66]. In the quasi-free region, one has small momentum transfers and typical neutrino energies of 0.2 GeV. The ratio of the proton to the neutron yield depends sensitively on G_1^s and F_2^s,

$$R = Y_p \,/\, Y_n = F(G_1^s,\, F_2^s)\ . \tag{3.25}$$

For example, if $F_2^s(0) = -0.22$, $R_{\bar\nu}$ varies between 0.85 and 1.9 for $G_1^s(0)$ between 0 and -0.2. For further details, see ref. [66]. Of course, it is mandatory to understand well the nuclear structure issues, in particular one assumes that the knock-out is involving only one nucleon. How good these assumptions are is still under debate. What is important is to realize that neutrino and pv electron scattering processes nicely complement each other and that many different experiments have to be performed to ultimately pin down the strength of the strange matrix elements in the proton.

3.6 Down and Dirty: QED, QCD and Heavy-Quark Corrections

Up to now, I have entirely worked at tree level. At low energies, radiative corrections and effects from heavy quarks are expected to be small. However, in certain cases the tree level couplings are suppressed or vanish (as it is the case with the isocalar axial current) or one is trying to extract small numbers. Therefore, it is mandatory to investigate the effects of QED and QCD corrections on the processes considered before. Since the neutrino has no direct QED and QCD couplings, let me start with the discussion of corrections to neutrino-hadron (quark) scattering. To be specific, consider the induced isoscalar coupling. At tree level it vanishes, but that does not mean that it is zero altogether. The basic Feynman diagram which induces such a coupling is the famous triangle diagram. On one end, the axial part $(\gamma_\mu\gamma_5)$ of the neutrino current couples and the other two corners are attached via gluon exchanges to the external light (u,d,s) quark. In the loop, quarks of all flavors run around. A simple one-gluon exchange is forbidden by color neutrality. This is the basic diagram to lowest order in the strong interactions. The contribution of such a type of diagram has to be finite since the sum over all quark flavors leads to anomaly cancellation,

$$\sum_i T_3^i = 0\ , \tag{3.26}$$

where T_3 denotes the weak isospin. However, we are interested at scales much below the intermediate vector boson and heavy quark masses. In this energy regime, the triangle diagram leads to an induced isoscalar current which is not suppressed by inverse powers of the heavy quark masses as naive decoupling would suggest. Indeed, a straightforward perturbative calculation of this two-loop diagram yields [67]

$$A_\mu^Z = \frac{1}{2}(A_u - A_d) + (A_u + A_d)\frac{1}{4}\Big(\frac{g^2}{4\pi}\Big)^2 \ln\frac{\Pi_i m_{i+}^2}{\Pi_i m_{i-}^2}\ , \tag{3.27}$$

where $m_{i\pm}$ refers to the masses of the quarks with weak isospin $\pm 1/2$, i.e. with charges 2/3 and -1/3, respectively, and g is the strong coupling constant. Notice that the induced isoscalar current $(A_u + A_d)$ depends logarithmically on the quark masses and not on inverse powers of them (as advertised). Collins, Wilczek and Zee [67] have shown how to do this in a more systematic fashion. Their method is based on integrating out quark doublets in succession. Denote by $H = A_t - A_b$ the axial current of the heaviest doublet. The idea is to recast H in the form

$$H = \sum_i A_i \bar{L}_i + \sum_i B_i \bar{H}_i \ , \tag{3.28}$$

where the light ($\bar{L}_i$) and the heavy ($\bar{H}_i$) particle operators exhibit decoupling. This means that the matrix elements of the $\bar{L}_i$ see the heavy quarks only via power law corrections. Repeating this exercise for $\tilde{H} = A_c - A_s$, the induced isoscalar current reads

$$\Delta A_\mu^Z = (A_u + A_d) \frac{1}{4} \frac{\alpha_s(m_s)}{\pi} \frac{\alpha_c(m_c)}{\pi} \ln \frac{m_c^2}{m_s^2} \simeq 0.05\,(A_u + A_d) \ , \tag{3.29}$$

where I have neglected the much smaller contribution from the (t,b) doublet. Of course, this approach is too bold since at the scale of the strange quark mass $\lambda = m_s \simeq 175$ MeV, this perturbative treatment cannot be justified any more. Therefore, Kaplan and Manohar [50] have generalized the method of ref. [67] to integrate out the various quarks separately. Their argument is based on the assumption $M_Z > m_t$. This simplifies the analysis but is not mandatory. In this case, integrating out the Z^0 one has an effective Lagrangian for neutrino-quark scattering,

$$\mathcal{L}_{\text{eff}} = -\frac{G_F}{\sqrt{2}} \sum_{u,d,s,c,b,t} \bar{\nu}\gamma^\mu(1-\gamma_5)\nu\left[\bar{q}(T_3 - 2s^2 Q)\gamma_\mu q - \bar{q}T_3\gamma_\mu\gamma_5 q\right] \ . \tag{3.30}$$

In principle, one should now calculate corrections to this effective Lagrangian. However, due to the absence of direct QED or QCD couplings of the neutrino, one can rewrite these corrections as corrections to the hadronic current. The procedure goes as follows. One uses the renormalization group to evolve (3.30) down to the scale of the top mass. At that energy, one integrates out the t quark. This induces two types of corrections, one class scaling like $g^2(m_t)/m_t^2$ (i.e. exhibiting decoupling). Since the t is removed from the theory, the Z^0 axial current has an anomaly because (3.26) is not fulfilled any more. This leads to a multiplicative renormalization of the singlet axial current, which is the most important effect as we will see. One then proceeds by scaling down to $\lambda = m_b$, integrating out the b and so on until one reaches the scale $\lambda = 1$ GeV (which is a typical energy in the elastic νp scattering process and large enough to justify the perturbative treatment). Putting the important pieces from the multiplicative renormalizations together, the relation between the axial current at $\lambda = M_Z$ and at $\lambda = 1$ GeV reads [50]

$$\begin{aligned} A_\mu^Z(\lambda = 1\,\text{GeV}) &= A_\mu^Z(\lambda = M_Z) + \frac{1}{2}(A_u + A_d)\Delta_A \\ \Delta_A &= \Lambda_{\mu c}\left[\frac{1}{2} + \frac{3}{10}\Lambda_{cb}(\Lambda_{bt} - 1)\right] \\ \log \Lambda(m_i, m_j, N_f) &= N_f \frac{\alpha_s(m_i)}{\pi}\frac{\alpha_s(m_j)}{\pi} \log\frac{m_i^2}{m_j^2} \ , \end{aligned} \tag{3.31}$$

which leads to $\Delta A = 0.02$ [50]. Obviously one recovers the result (3.29) if one were to integrate out the s quark too. Notice that the finite pieces due to the renormalization group evolution are negligible and that there is also an induced vector current with

$\Delta_V < 10^{-4}$. The main effect of the photons comes from the penguin diagram leading to a running of the weak mixing angle [68]

$$s^2(\lambda^2) = s^2(M_W^2) + \sum_j \frac{\alpha}{3\pi} \ln\left(\frac{M_W^2}{\lambda^2}\right) Q_j (T_{3j} - 2s^2 Q_j) , \tag{3.32}$$

where the index j extends over the quark flavours which are integrated out. In addition, there are pure next-to-leading order electroweak effects. These have been considered in ref. [69]. The radiatively induced isoscalar axial current has a very small coefficient,

$$-\frac{3\alpha}{16\pi s^2}\left[1 + \frac{1}{2c^2}(1 - 2s^2 + \frac{20}{9}s^4)\right] = -0.003 , \tag{3.33}$$

which is negligible compared to the heavy-quark effects. Therefore, if one restricts oneself to scales above $\lambda \simeq 1$ GeV, the QED and QCD corrections to neutrino-hadron scattering are known and under control. Matters become more messy if one attempts to go to lower energies and also if one considers electron-hadron scattering as I will do now. In this case, we have in addition the direct photon exchange which gives rise to further complications. What we are after can be summarized as follows. Denote by C the tree level value of any coupling between the lepton and the hadron (quark) currents. Due to the radiative corrections, these couplings are modified

$$C = C[1 + R] , \tag{3.34}$$

so that R gives the ratio of the corresponding value including radiative corrections to the tree level one (in case that C happens to vanish at tree level, one has to choose another coupling as the reference point). Naively, one expects the size of these corrections to be small, the typical scale being

$$G_F m^2 \frac{\alpha}{4\pi} \leq 10^{-8} . \tag{3.35}$$

This rather small number is sometimes considerably enhanced and, furthermore, the radiative corrections can in some circumstances compete with the effects induced by the strange quarks. Holstein and Musolf [70] have given the most detailed evaluation of these effects. Clearly, the results of their calculations should be considered indicative since considerable uncertainties are involved as will be discussed below. A very basic and introductory presentation concerning the calculation of radiative corrections for the pv processes under consideration has been given by Musolf [71] which the reader not familiar with the concepts of renormalization in the standard model certainly will appreciate. In essence, there are two classes of Feynman diagrams contributing at one loop order. The first class involves only one quark in the nucleon. Typical representatives are vertex or vector boson propagator corrections. More complicated are the diagrams involving two quarks, such as the γZ^0 box or exchange current type diagrams where a massive vector boson is exchanged between two quarks while the photon couples to one of them (many pictorials are displayed in ref. [71]). Let me first consider the one quark type diagrams. These depend on the not yet known masses of the t quark and the Higgs boson and, if there is physics beyond the standard model, implicitely on the parameters of this new physics. This latter dependence is parametrized in terms of S, T and U or ϵ_1, ϵ_2 and ϵ_3 (as discussed by Altarelli at this school [49]). There are two effects which can conspire to give much larger values for these loop corrections than the estimate (3.35) would suggest. These are large logarithms of the light fermion to the vector boson mass ratios and the suppression of tree level couplings. A particularly illustrative example is the scattering process $\nu_e\mu \to \nu_e\mu$. While the tree level $Z^0-\mu$ coupling is suppressed, the W-vertex correction to the $Z^0 - \nu_e$ coupling induces a $\log(m_e^2/M_W^2)$, with m_e the electron mass. The corresponding ratio follows to be [70]

$$R(\nu_e\mu \to \nu_e\mu) = \frac{\alpha}{4\pi}\frac{8}{3}\frac{\ln(m_e^2/M_W^2)}{4s^2-1} \simeq 0.5 \ . \tag{3.36}$$

This calculation is clean in that it involves only leptons. For the case of electron-hadron scattering, the quarks are bound within a nucleon of a size of about 1 fm. Using the uncertainty principle, this allows one to get an idea about the typical quark momenta. However, this is a very crude estimate and induces some uncertainty (for details, see ref. [70]). Furthermore, it is not obvious which quark mass values one should insert. When one considers quark loop corrections to the vector boson propagators, one obviously deals with the current quarks. In contrast, in the various box diagrams it is more justified to use the so-called constituent masses (~ 330 MeV for the u and d quarks) to account for the binding effects. These subtleties are also discussed in refs. [70,71]. For processes of the type $V(e) \times A(N)$ (which means that the vector current of the electron couples to the nucleonic axial current), one finds corrections $R^p = -0.65 \ldots - 0.28$ and $R^n = -0.56 \ldots - 0.06$ for the proton and the neutron, in order. These are clearly larger than the dimensional argument (3.35) suggests. Furthermore, we also have to consider the many quark diagrams. Here, the situation is less transparent, i.e. it is much more difficult to deal with this class. To proceed, one can resort to some kind of "hadronic duality" [70,71,72]. This means that one considers a meson cloud picture instead of the more complex quark diagrams. As examples, the excitation of quark-antiquark pairs with the quantum numbers of a pion probed by the photon and with a successive vector boson exchange translates to the photon coupling to a pion in flight, with one pion-nucleon coupling parity-conserving (strong) and the other pv (weak). Similarly, other diagrams translate into a $\gamma\rho^0$ conversion followed by a pv ρN coupling. Such diagrams contribute to the nucleon axial current via [70,72]

$$\begin{aligned} \delta A_N^\pi &= \frac{g_{\pi N} h_{\pi N}}{m^2}\left[\frac{\pi}{6}\frac{m}{M_\pi} + \ln(\frac{M_\pi}{m}) + \ldots\right] , \\ \delta A_N^\rho &= \frac{h_{\rho N}}{g_{\rho N}}\frac{1}{m_\rho^2} \ . \end{aligned} \tag{3.37}$$

The strong coupling constants ($g_{\pi N}$, $g_{\rho N}$) are fairly well known. Matters are different for the parity-violating couplings ($h_{\pi N}$, $h_{\rho N}$). Their calculation has been pioneered by Donoghue et al. [73], a recent update including additional experimental constraints and chiral soliton model calculations can be found in refs. [74.75]. The pion cloud contribution diverges in the chiral limit which is no surprise in the light of the discussion presented in section 2. Ultimately, CHPT methods might shed some light on these long-distance contributions. In ref. [70], the estimates for processes of the type $V(e) \times A(N)$ range from $-0.07 \ldots + 0.37$ for R^p and $-0.07 \ldots + 0.24$ for R^n.

Combining these numbers with the ones from the one quark diagrams, one sees that there are potentially large corrections. To make all this more transparent, let us consider again the determination of the strange anomalous magnetic moment from pv electron-proton scattering. In backward direction and at small Q^2 we have [76]

$$A_{ep} \simeq -\frac{G_F Q^2}{4\sqrt{2}\pi\alpha}\left\{\frac{\kappa_p}{\mu_p}[1+R^M_{\rm strange}+R^M_{\rm rad}] - \frac{2m(E'+E)}{Q^2}\frac{1-4s^2}{\mu_p}g_A(1+R^A_p)\right\}, \quad (3.38)$$

where $R^M_{\rm strange,\,rad}$ denotes the strangeness and radiative corrections to the magnetic moment and R^A_p the one to the proton axial current. Two important observations are in order. First, the radiative corrections to the magnetic asymmetry are comparable to the strange quark induced ones,

$$\frac{R^M_{\rm rad}}{R^M_{\rm strange}} \simeq \frac{0.1}{\mu_s} \, . \quad (3.39)$$

Second, there is a large correction to the background of the axial asymmetry which makes up roughly 30 per cent of the signal for the SAMPLE kinematics. These two effects severely constrain the accuracy for determining the strange anomalous magnetic moment. To overcome these problems, it is therefore mandatory to perform many complementary experiments like e.g. backward-angle $\vec{e}\,d$ scattering, pv quasi-elastic scattering and so on. For a nice overview about these topics, I refer to ref. [76]. Finally, let me stress again that more theoretical effort is needed to further tighten the limits on these radiative corrections so that an unambiguous interpretation of the experiments will be possible.

4 Summary and Outlook

The standard model of the strong and electroweak interactions is an embarrassingly successful theory. It is least understood at its extreme energy limits. First, at very high energies, it is not known what exactly triggers the spontaneous symmetry breaking of $SU(2)_L \times U(1)_Y \to U(1)_{\rm em}$ at a scale of $<\phi> = 250$ GeV. While the standard Higgs boson does the job, one is left with just too many free parameters to feel confident with such a scenario. At present, the physics beyond the standard model hides itself quite effectively as discussed by Altarelli at this school [49]. Future high-energy colliders like the LHC or the SSC will hopefully shed light on the electroweak symmetry breaking sector. May be less spectacular, but as challenging, is the problem of hadron structure in the non-perturbative regime. Chiral perturbation theory is a method which allows to work out systematically the consequences of the spontaneous chiral symmetry breaking in QCD. It is based on a simultaneous expansion in the (small) external momenta and (light) quark masses (u,d,s) and enjoys considerable success in the meson sector. In the first part of these lectures, I have mostly been concerned with the problems surrounding the inclusion of matter fields in the chiral expansion. While this is technically a straightforward procedure based on chiral counting rules, it is conceptually less transparent than in the meson sector due to the appearance of the nucleon mass term. This is a scale of the order of the chiral symmetry breaking scale also it is not related to the symmetry violation and, furthermore, it does not vanish in the chiral limit. I have concentrated here on a few selected processes in the two-flavor sector like nucleon Compton scattering or pion photo- and electroproduction. Since the u and d quarks are really light, the corresponding expansion parameters $M_\pi/4\pi F_\pi$ and $E_{\rm pion}/F_\pi$ are small and one has a better chance of a converging chiral expansion. However, as we have seen in the discussion of the LET (2.14), the appearance of the nucleon mass renders these dimensional arguments rather dangerous. It appears at this point that in the heavy mass formulation of baryon CHPT such terms should not show up since the baryon propagator does not include the nucleon mass. However, in that case the large new scale is hidden by means of the Goldberger-Treiman relation in large prefactors accompanying the axial-vector coupling strength g_A. All calculations performed so far have been carried out in the one-loop approximation indicating that it is mandatory to include at least the terms of order q^4 (which are beyond next-to-leading order). This argument is further strengthened when one considers the three-flavor sector. Here, one very often finds large kaon and eta loop corrections which make one feel uneasy about the validity of the chiral expansion [77]. The suggestion that these large loop effects are largely canceled by contributions from the spin-3/2 decuplet in the intermediate states [16,18] is appealing but has not yet been put on a firm basis, which means that one has to perform a full order q^4 (or higher) calculation. There are, however, on-going activities in this area and further progress can be expected soon.

The second topic of these lectures concerned the quark structure of the nucleon at low energies. I have discussed how particular combinations of the electromagnetic and weak form factors, which parametrize the structure of the nucleon and its flavor decomposition, allow to extract interesting matrix elements like the strange anomalous magnetic moment, $F_2^s(0)$, or the strange axial nucleon charge, $G_1^s(0)$. Parity-violating electron scattering (γZ^0-interference) and (anti)neutrino scattering off nucleons or nuclei are the tools to determine these and other matrix elements like e.g. the extension of the nucleon as given by the strange electric radius. However, and that was a major topic here, since in most cases one tries to extract small numbers, one also has to account for radiative and heavy quark corrections. While these are nominally small, they often tend to be enhanced by large factors way above the naive dimensional estimates. In the most extreme case (cf. the isoscalar axial coupling), the tree level couplings with which one works at leading order are vanishing. In neutrino-hadron scattering, if one does not go below a typical scale of say 1 GeV, these various corrections are under control. As a particular example, I have discussed the induced isoscalar axial coupling, which is not only empirically interesting but also shows how under certain circumstances effective field theory methods can and *should* be used to calculate QED, QCD and heavy quark corrections [50]. When it comes to electron-hadron scattering, the situation is much less satisfactory. First, as I discussed, in many experiments one wants to work at much smaller energy scales and therefore has to account for the effects of quark confinement. Second, the direct photon coupling to the hadrons (quarks) gives rise too much more diagrams than it is the case in neutrino scattering. The most systematic analysis of radiative corrections to parity-violating electron-hadron scattering [70] translates the Feynman diagrams involving more than one quark in the nucleon into "hadronic ones", i.e. making use of the meson cloud picture. This induces uncertainties which are of the order of the calculated effects themselves. Another source of uncertainty is the momentum distribution of the bound quarks which one can only account for in an approximate fashion. Obviously, the theorists are called for providing better calculations of this type so that the experimenters have a better chance to extract the various strange matrix elements. Further complications arise when one considers the scattering off nuclei [78]. As I pointed out, many complimentary experiments which are sensitive to different combinations of the form factors and have different next-to-leading order corrections have to and will be performed at MAMI, CEBAF, ... [79]. These experiments are eagerly awaited. To come back to the first part of the lectures, once we will have learned how to treat the three-flavor sector of baryon CHPT, one will also be able to discuss the strange matrix elements and confront the theoretical predictions with the data.

Acknowledgments

I would like to thank the organizers, in particular Professor W. Plessas, for their invitation and kind hospitality extended to me. This work was supported in part by the Deutsche Forschungsgemeinschaft through a Heisenberg fellowship and the Schweizerischer Nationalfonds.

References

[1] J. Goldstone, *Nuovo Cim.* **19** (1961) 154
[2] H. Leutwyler, in "Recent Aspects of Quantum Fields", eds. H. Mitter and H. Gausterer, Springer Verlag, Berlin, 1991
[3] Ulf-G. Meißner, Recent Developments in Chiral Perturbation Theory, preprint BUTP-93/01 (1993)
[4] S. Weinberg, *Phys. Rev. Lett.* **17** (1966) 616
[5] H. Lehmann, *Phys. Lett.* **B41** (1972) 529; *Acta Phys. Austriaca Suppl.* **11** (1973) 139
[6] S. Weinberg, *Physica* **96A** (1979) 327
[7] J. Gasser and H. Leutwyler, *Ann. Phys. (N.Y.)* **158** (1984) 142
[8] G. Ecker, J. Gasser, A. Pich and E. de Rafael, *Nucl. Phys.* **B321** (1989) 311; J. F. Donoghue, C. Ramirez and G. Valencia, *Phys. Rev.* **D39** (1989) 1947
[9] J. Gasser and Ulf-G. Meißner, *Phys. Lett.* **B258** (1991) 219
[10] J. Gasser and Ulf-G. Meißner, *Nucl. Phys.* **B357** (1991) 90
[11] S. Coleman, J. Wess and B. Zumino, *Phys. Rev.* **177** (1969) 2239; C. G. Callan, S. Coleman, J. Wess and B. Zumino, *Phys. Rev.* **177** (1969) 2247
[12] A. Krause, *Helv. Phys. Acta* **63** (1990) 3
[13] J. Gasser, M.E. Sainio and A. Švarc, *Nucl. Phys.* **B307** (1988) 779
[14] E. Jenkins and A.V. Manohar, *Phys. Lett.* **B255** (1991) 558
[15] S. Weinberg, *Nucl. Phys.* **B363** (1991) 3
[16] E. Jenkins and A.V. Manohar, in "Effective Field Theories of the Standard Model", ed. Ulf–G. Meißner, World Scientific, Singapore, 1992
[17] Ulf-G. Meißner, *Int. J. Mod. Phys.* **E1** (1992) 561
[18] E. Jenkins and A.V. Manohar, *Phys. Lett.* **B259** (1991) 353
[19] V. Bernard, N. Kaiser and Ulf-G. Meißner, in preparation
[20] N.M. Kroll and M.A. Ruderman, *Phys. Rev.* **93** (1954) 233
[21] V. Bernard, J. Gasser, N. Kaiser and Ulf-G. Meißner, *Phys. Lett.* **B268** (1991) 291
[22] V. Bernard, N. Kaiser and Ulf-G. Meißner, *Nucl. Phys.* **B383** (1992) 442
[23] A. M. Bernstein and B.R. Holstein, *Comm. Nucl. Part. Phys.* **20** (1991) 197

[24] D. Drechsel and L. Tiator, *J. Phys.* **G18** (1992) 449
[25] V. Bernard, N. Kaiser and Ulf-G. Meißner, *Phys. Lett.* **B282** (1992) 448
[26] J. H. Koch, S. Scherer and J. L. Friar, to be published, and S. Scherer and J. H. Koch, *Nucl. Phys.* **A534** (1991) 461
[27] T. P. Welch et al., *Phys. Rev. Lett.* **69** (1992) 2761
[28] V. Bernard, N. Kaiser, T.-S. H. Lee and Ulf-G. Meißner, *Phys. Rev. Lett.* **70** (1993) 387
[29] Y. Nambu and D. Lurié, *Phys. Rev.* **125** (1962) 1429; Y. Nambu and E. Shrauner, *Phys. Rev.* **128** (1962) 862
[30] V. Bernard, N. Kaiser and Ulf-G. Meißner, *Phys. Rev. Lett.* **69** (1992) 1877
[31] F.J. Federspiel et al., *Phys. Rev. Lett.* **67** (1991) 1511
[32] F.E. Low, *Phys. Rev.* **96** (1954) 1428; M. Gell-Mann and M.L. Goldberger, *Phys. Rev.* **96** (1954) 1433
[33] V. Bernard, N. Kaiser, J. Kambor and Ulf-G. Meißner, *Nucl. Phys.* **B388** (1992) 315
[34] R.L. Workman and R.A. Arndt, *Phys. Rev.* **D45** (1992) 1789
[35] S.D. Drell and A.C. Hearn, *Phys. Rev. Lett.* **16** (1966) 908; S.B. Gerasimov, *Sov. J. Nucl. Phys.* **2** (1966) 430.
[36] J. Ashman *et al.*, *Phys. Lett.* **B206** (1988) 364.
[37] V. Bernard, N. Kaiser and Ulf-G. Meißner, Bern University preprint BUTP-92/51, 1992
[38] V. Burkert and Z. Li, *Phys. Rev.* **D47** (1993) 46
[39] G. Höhler, in Landolt–Börnstein, vol.9 b2, ed. H. Schopper, Springer, Berlin, 1983
[40] H. Arenhövel, these proceedings; B. Schoch, these proceedings
[41] R.E. Taylor, *Rev. Mod. Phys.* **63** (1991) 573
[42] S. Platchkov et al., *Nucl. Phys.* **A510** (1990) 740
[43] M. Gari and W. Krümpelmann, Phys. Lett. **B274** (1992) 159; A. Lung et al., *Phys. Rev. Lett.* **70** (1993) 718
[44] S.J. Brodsky and G.P. Lepage, *Phys. Rev.* **D22** (1980) 2157 and references therein
[45] Ulf-G. Meißner, *Phys. Rep.* **161** (1988) 213
[46] D. Beck, *Phys. Rev.* **D39** (1989) 3248
[47] R. Cahn and F. Gilman, *Phys. Rev.* **D17** (1978) 1313
[48] D. Beck, University of Illinois preprint NPL-91-015, 1991
[49] G. Altarelli, these proceedings
[50] D. Kaplan and A.V. Manohar, *Nucl. Phys.* **B310** (1988) 527
[51] E. Reya, these proceedings
[52] J. Gasser, H. Leutwyler and M.E. Sainio, *Phys. Lett.* **253B** (1991) 252, 260
[53] C.Y. Prescott *et al.*, *Phys. Lett.* **B77** (1978) 1347; *Phys. Lett.* **B84** (1979) 524
[54] BATES experiment 89-06, see e.g. D. Beck in ref. [61]
[55] R.D. McKeown, *Phys. Lett.* **B219** (1989) 40
[56] N.W. Park and H. Weigel, *Nucl. Phys.* **A541** (1992) 453
[57] V. Bernard and Ulf-G. Meißner, *Phys. Lett.* **B216** (1989) 392; **B223** (1989) 439
[58] R.L. Jaffe, *Phys. Lett.* **B229** (1989) 275
[59] M.J. Musolf and T.W. Donnelly, *Nucl. Phys.* **A546** (1992) 509
[60] T.W. Donnelly et al., *Nucl. Phys.* **A541** (1992) 525
[61] "Parity Violation in Electron Scattering", eds. E. Beise and R. McKeown, World Scientific Publ., Singapore, 1990
[62] E. Beise and R. McKeown, *Comm. Nucl. Part. Phys.* **20** (1991) 105
[63] L.A. Ahrens et al., *Phys. Rev.* **D35** (1987) 785
[64] V. Bernard, N. Kaiser and Ulf-G. Meißner, *Phys. Lett.* **B237** (1990) 545
[65] G.T. Garvey, W.C. Louis and D.H. White, Los Alamos preprint LA-UR-93-0037, 1992

[66] G.T. Garvey et al., *Phys. Lett.* **B289** (1992) 249
[67] J. Collins, F. Wilczek and A. Zee, *Phys. Rev.* **D18** (1978) 242
[68] H. Georgi, "Weak Interactions and Modern Particle Physics", Benjamin/Cummings, Reading, MA, 1984
[69] R.N. Mohapatra and G. Senjanovic, *Phys. Rev.* **D19** (1979) 2165
[70] M.J. Musolf and B.R. Holstein, *Phys. Lett.* **B242** (1990) 461
[71] M.J. Musolf in ref. [61]
[72] W.C.Haxton, E.M.Henley and M.J. Musolf, *Phys. Rev. Lett.* **63** (1989) 949
[73] B. Desplanques, J.F. Donoghue and B.R. Holstein, *Ann. Phys.* **124** (1980) 449
[74] N. Kaiser and Ulf-G. Meißner, *Nucl. Phys.* **A499** (1989) 699
[75] Ulf-G. Meißner, *Mod. Phys. Lett.* **A5** (1990) 1703
[76] M.J. Musolf, MIT preprint CTP 2120, 1992
[77] J. Bijnens, H. Sonoda and M.B. Wise, *Nucl.Phys.* **B261** (1985) 185
[78] C.J. Horowitz and J. Piekarewicz, Florida State University preprint, FSU-SCRI-93-18, 1993
[79] Some examples are MAMI-A4, CEBAF 91-004, CEBAF 91-010 and CEBAF 91-017

Polarized Structure Functions and the Spin of the Proton

E. Reya

Institut für Physik, Universität Dortmund, D-44221 Dortmund, Germany

1 Introduction

During the past six years the interest in spin physics originated mainly from the theoretically not anticipated result obtained by the EMC experiment [1] for the (longitudinally) polarized proton structure function $g_1^p(x, Q^2)$. In the following I will give a general review of the status and the importance of the study of polarized structure functions and parton distributions. For obvious reasons I shall concentrate mainly on the developments during the past six years.

In the next Section I will first discuss the interesting aspects of g_1, remind you of the origin of the surprising longitudinal spin 'problem', which initiated all the interest and excitement during the past few years, and discuss comparatively two perturbative scenarios for its solution in terms of either a large total sea-quark polarization or a large total gluon polarization (or a suitable combination of both); then I shall use the operator language to formulate these two scenarios for the spin structure of the proton which leads, via the generalized $U(1)$ Goldberger-Treiman relation, to non-perturbative definitions (and estimates) of the total contributions of quark and gluon components to the proton spin. This Section will be concluded by a discussion of the Bjorken-x dependence of polarized parton distributions and by a brief review of the most important experiments to be performed during the next few years which will be of crucial importance for testing and discriminating between the various theoretical ideas concerning the partonic spin content of nucleons; it is finally pointed out that a recent Fermilab experiment on asymmetries in pion production from (doubly) longitudinally polarized (anti)proton-proton scattering is inconclusive concerning the size of the total gluon polarization in the proton.

In Section 3 I will consider $g_2(x, Q^2)$, related to the transverse spin structure function of the nucleon, which will serve as a probe of the quark-gluon bound-state dynamics ("higher twist"), but is so far unfortunately totally unmeasured. Finally, a new class of transverse polarization (twist 2 and 3) chiral-odd nucleon structure functions, the so-called "transversity" distributions $h_{1,2}(x, Q^2)$, which have received much attention recently, will be discussed in Section 4 and the conclusions will be summarized in Section 5.

2 The Structure Function g_1 and the Spin of the Nucleon

The excitement about spin physics was generated in 1987 by theoretically not anticipated data of the EMC collaboration [2] at CERN on the longitudinally polarized deep inelastic structure function $g_1^p(x,Q^2)$ of the proton which appears in the antisymmetric component of the hadronic tensor

$$\begin{aligned} W_{\mu\nu}(p,q) &= \int d^4x e^{iq\cdot x} < p,s|j_\mu(x)j_\nu(0)|p,s > \\ &= -g_{\mu\nu}F_1 + \frac{1}{p\cdot q}p_\mu p_\nu F_2 + i\frac{M}{p\cdot q}\epsilon_{\mu\nu\alpha\beta}q^\alpha s^\beta g_1 + \cdots \end{aligned} \tag{1}$$

with p^μ being the four-momentum of the (proton) target and s^β its spin, with $p^2 = M^2$ and $s^2 = -1$, and as usual $x = Q^2/2p\cdot q$ where $Q^2 \equiv -q^2$. Experimentally $g_1(x,Q^2)$ is obtained by measuring the longitudinal polarization asymmetry [1]

$$A_L \equiv \frac{\overrightarrow{\mu}\overleftarrow{p} - \overrightarrow{\mu}\overrightarrow{p}}{\overrightarrow{\mu}\overleftarrow{p} + \overrightarrow{\mu}\overrightarrow{p}} \simeq \frac{g_1^p(x,Q^2)}{F_1(x,Q^2)} \tag{2}$$

where the arrows refer always to longitudinally polarized beam and target. Furthermore, within the 'naive' quark-parton model,

$$2F_1(x,Q^2) = \sum_q e_q^2[q+\bar{q}] + O(\alpha_s) \simeq \frac{1}{x}F_2(x,Q^2) \tag{3}$$

$$2g_1(x,Q^2) = \sum_q e_q^2[\Delta q + \Delta\bar{q}] + O(\alpha_s) \tag{4}$$

where $q(x,Q^2) \equiv q_+ + q_-$ and $\Delta q(x,Q^2) \equiv q_+ - q_-$ with $q_+(x,Q^2)$ $[q_-(x,Q^2)]$ denoting the quark distributions of flavor q in the proton aligned [anti-aligned] with its spin. When taken together with previous data from SLAC [3], the more recent EMC results [1] lead to a reasonably accurate determination of $g_1^p(x,Q^2)$ for $0.01 \leq x \leq 0.7$, with an average $< Q^2 > \simeq 10$ GeV$^2 \equiv Q_0^2$ from EMC and $< Q^2 > \simeq 5$ GeV^2 from SLAC. The combined SLAC and EMC data [1,3] lead to a rather small value for the first moment of $g_1^p(x,Q_0^2)$:

$$g_1^p(Q_0^2) \equiv \int_0^1 g_1^p(x,Q_0^2)dx = 0.126 \pm 0.010 \pm 0.015\ . \tag{5}$$

It should be emphasized that almost all of $\int_0^1 g_1^p dx$ stems from the region [1] $x \gtrsim 0.01$, i.e.

$$\int_{0.01}^1 g_1^p(x,Q_0^2)dx \simeq 0.123\ . \tag{6}$$

This surprisingly <u>small</u> EMC result initiated all the interest in spin physics during the past few years (for recent reviews see, for example, Refs. 4-12), since it is almost half as large as "naively" expected theoretically by Gourdin and Ellis and Jaffe (EJ) [13]

$$g_1^p(Q_0^2)_{EJ} \simeq 0.19. \tag{7}$$

This latter so-called Ellis-Jaffe "sum rule" has been obtained, as we shall see, by assuming that the total polarization of strange sea quarks contributes insignificantly to the spin of the proton, as one nevertheless expects intuitively due to the fact that it is easier for a gluon to create a nonstrange pair than a heavier strange pair – a situation very similar to the unpolarized broken $SU(3)$ sea as observed by neutrino-nucleon scattering experiments [14]. (Note that the original 1983 SLAC measurement [3], 0.17±0.05, was consistent with the expectation in Eq. 7.) The experimental results in Eqs. 5 and 6 have been extracted from measurements of $g_1^p(x, Q^2)$ at $x \gtrsim 0.01$ shown in Fig 1.

The surprising implications of the 'small' EMC result in Eq. 5 become immediately transparent by noting that $\int_0^1 g_1 dx$ measures the matrix element of the flavor-$SU(3)$ axial vector currents $j_{5a}^{\mu} = \bar{q}\gamma^{\mu}\gamma_5(\lambda_a/2)q$ which in turn can be expressed by the total polarizations (first moments), carried by (anti)quarks and gluons,

$$\Delta q(Q^2) \equiv \int_0^1 [\Delta q(x, Q^2) + \Delta\bar{q}(x, Q^2)]dx, \quad \Delta g(Q^2) \equiv \int_0^1 [g_+(x, Q^2) - g_-(x, Q^2)]dx, \tag{8}$$

in the following general way:

$$< p, s|j_{5a}^{\mu}|p, s >= \Delta\tilde{q}_a s^{\mu} \equiv (\Delta q_a + \delta_{a0}\Delta C_g \frac{\alpha_s}{2\pi}\Delta g)s^{\mu} \tag{9}$$

where Δq and Δg denote from now on always the total polarizations defined in Eq. 8, unless otherwise stated. The detailed form of the matrix element on the r.h.s. of Eq. 9 is of course a matter of convention for choosing an appropriate factorization scheme, i.e. particular values of the gluonic Wilson coefficient ΔC_g which describes the coupling of $\Delta g(Q^2)$ to the photon induced by the γ_5-triangle anomaly. Most of the (partly) controversial discussions during the past few years were and still are related to this problem and I shall discuss them in detail in the subsequent subsections. According to Eq. 4 the first moment of g_1^p can be expressed in terms of the flavor non-singlet ($a \neq 0$) and singlet ($a = 0$) partonic contributions in Eq. 9 in the following way:

$$\begin{aligned} & \qquad\qquad\qquad \text{non-singlet} \qquad\qquad \text{singlet} \\ \int_0^1 g_1^p(x, Q^2)dx &= \left[\frac{1}{12}\Delta q_3 + \frac{1}{36}\Delta q_8 \;+\; \frac{1}{9}\Delta\tilde{\Sigma}(Q^2)\right](1 - \alpha_s(Q^2)/\pi) \\ &= \left[(0.123 \pm 0.001) \;+\; \frac{1}{9}\Delta\tilde{\Sigma}(Q^2)\right](1 - \alpha_s(Q^2)/\pi) \end{aligned} \tag{10}$$

which is valid up to $O(\alpha_s)$ and where we have included the fermionic Wilson coefficient [15] $\Delta C_q = \int_0^1 \Delta C_q(x, \alpha_s)dx = 1 - \alpha_s/\pi$ with the HO $\alpha_s(Q_0^2)/\pi = 0.061$ for $\Lambda_{\overline{MS}} =$ 0.2 GeV. The flavor non-singlet contribution $\Delta q_3/12 + \Delta q_8/36 = 0.123 \pm 0.001$ is fixed by the Gamov-Teller part of octet hyperon β-decays via the Bjorken sum rule [16] ($SU(2)_f$ symmetry)

$$\Delta q_3 = \Delta u - \Delta d = g_A \equiv F + D = 1.254 \pm 0.006 \tag{11}$$

and [17]

$$\Delta q_8 = \Delta u + \Delta d - 2\Delta s \equiv 3F - D = 0.68 \pm 0.04 \tag{12}$$

where a $SU(3)_f$ symmetry has been assumed for relating matrix elements of charged and neutral weak axial currents in order to extract Δq_8! (Note that $SU(3)_f$ breaking effects could be sizeable, which will be addressed later.)

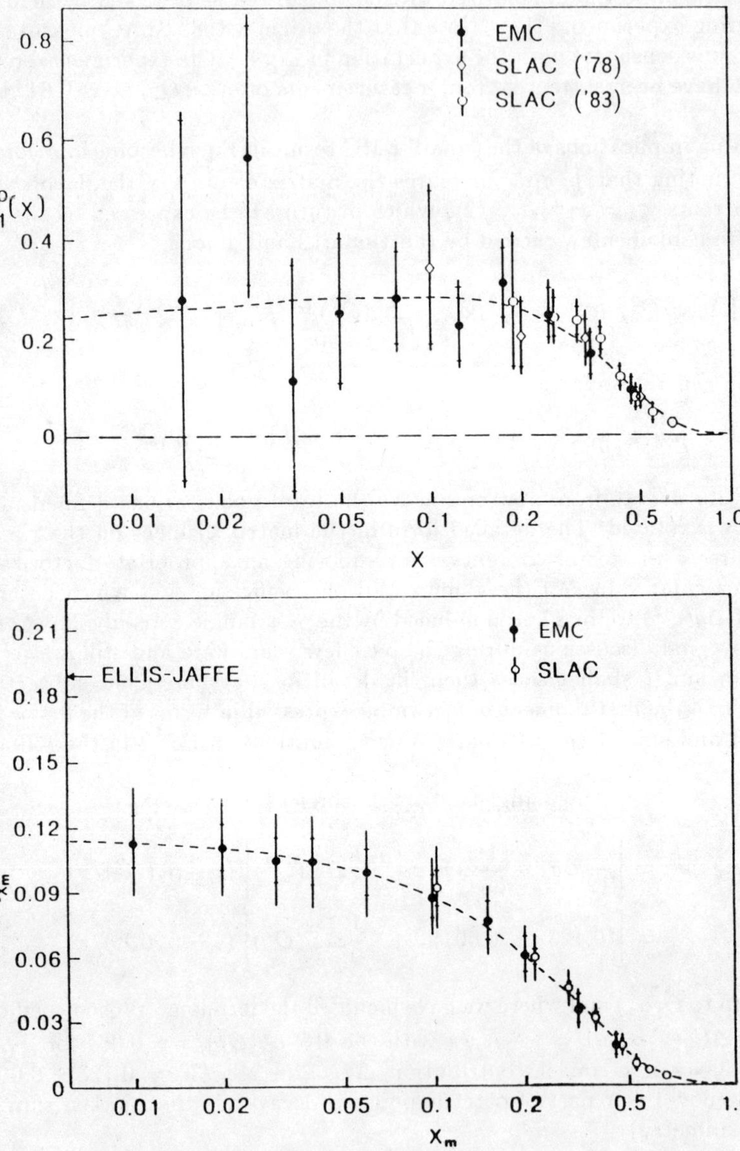

Fig. 1. EMC [1] and SLAC [3] measurements of $g_1^p(x, Q^2)$ and the appropriate integrated results as a function of the lower integration limit x_m. The dashed curve represents an optimized fit by EMC which gives rise to Eq. 5.

The surprising result now is that the non-singlet contribution in Eq. 10 alone already suffices to explain the 'small' EMC measurement in Eq. 5, in contrast to the larger EJ estimate (7), which implies an unexpectedly small or possibly even vanishing singlet component in Eq. 10:

$$\Delta\tilde{\Sigma}(Q_0^2) = 0.10 \pm 0.17 \tag{13}$$

where

$$\Delta\tilde{\Sigma}(Q^2) = \Delta\Sigma + 3\Delta C_g \frac{\alpha_s(Q^2)}{2\pi}\Delta g(Q^2) \tag{14}$$

according to Eq. 9 for $f = 3$ light quark flavors where

$$\Delta\Sigma \equiv \Delta q_0 = \Delta u + \Delta d + \Delta s = \Delta q_8 + 3\Delta s \tag{15}$$

with Δq_8 being given by Eq. 12. It is now obvious that the large EJ-estimate in Eq. 7 follows from the large Δq_8 contribution to $\Delta\tilde{\Sigma}$ in Eq. 10 by assuming $\Delta s \simeq 0$ in Eq. 15 and neglecting the $\alpha_s(Q^2)\Delta g(Q^2)$ term in Eq. 14. It should be furthermore noted that, in the 1-loop leading order (LO) of QCD, the total polarization of each quark flavor is conserved, i.e. Q^2-independent,

$$\Delta q(Q^2) = \text{ const.} \tag{16}$$

as a consequence of helicity conservation at the quark-gluon vertex, i.e. [18] $\Delta P_{qq}^{(0)} \equiv \int_0^1 \Delta P_{qq}^{(0)}(x)dx = 0$ and $\Delta P_{qg}^{(0)} = 0$. Furthermore [18,19] $\Delta P_{gq}^{(0)} = 2$ and since $\Delta P_{gg}^{(0)} = \beta_0/2$, where $\beta_0 = 11 - 2f/3$ in $\alpha_s = \alpha_s^{LO} = 4\pi/(\beta_0 \ln Q^2/\Lambda_{LO}^2)$, the polarized AP evolution equation [18] for $\Delta g(Q^2)$ implies that, apart from minor corrections, $\alpha_s(Q^2)\Delta g(Q^2) \simeq$ const. does not renormalize in LO,

$$\frac{d}{d\ln Q^2}\left[\alpha_s(Q^2)\Delta g(Q^2)\right] = 0 + O(\alpha_s^2), \tag{17}$$

similarly to $\Delta\Sigma(Q^2) = \Delta\Sigma=$ const. Therefore the combination $\alpha_s\Delta g$ should be treated at the same level as the contribution of the conserved quark polarizations to g_1 in Eq. 10 which has been originally emphasized in Refs. 20, 21.

The puzzling aspects and questions are now obvious:
(i) Why is the total polarization of the singlet combination $\Delta\tilde{\Sigma}(Q_0^2)$ in Eq. 13 vanishingly small as compared with the "naive" expectation $\Delta\tilde{\Sigma} \simeq \Delta\Sigma \simeq \Delta q_8 \simeq 0.68$ due to Eq. 15 for vanishing sea and gluonic contributions?
(ii)Why is there such a large difference in the flavor singlet sector between parton quarks and $SU(6)$ constituent quarks

$$\Delta\tilde{\Sigma}(Q_0^2) \simeq 0.1 << \Delta\tilde{\Sigma}_{SU(6)} = 1, \tag{18}$$

although there is only about a 30 % difference in the flavor non-singlet sector

$$\Delta q_3 \lesssim 1.25 \lesssim \Delta q_3^{SU(6)} = \frac{5}{3}, \quad \Delta q_8 \simeq 0.68 \lesssim \Delta q_8^{SU(6)} = 1 \tag{19}$$

due to the $SU(6)$ values $\Delta u = 4/3$, $\Delta d = -1/3$, $\Delta s = 0$? One expects that, for conserved quantities, constituent and parton results should coincide; the small differences in Eq. 19 are attributed to helicity non-conservation induced at low energy

scales by finite quark mass effects which break chirality and thus the symmetry in the non-perturbative region which creates a difference in the initial values for the perturbative QCD evolution [4]. Thus the huge difference in Eq. 18 appears to be too large to be explained in this way and demands a different explanation.
The explanation of the smallness of the total singlet contribution in Eq. 14 has been a matter of dispute during the past years: Whether a large negative Δs in Eq. 15 [or possibly some $SU(3)$ broken sea component] or a large Δg in Eq. 14 or, perhaps more realistically, a combination of both is responsible for the cancellation of the Δq_8 term in Eq. 15, is of course entirely a matter of the convention used in defining the polarized quark distributions due to the well known factorization scheme dependence of the higher order (HO) Wilson coefficient $\Delta C_g(x, \alpha_s)$. In other words, is it $\Delta\tilde{\Sigma}$ which is to be interpreted as the total quark polarization or is it $\Delta\Sigma$ in Eq. 15 with Δg as a separate independent gluon contribution? Let me now turn to a more detailed discussion of these two extreme alternative interpretations and whether physical and theoretical intuition might favor one over the other.

2.1 The Large Total Sea-Quark Polarization

Here all possible gluonic contributions in Eq. 14 are absorbed into the definition of quark polarizations, i.e. $\Delta q \to \Delta\tilde{q}$ according to Eq. 9. To simplify the notation let me just set $\Delta C_g = 0$, so that $\Delta\tilde{q} = \Delta q$ and thus $\Delta\tilde{\Sigma} \simeq \Delta\Sigma \simeq 0.1$ according to Eqs. 13 and 14 which give, using Δq_8 from (12) in Eq. 15 [22,23],

$$\Delta s = -0.19 \pm 0.06 \,. \tag{20}$$

This surprising result, implied by the new EMC measurement [1], of a large and negative polarization of the strange sea is in contrast to the intuitively reasonable ("canonical") expectation $\Delta s \simeq 0$ leading to Eq. 7 as discussed thereafter. The finite Δs in Eq. 20 implies significant changes of the total polarizations carried by u and d quarks:

$$\begin{array}{lllllllll} & & & & \text{canonical} & & & \text{new (EMC)} & \\ \Delta u & = & \frac{1}{2}(\Delta q_8 + \Delta q_3) + \Delta s & \simeq & 0.97 & + & \Delta s & = & 0.78 \pm 0.06 \\ \Delta d & = & \frac{1}{2}(\Delta q_8 - \Delta q_3) + \Delta s & \simeq & -0.29 & + & \Delta s & = & -0.48 \pm 0.06 \,. \end{array} \tag{21}$$

For comparison the $SU(6)$ static quark model predictions in Eq. 19 were $\Delta u = 4/3$ and $\Delta d = -1/3$. Apart from contributing differently to the nucleon's spin, such changes of Δu and Δd may be of astrophysical relevance [23] as well as of substantial consequences for, e.g., laboratory searches of supersymmetric dark matter (photino) candidates, and for the flux of neutrinos from dark matter (photino) annihilation in the sun which will be reduced due to the reduced photino trapping rate in the sun.

As discussed repeatedly, such a large total polarization of strange quarks is hard to swallow from a physics point of view and has therefore been repeatedly questioned [4-12]. It should be remembered that this result, Eq. 20, depends critically on the assumed $SU(3)_f$ symmetry between hyperon decay matrix elements of the flavor changing charged weak axial currents and the neutral axial currents in Eq. 9 in order to estimate Δq_8 in Eq. 12. There are serious obvious arguments against [17] a $SU(3)$ flavor symmetry, in contrast to the fundamental $SU(2)_f$ isospin rotation between matrix elements of charged and neutral axial currents which results in the Bjorken sum rule (11). An interesting alternative explanation of the EMC result has therefore been suggested by Lipkin [24] which is not based on a $SU(3)_f$ symmetry by assuming the sea to be only $SU(2)_f$ symmetric, with no $s\bar{s}$ pairs, i.e. $\Delta s \simeq 0$. In this case an additional assumption is necessary to enable one to extract the total helicity carried by valence quarks $\Delta q_v \equiv \Delta q - \Delta q_{sea}$ from hyperon β-decay data: It is assumed [24] that sea quarks do not participate in (flavor changing) β-decays which then determine $\Delta u_v \simeq 0.97$ and $\Delta d_v \simeq -0.29$ via Eqs. 11 and 12 (or Eq. 21). The EMC measurement (13) can then be explained entirely in terms of the total polarization of the light sea quarks [24].

$$\Delta u_{sea} = \Delta d_{sea} \simeq -0.29, \quad \Delta s \stackrel{!}{=} 0. \tag{22}$$

(Note that according to our definition (8), $\Delta u_{sea} \equiv 2\int_0^1 \Delta\bar{u}(x, Q^2)dx$.) This scenario has the welcome property that the $SU(3)_f$ symmetry is strongly broken with a marginal or possibly even vanishing Δs – a situation similar to the unpolarized broken $SU(3)$ sea [14] $\bar{u} = \bar{d} > s$. This scenario should, however, be considered as a toy model since the gluon component has been entirely disregarded which is responsible for the somewhat exaggerated $SU(3)_f$ breaking $\Delta q_8 \simeq 0.1$ as compared to the $SU(3)_f$ symmetric result $\Delta q_8 \simeq 0.68$ in Eq. 12. Even a small finite Δg contribution in Eq. 14 can already reduce substantially the large $SU(3)_f$ symmetry breaking.

There is an alternative proposal [25-27] that invokes the Skyrme model for an explanation of the small EMC result. Here one argues [25] that the quark singlet combination $\Delta\Sigma$ is suppressed by $1/N_c$ while $\Delta q = O(1)$ for each separate flavor, and a similar suppression should hold for $\Delta g(Q^2)$. Although it is somewhat unclear what precise relation there is between the Skyrme model and QCD, this explanation is even questionable [6,27] within the Skyrme model itself.

The characteristic feature of all these explanations of the small EMC result (13) is a small, or even vanishing, singlet quark contribution $\Delta\tilde{\Sigma} \equiv \Delta\Sigma = 0.10 \pm 0.17$. Therefore, these scenarios give rise to a "spin-surprise" (which is certainly not a "spin -crisis"!) by considering the helicity sum rule

$$\frac{1}{2} = \frac{1}{2}\Delta\Sigma + \Delta g + L_z \simeq \Delta g + L_z \tag{23}$$

i.e. the quarks do not sizeably contribute to the proton spin. Thus the spin of a nucleon is built up only from gluon-helicities and angular momentum L_z (which is the average value of the angular momentum of all quarks and gluons along the quantization z-axis due to the finite k_T carried by partons [6,17,28,29]) – a rather unexpected and curious result. It should be emphasized that the very reason for this surprise lies in the fact that for this kind of models $\Delta\Sigma$ is not constituent like, i.e. $\Delta\Sigma << \Delta\Sigma_{SU(6)} = 1$, as in Eq. 18. This is in contrast to the intuitively more reasonable pre-EMC Gourdin-Ellis-Jaffe expectation [13] $\Delta s \simeq 0$ which implies, due to Eqs. 15 and 12, $\Delta\Sigma \simeq \Delta q_8 \simeq 0.68$, i.e. about 70% of the proton's spin is carried by its fermionic constituents and the rest comes from $\Delta g(Q^2) + L_z(Q^2)$ in Eq. 23. It should be mentioned that, although the L_z contribution in (23) can be formally theoretically formulated in a consistent covariant way [6], there appears to be no direct experimental test of the size as well as the sign of $L_z(Q^2)$ – apart from azimuthal distributions [30] which are only sensitive to some average $< k_T^2 >$ of rotating constituents in a polarized nucleon target.

We turn now to an alternative explanation of the EMC result where the total gluon polarization plays a dominant rôle and which is not plagued by the aforementioned physical drawbacks.

2.2 The Large Total Gluon Polarization

Here the smallness of the singlet component in Eq. 13 is entirely described within the QCD improved parton model and is attributed to a large gluon helicity component Δg in Eq. 14 using [20] $\Delta C_g = -1$:

$$\Delta\tilde{\Sigma}(Q_0^2) = \Delta\Sigma - 3\frac{\alpha_s(Q_0^2)}{2\pi}\Delta g(Q_0^2) = 0.10 \pm 0.17 \tag{24}$$

where $\Delta\Sigma$ (not $\Delta\tilde{\Sigma}$!) is defined, via Eq. 15, as the total helicity of quarks. As an extreme assumption one takes, for definiteness, $\Delta s \simeq 0$ (although a more realistic finite but small value of Δs cannot be ruled out experimentally) which implies a large $\Delta\Sigma \equiv \Delta q_8 + 3\Delta s \simeq \Delta q_8 \simeq 0.68$ and thus Eq. 24 requires

$$\Delta g(Q_0^2) = 6.3 \pm 1.9 \tag{25}$$

in order to explain the small EMC result. The coupling of $\Delta g(Q^2)$ to the photon, induced by the γ_5-triangle anomaly [31], is described by the first moment of the Wilson coefficient $\Delta C_g = -1$ which has been defined relative to $\alpha_s/4\pi$ for each flavor of quarks and antiquarks separately [20,32]

$$\Delta C_g(\alpha_s) = \int_0^1 \Delta C_g(x, \alpha_s)dx = -\alpha_s/4\pi \tag{26}$$

i.e. $\Delta C_g(\alpha_s) \equiv (\alpha_s/4\pi)\Delta C_g$, which gives Eq. 24. Formally the gluon contribution in (24) appears as a non-leading correction at order α_s but, because of Eq. 17, $\alpha_s\Delta g$ is constant in Q^2 in the leading approximation and therefore should be treated at the same level as the Q^2-independent quark contributions to g_1 in the 'naive' LO parton model. It should be furthermore emphasized that the splitting of $\Delta\tilde{\Sigma}(Q^2)$ into a quark <u>and</u> gluon term depends on the definition of the quark which is formally the leading term: This partition (factorization scheme, $\Delta C_g = -1$) in Eq. 24 corresponds to $\Delta\Sigma$ defined in such a way as being <u>conserved</u> [4,33] also at the two-loop accuracy of massless QCD (i.e. $\Delta P_{qq}^{(1)} = 0$),

$$\frac{d}{d\ln Q^2}\Delta\Sigma(Q^2) = 0 + O(\alpha_s^3). \tag{27}$$

Since the present explanation of the low EMC measurement resorts to a large total quark polarization, $\Delta\Sigma \simeq \Delta q_8 \simeq 0.7$, we have

$$\Delta\Sigma \simeq 0.7 \leq \Delta\Sigma_{SU(6)} = 1 \tag{28}$$

i.e. $\Delta\Sigma$ is constituent-like, in contrast to Eq. 18, as one expects for conserved quantities. This is now in agreement with the flavor-nonsinglet sector in Eq. 19 and again the, say, 30% difference between partons and constituents in (28) represents the breaking from mass terms. Furthermore, in contrast to the previous explanation (scheme) based on a large sea polarization, we do not face a 'spin-surprise' here, since there is an (almost) "canonical" explanation of the nucleon spin

$$\frac{1}{2} = \frac{1}{2}\Delta\Sigma + \Delta g(Q^2) + L_z(Q^2) \tag{29}$$

where about 70% of the nucleon's spin is carried by its fermionic constituents and the rest comes from $\Delta g + L_z$ with $\Delta g(Q_0^2)$ being, however, sizeable according to Eq. 25.

Since the value of ΔC_g in Eq. 26 has been a matter of dispute during the past years [4-8,20,32-41], let me briefly comment on some details of its derivation. This is due to the fact, as has been already repeatedly emphasized, that the gluonic contribution in Eq. 24 is of course a matter of the convention used in defining the polarized quark distributions, due to the usual factorization scheme dependence of the HO Wilson coefficient $\Delta C_g(x, \alpha_s)$ – a situation very similar to unpolarized deep inelastic scattering beyond the leading order (for a detailed discussion of various schemes as well as of the related transformations of the 2-loop anomalous dimensions, the interested reader is referred to Refs. 33,42). Since a deep inelastic scattering process is supposed to probe the hard pointlike parton structure of nucleons, one can, however, arrive at the definite and finite result in Eq. 26 by resorting to the following two physically intuitive assumptions: (i) the gluonic term in (24) receives only hard, perturbatively calculable contributions and (ii) $\Delta\Sigma$ remains a conserved quantity. The gluonic contribution to g_1 is now calculated via the polarized subprocess $\gamma^* g_\pm \to q\bar{q}$ depicted in Fig. 2. The

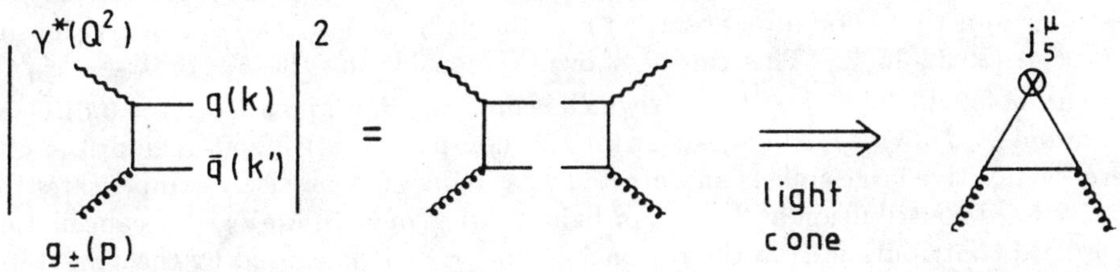

Fig. 2. The matrix element squared of the polarized subprocess $\gamma^* g \to q\bar{q}$, being related to the γ_5-triangle anomaly at the light-cone ($Q^2 \to \infty$), giving rise to the Wilson coefficient ΔC_g, i.e. to the coupling of Δg to g_1.

collinear divergences can be regularized in various different ways by giving, for example, a small mass to gluons ($p^2 < 0$) [32] or to quarks ($k^2 = k'^2 = m^2$) [20,35,37,10] or by using dimensional regularization ($n = 4 + 2\epsilon, \epsilon > 0$) [34,37,38,10] or by assuming a lower bound $k_{T,min}$ for the k_T of the outgoing quarks relative to the incoming gluon [36]. A straightforward calculation of the diagram in Fig. 2 then gives

$$g_1^{\gamma^* g \to q\bar{q}}(x, Q^2) =$$

$$\frac{\alpha_s}{4\pi} \times \left[\begin{array}{llll} 2\Delta P_{qg}^{(0)}(x)\left(\ln\frac{Q^2}{-p^2} + \ln\frac{1}{x^2} - 2\right) & + & 0 & p^2 < 0, m^2 = 0 \quad (30a) \\ 2\Delta P_{qg}^{(0)}(x)\left[\begin{array}{l} \ln Q^2/m^2 \\ -1/\hat{\epsilon} \end{array} + \ln\frac{1-x}{x} - 1\right] & + & 2(1-x) & \begin{array}{l} p^2 = 0, m^2 \neq 0 \\ \overline{\mathrm{MS}}(4+2\epsilon) \end{array} \quad (30b) \end{array}\right.$$

where $\Delta P_{qg}^{(0)}(x) = (2x-1)/2$ and $-1/\hat{\epsilon} \equiv -1/\epsilon + \gamma_E - \ln 4\pi + \ln Q^2/\mu^2$. When convoluted with $\Delta g(x, Q^2)$ in the usual way, this gives then the gluonic contribution to $g_1^p(x, Q^2)$ provided the collinear divergence terms proportional to $\ln Q^2/-p^2$, $\ln Q^2/m^2$ or $-1/\hat{\epsilon}$ are factorized, as usual, into $\Delta q(x, Q^2)$ and resummed by the evolution equations. The gluonic coefficient function $\Delta C_g(x, \alpha_s)$ is now determined by the remaining "finite" terms depending on the factorization scheme chosen, i.e. on how much of these "finite" terms is absorbed, together with the collinear singularities, into the definition of Δq. From Eq. 30a one might conclude

$$\Delta C_g^{(a)}(x, \alpha_s) = \frac{\alpha_s}{4\pi}(2x-1)(\ln\frac{1}{x^2} - 2) \tag{31a}$$

which, when integrated gives our advocated result in Eq. 26. On the other hand, from Eq. 30b one may (naively) conclude [10,37,38]

$$\Delta C_g^{(b)}(x, \alpha_s) = \frac{\alpha_s}{4\pi}\left[(2x-1)(\ln\frac{1-x}{x} - 1) + 2(1-x)\right] \tag{31b}$$

which has a vanishing first moment

$$\int_0^1 \Delta C_g^{(b)}(x, \alpha_s)dx = \frac{\alpha_s}{4\pi}[-1+1] = 0. \tag{32}$$

This vanishing result is obviously caused by the last positive term $+2(1-x)$ in Eq. 31b. A closer inspection [39,36,8] reveals, however, that this positive contribution $+2(1-x)$ in Eq. 30b derives from the soft non-perturbative collinear region where $k_T^2 \sim m_q^2 << \Lambda^2$ and thus should be absorbed into the definition of the light quark density $\Delta\Sigma(x, Q_0^2)$ [20,32-36,39]. This can be shown by considering the k_T, or $\lambda \equiv 4k_T^2/Q^2$ distribution [32,36,39] of $\int g_1^{\gamma^* g \to q\bar{q}} dx$ which is illustrated in Fig. 3: For $m = 0.01$ GeV, for example, $(d/d\lambda)\int g_1$ changes sign in the non-perturbative soft region $\lambda << 1$ where the positive large spike, concentrated at $\lambda \lesssim 4m^2/Q^2$, precisely compensates the negative hard contribution at $\lambda \gtrsim 1$. For light quarks ($m = 10$ MeV), this cancellation of the hard contribution from the region $k_T^2 \sim m^2 << \Lambda^2$ is caused by the same term in the $\gamma^* g_\pm \to q\bar{q}$ matrix-element-squared [39] which gives rise to the $+2(1-x)$ term in Eq. 30b. Absorbing this soft collinear contribution into the light quark density $\Delta\Sigma(x, Q_0^2)$ leaves us with the hard gluonic contribution to ΔC_g and Eq. 31b changes to

$$\Delta C_g^{(b')}(x, \alpha_s) = \frac{\alpha_s}{4\pi}(2x-1)\left(\ln\frac{1-x}{x} - 1\right) \tag{31b'}$$

which yields again the non-vanishing first moment in Eq. 26. (For heavy quarks ($Q = c, b, \cdots$), however, there are no contributions from the soft region and the positive contribution $+2(1-x)$ in Eq. 31b comes from the hard perturbative region [33,36,39] $\lambda \simeq 1$, i.e. around $k_T^2 \sim m_Q^2 >> \Lambda^2$, as illustrated by the short-dashed curve in Fig.3. Thus Eq. 31b is the appropriate expression for heavy quarks [43,44] with its vanishing first moment in Eq. 32, i.e. heavy quarks do not contribute to the total spin of the nucleon). It should be mentioned that different regularization procedures, for example introducing a 'brute force' lower bound for k_T with [36] $k_{T,min}^2 \sim O(\Lambda^2)$, result in the same hard gluon contribution to $\Delta C_g(x, \alpha_s)$ as in Eq. 31b'. For a gluon off-shellness $-p^2 = O(\Lambda^2)$ and $m^2 = 0$, as shown by the dotted curve in Fig. 3, one obtains again the same non-vanishing result for ΔC_g but with a different x-dependence for ΔC_g in Eq. 31a than the one in Eq. 31b'. Furthermore it is a peculiarity of the present case that the first moment ΔC_g in Eq. 26 can be calculated even without introducing any regulator at all [32,35], as shown by the long-dashed curve in Fig. 3 corresponding to $p^2 = 0 = m^2$, due to the vanishing of the first moment of $\Delta P_{qg}^{(0)}(x)$ in Eq. 30b.

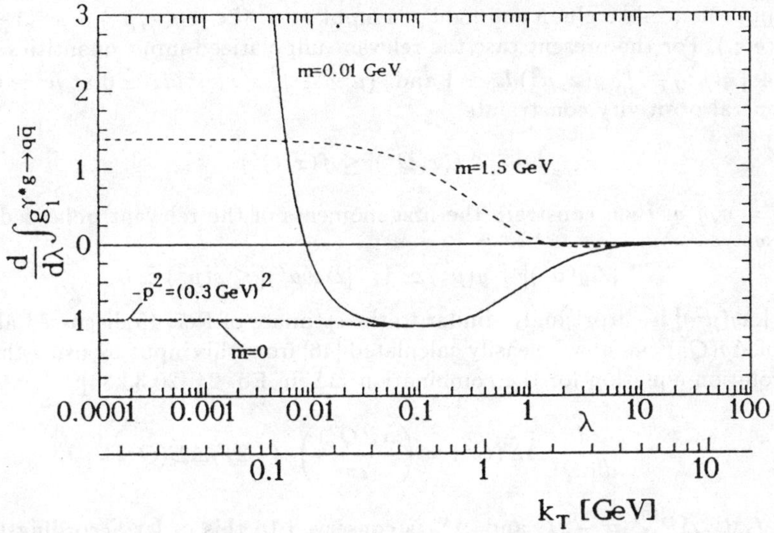

Fig. 3. The $\lambda \equiv 4k_T^2/Q^2$ distributions [39] of $\int g_1^{\gamma^* g\to q\bar{q}}dx$ in units of $e_q^2\alpha_s/8\pi$, using $Q^2 = 10\ \text{GeV}^2$, for various values of the quark mass parameter m. For comparison the dotted curve shows the analogous result for an off-shell gluon, $-p^2 \simeq \Lambda^2$, but with $m = 0$. Note that the λ-integral over $(d/d\lambda)\int g_1$ for this latter case has the value -2, which is also the case for the massless $m = 0$ distribution (long-dashed curve).

Having established that the finiteness of the gluonic Wilson coefficient is caused by the hard perturbatively calculable contributions of the triangle anomaly, the large total gluon polarization in Eq. 25, needed in addition for explaining the small EMC result, can of course not be calculated perturbatively and one has to resort to non-perturbative methods and/or to different hard processes where Δg can be measured independently. Before turning to a discussion of these topics let me mention that one expects, on rather general grounds [45] based on the (intrinsic) bound-state dynamics of the nucleon, counting rules, consistency constraints for g_-/g_+ and $\Delta g/g$ as $x \to 1$ and $x \to 0$, etc., the total gluon helicity to be sizeable, $\Delta g(\mu^2) \simeq 1.2$. Although the scale μ^2 for this prediction is unknown, one expects an intrinsic quantity to be relevant at a low bound-state scale, probably $\mu^2 \lesssim 1\ \text{GeV}^2$. Given the intrinsic polarization, one can obtain the "extrinsic" one by applying the evolution equations (radiative effects) starting at the bound-state scale μ^2 which increase [8,46] this intrinsic result to a value compatible with the experimental requirement at $Q_0^2 = 10\ \text{GeV}^2$ in Eq. 25:

$$\Delta g(Q_0^2) = 2 \text{ to } 4.5 \tag{33}$$

using $\mu^2 = (1 \text{ to } 0.3)\ \text{GeV}^2$ as the input scale for the solution (Eq. 37 below) of the evolution equation.

Along similar lines, the perturbatively defined $\Delta g(Q_0^2)$ can be dynamically estimated [46,42,47] using the renormalization group evolution equations supplemented by some non-perturbative valence or valence-like boundary conditions at a low bound-state resolution scale $Q^2 = \mu^2$. This radiative approach is very similar to the one for unpolarized nucleons [48] where the gluon and sea distributions, in particular in the unknown small-x regions, can be successfully generated from an experimental valence and finite valence-like gluon and light-sea input at μ^2 (i.e. $xg(x,\mu^2) \sim x^\alpha(1-x)^\beta$ with $\alpha > 0$, etc.). For the present case the relevant unpolarized input quantities are found to be [48] $g(\mu^2) \equiv \int_0^1 g(x,\mu^2)dx \simeq 1$ and $s(\mu^2) \equiv \int_0^1 s(x,\mu^2)dx \simeq 0$ at $\mu^2 \simeq 0.3$ GeV^2. The general positivity constraints

$$|\Delta f(x,Q^2)| \leq f(x,Q^2) \tag{34}$$

with $f = q, \bar{q}, g$, then constrain the first moments of the relevant polarized distributions by

$$|\Delta g(\mu^2)| \lesssim g(\mu^2) \simeq 1, \quad |\Delta s(\mu^2)| \leq s(\mu^2) \simeq 0 \tag{35}$$

where $|\Delta g(\mu^2)|$ is surprisingly similar to the estimate of Ref. 45 discussed above. The value of $\Delta g(Q_0^2)$ can now be easily calculated [46] from this input by using the relevant HO evolution equation for the combination $\Delta\Sigma$ in Eq. 24 [20,33,35]:

$$\frac{d}{d\ln Q^2}\Delta\tilde{\Sigma}(Q^2) = \left(\frac{\alpha_s(Q^2)}{2\pi}\right)^2 (-2f)\Delta\tilde{\Sigma}(Q^2) \tag{36}$$

where $f\Delta C_g \Delta P_{gq}^{(0)} = -2f$, and $\Delta\Sigma$ is conserved to this order according to Eq. 27. The solution of Eq. 36 is given by

$$\Delta g(Q^2) = \frac{\alpha_s(\mu^2)}{\alpha_s(Q^2)}\left[1 - \frac{2}{3}\frac{\alpha_s(\mu^2)-\alpha_s(Q^2)}{\pi}\right]\Delta g(\mu^2) + \frac{4}{9}\left[\frac{\alpha_s(\mu^2)}{\alpha_s(Q^2)} - 1\right]\Delta\Sigma \tag{37}$$

where I have used $\beta_0 = 11 - 2f/3 = 9$ for $f = 3$ light flavors. With $\Delta\Sigma \equiv \Delta q_8 + 3\Delta s \simeq \Delta q_8 \simeq 0.68$ and using the input in Eq. 35, this yields [46] $[\alpha_s(\mu^2)/\pi \simeq 0.216]$

$$-2.5 \lesssim \Delta g(Q_0^2) \lesssim 4 \tag{38}$$

in agreement with experiment, Eq. 25. It should be remarked that present EMC data [1], taken at face value, allow even for a negative [49] $\Delta g(Q_0^2)$ compatible with (38) which will be discussed in more detail later.

2.3 Non-Perturbative Operator/Current Algebra Approaches and Estimates

In the previous mechanism the effect of the γ_5-triangle anomaly is completely described within the context of the perturbative QCD improved parton model: Δg can be measured in hard processes and once the total singlet contribution $\Delta\tilde{\Sigma}$ is split into an anomalous gluon term and a quark term according to Eq. 24, $\Delta\Sigma$ approaches its naive value. In this sense the mechanism is "perturbative", although $\Delta\Sigma$ and Δg can certainly not be computed ab initio in perturbation theory, and the quark-like and gluon-like terms are, by construction, individually gauge invariant and free of any other theoretical ambiguities. This is in contrast to current algebra (CA) definitions [7,9,40,50,51] which in general include non-perturbative ($Q^2 \to 0$) contributions [5,52]. One can imagine that further non-perturbative contributions also due to the anomaly (e.g. instanton terms [53]) could be present in the conserved quark term making it different from the constituent limit even after subtraction of the anomalous gluon component. More explicitly, the gauge non-invariant topological gluonic current k_μ is usually employed for defining the total gluon helicity

$$< p, s|k_\mu|p, s > \equiv -\frac{\alpha_s}{2\pi}\Delta g^{CA}\bar{u}\gamma_\mu\gamma_5 u \tag{39}$$

with k_μ appearing in the anomalous divergence relation (neglecting the small light quark masses m_q)

$$\begin{aligned} \partial_\mu j_5^\mu &= f\frac{\alpha_s}{2\pi}\mathrm{tr}\left(F_{\alpha\beta}\tilde{F}^{\alpha\beta}\right) \\ &= f\partial_\mu\left[\frac{\alpha_s}{2\pi}\epsilon^{\mu\nu\rho\sigma}\mathrm{tr}A_\nu(F_{\rho\sigma} - \frac{2}{3}g_s A_\rho A_\sigma)\right] \equiv f\partial_\mu k^\mu \end{aligned} \tag{40}$$

which is due to the Adler-Bell-Jackiw γ_5-triangle anomaly [54]. Therefore the singlet axial vector current j_5^μ is not conserved even in the chiral limit for massless quarks. However, according to Eq. 40, a conserved (but gauge variant) quark singlet component $\Delta\Sigma^{CA}$ can be obtained by rewriting $j_5^\mu = (j_5^\mu - fk^\mu) + fk^\mu$ and identifying $< p, s|(j_5^\mu - fk^\mu)|p, s > \equiv s^\mu\Delta\Sigma^{CA}$. Furthermore, an explicit computation by Forte [53] of $< p, s|k_\mu|p, s >$ in Eq. 39 has revealed that $(\alpha_s/2\pi)\Delta g^{CA} = (\alpha_s/2\pi)\Delta g + \Omega$ where Δg corresponds to the gauge invariant hard piece of the anomaly which may be identified with our perturbative partonic definition used thus far; the remaining gauge variant Chern-Simons (instantonic) piece Ω is the soft non-perturbative contribution to the triangle-anomaly which does not correspond to a hard (large-k_T) subprocess (i.e. should be considered as part of the one-jet events [5,52] $\gamma^* q_\pm \to q_\pm$) and thus should be absorbed into the definition of the conserved total quark polarization:

$$\Delta\tilde{\Sigma} = [\Delta\Sigma^{CA} - f\Omega] - f\frac{\alpha_s}{2\pi}\Delta g \equiv \Delta\Sigma - f\frac{\alpha_s}{2\pi}\Delta g \tag{41}$$

leaving us with our original expression (24) with both terms on the r.h.s. of (41) being separately gauge invariant. The term $-f\Omega$ may be interpreted as an additional sea polarization induced by instantonic effects.

It was claimed that, by evaluating the matrix element $< p, s|k_\mu|p, s >$ on the lattice, one obtains the bound [55]

$$3\left|\frac{\alpha_s}{2\pi}\Delta g + \Omega\right| < 0.05. \tag{42}$$

This result is surprisingly small and, moreover, can obviously not be used to extract some information about Δg. It remains to be seen whether this result can be trusted in view of the approximations involved (quenched quarks, small $6^3 \times 10$ lattices and how to accommodate instantons, $\cdots$) and in view of the delicate problem of how to define the topological charge on a lattice.

The splitting of $\Delta\tilde{\Sigma}(Q^2)$ into a quark-like and a gluon-like term, both of them gauge invariant and in particular with the correct behavior under the renormalization group (RG), has been formulated and discussed in very general terms by Shore and Veneziano [40,51]. For this reason I will briefly outline their main results, although there is no general consensus on their interpretation [7,9,50].

Generalizing the classic results of current algebra and PCAC, the Goldberger-Treiman (GT) relation for the flavor non-singlet isotriplet channel ($F_\pi g_{\pi NN} = 2MG_A^{(3)}(0) \equiv Mg_A$ with $F_\pi \simeq 93$ MeV), to the flavor singlet channel one obtains [40,50,51] the so called "$U(1)$ Goldberger-Treiman" relation

$$F_{\eta 0}(0) g_{\eta_0 NN}(0) = 2MG_A(0). \tag{43}$$

The singlet axial-vector form factor $G_A(0)$ is defined from $< p,s|j_5^\mu(0)|p,s >$ which is related, via the OPE [40,51], to the large Q^2 limit of the matrix element $< p,s|j_\mu(q)j_\nu(-q)|p,s >$ in Eq. 1 measured indirectly by the polarized EMC/SLAC experiments [1-3]: $G_A(0)|_{Q_0^2=10\ \mathrm{GeV}^2} \equiv \Delta\tilde{\Sigma}(Q_0^2) = 0.10 \pm 0.17$ according to Eq. 13. It is important to note that η_0 appearing on the l.h.s. of the singlet GT-relation does not correspond to the physical $SU(3)$-singlet η' but instead to the unphysical would-be Nambu-Goldstone boson in the OZI limit of QCD [40] where the triangle anomaly is absent. This is so because $G_A(0)$ on the r.h.s. of (43) is not RG invariant, due to the non-conservation of the singlet axial-vector current in Eq. 40; therefore the l.h.s. of the singlet GT-relation must be scale dependent as well in order to be consistent with the RG, and so cannot simply correspond to the substitution of the η' for the π in the classical flavor non-singlet GT-relation where the anomaly is absent. This RG non-invariance resides in $F_{\eta_0}(0) \equiv 2f\sqrt{\chi'(0)}$, where $\chi'(0) \equiv \partial(i\int d^4x e^{iqx} < 0|T\partial\cdot k(x)\partial\cdot k(0)|0 >)/\partial q^2|_{q^2=0}$ with the topological charge density $\partial\cdot k$ defined in (40), being related to the topological susceptibility of massless QCD. (It would be tempting to relate $\chi'(0)$, using some smoothness assumption, to the topological susceptibility of a quarkless pure Yang-Mills theory, a quantity measured on the lattice [56] or by QCD sum rules [57] and typically found to be small implying the smallness of $G_A(0) \simeq 0.1$ via Eq. 43. The problem with this naive scenario, however, is that $\chi'(0)$ also controls the η' decay constant entering the calculation $\eta' \to \gamma\gamma$ which is well explained with an unsuppressed η' decay constant [40,58].) In order to gain more insight into the mechanism responsible for the smallness of the l.h.s. of the $U(1)$ GT-relation (43), it is useful to rewrite it in terms of the physical η' [51]:

$$F g_{\eta' NN}(0) + \frac{1}{2f} F^2 m_{\eta'}^2 g_{GNN}(0) = 2MG_A(0). \tag{44}$$

Here F is now a RG invariant process independent decay constant, related to the zero-momentum limit of the propagator for the composite operator $\bar{q}\gamma_5 q$, which is not the η' decay constant $F_{\eta'}$ defined analogously to F_π. Furthermore, g_{GNN} is the scale dependent coupling of the nucleon to the gluonic operator $\partial\cdot k$ in Eq. 38. From the requirement that the total quark polarization $\Delta\Sigma$ should be RG invariant, one may identify [40,51] in the $U(1)$ GT-relation (44)

$$2M\Delta\Sigma \equiv F g_{\eta' NN}\,, \quad 2M(-f)\frac{\alpha_s}{2\pi}\Delta g \equiv \frac{1}{2f} F^2 m_{\eta'}^2 g_{GNN} \tag{45}$$

as non-perturbative (gauge- and isospin-invariant) definitions of the "quark" and "gluon" components.

To go beyond these formal definitions one needs some guiding principle which allows to estimate the actual size of each individual term in the singlet GT-relation (44): One expects [51,58] the OZI approximation to apply to those quantities which are RG invariant (e.g. $Fg_{\eta' NN}$), i.e. conserved, and thus have the same RG behavior in QCD itself and in the OZI limit; for those quantities which are <u>not</u> RG invariant in QCD (e.g. g_{GNN}, G_A), but which are RG invariant in the OZI limit, one expects the OZI approximation to be strongly <u>violated</u> and therefore not applicable. Thus [51]

$$\Delta\Sigma \equiv \frac{1}{2M} F g_{\eta' NN} \overset{OZI}{\simeq} \frac{1}{2M}\left(\sqrt{6}F_\pi\right)\left(\sqrt{2}g_{\eta_8 NN}\right) \simeq 0.7 \tag{46a}$$

$$\Delta g \not\simeq \Delta g^{OZI} \sim g_{GNN}^{OZI} = 0 \tag{46b}$$

$$G_A \not\simeq G_A^{OZI} = \Delta q_8 \simeq 0.68 \tag{46c}$$

which implies that the quark component $\Delta\Sigma$ of the proton spin is <u>not</u> small, and, moreover, the observed small EMC value of $G_A \simeq 0.1$ would then be due to an approximate cancellation in (44) between $\Delta\Sigma$ and a <u>large</u> gluon component Δg, defined in (45), which must therefore display large OZI violations according to (46b).

This interpretation is suggested (but certainly not required) by the success of the OZI approximation for the decay $\eta' \to \gamma\gamma$ where the analogous relation to (44) is given by [58]

$$F g_{\eta'\gamma\gamma}(0) + \frac{1}{2f}F^2 m_{\eta'}^2 g_{G\gamma\gamma}(0) = \frac{4N_c}{3\pi}\alpha. \tag{47}$$

Here each term is separately RG invariant and thus the OZI approximation should apply reliably which gives simply

$$\sqrt{6}F_\pi g_{\eta'\gamma\gamma} \simeq \frac{4N_c}{3\pi}\alpha \tag{47'}$$

since $g_{G\gamma\gamma} \simeq 0$ in the OZI limit. This latter relation is quite accurately satisfied experimentally [58] which supports the reliability of the above arguments.

It is certainly interesting to have rather general <u>non</u>-perturbative arguments in favor of a large, conserved and thus constituent-like quark component $\Delta\Sigma$ which, on account of the small EMC measurement [1], implies a large gluon polarization Δg. Such a distinction between $\Delta\Sigma$ and Δg cannot be obtained from more recent lattice QCD estimates with dynamical staggered fermions [59] which yielded $G_A(0) \equiv \Delta\tilde{\Sigma} = 0.18 \pm 0.02$, in agreement with the EMC expectation in Eq. 13, but does not allow for extracting the separate quark and gluon contributions to $\Delta\tilde{\Sigma} = \Delta\Sigma - 3(\alpha_s/2\pi)\Delta g$. The same holds also for the postdictions of the EMC effect derived from QCD sum rules [60] which result also in a similarly small [61] <u>total</u> singlet spin contribution $\Delta\tilde{\Sigma}$. Nevertheless it is clear that only future (dedicated) experiments can ultimately decide about the physical reality of the various theoretical ideas and scenarios discussed so far. Since all realistic measurements are of course sensitive to the explicit x-dependence of the polarized parton distributions, $\Delta f(x, Q^2)$, and not just to their first moments, I will first discuss the various constraints and parametrizations for the Bjorken-x dependence of $\Delta f(x, Q^2)$, and afterwards turn to a discussion of phenomenological signatures for Δg and $\Delta\bar{q}$.

2.4 Bjorken-x Dependence of Polarized Parton Distributions

The directly measured structure function $g_1(x,Q^2)$ in Eq. 2, and not just its first moment in Eq. 10, is to $O(\alpha_s)$ in general given by

$$g_1(x,Q^2) = \frac{1}{2}\sum_{u,d,s} e_q^2\,[\Delta q + \Delta\bar{q}] * \Delta C_q + \frac{1}{3}\frac{\alpha_s(Q^2)}{2\pi}\Delta C_g * \Delta g \tag{48}$$

with the convolutions being defined by

$$\Delta C * \Delta f \equiv \int_x^1 (dz/z)\Delta C(z)\Delta f(x/z,Q^2) \tag{49}$$

and the fermionic Wilson coefficient is given by [34]

$$\begin{aligned}\Delta C_q(x,\alpha_s) = \delta(1-x) + \frac{\alpha_s}{2\pi}\frac{4}{3}\Bigg[&(1+x^2)\left(\frac{\ln(1-x)}{1-x}\right)_+ - \frac{3}{2}\frac{1}{(1-x)_+} - \frac{1+x^2}{1-x}\ln x \\ &+2+x-\left(\frac{9}{2}+\frac{\pi^2}{3}\right)\delta(1-x)\Bigg]\end{aligned} \tag{50}$$

which integrates to $\int_0^1 \Delta C_q(x,\alpha_s)dx = 1-\alpha_s(Q^2)/\pi$ as used in Eq. 10. The gluonic Wilson coefficient in (48) is, according to Eq. 31b', given by

$$\Delta C_g(x) = (2x-1)(\ln\frac{1-x}{x} - a) \tag{51}$$

where one usually chooses $a = 1$ although this choice is a mere matter of convention corresponding to factorizing the collinear piece in Eq. 30b proportional to $2\Delta P_{qg}^{(0)}\ln Q^2/m^2$ (or to $-1/\hat{\epsilon}$) into $\Delta\Sigma(x,Q^2)$. Since m^2 (or μ^2) is an entirely arbitrary regulator mass parameter, any other m'^2 could do as well, leaving us with a term $2\Delta P_{qg}^{(0)}\ln\, m'^2/m^2$ in (30b) which gives rise to $a \neq 1$ in Eq. 51. Note that this additional factorization convention ambiguity disappears for the first moment since it integrates to zero, $\int_0^1 \Delta C_g(x)dx = -1 - 0\cdot a$ as has been used in Eq. 24 and thereafter. It should furthermore be noted that a theoretically consistent Q^2-dependent next-to-leading order (NLO) analysis in Bjorken-x space of $g_1(x,Q^2)$ in Eq. 48 is not possible at present since the polarized 2-loop splitting functions $\Delta P_{ij}^{(1)}(x)$, required for the NLO Q^2-evolutions of $\Delta f(x,Q^2)$, are not available yet, in contrast to their first moments discussed up to now.

In view of the present scarce data [1-3], shown in Fig. 1, very little is known about the detailed x-dependence of $\Delta q(x,Q^2)$, $\Delta\bar{q}(x,Q^2)$ and $\Delta g(x,Q^2)$. They have only to satisfy the general positivity constraints (34)

$$|\Delta f(x,Q^2)| \leq f(x,Q^2) \tag{52}$$

with $f = q,\bar{q},g$, as well as the constraints (11) and (12) for their first moments

$$\Delta u_v(Q^2) = 2F \simeq 0.97, \quad \Delta d_v(Q^2) = F - D \simeq -0.29 \tag{53}$$

where the valence densities are as usual defined as $\Delta q(x,Q^2) = \Delta q_v + \Delta\bar{q}$ with $\Delta\bar{q} \equiv \Delta\bar{u} = \Delta\bar{d} = \Delta s = \Delta\bar{s}$, in contrast to the first moment definition in Eq. 8.

Furthermore, the small-x behavior of $g_1(x, Q^2)$ is, according to standard Regge lore, expected to be [25]

$$g_1(x, Q^2) \sim x^{-\alpha_{A_1}(0)} \quad \text{as } x \to 0 \ \text{(at } Q^2 = ?) \tag{54}$$

with the scale Q^2, where this relation is supposed to hold, being unknown, and where $\alpha_{A_1}(0) \in (-0.5, 0)$ is the intercept of the degenerate $A_1 \equiv a_1(1260)$, $D \equiv f_1(1285)$ and $E \equiv f_1(1420)$ Regge trajectories. From this we conclude that $g_1 \sim x^0$ as $x \to 0$, to be compared with a fit [25] (for $x < 0.2$) to the EMC data [1] $g_1^p(x, Q^2) \sim x^{0.07\,{}^{+0.32}_{-0.42}}$. The main uncertainty lies of course in the value of Q^2 where this Regge behavior is implemented. If, for example, we use (54) at a scale $Q^2 \simeq 1$ GeV^2 then, according to the QCD evolution, $g_1^p(x, Q_0^2)$ as well as $\Delta q(x, Q_0^2), \Delta\bar{q}(x, Q_0^2)$ and $\Delta g(x, Q_0^2)$ derived from it will be rather steeper at $x \to 0$ than in Eq. 54 at the average scale $Q_0^2 \simeq 10$ GeV^2. The only somewhat reliable conclusion we can draw from this is that Δq_v, $\Delta\bar{q}$ and Δg are <u>less singular</u> as $x \to 0$ than the unpolarized distributions $q_v \sim x^{-\alpha_\rho(0)} \sim x^{-1/2}$ and $\bar{q}, g \sim x^{-\alpha_{\mathcal{P}}(0)} \sim x^{-1}$.

There have been many attempts and suggestions ('educated guesses') [35,52,62-69] for parametrizations of $\Delta f(x, Q^2)$ which satisfy the above constraints and requirements. A compact and very elegant version of such parametrizations have been proposed recently [69]:

$$\Delta u_v(x, Q_0^2) = \alpha(x) u_v(x, Q_0^2), \quad \Delta d_v(x, Q_0^2) = \beta(x) d_v(x, Q_0^2) \tag{55}$$

with $\alpha, \beta \to 1$ as $x \to 1$ which reflects the argument [70] that the valence quark at $x = 1$ remembers the spin of the parent proton as supported also by perturbative QCD (i.e. [18] $P_{i_+ j_+} \sim$ const., $P_{i_- j_+} \to 0$ as $x \to 1$). This constraint is usually referred to as an 'optimal dilution' of the unpolarized valence distributions, i.e. assuming that the valence quarks carry most of the proton helicity only at large values of x. On the other hand, the region near $x = 0$ is expected to be dominated by the sea ($\Delta g, \Delta\bar{q}$) so that the spin of the parent proton would no longer be reflected in the valence quarks. This suggests that $\alpha, \beta \to 0$ as $x \to 0$. A representative example of these functions which reproduce present measurements is [69]

$$\alpha(x) = x^{0.3}, \quad \beta(x) = \left(\frac{x - x_0}{1 - x_0}\right) x^b \tag{56}$$

with $x_0 \simeq 0.35 - 0.75$ and $b = 0.03 - 0.76$. These valence contributions alone can obviously not account for the EMC data [1] which require a large polarized sea of the form

$$\Delta\bar{q}(x, Q_0^2) = -A_s x^{\delta_s} (1 - x)^p \bar{q}(x, Q_0^2) \text{ with } \delta_s > 0,\ p >> 1 \tag{57}$$

or/and large polarized gluon densities typically parametrized by

$$\Delta g^a(x, Q_0^2) = \pm A_a x^{\delta_g} (1 - x)^p g(x, Q_0^2) \ \text{ with } \ \delta_g \simeq 1,\ p \gtrsim 10 \tag{58a}$$

or

$$\begin{aligned} \Delta g^b(x, Q_0^2) &= \pm g(x, Q_0^2) && \text{for } x > x_c \simeq 0.08 \\ &= \pm A_b x g(x, Q_0^2) && \text{for } x \leq x_c \end{aligned} \tag{58b}$$

where the latter polarized gluon distribution, saturated for $x > x_c$ as maximally allowed by Eq. 52, is somewhat disfavored [71] by recent Fermilab measurements of the longitudinal spin asymmetry for π^0 production [72]. Typical examples of these two extreme (positive) polarized gluon distributions $\Delta g^{a,b}$ are shown in Fig. 4 and compared with a representative unpolarized gluon density. It should be furthermore emphasized that present data, taken at face value, do not even constrain [49] the sign of Δg as will be discussed in more detail below.

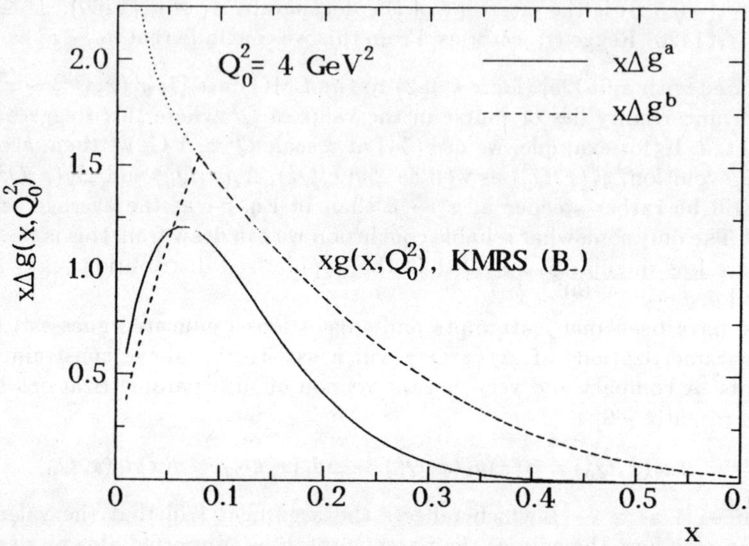

Fig. 4. Representative examples for the two large (positive) gluon scenarios in Eq. 58a,b which are taken from Ref. 71 and correspond to $\int_0^1 \Delta g(x, Q_0^2)dx \simeq 4$. The dotted curve corresponds to the unpolarized KMRS(B_-) gluon distribution [73] which satisfies the positivity constraint (52).

In Fig. 5 we show the representative results [65] of the extreme scenario where the EMC effect is explained by the $\Delta\bar{q}$ term in Eq. 48, assuming $\Delta g \equiv 0$: The fit has been based on the ansatz (57) with the result

$$x\Delta\bar{q}(x, Q_0^2) = -0.22x^{0.5}(1-x)^{14} \tag{59}$$

which integrates to $\Delta\bar{q}(Q_0^2) \equiv \int_0^1 \Delta\bar{q}(x, Q_0^2)dx = -0.1$ in agreement with Eq. 20 where Δs corresponds to $2\Delta\bar{q}$. It should be noted that in Fig. 5 the valence contribution alone would, when integrated, imply the Ellis-Jaffe result (7) as can be inferred from Eq. 10 (or Eq. 48) using the constraint (53). Alternatively, the required large and negative sea contribution in Fig. 5 can be attributed to a large gluon component in Eq. 48, assuming $\Delta\bar{q} \equiv 0$,

$$\frac{2}{3}\frac{\alpha_s(Q^2)}{4\pi}\int_x^1 \frac{dz}{z}\Delta C_g(z)\Delta g\left(\frac{x}{z}, Q_0^2\right) \tag{60}$$

with a large total gluon helicity $|\Delta g(Q_0^2)| \simeq 4-6$ according to Eq. 25. The actual form of $\Delta g(x, Q_0^2)$ becomes now strongly dependent on the chosen x-dependence of $\Delta C_g(x)$ in Eq. 51. Due to these ambiguities, present data ($x \geq 0.01$) allow even for a negative [49] $\Delta g(x, Q_0^2)$ and cannot distinguish between $\Delta C_g \equiv \int_0^1 \Delta C_g(x)dx = -1$

and $\Delta C_g = 0$ where the latter choice corresponds to keeping the soft $2(1-x)$ term in the 'hard' coefficient function as done in Eq. 31b and discussed thereafter. This is so because the 'truncated' moment ($x \geq 0.01$) of Eq. 60,

$$\frac{2}{3}\frac{\alpha_s(Q^2)}{4\pi}\int_{0.01}^{1} dx \int_x^1 \frac{dz}{z}\Delta C_g(z)\Delta g\left(\frac{x}{z}, Q_0^2\right) \tag{61}$$

does not factorize anymore into $\Delta C_g \cdot \Delta g(Q_0^2)$, in contrast to considering the (full) first moment $\int_0^1 g_1^p(x, Q_0^2)dx$ which has been done thus far (cf. Eq. 14). For the two choices, according to Eqs. 51 and 31b,

$$\Delta C_g^a(x) = (2x-1)\ln\frac{1-x}{x} - a(2x-1) \tag{62a}$$

$$\Delta C_g^b(x) = (2x-1)\ln\frac{1-x}{x} - b(2x-1) + 2(1-x) \tag{62b}$$

corresponding to $\Delta C_g^a \equiv \int_0^1 \Delta C_g^a(x)dx = -1$ and $\Delta C_g^b = 0$, representative results are shown in Fig. 6 for $a \simeq b_+ \simeq 2$ (which require a positive $\Delta g(x, Q_0^2)$ as illustrated in Fig. 6) and for $b_- \simeq -3$ (which requires a negative polarized gluon density). The polarized gluon densities shown in Fig. 6 integrate to $|\int_{0.01}^1 \Delta g^{a,b}(x, Q_0^2)dx| \simeq 4.5$ and give a fair description of the EMC data in each case.

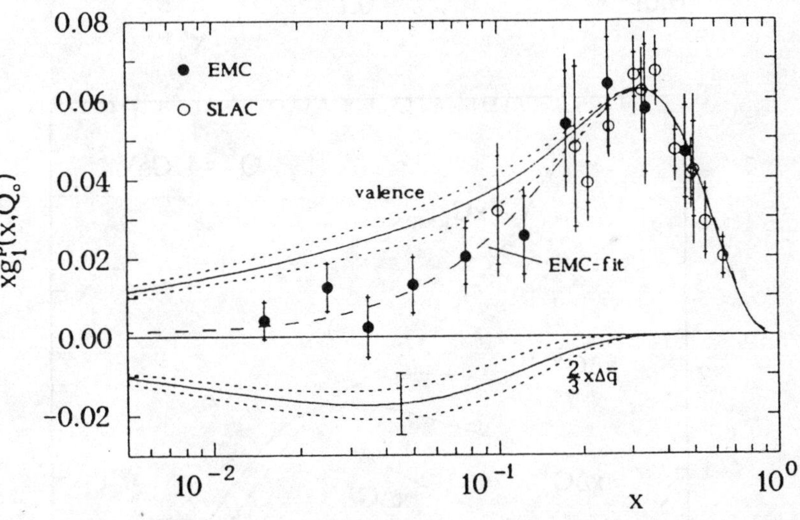

Fig. 5. Extraction [65] of the polarized valence and sea $\Delta\bar{q}$ distribution (assuming $\Delta g \equiv 0$) from the data [1,3] on g_1^p. The central EMC fit [1] is shown by the long-dashed curve. The short-dashed curves correspond to the allowed range of $3F - D$ in Eq. 12. The vertical bar indicates the typical uncertainty for $2x\Delta\bar{q}/3$ due to the EMC data. Note that the valence contribution alone integrates to about 0.19, i.e. the Ellis-Jaffe result (7).

Nevertheless, it is clear that only future (dedicated) experiments can ultimately decide about such ambiguities and about the physical reality of the various theoretical scenarios discussed so far. In particular, the polarized gluon distribution $\Delta g(x, Q^2)$, about which no detailed experimental information whatsoever exists for the time being, is here of crucial importance. It is thus clear that we need much MORE information from polarized experiments in the (hopefully near) future.

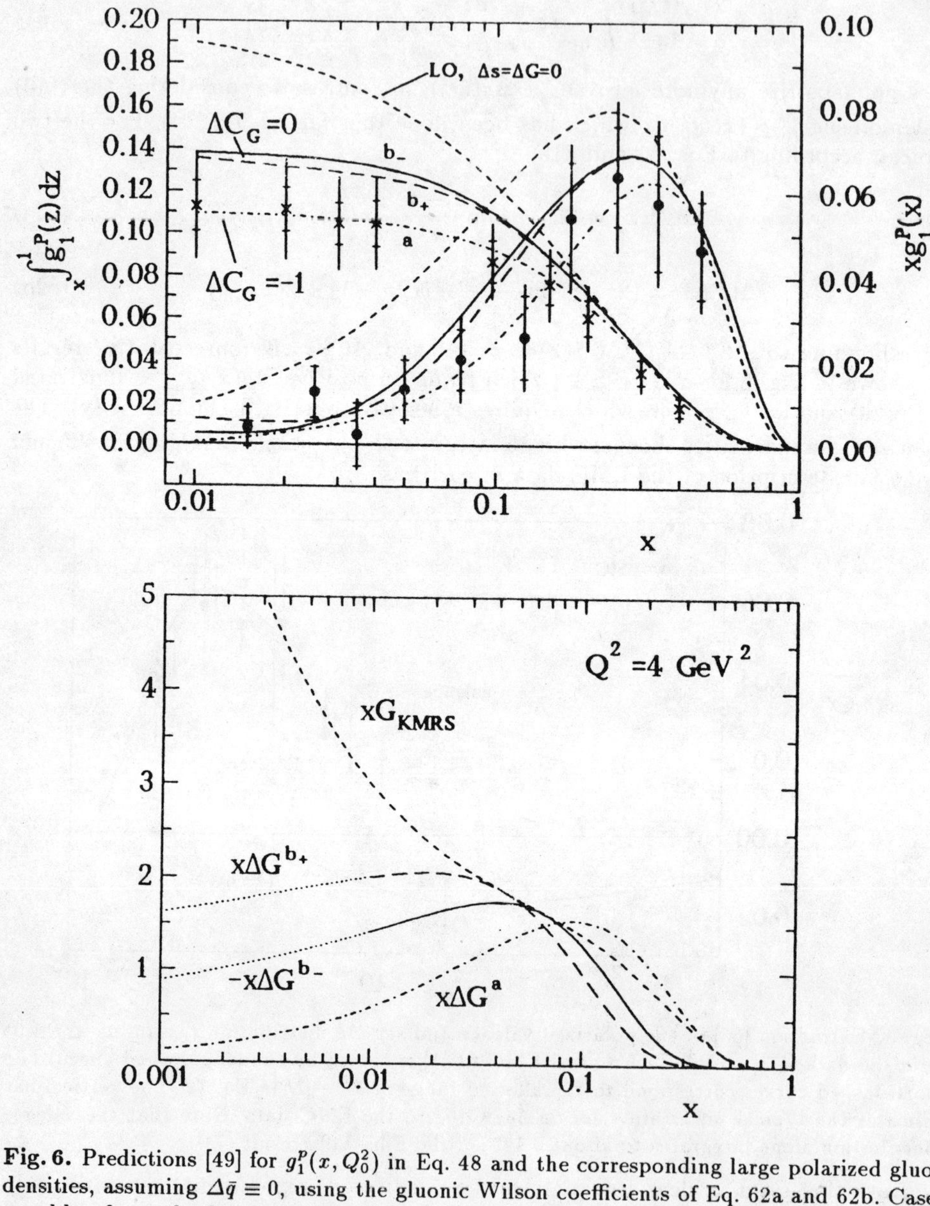

Fig. 6. Predictions [49] for $g_1^p(x, Q_0^2)$ in Eq. 48 and the corresponding large polarized gluon densities, assuming $\Delta\bar{q} \equiv 0$, using the gluonic Wilson coefficients of Eq. 62a and 62b. Cases a and b refer to the first moments $\Delta C_g = -1$ and $\Delta C_g = 0$, respectively. The data are taken from Ref. 1.

2.5 Phenomenological Signatures of Δg and $\Delta\bar{q}$

There have been many suggestions for measuring polarized parton distributions with hard processes [4,5,7,8] but it appears to be particularly difficult to extract information about $\Delta g(x,Q^2)$. Since it is beyond the scope of this talk to discuss all polarized processes suggested so far, I will concentrate on those which I believe are the most important ones for delineating Δg. (All realistic measurements are of course sensitive to $\Delta f(x,Q^2)$ and not just to their first moments. The interested reader is referred to the reviews in Refs. 5,7,8,35,52,74 and the references therein, for a discussion of the detailed x- and Q^2-dependence of polarized parton distributions.) As in the unpolarized case, only processes where Δg occurs directly already in the LO (with no Δq and $\Delta\bar{q}$ contributions present) appear to be the most promising source for measuring $\Delta g(x,Q^2)$. This is the case for

(i) deep inelastic (or photon) production of heavy quarks $Q=c,b$ via the fusion process $\gamma^*(\gamma)\Delta g \to Q\bar{Q}$ responsible for open charm [22,35,44] or J/ψ production [75,76]. Although this production mechanism for open charm production is theoretically clean with a sizeable polarization asymmetry [44] $A^c = \Delta\sigma^c/\sigma^c = 20-30\%$, it will be hard for the presently ongoing polarized fixed target DIS experiments [77,78] to observe a sufficient amount of events since $\sigma^c \simeq 400$ nb for $E_\gamma \gtrsim 100$ GeV. Here, I believe, because of the high-intensity $\vec{e}$-beam, the SLAC experiment [78] is in a much better shape than the SMC $\vec{\mu}\vec{p}$ experiment [77] at CERN. The situation is similar for J/ψ production [75,76] but here, however, one faces the additional model dependence (duality [79] or color singlet [80] model) for bound-state production. Anyway, the hope is [76] that SMC could measure $\Delta g(x, m^2_{J/\psi})$ in the region $0.03 \lesssim x \lesssim 0.25$;

(ii) hadronic heavy quark production [81,82], via the LO subprocesses

$$\Delta g\Delta g \to Q\bar{Q}, \qquad \Delta q\Delta\bar{q} \to Q\bar{Q} \tag{63}$$

which depends "quadratically" on Δg and where the $\Delta q\Delta\bar{q}$ contribution is small [82], appears to be a very sensitive and presumably the best (most realistic!?) test of Δg. Here the polarized $\vec{p}\,\vec{p}$ RHIC collider ($\sqrt{s} = 50-500$ GeV) with high luminosity ($L \gtrsim 10^{32}$ cm^{-2}sec^{-1}) will play a decisive role provided, of course, our experimental colleagues succeed in polarizing longitudinally both proton beams;

(iii) more recently it has been emphasized that singly polarized hadron-hadron collisions [83-86], i.e. where only one incoming (possibly fixed target) hadron is polarized, provide access to $\Delta g(x,Q^2)$ as well which appears to be experimentally easier than working with both initial hadrons polarized. However, the prize one has to pay here is that for Drell-Yan dilepton production [83,84] either the polarization of one of the final leptons has to be measured [83],

$$p\vec{p} \to \gamma^* X \to \mu^+\vec{\mu}^- X$$

or the angular distribution of the produced lepton pair [84] as a polarimeter for the virtual photon. Alternatively one could also consider the production of a polarized direct-photon [85,86] via $p\vec{p} \to \vec{\gamma}X$. It should be emphasized that one could use here the energetic unpolarized proton beams of the TEVATRON $\bar{p}p$ or HERA ep colliders ($E_p \simeq 1$ TeV) to be scattered off a polarized fixed p-target [85], still giving rise to a sizeable $\sqrt{s} \simeq 43$ GeV.

There have been many further suggestions and studies to measure $\Delta g(x,Q^2)$ and/or $\Delta\bar{q}(x,Q^2)$ in other suitable hard processes initiated by doubly polarized hadron-hadron collisions such as the production of large-p_T photons [67,86-90], Drell-Yan dimuons [34,67,86,89,90], large-p_T jet asymmetries of final $jj, jjj, jj\gamma$ events [87,62,68,91-94], etc. Again, the future high luminosity RHIC and UNK colliders will be essential for performing such experiments. Recently, polarization asymmetries for semi-inclusive pion production have been measured from doubly longitudinally polarized (anti)proton-proton collisions at the Fermilab SPF [72] resulting in $A_{LL}^{\pi^0}(\overset{(-)}{p}p) \simeq 0$ for $1 \lesssim p_T^{\pi^0} \lesssim 4$ GeV at $E_{beam}^{\overset{(-)}{p}} = 200$ GeV, i.e. $\sqrt{s} = 20$ GeV. It should be pointed out that this result does <u>not</u> necessarily imply a vanishing [72,93] gluon polarization Δg, but is equally consistent with a large $\Delta g \simeq 3-6$ [71]. A clean distinction between a large and a small Δg scenario could be achieved, if it were possible to perform such a semi-inclusive experiment at, say, $\sqrt{s} \simeq 100$ GeV with $p_T^{\pi^0} \gtrsim 5$ GeV [71].

Furthermore, measurements of $g_1(x,Q^2)$ at various fixed values of Q^2 would be very desirable in order to extract from measured scaling-<u>violations</u> $\Delta g(x,Q^2)$ and $\Delta\bar{q}(x,Q^2)$ using, as usual, the polarized evolution equations [18]. Strictly speaking, this would be possible only in the LO at present [35,52,64-69,89]. Since Δg enters Eq. 14 formally in HO, a theoretically consistent analysis in Bjorken-x space requires, however, the polarized 2-loop splitting functions $\Delta P_{ij}^{(1)}(x)$, which are not yet available, in order to allow for a renormalization/factorization scheme invariant analysis of $g_1^p(x,Q^2)$ in Eq. 48. It should also be noted that the inclusion of neutral (Z^0) and charged ($W^{\pm}$) weak currents gives rise to additional parity-violating polarized structure functions [95] $g_{3,4}$ which could reveal some direct information about the polarized valence distributions in the proton, provided $Q^2 \gtrsim 10^3$ GeV^2.

Finally, I remind you that the small EMC result (5) implies further dramatic consequences for the polarized <u>neutron</u> structure function $g_1^n(x,Q_0^2)$ by making use of the "sacred" (isospin invariance) Bjorken sum rule [16] in Eq. 11,

$$\int_0^1 [g_1^p(x,Q^2) - g_1^n(x,Q^2)]\, dx = \frac{1}{6} g_A (1 - \alpha_s(Q^2)/\pi) \tag{64}$$

where I have included the small α_s/π correction [15]. This gives, using Eq. 5,

$$\int_0^1 g_1^n(x,Q_0^2)dx = -0.07 \pm 0.02 \tag{65}$$

which is about ten (!) times larger than the pre-EMC Ellis-Jaffe expectation $g_1^n(Q_0^2)_{EJ} \simeq -0.006$ based on Eq. 7. These predictions for g_1^n can be straightforwardly translated into Bjorken-x space,

$$g_1^n(x,Q_0^2) = g_1^p(x,Q_0^2) - \frac{1}{6}\left[\Delta u_v(x,Q_0^2) - \Delta d_v(x,Q_0^2)\right], \tag{66}$$

where the insignificant HO contribution due to ΔC_q in Eq. 64 has been suppressed, which are shown in Fig. 7 for two different sets of polarized valence densities [63,65] with $g_1^p(x,Q_0^2)$ taken from EMC/SLAC measurements [1,3].

Due to the large negative area of g_1^n in Eq. 65, the predictions in the small-x region ($x \lesssim 0.2$) differ of course significantly from the original pre-EMC Gourdin-Ellis-Jaffe

[13] estimate based on assuming $\Delta\bar{q} = \Delta g = 0$. Since this latter estimate disagrees with the more recent EMC measurement [1] in Eq. 5, a confirmation of the large and negative predictions for $g_1^n(x, Q_0^2)$ in Fig. 7 could serve also as a sensitive cross-check of the EMC data on $g_1^p(x, Q_0^2)$ in the small-x region. Such neutron measurements will be available soon [77, 78] which, furthermore, will serve as a fundamental test of the crucial Bjorken sum rule and of the QCD-improved parton model! Any significant deviation from these predictions would indicate the importance of non-perturbative higher-twist contributions invalidating Eq. 66.

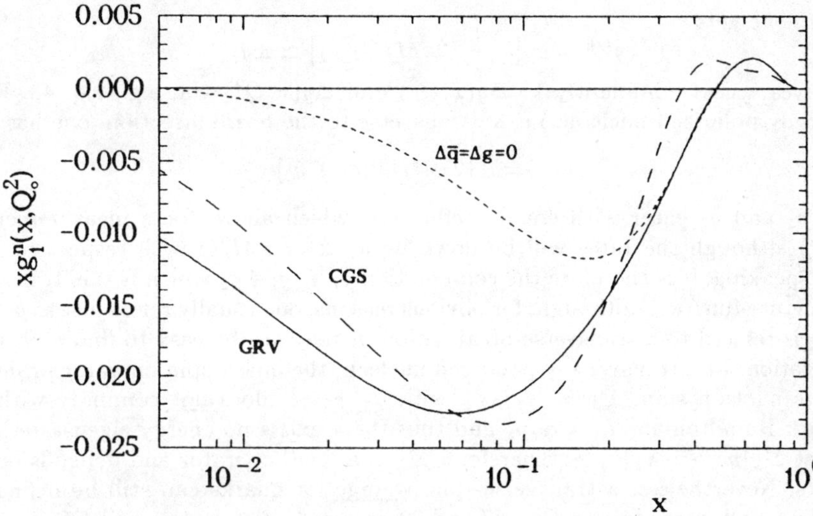

Fig. 7. Predictions for $g_1^n(x, Q_0^2)$ according to Eq. 66, where $g_1^p(x, Q_0^2)$ has been taken from the central EMC fit [1]. The solid (GRV [65]) and dashed (CGS model I [63]) curves refer to two different polarized valence quark densities. For illustration the short-dashed curve shows the expectation for g_1^n based on the original estimate of Gourdin, Ellis and Jaffe [13] (i.e. $\Delta\bar{q} = \Delta g = 0$) using the above valence distributions of GRV [65].

Alternatively, there are non-QCD models that contemplate a violation [96] of the Bjorken sum rule, where the r.h.s. of (64) is expected to be about half as large ($\simeq 0.1$) due to physically motivated final states ("fire-strings") as opposed to quarks and gluons; this violation, being of non-partonic origin, is thus not related to a breakdown of isospin symmetry. On the other hand it has been speculated that the non-perturbative instanton contribution Ω to Eq. 39, as given in Eq. 41, could give rise to isospin violations of the order [97] $\Omega^p/\Omega^n \sim m_d/m_u$ which again might suppress the r.h.s. of the Bjorken sum rule (64) by about a factor of 2.

3 The Transverse Spin Structure Function g_2

The second spin-dependent deep inelastic structure function g_2, which is kinematically suppressed, appears in the antisymmetric component of the hadronic tensor in Eq. 1 in the following way (for recent reviews, see Refs. 98, 99, 4):

$$W_{\mu\nu} = \cdots + i\epsilon_{\mu\nu\alpha\beta}q^\alpha \frac{M}{p\cdot q}\left[s^\beta g_1 + \frac{p\cdot q s^\beta - s\cdot q p^\beta}{p\cdot q} g_2\right] + \cdots \tag{67}$$

For the longitudinal polarization considered so far, i.e. s^β (anti)parallel to the beam direction, one gets

$$W^{\parallel} = \pm\left[g_1 - (2xM/Q)^2 g_2\right] \simeq \pm g_1 \tag{68}$$

which gives access dominantly to $\Delta q(x,Q^2)$ and $\Delta g(x,Q^2)$ via, e.g., Eq. 48. For a transversely polarized nucleon, i.e. s^β transverse to the beam direction, one has

$$W^{\perp} = \pm(2xM/Q)[g_1 + g_2] \tag{69}$$

so that g_1 and g_2 enter with equal coefficients, which allows for a measurement of $g_2(x,Q^2)$ although the rates will be down by a factor xM/Q with respect to (68). Strictly speaking, it is therefore the combination $g_T \equiv g_1 + g_2$ which is the 'transverse spin structure function' although, for obvious reasons, one usually refers just to g_2 [98]. Since g_2 is related to a transverse polarization, it may not be easy to find a partonic interpretation: In a transversely polarized nucleon, the quark-spin operator projected along the nucleon spin, $\Sigma_\perp = \gamma_0\gamma_5 \not{s}_\perp$ with $\not{s}_\perp \sim \gamma^1$, does <u>not</u> commute with the free quark Hamiltonian $H_0 = \alpha_z p_z$ and thus there exists no energy eigenstate $|p_z>$ such that $\Sigma_\perp |p_z> = \lambda_\perp |p_z>$. Therefore, $\Sigma_\perp$ is a 'bad' operator and depends on the dynamics. Nevertheless, a transverse-spin <u>average</u> for quarks can still be defined in the nucleon and, according to Eqs. 67 and 69, it is just $g_T \equiv g_1 + g_2$ which is sensitive to the quark-gluon interactions – a clear sign that no simple parton interpretation can be made for it [98,100]. (This is in contrast to the longitudinally polarized nucleon in (68), where the quark helicity operator $\Sigma_\parallel = \gamma_0\gamma_5 \not{s}_\parallel$, with $\not{s}_\parallel \sim \gamma^3$, commutes with H_0 and thus $g_1(x)$ measures directly the quark helicity distribution.) Furthermore, g_2 vanishes [100] for a free (massless or massive) quark, i.e. for a pointlike nucleon, and thus g_2 cannot be expressed as an incoherent sum over free on-shell partons. The partons must be interacting (quark-gluon interactions) and/or virtual in order to contribute to g_2. Therefore g_2 will serve as a unique probe of 'higher twist' (twist $\equiv$ dim. – spin = 3) as well.

Regardless of the difficulties with a partonic interpretation, one usually writes

$$g_2 = g_2^{WW} + \bar{g}_2 \tag{70}$$

where the twist-2 'Wandzura-Wilczek' piece [101] g_2^{WW} receives contributions from the same class of twist-2 light-cone operators which determine g_1, giving rise to the so-called Wandzura-Wilczek relation

$$g_2^{WW}(x,Q^2) = -g_1(x,Q^2) + \int_x^1 \frac{dz}{z} g_1(z,Q^2), \tag{71}$$

and where $\bar{g}_2(x,Q^2)$ refers to the twist-3 contribution which can be non-vanishing due to quark-gluon correlations [102,98,100] in the target nucleon and due to quark masses. In other words, $\bar{g}_2$ is determined by the partons' transverse momenta and their off-shellness ($\sim \Lambda^2$) which are unknown in the parton model. The origin of the WW-relation (71) can be easily understood in terms of a light-cone operator expansion of (67) where two classes of operators occur [15,98,100]. One class is represented by totally symmetric operators of twist-2 (spin $= n+1$)

$$O_2^n : \bar{q}\gamma_5\gamma^{\{\sigma} D^{\mu_1} D^{\mu_2} \dots D^{\mu_n\}} q, \; etc. \tag{72}$$

where $n \geq 0$ and the appropriate subtraction of trace-terms is always implied. To the second class belong twist-3 operators (spin $= n$) with mixed symmetry that need at least two (antisymmetrized) indices ($n > 0$)

$$\begin{aligned} O_3^n : &\bar{q}\gamma_5\gamma^{[\sigma} D^{\{\mu_1]} D^{\mu_2} \dots D^{\mu_n\}} q \\ &\bar{q}\gamma_5 m \left[\gamma^\sigma, \gamma^{\{\mu_1}\right] D^{\mu_2} \dots D^{\mu_n\}} q, \; etc. \end{aligned} \tag{73}$$

The (Mellin) moments of g_1 and g_2 are then given by

$$\int_0^1 x^n g_1(x, Q^2) dx = a_n, \quad \int_0^1 x^n g_2(x, Q^2) dx = \frac{n}{n+1}(d_n - a_n) \tag{74}$$

where a_n and d_n denote the nucleon matrix elements of O_2^n and O_3^n, respectively. If the twist-3 operators are negligible, i.e. all $d_n = 0$, then Eq. 74 immediately gives the WW relation (71) [101,98,103] which completely determines the twist-2 contribution g_2^{WW} to g_2 in Eq. 70 in terms of the measured $g_1(x, Q^2)$: The expected shape of $g_2^{WW}(x, Q_0^2)$ is shown in Fig. 8 derived from the existing EMC data [1] for $g_1(x, Q_0^2 \simeq 10 \text{ GeV}^2)$. Note that an experimental confirmation of the prediction for $g_2^{WW}(x, Q_0^2)$ in Fig. 8 (dashed curve) would imply the absence of any twist-3 $\bar{g}_2$ contribution to g_2 in Eq. 70. The expectation that $\bar{g}_2$ should be small is based on the fact that it vanishes for ultra-relativistic on-shell quarks where $s^\beta \sim p^\beta$ (i.e. in this case there are not enough four-vectors to form an antisymmetric combination in Eq. 67) and that non-relativistic corrections, being of the order m_q/M or m_q/Λ, are small for light quarks. In general, however, this twist-3 piece $\bar{g}_2$, being the contribution from the d_n sequence in Eq. 74, is not expected to vanish due to the off-shellness of interacting quarks [4,98,100,102] with virtuality k^2 where k^2/Λ^2 is not small.

Within the covariant relativistic parton model approach it has, however, been argued [103] that the quark virtuality k^2 is expected to be not too sizeable and therefore deviations ($\bar{g}_2$) from the Wandzura-Wilczek contribution to g_2 should be small, i.e. $g_2 \simeq g_2^{WW}$ with g_2^{WW} given in Eq. 71. This can be seen as follows. As a straightforward generalization of the unpolarized case [104], the covariant parton model expression for the asymmetric hadronic tensor in (67) is given by a convolution over the struck parton's momentum,

$$W_{\mu\nu}^{(A)}(q, p, s) = \sum_{s'=\pm 1} \int d^4k f_{s'}(p, k, s) w_{\mu\nu, s'}^{(A)}(q, k) \delta\left[(k+q)^2 - m^2\right] \tag{75}$$

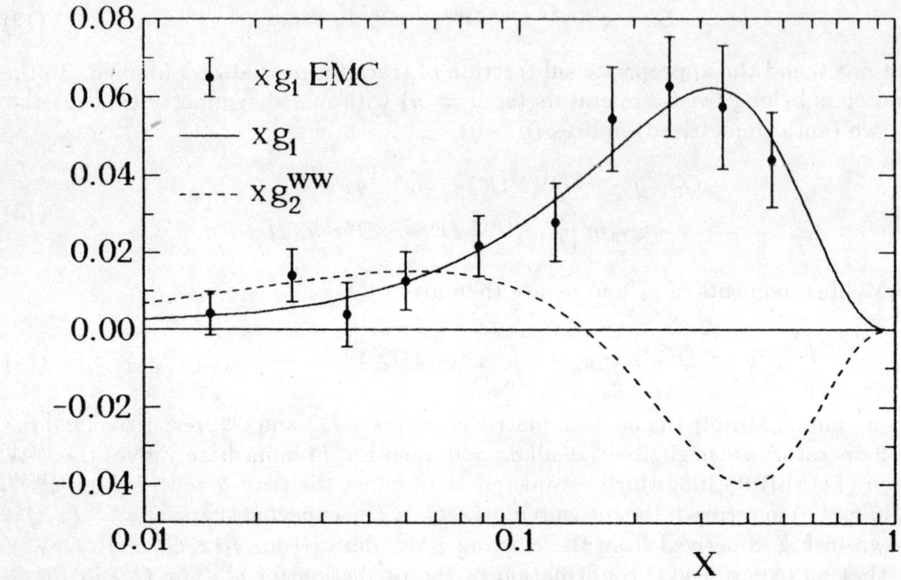

Fig. 8. Expectations for the twist-2 contribution $g_2^{WW}(x, Q_0^2)$ to g_2 as computed from the WW relation (71) using a parametrization (solid curve) of the EMC data [1] for g_1^p at $Q_0^2 \simeq 10$ GeV^2.

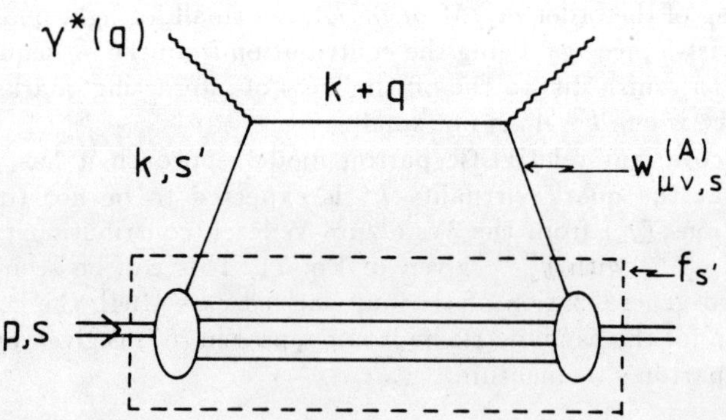

Fig. 9. Parton model 'hand bag' diagram for deep inelastic scattering.

which is illustrated by the diagram in Fig. 9 and $f_{s'}$ being some Lorentz-invariant hadronic wave function. The antisymmetric tensor for the parton-photon interaction is given by

$$\begin{aligned} w^{(A)}_{\mu\nu,\pm 1} &= \frac{1}{4}\mathrm{tr}\left[(1 \mp \gamma_5 \not{n})(\not{k} + m)\gamma_\mu(\not{k} + \not{q} + m)\gamma_\nu\right] - (\mu \leftrightarrow \nu) \\ &= \pm im\epsilon_{\mu\nu\alpha\beta} q^\alpha n^\beta \end{aligned} \tag{76}$$

with the off-shell parton spin vector ($k \cdot n = 0,\ n^2 = -1$)

$$n^\beta = \left(k^\beta - \frac{k^2}{p \cdot k} p^\beta\right) p \cdot k / \sqrt{(p \cdot k)^2 k^2 - M^2 k^4}.$$

Defining

$$(f_+ - f_-)m / \left[k^2 - M^2 k^4 / (p \cdot k)^2\right]^{1/2} \equiv -\frac{k \cdot s}{M} \tilde{f}(p \cdot k, k^2),$$

a comparison of Eq. 75 with 67 yields [103]

$$\begin{aligned} g_1(x) = &\frac{\pi x}{8} \int dk^2 dk_T^2 \tilde{f}\left(x + \frac{k^2 + k_T^2}{xM^2}, k^2\right) \times \\ &\times \left(1 - \frac{k^2 + k_T^2}{x^2 M^2}\right)\left(1 - \frac{2k^2}{x^2 M^2 + k^2 + k_T^2}\right) \end{aligned} \tag{77}$$

$$g_T(x) \equiv g_1(x) + g_2(x) = \frac{\pi x}{8} \int dk^2 dk_T^2 \tilde{f}\left(x + \frac{k^2 + k_T^2}{xM^2}, k^2\right)(k_T^2 / x^2 M^2) \tag{78}$$

for a longitudinally and transversely polarized proton, respectively, and where the scale $Q^2 \equiv -q^2$ at which these results are supposed to hold remains, of course, undetermined. Eq. 78 immediately tells us that a finite transverse polarization result g_T can arise only for a non-vanishing parton transverse momentum, $k_T^2 \neq 0$! Therefore the EMC observation [1] of a non-vanishing value of $\int_0^1 g_1(x)dx$ in Eq. 5 is direct evidence for $k_T^2 \neq 0$ (via the Burkhardt-Cottingham sum rule $\int_0^1 g_2(x)dx = 0$, to be discussed below). Furthermore, the second factor in parentheses in the integrand of Eq. 77 proportional to $2k^2$ violates the Wandzura-Wilczek sum rule and explicitly describes the corrections ($\bar{g}_2$) in Eq. 70, depending on the value of k^2. In the absence of this term one easily derives from Eqs. 77 and 78

$$\frac{d}{dx}[g_1(x) + g_2(x)] = -\frac{1}{x} g_1(x) \tag{79}$$

and hence the sum rule (71), which proves our above statement $g_2 = g_2^{WW}$ for $k^2 = 0$. There seems to be not much room for a sizeable $k^2 \neq 0$ since the first term in parentheses in the integrand of Eq. 77 gives rise to a zero for $g_1(x)$ if $k^2 + k_T^2 \simeq x^2 M^2$, the lack of experimental evidence for which (Fig. 1) suggesting that $k^2 + k_T^2 << x^2 M^2$. Ignoring the possibility that $k^2 < 0$ contributes appreciably, the second factor in parentheses in Eq. 77 is essentially unity, giving thus only small corrections to the WW sum rule. It should be noted that $g_T(x)$ in Eq. 78 gets no additional off-mass-shell correction, besides from the one entailed in $\tilde{f}$, and that a measurement of g_1 gives a direct estimate [103] of the mean (intrinsic) k_T of partons as a function of x, in complete analogy to the unpolarized case [104].

Unfortunately, the virtuality k^2 alone is not a reliable measure for the importance of all possible twist-3 contributions to $\bar{g}_2$ in Eq. 70, in particular not for quark-gluon correlations and mass operators [100,102] as depicted in (73). Indeed, the first of the twist-3 operators in (73) can be shown [102] to be an operator of order g_s involving the gluon field strength tensor, $g_s\bar{q}\tilde{F}^{\sigma\mu_1}\gamma^{\mu_2}\cdots q$. (Note that even for a free quark, where g_2 vanishes, g_2^{WW} has to be exactly cancelled by $\bar{g}_2$ in Eq. 70; in this case the contribution to $\bar{g}_2$ is entirely given by the mass operator in Eq. 73.) In the chiral bag model, for example, the bag boundary simulates [100,105] the confinement effects, which arise from quark-gluon interactions, giving rise to large twist-3 contributions $< O_3^n > \sim < O_2^n >$ i.e. $|\bar{g}_2| \sim |g_2^{WW}|$ and thus to sizeable departures from the WW prediction of g_2 in Eq. 71 as can be seen from Fig. 10. It should be noted that such large twist-3 effects are mainly caused by the non-covariance of the relativistic bag model, which originates from the implementation of the bag boundary (surface) in the equation of motion $i\not{\partial}\psi(\xi) = \delta(|\vec{\xi}\,| - R)\psi(\xi)$, although the average virtuality of the quark in the bag is small [106], $< k^2 >_{bag} \simeq 0.02$ GeV^2, from which one might (perhaps erroneously) conclude that $\bar{g}_2 \simeq 0$ according to the covariant parton model discussed above. The predictions [106] for $g_{1,2}$ shown in Fig. 10 correspond to an improved and extended (one-gluon exchange) version [105] of the MIT bag model [100]. It should, however, be emphasized that the results of such bound-state models are expected to hold at some non-perturbative bound-state scale, typically $Q^2 = O(\Lambda^2)$. Strictly speaking it would therefore not even be possible to use these

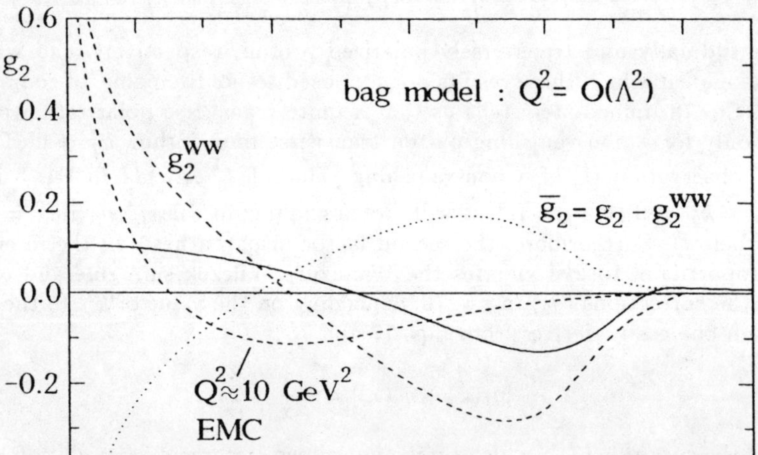

Fig. 10. Predictions [106] for g_2 of an improved and extended (one-gluon exchange) version [105] of the MIT bag model using a bag radius $R = 0.7$ fm and for the spin-singlet (triplet) intermediate diquark masses 0.56 (0.8) GeV. g_2^{WW} has been calculated according to Eq. 71. The EMC prediction for g_2^{WW} at $Q^2 \simeq 10$ GeV^2 is the same as in Fig. 8.

predictions as an input for an evolution to a larger, experimentally relevant scale $Q^2 \gtrsim 1$ GeV^2, unless one arbitrarily chooses the bag bound-state scale $Q^2 >> \Lambda^2$

from where a perturbative RG evolution could be started. In order to demonstrate the importance of different scale effects, the EMC prediction for g_2^{WW} at $Q^2 \simeq 10$ GeV^2 is shown in Fig. 10 as well (this dashed curve is the same as the one in Fig. 8). The difference between the bag and EMC prediction is indeed very large, indicating that expectations from bound-state models are not very relevant for actual deep inelastic measurements. This is not very surprising since the bag predictions [100,105] for g_1 (as well as for unpolarized structure functions) disagree with actual measurements. Therefore the large bag-model prediction for $\bar{g}_2$ in Fig. 10 appears to be also not too relevant for future experiments. This is here particularly disturbing because the Q^2-evolution of $\bar{g}_2(x, Q^2)$ is theoretically unknown and in general an unsolved intricate problem, because the number of independent twist-3 operators O_3^n in (73) increases with n [107]. Thus there exist no Altarelli-Parisi type evolution equations and any extrapolation to the interesting small-x region is entirely unknown. There have been attempts [108] to construct practical approximations to the Q^2-evolution in the limit $x \longrightarrow 1$ (large n), but the very large-x region is experimentally (and theoretically) less relevant, as well as in the limit $N_c \longrightarrow \infty$. Provided this latter large N_c limit has any physical relevance, the effect of the Q^2-evolution of $\bar{g}_2(x, Q^2)$ is large throughout the whole x-region [106].

Other models [109], based on the light-cone quark model [110], give substantially different predictions for $g_2(x)$ than the ones shown in Fig. 10: g_2 and $\bar{g}_2$ could also become strongly negative (of the order of -1) for $x < 0.1$.

It is therefore very important to check experimentally first the WW relation (71) [i.e. is $g_2 \neq g_2^{WW}$?] and then to extract $\bar{g}_2 = g_2 - g_2^{WW}$. No measurements of g_2 exist so far. A measurement of $\bar{g}_2(x, Q^2)$ which appears to be easily accessible to experiment, will provide, for the first time, important information about the nucleon matrix elements of specific interaction-dependent quark-gluon operators of higher-twist and thus further insight into the bound-state dynamics involved.

A further important issue to be tested experimentally is the validity of the Burkhardt-Cottingham (BC) sum rule [111,98]

$$\int_0^1 g_2(x, Q^2)dx = 0 \tag{80}$$

(Note that g_2^{WW} in Eq. 71 obeys, on rather general grounds, this sum rule $\int_0^1 g_2^{WW}(x, Q^2)dx = 0$). This sum rule is a sufficiently general result of a superconvergence relation based on Regge asymptotics [$\alpha(0) < 0$]. It turns out to be very robust and is most probably true [98,100,112] (unless there are Regge cuts with branch points at $\alpha(0) \geq 0$ or specific non-polynomial $J = 0$ fixed poles in Compton amplitudes that could invalidate the BC sum rule only by terms of order at most $\frac{1}{Q^2}$ [113]). It should be pointed out that the light-cone approach does <u>not</u> predict the first moment in Eq. 80, since the n-moment of g_2 in Eq. 74 is only valid for $n > 0$. Although the presence of the factor n in Eq. 74 indicates the validity of the BC sum rule, we would need the $n \longrightarrow 0$ continuation of the twist-3 matrix element d_n: If $d_n < 1/n$ as $n \longrightarrow 0$, the BC sum rule (80) would hold; it would not hold if $d_n \sim 1/n$ for $n \longrightarrow 0$ (but we know that the twist-2 matrix element a_0 is not singular because the first moment of g_1 is finite). It is important to check the validity of the BC sum rule by experiment

– as far as possible. It would be very interesting if it were not true, because this would imply a non-conventional behavior of twist-3 operators ($d_n \sim 1/n$) or the importance of long-range effects [100,109].

4 Transverse Chiral-Odd ('Transversity') Structure Functions

For completeness let me finally mention the chiral-odd "transversity" distributions [114-116,99,15,107] $h_i(x, Q^2)$ which received some interest recently. A transversity distribution is chiral-odd in the sense that it measures the correlation between left- and right-handed quarks ($L \leftrightarrow R$) with $q_{R,L} \equiv \frac{1}{2}(1 \pm \gamma_5)q$ and the chirality being defined via $\gamma_5 q_{R,L} = \pm q_{R,L}$. This is in contrast to the deep inelastic structure functions $g_{1,2}$ discussed thus far which, for obvious reasons, are always chiral even ($L \to L, R \to R$), apart from small quark mass corrections, as illustrated in Fig. 11: In a (fully) deep inelastic scattering the quark lines leaving and entering the nucleon are of a single chirality since the photon (gluons) participating in the hard scattering process cannot change the chirality of the quark. Thus for each flavor, only two independent quark-nucleon amplitudes (quark distributions) enter in LO the description of deep inelastic scattering as shown in Fig. 12. The average over chirality (or helicity) gives the (unpolarized) quark momentum distributions in $F_1(x, Q^2)$ in Eq. 3, whereas the difference gives the chirality (or helicity) weighted (longitudinally polarized) quark momentum distributions in $g_1(x, Q^2)$ in Eq. 4. On the other hand we have seen in Eq. 69 that the spin asymmetry from a transversely polarized nucleon is suppressed by a factor $M/\sqrt{Q^2}$ since it measures an amplitude in which the nucleon helicity changes but in whose hard scattering the quark chirality does not change [115,116] (remember that a transversely polarized nucleon can be regarded as a superposition of longitudinally polarized states, $|p_z, \pm e_x\rangle = (|p_z, e_z\rangle \pm |p_z, -e_z\rangle)/\sqrt{2}$). As we have seen in Section 3, there are two main effects that cause the transverse asymmetry, i.e. $g_T = g_1 + g_2$, not to vanish: These are the (possibly) small effects proportional to m_q that allow the quark helicity to flip without changing chirality, and quark-gluon interactions that allow the quark helicity to be transferred to gluons (which connect a hard quark line in Fig. 11 directly with the nucleon) while preserving chirality.

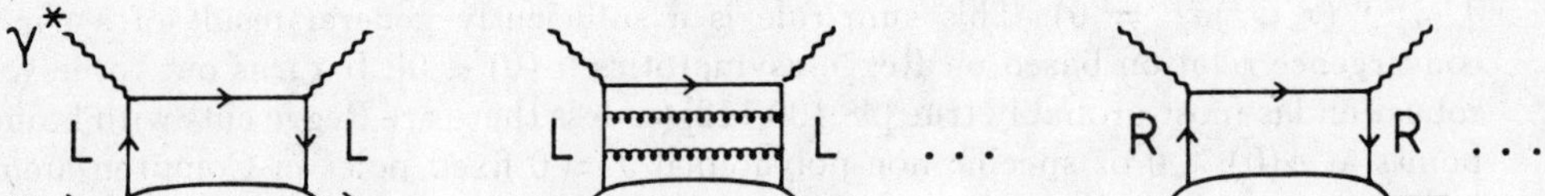

Fig. 11. Chiral (even) structure of deep inelastic scattering.

Fig. 12. Left- and right-handed chiral even quark distributions whose sum and difference give $F_1(x,Q^2)$ and $g_1(x,Q^2)$, respectively.

In general, however, the chirality of the quark lines leaving and entering a given nucleon need not be the same, as illustrated in Fig. 13. The reason being that chirality is not a symmetry of QCD, even when the quarks are massless; it is broken by non-perturbative condensates $\langle \bar{q}_L q_R \rangle \neq 0$ in the QCD vacuum and it fails in the nucleon for the same reason of bound state effects [115,116]. Since the quark chirality (= helicity + $O(m_q)$) flips, this new leading twist-2 structure function is sometimes [115,116] called $h_1(x,Q^2)$. Its interpretation is obscure in the chiral basis, but is revealed [115] by using the transverse projection operators $P^{\uparrow\downarrow} = \frac{1}{2}(1 \pm \gamma_5 \not{s}_\perp)$ instead of $P_{R,L} = \frac{1}{2}(1 \pm \gamma_5)$. Here the transversely polarized nucleon and quarks are classified as eigenstates of the transversely projected Pauli-Lubanski spin operator $\gamma_5 \not{s}_\perp$ (which is the more familiar form proportional to $W_\mu = -\frac{1}{2}\epsilon_{\mu\nu\rho\sigma} J^{\nu\rho} P^\sigma$), i.e. $\gamma_5 \not{s}_\perp u(p_z, s_\perp) = \pm u(p_z, s_\perp)$. It should be noted that $\gamma_5 \not{s}_\perp$ commutes with the free-quark Hamiltonian $H_0 = \alpha_z p_z$, in contrast to the transverse quark spin operator $\Sigma_\perp = \gamma_0 \gamma_5 \not{s}_\perp$ relevant for g_T in Section 3, and therefore a simple parton model can be made to interpret $h_1(x,Q^2)$. This basis was originally introduced in hadron-hadron scattering [117], and is known as the 'transversity' basis. The operator $\gamma_5 \not{s}_\perp = \gamma_0 \Sigma_\perp$ is called 'transversity' operator and the corresponding eigenstates the 'transversity' eigenstates. Hence one names h_1 the quark transversity distribution: It measures, for each quark flavor, the probability in a transversely polarized nucleon to find a quark in an eigenstate of the transverse Pauli-Lubanski operator with eigenvalue +1/2, i.e. aligned with the nucleon polarization, minus the probability to find it oppositely polarized (eigenvalue −1/2):

$$h_1^q(x,Q^2) = q^\uparrow(x,Q^2) - q^\downarrow(x,Q^2) \tag{81}$$

and there are many different notations for h_1 in the literature [107,114,118], such as $h^T, \Delta_1 q, \Delta_T q$, etc. Furthermore this twist-2 distribution $h_1(x,Q^2)$, where only fermions contribute, obeys a simple Altarelli-Parisi type of evolution equation [118]. It should be reemphasized that the 'transversity' distribution h_1 must not be confused with transverse spin: Spin is measured, as discussed previously ($g_1, g_T = g_1 + g_2$), by the operator $\bar{q}\gamma_\mu\gamma_5 q = \bar{q}_L\gamma_\mu\gamma_5 q_L + \bar{q}_R\gamma_\mu\gamma_5 q_R$ which preserves quark chirality and has nothing to do with h_1 which flips it.

There are further subleading (twist-3) transversity distributions [115] such as h_2 which is the chiral-odd analogue to g_2. In general the parton distributions in QCD can be defined by the light-cone Fourier transform of field operator products and the simplest distributions are related to the target matrix elements of bilinear quark operators $< ps|\bar{q}(0)\Gamma q(\xi)|ps >$ [115]:

$$\begin{array}{rlll} \Gamma = & \gamma_\mu & : & F_1 \\ & \gamma_\mu\gamma_5 & : & g_1, g_T \equiv g_1 + g_2 \\ & i\sigma_{\mu\nu}\gamma_5 & : & h_1, h_L \equiv \frac{1}{2}h_1 + h_2 \\ & 1 & : & e \end{array} \tag{82}$$

where twist-4 distributions have been suppressed and the last twist-3 chiral-odd scalar distribution $e(x)$ has the interesting property of being related, via its first moment, to the nucleon σ-term but it appears to be very difficult to be measurable in a quark chirality flip from a spin-averaged target [115]. The transversity densities $h_{1,2}$ measure the tensor current of the nucleon.

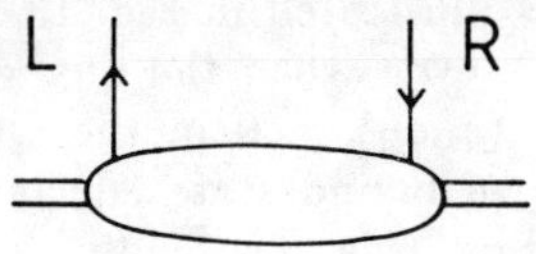

Fig. 13. Chiral (odd) flip distribution which gives the 'transversity' structure function $h_1(x, Q^2)$.

The chiral-odd h_1 distributions, Fig. 13, cannot be measured in a fully inclusive deep inelastic *ep* scattering process (where the chirality is always conserved, Fig. 11, apart from small quark mass corrections), but it appears in a transversely polarized hadron-hadron initiated reaction, e.g. Drell-Yan dilepton production, at the dominant 'scaling' order since the chirality of a quark can be flipped here without suppression as illustrated in Fig. 14. Note that the perturbative hard subprocess conserves, of course, always chirality. There have been various suggestions [115-121] to measure the quark transverse polarization in doubly transversely polarized hadron-hadron initiated reactions using the azimuthal dependence of the produced particles, like Drell-Yan dileptons, direct photons, heavy quarks and quark/gluon jets. For example, for the case of dilepton production [115] the spin asymmetry A_{TT} for transverse-transverse collisions measures directly h_1, whereas for longitudinal-transverse collisions A_{LT} gives access to h_L provided g_1 and $g_T = g_1 + g_2$ are known; longitudinal-longitudinal collisions obviously give g_1 via A_{LL}. Such measurements could, at least in principle, shed light on the nucleon's "tensor charge" $\delta q(Q^2)$,

$$\int_0^1 dx \left[h_1^q(x, Q^2) - h_1^{\bar{q}}(x, Q^2)\right] = \delta q(Q^2) \tag{83}$$

with

$$< ps|\bar{q}i\sigma^{\mu\nu}\gamma_5 q|ps >= \frac{2}{M}(s^\mu p^\nu - s^\nu p^\mu)\delta q \tag{84}$$

for each quark flavor. Therefore, δq counts the number of valence quarks (quarks <u>minus</u> antiquarks) of opposite transversity – a quantity which is Q^2-dependent [118] and related to an operator ($\bar{q}i\sigma^{\mu\nu}\gamma_5 q$) which is odd under charge conjugation. This is in contrast to the quark spin operator $\bar{q}\gamma^\mu\gamma_5 q$ which is even under charge conjugation and therefore measures the total (Q^2-independent) quark helicity contribution (quarks plus antiquarks) Δq to the nucleon spin as defined in Eq. 8 and discussed thereafter. To understand the physical meaning of the tensor charge, it is helpful to make a comparison between δq and Δq in the rest frame of the nucleon, $p^\mu = (M, \vec{0})$ and $s^\mu = (0, \vec{s})$:

$$< ps|\bar{q} i\sigma^{i0}\gamma_5 q|ps > = < ps|\bar{q}\Sigma^i q|ps > = 2s^i \delta q \quad (85)$$

$$< ps|\bar{q}\gamma^i\gamma_5 q|ps > = < ps|\bar{q}\gamma_0\Sigma^i q|ps > = 2s^i \Delta q \quad (86)$$

i.e. the spin (helicity) operator related to axial currents differs just by an additional γ_0 matrix from the (Pauli-Lubanski) 'transversity' related to tensor currents. (Note that $\bar{q}\gamma_0\Sigma^i q = q^+\Sigma^i q$, and hence its name spin operator.) Although model calculations indicate that $h_1(x)$ is large [115], $h_1 \simeq g_1$, the experimental feasibility of the suggested doubly polarized Drell-Yan reactions seems to be limited both by the low rates [121], even at RHIC [116], and by the difficulty to produce intense high energy polarized colliding hadron (in particular antiproton [116]) beams.

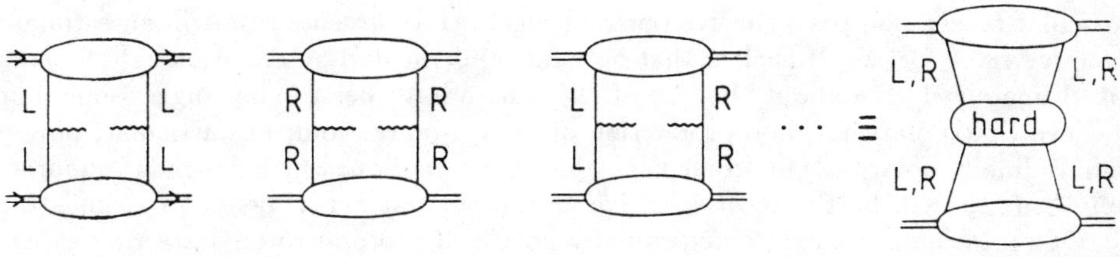

Fig. 14. Chiral structure of hadronic Drell-Yan production of lepton pairs with invariant mass $Q^2 \equiv q^2 > 0$ produced by the virtual photons. Chirality flip ($R \leftrightarrow L$) occurs without suppression due to the two initial hadrons (non-perturbative bound states) involved in the process.

For this reason it has been suggested more recently to use single transverse polarization experiments where only one initial hadron is polarized, with the other one being unpolarized, and the polarization of the final state particle is observed instead [122-124]. Alternatively, transversity distributions can also be measured by a semi-inclusive deep inelastic $e\vec{p}$ scattering experiment, with an unpolarized electron and a transversely polarized proton target, producing a polarized quark jet, for example [123,125,126]. It will not be easy to measure the transverse handedness of the quark in the final state and old ideas [127] have been recently revived [123,126,128] for determining the polarization of an outgoing quark (or gluon) via the hadron distribution in the jet. Since transverse and "transversity" distributions are experimentally entirely unexplored, we hopefully can look forward to many new results in the years to come.

5 Summary and Conclusions

The explanation of the spin content of the nucleon in terms of a large quark-sea or gluon component (or a combination of both) is, strictly speaking, a matter of the (theoretical factorization) convention used in defining polarized quark distributions. If the gluon component is entirely absorbed into the definition of total quark polarizations, a large negative strange sea polarization is required for explaining the "small" EMC measurement of $\int_0^1 g_1^p dx$ at $Q^2 \simeq 10\ \mathrm{GeV}^2$, or a $SU(3)_f$ broken sea polarization with $\Delta\bar{u} = \Delta\bar{d} >> \Delta s \simeq 0$ can do as well. However, a sizeable gluon contribution Δg appears to be intuitively more plausible since its coupling to the photon, induced by the triangle anomaly, is non-vanishing if only hard perturbative contributions, appropriate for a deep inelastic process, are taken into account; the soft non-perturbative pieces are absorbed into the definition of quark polarizations which corresponds to a large and conserved total polarization of the flavor singlet quark component $\Delta\Sigma \equiv \Delta u + \Delta d + \Delta s$. Thus $\Delta\Sigma$ is constituent-like, as are the conserved flavor non-singlet quark polarizations fixed by hyperon β-decays. The large total gluon polarization Δg, needed in addition for explaining the "small" EMC measurement, cannot be calculated perturbatively and one has to resort to non-perturbative methods and/or to different hard processes where Δg can be measured independently.

Various theoretical arguments have been given in favor of a sizeable Δg and in particular recent non-perturbative (current algebra) arguments result in an estimate of a large value for Δg. It is clear that only future dedicated polarization experiments can ultimately decide about the size of Δg: There have been many suggestions, but I believe that polarized photoproduction of heavy quarks and, in particular, purely (longitudinally polarized) hadronic heavy quark production, which depends "quadratically" on Δg, will be the most sensitive and important experiments for delineating $\Delta g(x, Q^2)$. Perhaps a singly longitudinally polarized proton-proton scattering experiment, using the 1 TeV unpolarized proton beams of TEVATRON or HERA to be scattered off a polarized fixed target, could do the job as well provided the experimentalists manage to measure the polarization of one of the final leptons of a produced Drell-Yan lepton pair, say. In addition, of course, the EMC result has to be checked by remeasuring g_1^p and a measurement of the polarized neutron structure function $g_1^n(x, Q^2)$, hopefully available soon, will serve as a fundamental test of the crucial (isospin symmetry) Bjorken sum rule and of the QCD-improved parton model.

A measurement of the transverse spin structure function $g_2(x, Q^2)$, or more precisely $g_1 + g_2$, will provide important information about specific quark-gluon operators of higher twist and thus further insight into the bound-state dynamics of nucleons. This will be achieved by looking for deviations of g_2 from its twist-2 Wandzura-Wilczek contribution, uniquely determined by g_1. A further important issue to be tested experimentally is the validity of the Burkhardt-Cottingham sum rule for $g_2(x, Q^2)$ which is expected to hold on rather general grounds.

Finally, the so far entirely unknown chiral-odd "transversity" distributions, although not related to the spin structure of nucleons, should be experimentally explored either in doubly transversely polarized hadron-hadron scattering experiments or in semi-inclusive deep inelastic reactions with just the proton target being transversely polarized and by measuring the polarization of the produced outgoing quark.

Note added: While this review was being completed, the SMC collaboration just published [129] their recent measurements of $g_1^d(x, Q^2)$ which confirm the Bjorken sum rule (64) and thus the (large) expectation for $g_1^n(x, Q^2)$ in Eq. 65, using the EMC result for g_1^p.

Acknowledgments

I am indebted to M. Glück for numerous discussions and for fruitful collaborations on various topics discussed here. I also thank M. Stratmann and W. Vogelsang for their collaboration and help in preparing some of the figures presented. Finally I thank G. Altarelli for some helpful and clarifying comments. This work has been supported in part by the 'Bundesministerium für Forschung and Technologie', Bonn.

References

[1] J. Ashman et al., EMC collab., Phys. Lett. **B206** (1988) 364; Nucl. Phys. **B328** (1989) 1 and references therein.
[2] V. Papavassiliou, EMC collab., Proceedings of the EPS Conf. on High Energy Physics, Uppsala 1987, ed. O. Bartner, vol. I, p. 441.
[3] M.J. Alguard et al., SLAC-Yale collab., Phys. Rev. Lett. **37** (1976) 1261; G. Baum et al., SLAC-Yale collab., Phys. Rev. Lett. **45** (1980) 2000; **51** (1983) 1135.
[4] G. Altarelli, Proceedings of the 27th Int. E. Majorana Summer School of Subnuclear Physics, Erice 1989, ed. A. Zichichi, Plenum Press [CERN-TH.5675/90]; and Proceedings of the HERA-Workshop, Hamburg 1991, eds. W. Buchmüller and G. Ingelmann (DESY, 1992), vol. I, p. 379 [CERN-TH.6340/91].
[5] G.G. Ross, Proceedings of the 1989 Int. Symposium on Lepton and Photon Interactions at High Energies, Stanford, ed. M. Riordan (World Scientific, 1990), p. 41.
[6] R.L. Jaffe and A. Manohar, Nucl. Phys. **B337** (1990) 509.
[7] H.Y. Cheng, Chin. J. Phys. **29** (1991) 67.
[8] E. Reya, Proceedings of the Int. Workshop on High Energy Physics Phenomenology II "Particle Phenomenology in the 90's", Calcutta, 1991, eds. A. Datta, P. Ghose and A. Raychaudhuri (World Scientific, 1992) p. 109.
[9] A.V. Efremov, Adriatico Research Conf. on Polarization Dynamics in Nuclear and Particle Physics, Trieste 1992 [CERN-TH.6466/92].
[10] G.T. Bodwin and J. Qiu, Proceedings of the Polarized Collider Workshop, Penn. State Univ., 1990, AIP Conf. Proc. **233** (1991) 285.
[11] A.V. Manohar, Lectures at the 1992 Lake Luise Winter Institute, UC San Diego report UCSD/PTH 92-10.
[12] R. Decker, M. Nowakowski and U. Wiedner, Fortsch. Phys. **41** (1993) 87.
[13] M. Gourdin, Nucl. Phys. **B38** (1972) 418; J. Ellis and R.L. Jaffe, Phys. Rev. **D9** (1974) 1444; Erratum **D10** (1974) 1669.
[14] H. Abramowicz et al., CDHS collab., Z. Phys. **C15** (1982) 19; C. Foudas et al., CCFR collab., Phys. Rev. Lett. **64** (1990) 1207.
[15] J. Kodaira, S. Matsuda, T. Muta, K. Sasaki and T. Uematsu, Phys. Rev. **D20** (1979) 627; Nucl. Phys. **B159** (1979) 99.
[16] J.D. Bjorken, Phys. Rev. **148** (1966) 1467; ibid. **D1** (1970) 1376.
[17] L.M. Sehgal, Phys. Rev. **D10** (1974) 1663; Erratum **D11** (1975) 2016; D.B. Kaplan and A. Manohar, Nucl. Phys. **B310** (1988) 527; M. Anselmino, B.L. Ioffe and E. Leader, Sov. J. Nucl. Phys. **49** (1989) 136; H.J. Lipkin, Phys. Lett. **B230** (1989) 135.

[18] G. Altarelli and G. Parisi, Nucl. Phys. **B126** (1977) 298.
[19] C.S. Lam and B.A. Li, Phys. Rev. **D25** (1982) 683.
[20] G. Altarelli and G.G. Ross, Phys. Lett. **B212** (1988) 391.
[21] A.V. Efremov and O.V. Teryaev, Dubna report E2-88-287, 1988 [published in the Proceedings of the Int. Hadron Symposium, Bechyne, Czechoslovakia, 1988, eds. X. Fischer et al. (Czech. Academy of Sciences, Prague, 1989), p. 302].
[22] M. Glück and E. Reya, Univ. Dortmund report DO-TH 87/14; Z. Phys. **C39** (1988) 569.
[23] J. Ellis, R.A. Flores and S. Ritz, Phys. Lett. **B198** (1987) 393; J. Ellis and R.A. Flores, Nucl. Phys. **B307** (1988) 883.
[24] H.J. Lipkin, Phys. Lett. **B256** (1991) 284.
[25] S.J. Brodsky, J. Ellis and M. Karliner, Phys. Lett. **B206** (1988) 309; J. Ellis and M. Karliner, ibid. **B213** (1988) 73.
[26] G. Clément and J. Stern, Phys. Lett. **B220** (1989) 238.
[27] Z. Ryzak, Phys. Lett. **B217** (1989) 325; V. Bernard and U.G. Meissner, ibid. **B223** (1989) 439; J. Stern and G. Clément, ibid. **B231** (1989) 471.
[28] P. Ratcliffe, Phys. Lett. **B192** (1987) 180.
[29] J.D. Jackson, G.G. Ross and R.G. Roberts, Phys. Lett. **B226** (1989) 159.
[30] T. Meng, J. Pan, Q. Xie and W. Zhu, Phys. Rev. **D40** (1989) 769.
[31] R.L. Jaffe, Phys. Lett. **B193** (1987) 101.
[32] R.D. Carlitz, J.C. Collins and A.H. Mueller, Phys. Lett. **B214** (1988) 229.
[33] G. Altarelli and B. Lampe, Z. Phys. **C47** (1990) 315.
[34] P. Ratcliffe, Nucl. Phys. **B223** (1983) 45.
[35] G. Altarelli and W.J. Stirling, Particle World **1** (1989) 40.
[36] L. Mankiewicz and A. Schäfer, Phys. Lett. **B242** (1990) 455; L. Mankiewicz, Phys. Rev. **D43** (1991) 64.
[37] G.T. Bodwin and J. Qiu, Phys. Rev. **D41** (1990) 2755.
[38] A.V. Manohar, Phys. Rev. Lett. **66** (1991) 289; U. Ellwanger, Phys. Lett. **B259** (1991) 469.
[39] W. Vogelsang, Z. Phys. **C50** (1991) 275.
[40] G. Veneziano, Mod. Phys. Lett. **A4** (1989) 1605; G.M. Shore and G. Veneziano, Phys. Lett. **B244** (1990) 75.
[41] S.D. Bass, N.N. Nikolaev and A.W. Thomas, Adelaide Univ. report ADP-133-T80 (1990); S.D. Bass, B.L. Ioffe, N.N. Nikolaev and A.W. Thomas, J. Moscow Phys. Soc. **1** (1991) 317.
[42] M. Glück and E. Reya, Z. Phys. **C43** (1989) 679.
[43] A.D. Watson, Z. Phys. **C12** (1982) 123.
[44] M. Glück, E. Reya and W. Vogelsang, Nucl. Phys. **B351** (1991) 579.
[45] S.J. Brodsky and I. Schmidt, Phys. Lett. **B234** (1990) 144.
[46] M. Glück and E. Reya, Phys. Lett. **B270** (1991) 65.
[47] J. Kunz, P.J. Mulders and S. Pollock, Phys. Lett. **B222** (1989) 481.
[48] M. Glück, E. Reya and A. Vogt, Z. Phys. **C53** (1992) 127; Phys. Lett. **B306** (1993) 391.
[49] M. Glück, E. Reya and W. Vogelsang, Phys. Rev. **D45** (1992) 2552.
[50] T.P. Cheng and L.F. Li, Phys. Rev. Lett. **62** (1989) 1441; H. Fritzsch, Phys. Lett. **B329** (1989) 122; ibid. **B242** (1990) 451; ibid. **B256** (1991) 75; T. Hatsuda, Nucl. Phys. **B329** (1990) 376; A.V. Efremov, J. Soffer and N.A. Törnqvist, Phys. Rev. Lett. **64** (1990) 1495; Phys. Rev. **D44** (1991) 1369; A.V. Efremov, J. Soffer and O.V. Teryaev, Nucl. Phys. **B346** (1990) 97; J. Bartelski and S. Tatur, Phys. Lett. **B265** (1991) 192; U. Ellwanger and B. Stech, Phys. Lett. **B241** (1990) 409; Z. Phys. **C49** (1991) 683; J.H. Kühn and V.I. Zakharov, Phys. Lett. **B252** (1990) 615.

[51] G.M. Shore and G. Veneziano, Nucl. Phys. **B381** (1992) 23.
[52] G.G. Ross and R.G. Roberts, Rutherford Lab. RAL-90-062.
[53] S. Forte, Phys. Lett. **B224** (1989) 189; Nucl. Phys. **B331** (1990) 1.
[54] S.L. Adler, Phys. Rev. **177** (1969) 2426; J.S. Bell and R. Jackiw, Nuovo Cimento **A51** (1969) 47.
[55] J.E. Mandula, Phys. Rev. Lett. **65** (1990) 1403.
[56] G. Briganti, A. Di Giacomo and H. Panagopoulos, Phys. Lett. **B253** (1991) 427; A. Di Giacomo, E. Meggiolaro and H. Panagopoulos, Phys. Lett. **B291** (1992) 147.
[57] S. Narison, Phys. Lett. **B255** (1991) 101.
[58] G.M. Shore and G. Veneziano, Nucl. Phys. **B381** (1992) 3.
[59] R. Altmeyer, M. Göckeler, R. Horseley, E. Laermann and G. Schierholz, DESY 92-187/HLRZ 92-103.
[60] V.M. Belyaev. B.L. Ioffe and Ya.I. Kogan, Phys. Lett. **151B** (1985) 290; C.B. Chiu, J. Pasupathy and S.L. Wilson, Phys. Rev. **D32** (1985) 1786.
[61] S. Gupta, M.V.N. Murthy and J. Pasupathy, Phys. Rev. **D39** (1989) 2547; E.M. Henley, W.Y.P. Hwang and L.S. Kisslinger, Phys. Rev. **D46** (1992) 431.
[62] J. Babcock, E. Monsay and D. Sivers, Phys. Rev. **D19** (1979) 1483.
[63] P. Chiappetta, J. Ph. Guillet and J. Soffer, Nucl. Phys. **B262** (1985) 187.
[64] J. Qiu, G.P. Ramsey, D. Richards and D. Sivers, Phys. Rev. **D41** (1990) 65.
[65] M. Glück, E. Reya and W. Vogelsang, Nucl. Phys. **B329** (1990) 347.
[66] S. Gupta, J. Pasupathy and J. Szwed, Z. Phys. **C46** (1990) 111.
[67] H.Y. Cheng and S.N. Lai, Phys. Rev. **D41** (1990) 91.
[68] P. Chiappetta and G. Nardulli, Z. Phys. **C51** (1991) 435.
[69] H.Y. Cheng and C.F. Wai, Phys. Rev. **D46** (1992) 125.
[70] R.P. Feynman, "Photon Hadron Interactions" (Benjamin, New York, 1972).
[71] W. Vogelsang and A. Weber, Phys. Rev. **D45** (1992) 4069.
[72] D.L. Adams et al., E-581/704 collab., Phys. Lett. **B261** (1991) 197.
[73] J. Kwiecinsky, A.D. Martin, R.G. Roberts and W.J. Stirling, Phys. Rev. **D42** (1990) 3645.
[74] W. Vogelsang, Nucl. Phys. B (Proc. Suppl.) **29A** (1992) 113.
[75] J.P. Guillet, Z. Phys. **C39** (1988) 75; P. Kalyniak, M.K. Sundaresan and P.J.S. Watson, Phys. Lett. **B216** (1989) 397.
[76] R.M. Godbole, S. Gupta and K. Sridhar, Phys. Lett. **B255** (1991) 120.
[77] SMC collab. (CERN), V.W. Hughes et al.
[78] E142/143 collab. (SLAC), E.W. Hughes et al.
[79] H. Fritzsch, Phys. Lett. **67B** (1977) 217; L.M. Jones and H.W. Wyld, Phys. Rev. **D17** (1978) 2332; M. Glück and E. Reya, Phys. Lett. **79B** (1978) 453.
[80] E.L. Berger and D. Jones, Phys. Rev. **D23** (1981) 1521, R. Baier and R. Rückl, Nucl. Phys. **B218** (1983) 289; Z. Phys. **C19** (1983) 251.
[81] J. Cortes and B. Pire, Phys. Rev. **D38** (1988) 3586; M.A. Doncheski and R.W. Robinett, Phys. Lett. **B248** (1990) 188.
[82] A.P. Contogouris, S. Papadopoulos and B. Kamal, Phys. Lett. **B246** (1990) 523.
[83] A.P. Contogouris and S. Papadopoulos, Phys. Lett. **B260** (1991) 204.
[84] R.D. Carlitz and R.S. Willey, Phys. Rev. **D45** (1992) 2323.
[85] M. Glück and W. Vogelsang, Phys. Lett. **B277** (1992) 515.
[86] P. Mathews and R. Ramachandran, Z. Phys. **C53** (1992) 305.
[87] M.B. Einhorn and J. Soffer, Nucl. Phys. **B274** (1986) 714; C. Bourrely, J. Soffer and P. Taxil, Phys. Rev. **D36** (1987) 3373.
[88] E.L. Berger and J. Qiu, Phys. Rev. **D40** (1989) 778.
[89] S. Gupta, D. Indumathi and M.V.N. Murthy, Z. Phys. **C42** (1989) 493; Erratum **C44** (1989) 356.

[90] C. Bourrely, J.Ph. Guillet and J. Soffer, Nucl. Phys. **B361** (1991) 72; P. Chiappetta, P. Colangelo, J.Ph. Guillet and G. Nardulli, Marseille CPT-92/PE.2796 (1992).
[91] G. Ranft and J. Ranft, Nucl. Phys. **B165** (1980) 395.
[92] Z. Kunszt, Phys. Lett. **B218** (1989) 243.
[93] G.P. Ramsey, D. Richards and D. Sivers, Phys. Rev. **D37** (1988) 3140; G.P. Ramsey and D. Sivers, ibid. **D43** (1991) 2861.
[94] M.A. Doncheski, R.W. Robinett and L. Weinkauf, Phys. Rev. **D44** (1991) 2717.
[95] W. Vogelsang and A. Weber, Nucl. Phys. **B362** (1991) 3; P. Mathews and P. Ravindran, Phys. Lett. **B278** (1992) 175; V. Ravishankar, Nucl. Phys. **B374** (1992) 309.
[96] A. Giannelli, L. Nitti, G. Preparata and P. Sforza, Phys. Lett. **150B** (1985) 214; G. Preparata and J. Soffer, Phys. Rev. Lett. **61** (1988) 1167; G. Preparata, Univ. Milano MITH 91/18.
[97] S. Forte, Phys. Lett. **B309** (1993) 174.
[98] R.L. Jaffe, Comm. Nucl. Part. Phys. **14** (1990) 239.
[99] P.G. Ratcliffe, Univ. Milano MITH 92/8.
[100] R.L. Jaffe and X. Ji, Phys. Rev. **D43** (1991) 724.
[101] S. Wandzura and F. Wilczek, Phys. Lett. **72B** (1977) 195.
[102] E.V. Shuryak and A.I. Vainshtein, Nucl. Phys. **B201** (1982) 141.
[103] J.D. Jackson, G.G. Ross and R.G. Roberts, Phys. Lett. **B226** (1989) 159.
[104] M. Glück and E. Reya, Nucl. Phys. **B145** (1978) 24.
[105] A.W. Schreiber, A.I. Signal and A.W. Thomas, Phys. Rev. **D44** (1991) 2653.
[106] M. Stratmann, Univ. Dortmund DO-TH 93/06 (Phys. Rev. D).
[107] A.P. Bukhvostov, A.A. Kuraev and L.N. Lipatov, JETP Lett. **37** (1983) 482; Sov. Phys. JETP **60** (1984) 22; P. Ratcliffe, Nucl. Phys. **B264** (1986) 493; I. Antoniadis and C. Kounnas, Phys. Rev. **D24** (1981) 505.
[108] X. Ji and C. Chou, Phys. Rev. **D42** (1990) 3637; A. Ali, V.M. Braun and G. Hiller, Phys. Lett. **B266** (1991) 117.
[109] L. Mankiewicz and Z. Ryzak, Phys. Rev. **D43** (1991) 733; L. Mankiewicz and A. Schäfer, Phys. Lett. **B265** (1991) 167; A. Schäfer, Z. Phys. **C56** (1992) S179.
[110] Z. Dziembowski and J. Franklin, Phys. Rev. **D42** (1990) 905, and references therein.
[111] H. Burkhardt and W.N. Cottingham, Ann. Phys. (N.Y.) **56** (1970) 453.
[112] R.L. Heimann, Nucl. Phys. **B64** (1973) 429.
[113] T.P. Cheng and W.K. Tung, Phys. Rev. Lett. **24** (1970) 851.
[114] J.P. Ralston and D.E. Soper, Nucl. Phys. **B152** (1979) 109.
[115] R.L. Jaffe and X. Ji, Phys. Rev. Lett. **67** (1991) 552; Nucl. Phys. **B375** (1992) 527.
[116] G. Bunce et al., Particle World **3** (1992) 1.
[117] G.R. Goldstein and M.J. Moravcsik, Ann. Phys. (N.Y.) **98** (1976) 128; **142** (1982) 219; **195** (1989) 213.
[118] X. Artru and M. Mekhfi, Z. Phys. **C45** (1990) 669.
[119] J.L. Cortes, B. Pire and J.P. Ralston, AIP Conf. Proc. **233** (1991) 184; Z. Phys. **C55** (1992) 409.
[120] R.W. Robinett, Phys. Rev. **D45** (1992) 2563.
[121] X. Ji, Phys. Lett. **B284** (1992) 137.
[122] J. Qiu and G. Sterman, Phys. Rev. Lett. **67** (1991) 2264; SUNY Stony Brook ITP-SB-91-9.
[123] B. Carlitz, J. Collins, S. Heppelmann, G. Landinsky, R.L. Jaffe and X. Ji, Penn. State Univ. PSU/TH/101 (1992).
[124] M. Stratmann and W. Vogelsang, Phys. Lett. **B295** (1992) 277.
[125] X. Artru, AIP Conf. Proc. **223** (1991) 176.
[126] J. Collins, Penn. State Univ. PSU/TH/102 (1992), Nucl. Phys. B.

[127] O. Nachtmann, Nucl. Phys. **B127** (1977) 314; A.V. Efremov, Sov. J. Nucl. Phys. **28** (1978) 83.
[128] A.V. Efremov, L. Mankiewicz and N.A. Törnqvist, Phys. Lett. **B284** (1992) 394.
[129] B. Adeva et al., SMC collab., Phys. Lett. **B302** (1993) 533.

Small x Physics

J. Kwieciński

Department of Theoretical Physics,
H. Niewodniczański Institute of Nuclear Physics,
ul. Radzikowskiego 152, PL-31-342 KRAKÓW, Poland.

Abstract. The QCD expectations concerning the small x limit of parton distributions, where x is the Bjorken scaling variable, are reviewed. This includes discussion of the evolution equations in the small x region, the Lipatov equation, which sums the leading powers of $\ln(1/x)$, and the shadowing effects. Phenomenological implications of the theoretical expectations for the deep inelastic lepton-hadron scattering in the small x region, which will be accessible at the HERA ep collider, are described. We give predictions for structure functions F_2 and F_L and discuss specific processes sensitive to the small x physics such as heavy quark production, deep inelastic diffraction and jet production in deep inelastic lepton scattering. A brief review of nuclear shadowing in the inelastic lepton nucleus scattering at small x is also presented.

1 Introduction

Small x physics is very interesting both theoretically and phenomenologically. We shall be primarily concerned with the deep-inelastic lepton-hadron scattering in the small x limit, where x is the Bjorken scaling variable. The variable x is defined as :

$$x = \frac{Q^2}{2pq}, \tag{1.1}$$

where

$$Q^2 = -q^2 \tag{1.2}$$

and p and q are the four momenta of a hadron and of the virtual photon, respectively [1]. For large Q^2 we expect that the deep inelastic lepton - hadron scattering can be described within the QCD improved parton model [1-5]. The deep inelastic scattering at small x is then probing the parton distributions in a limit when the momentum fraction x carried by a parton is very small. QCD predicts several novel phenomena to occur in the small x limit of parton distributions such as the very strong increase of parton densities, parton screening effects possibly leading to parton saturation [6-11]. Understanding of parton distributions in this new regime is one of the most interesting and challenging theoretical problems in QCD. Moreover, the HERA ep collider will be able to probe the very small x region for reasonably high values of Q^2 for the parton model ideas to be applicable [11-13]. It is therefore important to know whether various QCD expectations at small x can be tested experimentally.

The main aim of these lectures is to summarize the QCD expectations concerning the small x behaviour of parton distributions and to discuss their possible phenomenological implications for HERA. We shall also discuss nuclear shadowing in inelastic lepton-nucleus scattering at small x.

2 Small x Limit of Parton Distributions in Perturbative QCD

2.1 Altarelli-Parisi Evolution Equations in the Small x Limit

The Altarelli-Parisi evolution equations describe the evolution of quark and gluon densities with Q^2 [1-5,14]. In the leading logQ^2 approximation LLQ^2 they have the following form:

$$Q^2\frac{\partial q_i(x,Q^2)}{\partial Q^2} = \frac{\alpha_s(Q^2)}{2\pi}[P_{qq}\otimes q_i + P_{qg}\otimes g]$$

$$Q^2\frac{\partial g(x,Q^2)}{\partial Q^2} = \frac{\alpha_s(Q^2)}{2\pi}[P_{gq}\otimes \Sigma + P_{gg}\otimes g], \tag{2.1.1}$$

where

$$P_{ij}\otimes f = \int_x^1 \frac{dx'}{x'}P_{ij}(\frac{x}{x'})f(x',Q^2) \tag{2.1.2}$$

and

$$\Sigma(x,Q^2) = \sum_{i=1}^{N_f}[q_i(x,Q^2)+\bar{q}_i(x,Q^2)]. \tag{2.1.3}$$

The functions $g(x,Q^2)$ and $q_i(x,Q^2)$ and $\bar{q}_i(x,Q^2)$ are the gluon, quark and antiquark distributions where the index i denotes the quark flavour. $\alpha_s(Q^2)$ is the running QCD coupling [1-5] and N_f denotes the number of active flavours. The quark distributions $q_i(x,Q^2)$ are equal to the sum of the valence and sea quark distributions i.e.:

$$q_i(x,Q^2) = q_i^v(x,Q^2) + q_i^s(x,Q^2) \tag{2.1.4}$$

and

$$q_i^s(x,Q^2) = \bar{q}_i(x,Q^2). \tag{2.1.5}$$

The functions $P_{ij}(z)$ are the familiar splitting functions [1-5]. The valence quark distributions $q_i^v(x,Q^2)$ satisfy the following equation:

$$Q^2\frac{\partial q_i^v(x,Q^2)}{\partial Q^2} = \frac{\alpha_s(Q^2)}{2\pi}P_{qq}\otimes q_i^v. \tag{2.1.6}$$

The structure function $F_2(x,Q^2)$ of deep inelastic (charged) lepton scattering is related in the standard way to the quark and antiquark distributions:

$$F_2(x,Q^2) = \sum_{i=1}^{N_f} e_i^2 x[q_i(x,Q^2)+\bar{q}_i(x,Q^2)], \tag{2.1.7}$$

where e_i is the electric charge of the quark having a flavour i. Both the evolution equations as well as the relation (2.1.7) between the structure function F_2 and the quark distributions acquire additional higher order terms in $\alpha_s(Q^2)$ in the next-to-leading approximation (NLLQ2) [1-5].

In physical gauges the Altarelli-Parisi equations in the LLQ^2 approximation correspond to ladder diagrams with ordered longitudinal momenta and strongly ordered transverse momenta along the chain (see Fig. 1)

$$x < x_n < ..... < x_1 < 1$$
$$Q^2 >> \kappa^2 >> k_n^2 >> .....k_1^2. \tag{2.1.8}$$

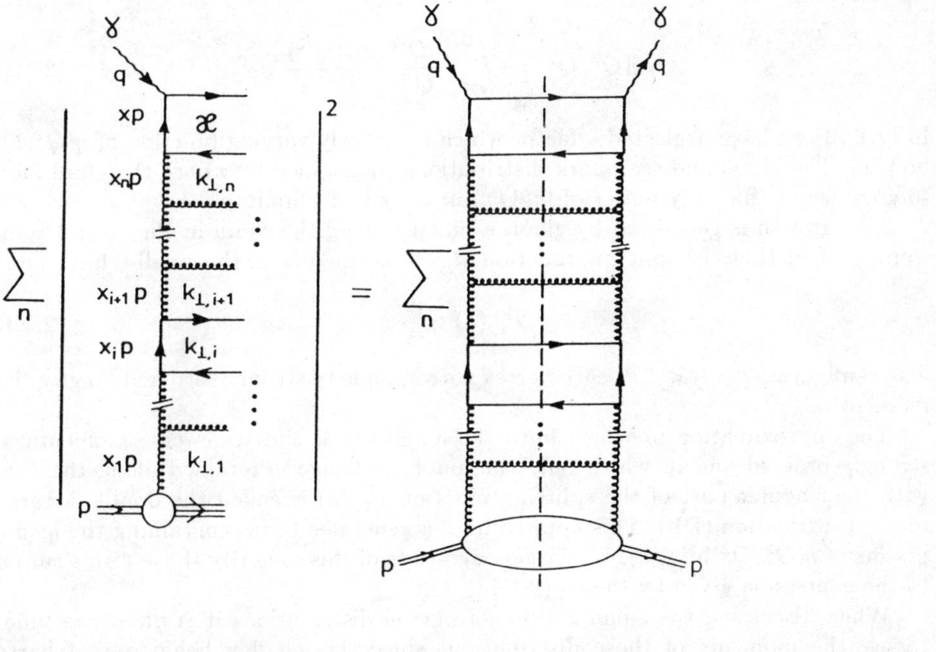

Fig. 1. The ladder diagram for the deep inelastic scattering in the LLQ^2 approximation.

In order to solve the evolution equations (2.1.1) and (2.1.6) one has to provide the input parametrisation of those distributions at some reference scale Q_0^2 which should be large enough for the QCD improved quark parton model to be applicable. The small x behaviour of the input parton distributions is usually obtained from Regge theory [1,9]:

$$xq_i^v(x, Q_0^2) \sim x^{1-\alpha_R} \tag{2.1.9}$$
$$xq_i^s(x, Q_0^2) \sim x^{1-\alpha_P}$$
$$xg(x, Q_0^2) \sim x^{1-\alpha_P}, \tag{2.1.10}$$

where α_R and α_P denote the reggeon and pomeron intercepts, respectively. Assuming that $\alpha_R \simeq 1/2$ and $\alpha_P \simeq 1$ we get the following behaviour for the valence quark, sea quark and gluon distributions at small x:

$$xq_i^v(x,Q_0^2) \sim x^{1/2} \tag{2.1.11}$$

$$xq_i^s(x,Q_0^2) \sim x^0$$

$$xg(x,Q_0^2) \sim x^0 \tag{2.1.12}$$

The $x^{1/2}$ type behaviour of the valence quark distributions is stable against the QCD evolution generated by the eq. (2.1.6). The x^0 behaviour of the sea quark and gluon distributions is, however, unstable and for $Q^2 > Q_0^2$ the evolution equations (2.1.1) generate a steeper behaviour, i.e.:

$$xg(x,Q_0^2) \sim \exp 2[\eta(Q^2,Q_0^2)\ln(1/x)]^{1/2}\}(2.1.13)$$

with a similar behaviour for the sea quarks, where the function $\eta(Q^2,Q_0^2)$ is given by the following formula:

$$\eta(Q^2,Q_0^2) = \int_{Q_0^2}^{Q^2} \frac{dQ'^2}{Q'^2}\frac{3\alpha_s(Q'^2)}{\pi}\,. \tag{2.1.14}$$

In (2.1.13) we have neglected a factor which is a slowly varying function of $\eta(Q^2,Q_0^2)$ $\ln(1/x)$. The gluon and sea quark distributions (multiplied by x) are therefore found to grow faster than any power of $\ln(1/x)$ in the small x limit.

This growth is generated by gluon exchange along the chain in Fig. 1 and comes from the fact that the splitting function $P_{gg}(z)$ is singular in the small z limit i.e.:

$$P_{gg}(z) \simeq \frac{6}{z}\,. \tag{2.1.15}$$

The configuration $z << 1$ at each vertex corresponds to strongly ordered longitudinal momenta.

The approximation in which both the longitudinal and transverse momenta are strongly ordered and in which only the gluon exchange is retained along the chain with the singular part of the splitting function $P_{gg}(z)$ is called the double logarithmic approximation (DL). This approximation generates terms containing the leading powers of $\eta(Q^2,Q_0^2)\ln(1/x)$. For large magnitude of this quantity those terms sum up to the expression given by the eq.(2.1.13).

When discussing the small x limit of parton distributions it is often convenient to use the moments of those distributions since the small x behaviour of parton distributions is controlled by the singularities of their moments in the moment index plane. The moment $g_n(Q^2)$ of the gluon distribution $g(x,Q^2)$ is defined as

$$g_n(Q^2) = \int_0^1 dx x^{n-1} g(x,Q^2) \tag{2.1.16}$$

and similarly for other parton distributions. The gluon distribution $g(x,Q^2)$ is related to the moments $g_n(Q^2)$ through the inverse Mellin transform:

$$g(x,Q^2) = \frac{1}{2\pi i}\int_{c-i\infty}^{c+i\infty} dn x^{-n} g_n(Q^2) \tag{2.1.17}$$

where the integration contour is located to the right of the singularities of the moment $g_n(Q^2)$ in the n plane. It is therefore clear from the eq. (2.1.17) that the (leading) small x behaviour of the gluon distributions is controlled by the leading (i.e. rightmost) singularity of $g_n(Q^2)$ in the n plane.

The evolution equations for moments can be solved in a closed form [1-5] and if we keep only the splitting function P_{gg} in the eq. (2.1.1), which is most relevant at small x, this solution is:

$$g_n(Q^2) = g_n(Q_0^2)exp[\gamma_{gg}(n)\eta(Q^2, Q_0^2)], \tag{2.1.18}$$

where

$$\gamma_{gg}(n) = \frac{1}{6}\int_0^1 dz z^{n-1} P_{gg}(z) \tag{2.1.19}$$

and where $g_n(Q_0^2)$ is the moment of the starting distribution. If we keep only the singular term $6/z$ in P_{gg} (i.e. if we restrict ourselves to the double logarithmic approximation) then $\gamma_{gg}(n)$ is equal to

$$\gamma_{gg}^{DL}(n) = \frac{1}{(n-1)}. \tag{2.1.20}$$

It follows from eqs. (2.1.18) and (2.1.20) that the moment $g_n(Q^2)$ has an essential singularity at $n = 1$. This is the leading singularity if the starting distribution has the behaviour (2.1.12) and it generates the small x behaviour given by the formula (2.1.13).

The solution of the evolution equations for the moments $q_{in}^v(Q^2)$ of the valence quark distributions is:

$$q_{in}^v(Q^2) = q_{in}^v(Q_0^2)exp[\gamma_{qq}(n)\eta(Q^2, Q_0^2)], \tag{2.1.21}$$

where

$$\gamma_{qq}(n) = \frac{1}{6}\int_0^1 dz z^{n-1} P_{qq}(z) \tag{2.1.22}$$

and the leading pole of the function $\gamma_{qq}(n)$ is at $n = 0$ [1,3-5]. The moment $q_{in}^v(Q_0^2)$ of the starting valence quark distribution will have a pole at $n = 1/2$ if this distribution has its small x behaviour given by the eq. (2.1.11). This pole will remain the leading singularity of the moment $q_{in}^v(Q^2)$ and in consequence the small x behaviour (2.1.11) of the valence quark distributions will remain unchanged by the QCD evolution.

The fact that the "x^0" behaviour of gluon and sea quark distributions is unstable against the QCD evolution suggests that the pomeron in QCD may be more complicated than just a simple Regge pole of intercept equal to 1. Moreover it should be realised that the LLQ^2 or DL approximations are incomplete in the small x limit. The LLQ^2 approximation sums by definition only those terms in the perturbative expansion which contain the leading powers of $\ln(Q^2)$ at each order of the perturbative expansion. In the small x limit, however, we should sum at least those terms which contain the leading powers of $\ln(1/x)$ and retain the full Q^2 dependence i.e. to consider the leading $\log(1/x)$ (LL$1/x$) approximation. This approximation leads to the Lipatov equation which will be described in the next Section.

2.2 The Leading Log$(1/x)$ Approximation and the Lipatov Equation

The leading $\log(1/x)$ approximation corresponds by definition to the sum of those terms in the perturbative expansion which contain the maximal powers of $\ln(1/x)$ at each order of the perturbative expansion. The basic quantity in this approximation is the unintegrated gluon distribution $f(x,k^2)$ defined as:

$$f(x,k^2) = Q^2 \frac{\partial x g(x,Q^2)}{\partial Q^2}|_{Q^2=k^2} \ , \tag{2.2.1}$$

where $g(x,Q^2)$ is the gluon distribution in a hadron. In the LL$1/x$ approximation the unintegrated gluon distribution satisfies the following equation [15-22]:

$$-x\frac{\partial f(x,k^2)}{\partial x} = \frac{3\alpha_s(k^2)}{\pi}k^2 \int_{k_0^2}^{\infty} \frac{dk'^2}{k'^2}\left\{\frac{f(x,k'^2)-f(x,k^2)}{|k'^2-k^2|} + \frac{f(x,k^2)}{[4k'^4+k^4]^{1/2}}\right\}$$
$$\equiv K_L \otimes f. \tag{2.2.2}$$

This equation is called in the literature the Balitzkij-Fadin-Kuraev-Lipatov equation or, in short, the Lipatov equation. It corresponds again to the sum of ladder diagrams (see Fig. 2) but unlike the LLQ^2 approximation the transverse momenta of the gluons are not ordered. The kernel K_L of the eq. (2.2.2) contains both the real gluon emission terms as well as the virtual corrections. The former correspond to terms proportional to $f(x,k'^2)$ while the latter to terms proportional to $f(x,k^2)$ in the rhs. of the eq. (2.2.2). The virtual corrections correspond to the "gluon reggeisation" [6,15-18] (or to the "non-Sudakov" form factor [20-22]). The variable(s) $k^2(k'^2)$ correspond to the transverse momenta of the gluons. The parameter k_0^2 is the infrared cut-off which is necessary if the running coupling constant effects are taken into account. (Strictly speaking, within the genuine LL$1/x$ approximation, the QCD coupling α_s should be set to be k^2 independent.)

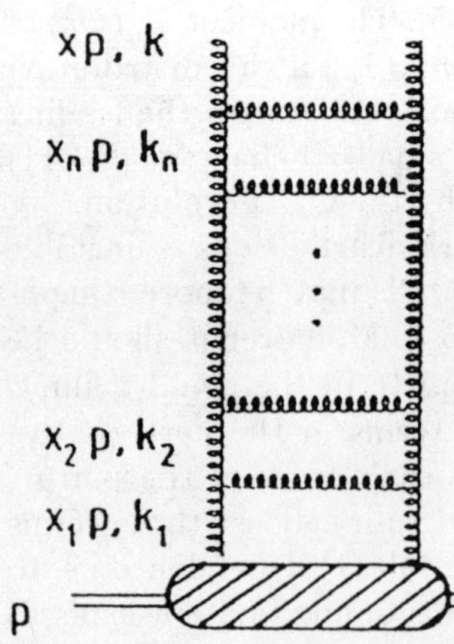

Fig. 2. The ladder diagram for the unintegrated gluon distribution $f(x,k^2)$ in the LL$1/x$ approximation.

When the running coupling effects are neglected (i.e. when one sets $\alpha_s(k^2) = \bar{\alpha}_s$) and when $k_0^2 = 0$ then the equation (2.2.2) can be solved analytically. To this aim it is convenient to introduce the moment $f_n(k^2)$ of the function $f(x,k^2)$:

$$f_n(k^2) = \int_0^1 dx x^{n-2} f(x,k^2) \tag{2.2.3}$$

which satisfies the following integral equation:

$$f_n(k^2) = \frac{f_0(k^2)}{(n-1)} + \frac{3\bar{\alpha}_s}{\pi(n-1)} k^2 \int_0^\infty \frac{dk'^2}{k'^2} \{ \frac{f_n(k'^2) - f_n(k^2)}{|k'^2 - k^2|} + \frac{f_n(k^2)}{[4k'^4 + k^4]^{1/2}} \} . \tag{2.2.4}$$

This equation can be solved by taking the Mellin transform $\tilde{f}_n(r)$ of $f_n(k^2)$

$$\tilde{f}_n(r) = \int_0^\infty dk^2 (k^2)^{-r-1} f_n(k^2) \tag{2.2.5}$$

which gives:

$$\tilde{f}_n(r) = \frac{\tilde{f}_0(r)}{n - 1 - \frac{3\bar{\alpha}_s}{\pi} \tilde{K}(r)} , \tag{2.2.6}$$

where $\tilde{K}(r)$ is the Mellin transform of the kernel of the integral equation (2.2.4), that is [6,16,17]

$$\tilde{K}(r) = -[\Psi(r) + \Psi(1-r) - 2\Psi(1)], \tag{2.2.7}$$

where $\Psi(z)$ is the logarithmic derivative of the Euler gamma function: $\Psi(z) \equiv \Gamma'(z)/\Gamma(z)$. The function $\tilde{f}_0(r)$ is the Mellin transform of the driving term $f_0(k^2)$. The moment $f_n(k^2)$ is given by the inverse Mellin transform

$$f_n(k^2) = \frac{1}{2\pi i} \int_{1/2-i\infty}^{1/2+i\infty} dr (k^2)^r \frac{\tilde{f}_0(r)}{n - 1 - \frac{3\bar{\alpha}_s}{\pi} \tilde{K}(r)} . \tag{2.2.8}$$

The integrand in (2.2.8) has poles at the zeros of its denominator. They define the anomalous dimensions in the LL($1/x$) approximation [23]. Note that for $r = 1/2$ the two poles pinch the integration contour in (2.2.8) leading to the (branch point) singularity in the n plane at $n = 1 + \lambda$ where

$$\lambda = \frac{3\alpha_s}{\pi} \tilde{K}(1/2) \equiv \frac{12 ln(2)}{\pi} \bar{\alpha}_s . \tag{2.2.9}$$

The leading small x behaviour is then found to be given by the following formula:

$$f(x,k^2) \sim const (k^2)^{1/2} \frac{x^{-\lambda}}{[ln(1/x)]^{1/2}} [1 + O(1/ln(1/x))]. \tag{2.2.10}$$

The parameter λ is equal to the maximal eigenvalue of the kernel K_L. The relatively unimportant factor $[\ln(1/x)]^{-1/2}$ in the eq. (2.2.10) comes from the fact that for the fixed coupling $\bar{\alpha}_s$ the eigenvalue spectrum of the kernel K_L is continuous. If the running coupling effects are included in the Lipatov equation as displayed explicitly in the eq.(2.2.2)) then the eigenvalue spectrum of K_L becomes discrete. The leading small x behaviour is then:

$$f(x,k^2) \simeq x^{-\bar{\lambda}} . \tag{2.2.11}$$

The maximal eigenvalue $\bar{\lambda}$ has now to be calculated numerically. It has been found to be dependent upon the infrared cut-off k_0^2 [24-28].

We found in this way that the gluon distributions in the LL$1/x$ approximation (multiplied by x) grow as the negative power of x in the small x limit. It should be noted that the exponent λ can have potentially large magnitude $\simeq 1/2$ or so.

This singular behaviour cannot hold forever and for sufficiently small values of x it has to be tamed by the screening corrections which will be described in the next Section.

2.3 Parton Screening

We showed in the previous Section that the gluon distributions (multipled by x), i.e. the functions $xg(x, Q^2)$, can grow indefinitely in the small x limit. The gluon distribution $g(x, Q^2)$ is equal by definition to the number density of gluons having transverse size $\sim 1/Q$ and carrying the fraction x of the hadron momentum in the infinite momentum frame. The quantity $xg(x, Q^2)$ is then equal to the number of those gluons within a bin $\Delta x \simeq x$ or equivalently to the number of gluons per unit rapidity $\ln(1/x)$. The transverse area $\tilde{S}(x, Q^2)$ occupied by those gluons is then given by the following formula:

$$\tilde{S}(x, Q^2) = xg(x, Q^2)\tilde{S}_g(Q^2), \tag{2.3.1}$$

where $\tilde{S}_g(Q^2)$ is the transverse area of a single gluon. Let us assume that it is equal to the (total) cross-section corresponding to the interaction of the hard probe gluon of virtuality Q^2 with a gluon in a hadron i.e.:

$$\tilde{S}_g(Q^2) = const\frac{\alpha_s(Q^2)}{Q^2} . \tag{2.3.2}$$

The total area occupied by the small x gluons is then given by:

$$\tilde{S}(x, Q^2) = const\ xg(x, Q^2)\frac{\alpha_s(Q^2)}{Q^2} . \tag{2.3.3}$$

This formula describes at the same time the cross-section corresponding to the interaction of the probe with a hadron within a QCD improved parton model. We showed in the preceding Sections that the quantity $xg(x, Q^2)$ can grow indefinitely with decreasing x. For sufficiently small values of x the transverse area occupied by the small x gluons can, for fixed Q^2, become comparable to the transverse area of a hadron $S_H = \pi R_H^2$ where R_H is the hadronic radius. When this happens (and in fact before this happens), the gluons can no longer be treated as free partons. They begin to screen each other and interact. This interaction leads to screening (or shadowing) effects. The main effect of shadowing is to tame the indefinite increase of parton distributions. One finds instead that at the sufficiently small values of x and/or Q^2 the gluon distributions approach the so-called saturation limit $xg_{sat}(x, Q^2)$ [6]

$$xg_{sat}(x, Q^2) = \frac{const}{\alpha_s(Q^2)} R_H^2 Q^2 . \tag{2.3.4}$$

In some models [29] the saturation limit contains some remnant weak x dependence. The most dramatic effect is the linear scaling violation exhibited by $xg_{sat}(x,Q^2)$.

If one assumes that the gluons are not distributed uniformly within a hadron but are concentrated around the "hot-spots" [30,31] having their radius $R_{h.s.}$ much smaller than the hadronic radius R_H then the shadowing effects are expected to be stronger. The saturation limit $xg_{sat}(x,Q^2)$ is then controlled by the radius $R_{h.s.}$ and not by R_H.

The shadowing effects modify the evolution equations as well as the Lipatov equation by the non-linear terms. Thus the Lipatov equation with shadowing has the following form [6,25]:

$$-x\frac{\partial f(x,k^2)}{\partial x} = \frac{3\alpha_s(k^2)}{\pi}k^2\int_{k_0^2}^{\infty}\frac{dk'^2}{k'^2}\{\frac{f(x,k'^2)-f(x,k^2)}{|k'^2-k^2|}+\frac{f(x,k^2)}{[4k'^4+k^4]^{1/2}}\}$$

$$-\frac{81\alpha_s^2(k^2)}{16R^2k^2}[xg(x,k^2)]^2. \tag{2.3.5}$$

This equation is called in the literature the Gribov, Levin, Ryskin (GLR) equation.

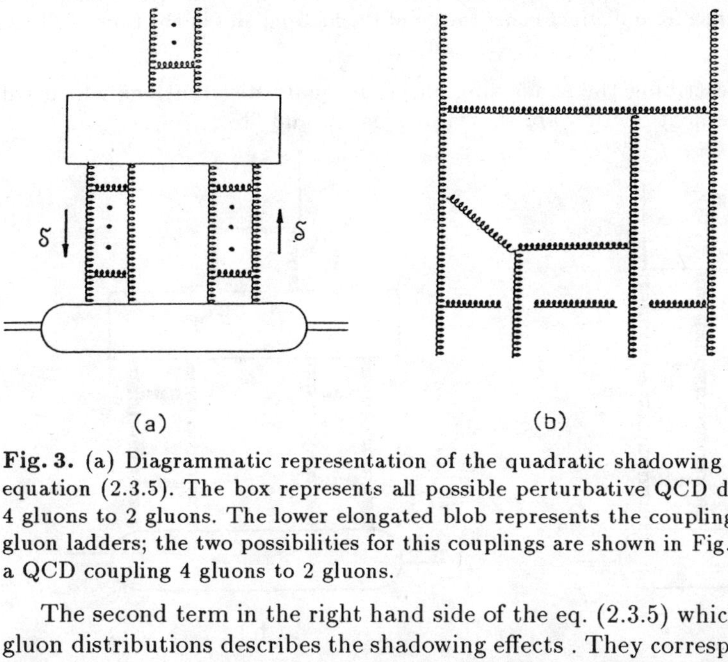

Fig. 3. (a) Diagrammatic representation of the quadratic shadowing term in the evolution equation (2.3.5). The box represents all possible perturbative QCD diagrams which couple 4 gluons to 2 gluons. The lower elongated blob represents the coupling of the hadron to the gluon ladders; the two possibilities for this couplings are shown in Fig. 4. (b) An example of a QCD coupling 4 gluons to 2 gluons.

The second term in the right hand side of the eq. (2.3.5) which is quadratic in the gluon distributions describes the shadowing effects . They correspond to the diagrams shown in Fig. 3. The parameter R describes the size of the region within which the gluons are concentrated. It arises from the integration over the 4-momentum δ flowing along the gluon ladders in Fig. 3. Specifically R comes from the integration over the transverse components of δ,

$$\frac{1}{R^2}\sim\int d\delta_T^2[F(-\delta_T^2)]^2 \tag{2.3.6}$$

and its value depends on exactly how the gluon ladders couple to the hadron. If the ladders couple to different partons (see Fig. 4a), then the form factor F is characterised by the hadronic radius, that is $R \sim 5$ GeV^{-1}. If we include a possibility that the ladders couple to the same constituent of the hadron (Fig. 4b) then it is appropriate to take R to be the radius of a valence quark, that is $R \sim 2$ GeV^{-1}. It is in the latter case when we speak about "hot-spots" [30,31].

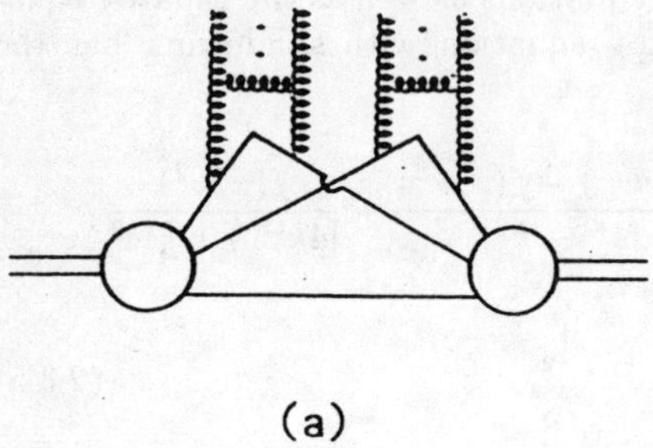

Fig. 4. Two possibilities for the coupling of the hadron (i.e. proton) to the gluon ladders. In Fig. (a) the ladders arise from distinct constituents of the hadron, and in (b) from the same constituent.

The diagrams describing the shadowing effects in quark distributions which will be introduced in the next Section are shown in Figs. 5a and 5b.

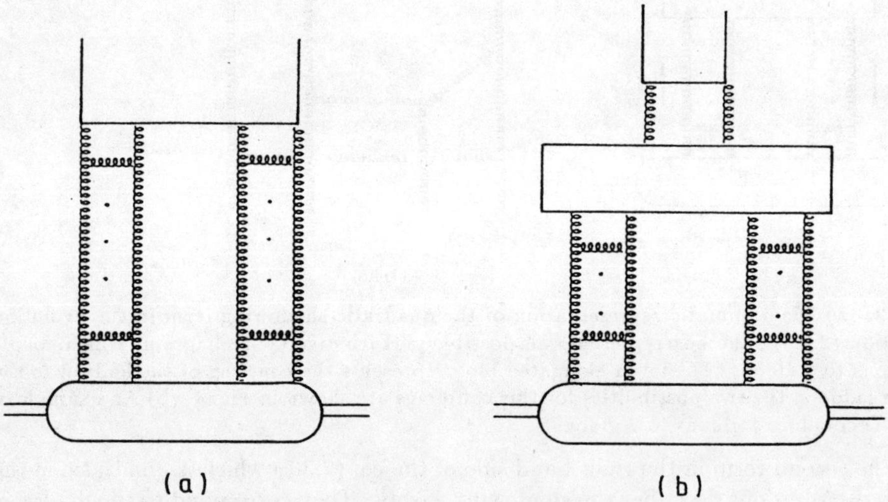

Fig. 5. Diagrams giving rise to shadowing terms in the evolution equations, (3.1.5 a-c), for the sea quark distributions.

Analysis of the equation (2.3.5) permits to study the onset of the singular $x^{-\lambda}$ type of behaviour generated by the Lipatov kernel and a simultaneous taming of this behaviour by the non-linear shadowing term. This analysis has been performed in [25,32] where the eq. (2.3.5) was solved by evolving in x downwards starting from the phenomenological boundary conditions at $x = x_0 = 10^{-2}$. The gluon distributions which come out from this analysis are presented in Figs. 6a and 6b. It may be seen that in the case when the shadowing term is neglected the $x^{-\lambda}$ behaviour emerges very clearly. It may also be seen that the shadowing corrections are rather small in the region of moderately small values of x ($x > 10^{-4}$ or so) at least for $R = 5$ GeV^{-1}.

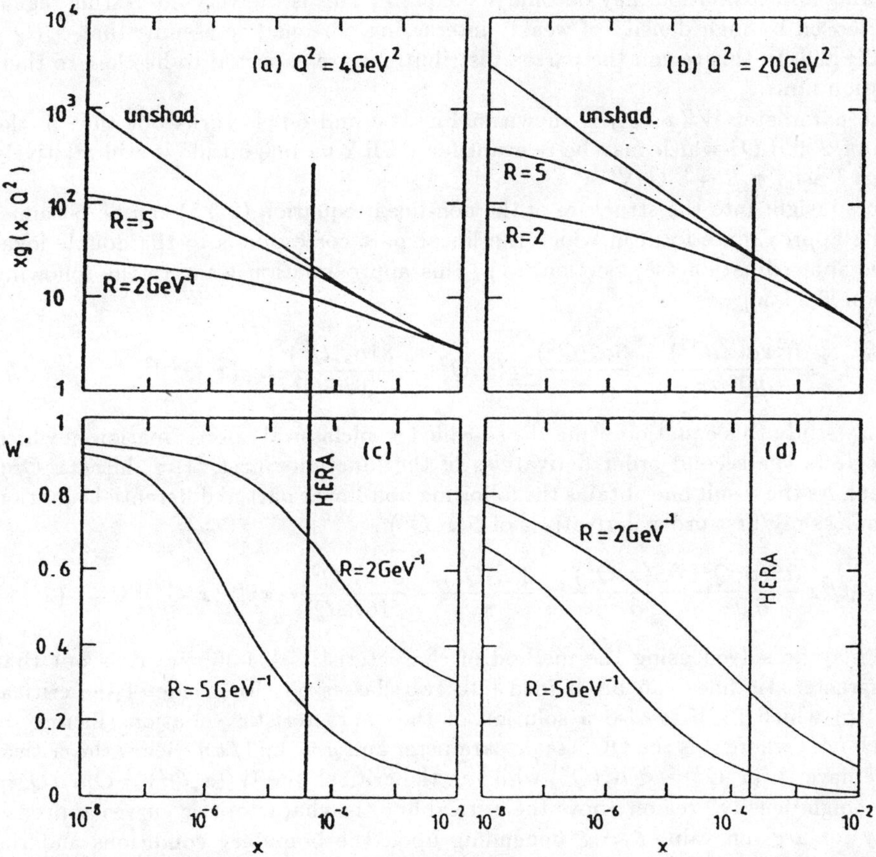

Fig. 6. (a),(b): The values of $xg(x, Q^2)$ at (a) $Q^2 = 4$ GeV^2 and (b) $Q^2 = 20$ GeV^2 calculated from the solution $f(x, k^2)$ of the integro-differential Lipatov equation (2.3.5). The three curves are, in descending order, the solution with shadowing neglected and the solutions with the shadowing term included with $R = 5$ GeV^{-1} and $R = 2$ GeV^{-1}. (c) and (d) show the values of W' defined as the ratio of the quadratic to the linear term in the rhs. of the eq. (2.3.5). The values of shadowing correspond to the solutions with shadowing shown in (a) and (b) (from [25]).

We can quantify the role of the shadowing effects introducing the parameter $W'(x, Q^2)$ equal to the ratio of the quadratic term to the linear term in the rhs.

of the eq. (2.3.5). One may consider three regions in the (x, Q^2) plane depending upon the magnitude of $W'(x, Q^2)$.
(i) In the region where $W'(x, Q^2) << \alpha_s(Q^2)$ the shadowing is negligible.
(ii) For $W'(x, Q^2) \simeq \alpha_s(Q^2)$ the shadowing becomes significant. It is expected to be described by the "leading" term which is quadratic in the gluon distribution as in the eq. (2.3.5).
(iii) In the region where $W'(x, Q^2) >> \alpha_s(Q^2)$ (i.e. $W'(x, Q^2) = O(1)$) the "leading" shadowing approximation may become inadequate. This is the very interesting region characterised by high density of weakly interacting partons (we assume that $\alpha_s(Q^2)$ is small) [29]. In this region the parton distributions are expected to be close to their saturation limit.

The parameter $W'(x, Q^2)$ is shown in Figs. 6c and 6d. It turns out that in the region of x and Q^2 which may be relevant for HERA its magnitude is still relatively small at least for $R \sim 5\ \mathrm{GeV}^{-1}$.

More insight into the structure of the non-linear equation (2.3.5) can be obtained from its approximate form in which the linear part corresponds to the double logarithmic approximation (see Section 2.1). This approximation leads to the following equation [6,33-36]:

$$-Q^2 x \frac{\partial^2 xg(x, Q^2)}{\partial Q^2 \partial x} = \frac{3\alpha_s(Q^2)}{\pi} xg(x, Q^2) - \frac{81\alpha_s(Q^2)^2}{16R^2Q^2}[xg(x, Q^2)]^2. \tag{2.3.7}$$

One may study this equation using the so-called semiclassical approximation in which one neglects the second order derivatives of the function $S(x, Q^2) = \ln[xg(x, Q^2)]$ [6,35,36]. As the result one obtains the following non-linear partial differential equation containing only first order derivatives of $S(x, Q^2)$:

$$-Q^2 x \frac{\partial S(x, Q^2)}{\partial Q^2} \frac{\partial S(x, Q^2)}{\partial x} = \frac{3\alpha_s(Q^2)}{\pi} - \frac{81\alpha_s(Q^2)^2}{16R^2Q^2} exp[S(x, Q^2)] \tag{2.3.8}$$

which may be solved using the method of characteristics [34-36]. It turns out that the characteristic lines can be divided into two classes on either side of the critical line $y_c(\xi)$ which itself is also a solution of the characteristic equations [here $\xi = \ln[\ln(Q^2/\Lambda^2)$ where Λ is the QCD scale parameter and $y = \ln(1/x)$]. Below the critical line we have $W'(x, Q^2) << \alpha_s(Q^2)$ while on the critical line $W'(x, Q^2) = O[\alpha_s(Q^2)]$. In the "high density" region above the critical line the characteristic curves approach infinity for a given value $\xi = \tilde{\xi}$ depending upon the boundary conditions and the parameters defining the shadowing term. The limit $y(\xi) \to \infty$ for $\xi \to \tilde{\xi}$ corresponds to $W'(x, Q^2) \to 1$ and $xg(x, Q^2) \to xg_{sat}(x, Q^2)$. The saturation limit in this case corresponds to the gluon distribution for which the right hand side of the eq. (2.3.7) vanishes i.e.:

$$xg_{sat}(x, Q^2) = \frac{16}{27\pi\alpha_s(Q^2)} R^2 Q^2. \tag{2.3.9}$$

The pattern of the characteristic lines described above is displayed in Fig.7

It turns out, however, that the parameter $W'(x, Q^2)$ obtained from the exact solution of the eq. (2.3.5) can have acceptable small values in regions above the semiclassical critical line. This in particular takes place in the region of (x, Q^2) which may be accessible at HERA. The semiclassical critical line should not therefore be regarded in general as the genuine boundary between the regions of low and high gluon densities.

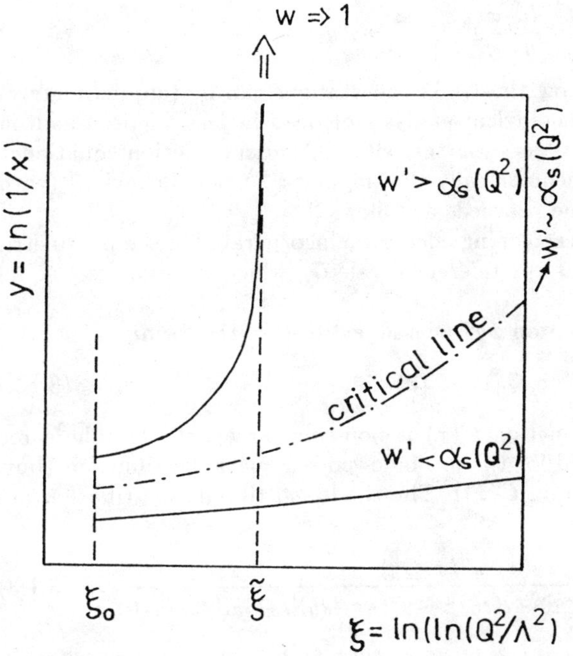

Fig. 7. A sketch of the characteristic curves, $y(\xi)$, obtained by solving the characteristic equations corresponding to the eq. (2.3.8), using input at $\xi = \xi_0$. In this approximation the gluon packing fraction parameter $W' \simeq \alpha_s(Q^2)$ on the critical line. In regions where shadowing is negligible the characteristic curves are straight lines (from [25]).

It is also possible to formulate the semiclassical approximation of the original equation (2.3.5) [25,37].

3 Small x Physics in Deep-Inelastic Lepton-Hadron Scattering

3.1 The QCD Predictions for the Deep-Inelastic Structure Functions at Small x

In the previous Sections we have described the QCD expectations concerning the small x behaviour of parton distributions. Our discussion focussed entirely on gluon distributions which dominate at small x. The singular small x behaviour, shadowing effects etc. should also manifest themselves in the sea quark distributions which at small x are predominantly driven by the gluons. It is the sea quark distributions which dominate in the structure function F_2 at small x. The longitudinal structure function $F_L(x, Q^2)$ is at small x also dominated by the gluons and in perturbative QCD one gets the following formula for $F_L(x, Q^2)$ [1]:

$$F_L(x,Q^2) = \frac{\alpha_s(Q^2)}{\pi}\{\int_x^1 \frac{dy}{y}(\frac{x}{y})^2 F_2(y,Q^2)$$

$$+2\sum e_q^2 \int_x^1 \frac{dy}{y}(\frac{x}{y})^2(1-\frac{x}{y})yg(y,Q^2)\}. \tag{3.1.1}$$

We shall now briefly discuss how the QCD expectations can be (approximately) incorporated within the phenomenological analysis of deep inelastic lepton-hadron scattering and of related processes based on the Altarelli-Parisi evolution equations.

In order to implement phenomenologically the singular $x^{-\lambda}$ type behaviour (with $\lambda \simeq 1/2$) and shadowing effects one proceeds as follows [38]:

(i) The singular behaviour and screening effects are incorporated in the parametrisation of the input distributions at the reference scale Q_0^2 which is usually set equal 4 GeV^2.

The unshadowed gluon distribution xg^u is assumed to take the form:

$$xg^u(x,Q_0^2) = C(x)x^{-\lambda}, \tag{3.1.2}$$

where $\lambda \simeq 1/2$ and where the function $C(x)$ is non-singular at $x = 0$. The same parametrisation is assumed to hold for the unshadowed sea quark distributions (but, of course, with the different function $C(x)$). The shadowed (input) distribution are modified as below:

$$xg(x,Q_0^2) = \frac{xg^u(x,Q_0^2)}{\{1+\theta(x_0-x)[xg^u(x,Q_0^2)-xg^u(x_0,Q_0^2)]/xg_{sat}(x,Q_0^2)\}} \tag{3.1.3}$$

$$xq_i^s(x,Q_0^2) = xq_i^{s(u)}(x,Q_0^2)\frac{xg(x,Q_0^2)}{xg^u(x,Q_0^2)}, \tag{3.1.4}$$

where $xg_{sat}(x,Q_0^2)$ is defined by the eq.(2.3.9). The shadowing effects are in this way switched on only for $x < x_0$ where x_0 is usually set equal 10^{-2}.

(ii) The input parton distributions are then evolved in Q^2 using the QCD evolution equations with the non-linear terms describing shadowing effects both for gluons as well as for the sea quarks. Those evolution equations have the following form [33]:

$$Q^2\frac{\partial xg(x,Q^2)}{\partial Q^2} = \frac{\alpha_s(Q^2)}{2\pi}x[P_{gq}\otimes\Sigma + P_{gg}\otimes g]$$

$$-\frac{81\alpha_s^2(Q^2)}{16R^2Q^2}\theta(x_0-x)\int_x^{x_0}\frac{dx'}{x'}[x'g(x',Q^2)]^2 \tag{3.1.5a}$$

$$Q^2\frac{\partial xq_i(x,Q^2)}{\partial Q^2} = \frac{\alpha_s(Q^2)}{2\pi}x[P_{qq}\otimes q_i + P_{qg}\otimes g]$$

$$-\frac{27\alpha_s^2(Q^2)}{160R^2Q^2}[xg(x,Q^2)]^2 + \frac{\alpha_s(Q^2)}{\pi Q^2}\theta(x_0-x)\int_x^{x_0}\frac{dx'}{x'}\frac{x}{x'}\gamma(\frac{x}{x'})x'G_H(x',Q^2) \tag{3.1.5b}$$

$$Q^2\frac{\partial xG_H(x,Q^2)}{\partial Q^2} = -\frac{81\alpha_s^2(Q^2)}{16R^2Q^2}\theta(x_0-x)\int_x^{x_0}\frac{dx'}{x'}[x'g(x',Q^2)]^2 \tag{3.1.5c}$$

where

$$\gamma(z) = -2z + 15z^2 - 30z^3 + 18z^4. \tag{3.1.6}$$

(iii) In the region of "large" x ($x > 10^{-2}$ or so) the parton distributions are constrained phenomenologically by the available data on deep inelastic lepton-hadron scattering and on other hard processes.

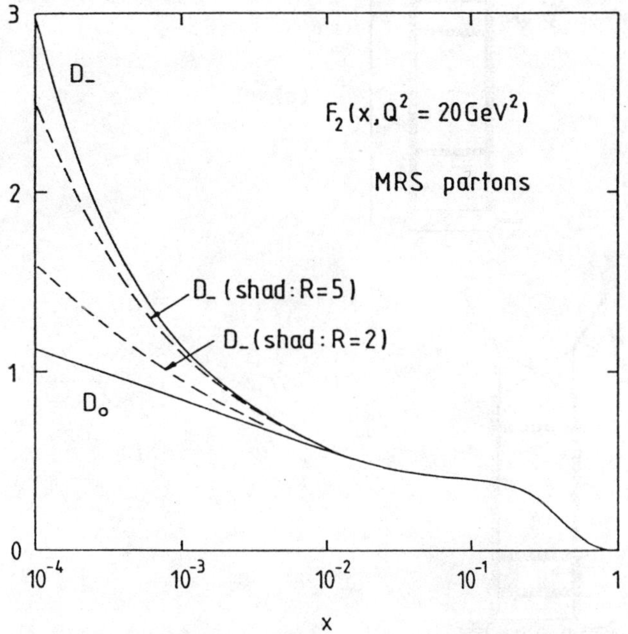

Fig. 8. The structure function F_2 as a function of x at $Q^2 = 20$ GeV^2 based on MRS partons [39]. Sets D_- and D_0 have the input gluon and sea quark distributions (multiplied by x) which behave respectively as $x^{-1/2}$ and x^0 at small x. The dashed curves show the effect of parton shadowing for two different choices of the parameter R.

In Fig. 8 we show the recent results of this analysis for the structure function $F_2(x, Q^2)$ [39]. This figure shows predictions concerning the behaviour of the structure function $F_2(x, Q^2)$ based on two sets of parton distributions, D_- and D_0, which respectively include and omit an $x^{-1/2}$ factor in the input distributions. Both sets describe the existing "large" x ($x > 0.05$ or so) data equally well. In the D_- case we also show results with and without shadowing corrections for two values of the radius parameter R i.e. for $R = 5$ GeV^{-1} (weak shadowing) and for $R = 2$ GeV^{-1} (strong shadowing motivated by the "hot spots" scenario), respectively. It can be seen that the shadowing effects are expected to be small in the small x region relevant for HERA (i.e. for $x \sim 10^{-3}$).

A missing ingredient in the procedure described above is any constraint on the size of the Lipatov component i.e. on that part of the gluon and sea quark distributions which has the $x^{-\lambda}$ behaviour with $\lambda \simeq 1/2$. The gluon distribution has not been required to satisfy the Lipatov equation (see Sec. 2.2); simply the $x^{-1/2}$ behaviour has been imposed on the input distribution at the reference scale $Q^2 = Q_0^2$, and also on the sea quark distributions which are themselves driven by the gluons.

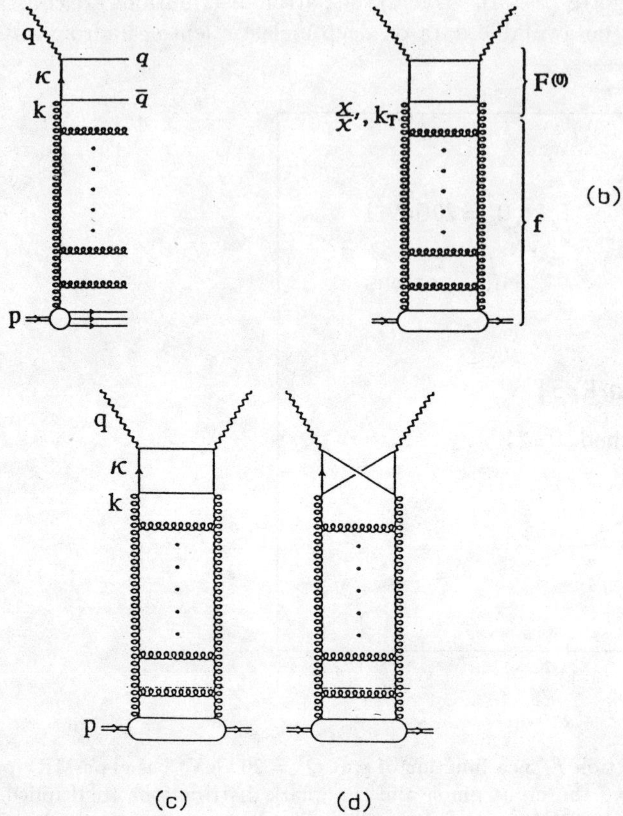

Fig. 9. (a) Diagrammatic representation of a gluon ladder contribution to the deep inelastic structure functions of the proton. q, κ, k and p denote the particle 4-momenta. (b) Diagrammatic representation of the factorisation formula of eq. (3.1.7). (c) Diagrams giving the small x behaviour of the structure functions for deep-inelastic electron-proton scattering.

We shall now discuss more quantitative predictions for the behaviour of the structure functions $F_2(x, Q^2)$ and $F_L(x, Q^2)$ at small x which will be directly based on the Lipatov equation [40]. The relevant diagrams are shown in Fig. 9 and the contribution to the (transverse and longitudinal) deep inelastic structure functions can be written in the following factorizable form [41-43]:

$$F_{T,L}(x, Q^2) = \int_x^1 \frac{dx'}{x'} \int \frac{dk_T^2}{k_T^4} f(\frac{x}{x'}, k_T^2) F_{T,L}^{(0)}(x', k_T^2, Q^2), \tag{3.1.7}$$

where x/x' is the longitudinal momentum fraction carried by the gluon which couples to the $q\bar{q}$ pair. The functions $F_{T,L}^{(0)}$ denote the quark box (and crossed box) approximation to the photon-gluon subprocess shown in the upper part of the diagrams shown in Fig. 9. In other words $F_{T,L}^{(0)}$, or to be dimensionally correct $F_{T,L}^{(0)}/k_T^2$, may be regarded as the structure functions of a gluon of approximate virtuality k_T^2. The function $f(x/x', k_T^2)$ is the unintegrated gluon density corresponding to the sum of ladder diagrams in the lower part of the diagrams shown in Fig. 9. In the leading $\log(1/x)$ approximation or, to be precise in the leading $\log(x'/x)$ approximation, the function f is given as the solution of the Lipatov equation (see Sections 2.2 and 2.3).

It may be instructive to see how the general factorizable form (3.1.7) reduces to the Altarelli-Parisi evolution of $q\bar{q}$ radiation from a gluon. In the Altarelli-Parisi treatment in the leading $\log Q^2$ approximation the integrations over the transverse momenta are dominated by the contributions from the strongly ordered configuration $k_T^2 << \kappa_T^2 << Q^2$ (see Sec. 2.1). In this limit there is no contribution to F_L and so we need only to consider the structure function $F_2(x, Q^2)$. If we keep only the strongly ordered contribution and we recall that x' is the momentum fraction of the gluon carried by the quark (or antiquark) which is struck by the photon, then F_2^0, which is equal to F_T^0 of (3.1.7) in the LLQ^2 approximation, is given by

$$F_2^{(0)}(x', k_T^2, Q^2) = \int_{k_T^2}^{Q^2} \frac{d\kappa_T^2}{\kappa_T^2} 2 \sum_q e_q^2 \frac{\alpha_s(\kappa_T^2)}{2\pi} x' P_{qg}(x'), \tag{3.1.8}$$

where $P_{qg}(x')$ is the Altarelli-Parisi splitting function. Thus (3.1.7) becomes

$$F_2(x, Q^2) = \int_x^1 dx' \int^{Q^2} \frac{d\kappa_T^2}{\kappa_T^2} \int^{\kappa_T^2} \frac{dk_T^2}{k_T^2} f(\frac{x}{x'}, k_T^2) 2 \sum_q e_q^2 \frac{\alpha_s(\kappa_T^2)}{2\pi} P_{qg}(x') \tag{3.1.9}$$

and hence, using (2.2.1), we have

$$Q^2 \frac{\partial F_2(x, Q^2)}{\partial Q^2} = 2 \sum_q e_q^2 \frac{\alpha_s(Q^2)}{2\pi} \int_x^1 dx' P_{qg}(x') \frac{x}{x'} g(\frac{x}{x'}, Q^2) \tag{3.1.10}$$

that is the conventional Altarelli-Parisi evolution of F_2 driven by $g \rightarrow q\bar{q}$.

There are at least two reasons why the formulae (3.1.9) and (3.1.10) based on the leading $\log(Q^2)$ approximation are inadequate in the small x region. The first is due to the Lipatov effect and the second arises from shadowing. We discuss these effects in turn. A crucial observation is that it is the dominance of the strong ordering of transverse momenta which leads to the nested integrations of (3.1.9). However, at small x where the leading $\log(1/x)$ terms dominate it is important to retain the full Q^2 dependence and not just the leading $\log(Q^2)$ terms. This is accomplished by the Lipatov equation for the unintegrated gluon distribution f which sums the ladder diagrams over the full phase space of the transverse momenta and not simply the strongly ordered part. In the case of fixed $\bar{\alpha}_s$ the Lipatov equation can be solved analytically and the leading small x behaviour of $f(x, k_T^2)$ is given by the formula (2.2.10). We note, in particular, the factor $(k_T^2)^{1/2}$ which may be traced to the anomalous dimension having magnitude 1/2 [23]. Due to this factor the region of strongly ordered transverse momenta is no longer dominant. The integrals are no longer of logarithmic dk_T^2/k_T^2 form and we must use the exact k_T^2 dependence of $F_2^{(0)}$ as well as integrating over the full region of phase space of transverse momenta.

At the very small x region the singular $x^{-\lambda}$ behaviour of f will be tamed by shadowing effects which lead to parton saturation (see Sec. 2.3). The saturation limit

f_{sat} of f grows linearly with k_T^2 and so, as before, the k_T^2 behaviour requires that we must integrate over full phase space of transverse momenta [42].

Both the analytic solution of the Lipatov equation given by the formula (2.2.10) as well as the saturation limit overestimate their respective effects [25,26]. The numerical solution of the Lipatov equation shows, particularly when the effects of the running QCD coupling are included, that the approximate analytic form considerably overestimates the actual solution [26]. Secondly, the numerical solution of the Lipatov equation with the non-linear shadowing term included (see eq. (2.3.5)) shows that the saturation limit is approached rather slowly [26] and is irrelevant for $x > 10^{-4}$ which will be probed at HERA (see Sec. 2.3). In order to obtain a reliable prediction it is therefore necessary to use the "exact" (numerical) solution f of the Lipatov equation in the formula (3.1.7) and this calculation has been performed in Ref. [40].

In the analysis of the diagrams shown in Fig. 9 it is convenient to use the basic light-like momenta q' and p where $q' = q + xp$. We decompose the gluon and quark 4-momenta k and κ, respectively, in terms of the Sudakov variables

$$k = ap - bq' + \mathbf{k_T}$$

$$\kappa = \alpha p - \beta q' + \boldsymbol{\kappa}_T. \tag{3.1.11}$$

We must carry out the integration over the box diagrams subject to quark mass-shell constraints which, in terms of the Sudakov variables, are of the form:

$$(\alpha - x)(1-\beta)2pq - \kappa_T^2 = m_q^2$$

$$(a-\alpha)\beta 2pq - (\boldsymbol{\kappa}_T - \mathbf{k})^2 = m_q^2. \tag{3.1.12}$$

The contributions of the diagrams of Fig. 9 to the functions $F_{T,L}(x, Q^2)$ are expressed in the form [40,41,43]:

$$F_T^0(x', k_T^2, Q^2) = 2\sum e_q^2 \frac{Q^2}{4\pi^2} \int_0^1 d\beta \int d^2\boldsymbol{\kappa}'_T \alpha_s(\kappa_T'^2 + m_{q0}^2)$$

$$*\{[\beta^2 + (1-\beta)^2][\frac{\kappa_T^2}{D_1^2} - \frac{\boldsymbol{\kappa}_T \cdot (\boldsymbol{\kappa}_T - \mathbf{k}_T)}{D_1 D_2}] + \frac{m_q^2}{D_1^2} - \frac{m_q^2}{D_1 D_2}]\}$$

$$*x'\delta\{x' - [1 + \frac{\kappa_T'^2}{Q^2}\beta(1-\beta) + \frac{k_T^2}{Q^2}]^{-1}\} \tag{3.1.13}$$

$$F_L^0(x', k_T^2, Q^2) = 2\sum e_q^2 \frac{Q^4}{\pi^2} \int_0^1 d\beta \int d^2\boldsymbol{\kappa}'_T \alpha_s(\kappa_T'^2 + m_{q0}^2)$$

$$*\beta^2(1-\beta)^2[\frac{1}{D_1^2} - \frac{1}{D_1 D_2}]x'\delta\{x' - [1 + \frac{\kappa_T'^2}{Q^2}\beta(1-\beta) + \frac{k_T^2}{Q^2}]^{-1}\}, \tag{3.1.14}$$

where

$$D_1 = \kappa_T^2 + \beta(1-\beta)Q^2 + m_q^2 \tag{3.1.15}$$

$$D_2 = (\boldsymbol{\kappa}_T - \mathbf{k}_T)^2 + \beta(1-\beta)Q^2 + m_q^2 \tag{3.1.16}$$

$$\boldsymbol{\kappa}_T = \boldsymbol{\kappa}'_T + (1-\beta)\mathbf{k_T} \tag{3.1.17}$$

and $x' \geq x$. The parameter m_q denotes the quark mass while the "mass" m_{q0} serves as the regulator which "freezes" the coupling at $\alpha_s(m_{q0}^2)$ for small $\kappa_T'^2$. For the heavy quark contribution it may be set $m_{q0} = m_h$ where m_h is the mass of the heavy quark h.

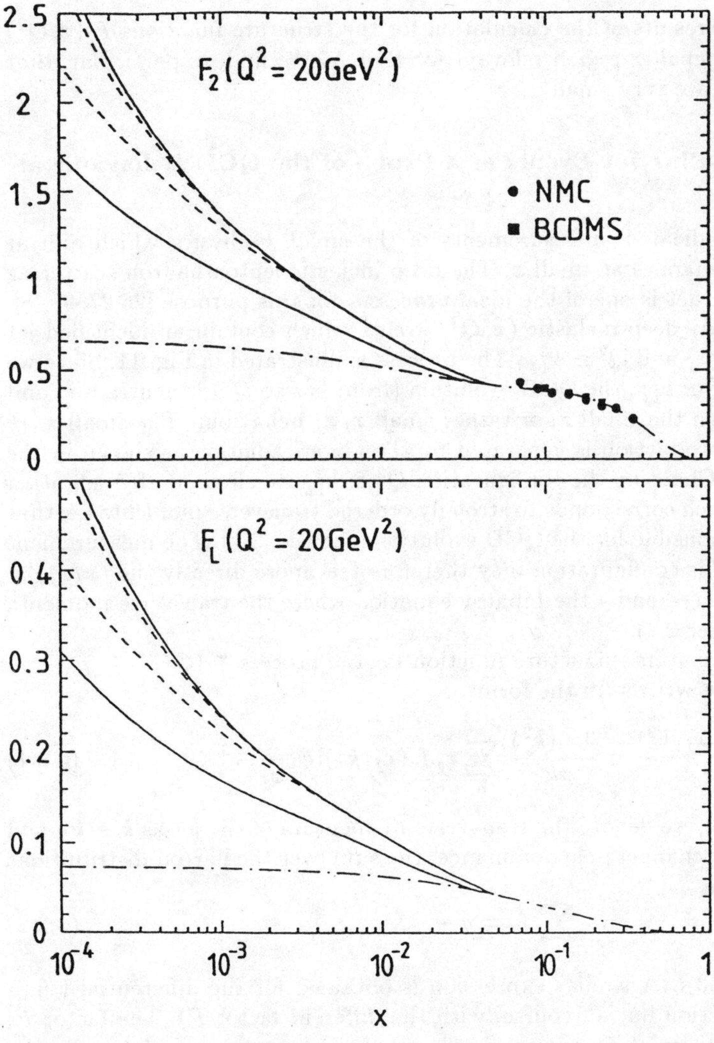

Fig. 10. Perturbative QCD predictions of the behaviour of the structure functions $F_2(x, Q^2)$ and $F_L(x, Q^2)$ at $Q^2 = 20$ GeV^2 and small x [40]. The continuous curves are with shadowing neglected, while the upper (lower) dashed curves have shadowing effects included with $R =$ 5 GeV^{-1} ($R = 2$ GeV^{-1}). For the upper three curves the infrared cut-off in (2.2.2) and in (3.1.7) is chosen to be $k_0^2 = 1$ GeV^2 while the lower of the two continuous curves give the unshadowed result for $k_0^2 = 2$ GeV^2. The dot-dashed curves are the background contributions.

For heavy quarks the formulas (3.1.7), (3.1.13) and (3.1.14) describe the photon-gluon fusion process which is expected to be the dominant one at small x. Those formulae are valid for arbitrary Q^2 including the region of small $Q^2 << m_h^2$. The cross-section for electroproduction of heavy quarks is in fact dominated by this "photoproduction" region.

In Fig. 10 we give results of the calculation for the structure functions $F_2(x, Q^2)$ and $F_L(x, Q^2)$ in the small x region relevant for HERA. We find, in particular, that the shadowing effects are very small.

3.2 Deep Inelastic Plus Jet Events as a Probe of the QCD Behaviour at Small x

There are several "dedicated" measurements of the small x physics which aim at revealing the QCD dynamics at small x. The deep inelastic lepton-hadron scattering containing a measured jet is one of the ideal processes for this purpose [26,27,44-48].

The idea is to study deep inelastic (x, Q^2) events which contain an identified jet (x_j, k_{1T}^2) where $x << x_j$ and $Q^2 \simeq k_{1T}^2$. The process is illustrated in Fig. 11. Since we choose events with $Q^2 \simeq k_{1T}^2$, the QCD evolution (from k_{1T}^2 to Q^2) is neutralised and attention is focussed on the small x, or rather small x/x_j behaviour. The small x/x_j behaviour of the jet production is generated by the gluon radiation as shown in the diagram of Fig. 11a . Choosing the configuration $Q^2 \simeq k_{1T}^2$ we eliminate by definition the gluon emission which corresponds to strongly ordered transverse momenta i.e. that emission which is responsible for the QCD evolution (see Sec. 2.1). The measurement of jet production in this configuration may therefore test more directly the $(x/x_j)^{-\lambda}$ behaviour which is generated by the Lipatov equation where the transverse momenta are not ordered (see Sec.2.2).

The differential transverse structure function for the process "$\gamma(Q^2)$" $+ p \rightarrow$ jet $(x_j, k) + X$ may be written in the form:

$$x_j \frac{\partial^2 F_T(x, Q^2; x_j, k^2)}{\partial x_j \partial k^2} = \frac{3\alpha_s(k^2)}{\pi k^4} [\sum_a x_j f_a(x_j, k^2)] F(\frac{x}{x_j}, k^2, Q^2), \qquad (3.2.1)$$

where, for convenience, we denote the transverse momentum of the jet as $k \equiv k_T$ and where, assuming the t-channel pole dominance, the sum over the parton distributions is:

$$\sum_a f_a = g + \frac{4}{9}\Sigma \qquad (3.2.2)$$

with Σ given by (2.1.3). (A similar expression is obtained for the differential longitudinal structure function but, of course, with the different factor F). The factor F, which has the dimensions of k^2 represents the photon-gluon process shown by the upper blob in Fig. 11b; that is F/k^2 can be identified with the gluon structure function integrated over the longitudinal momentum of the gluon. At small $z = x/x_j$ the function F is given by the sum of ladder diagrams shown in Fig. 12. The function F satisfies the Lipatov equation of the following form:

$$-z \frac{\partial H(z, k^2, Q^2)}{\partial z} = K_L \otimes H \qquad (3.2.3)$$

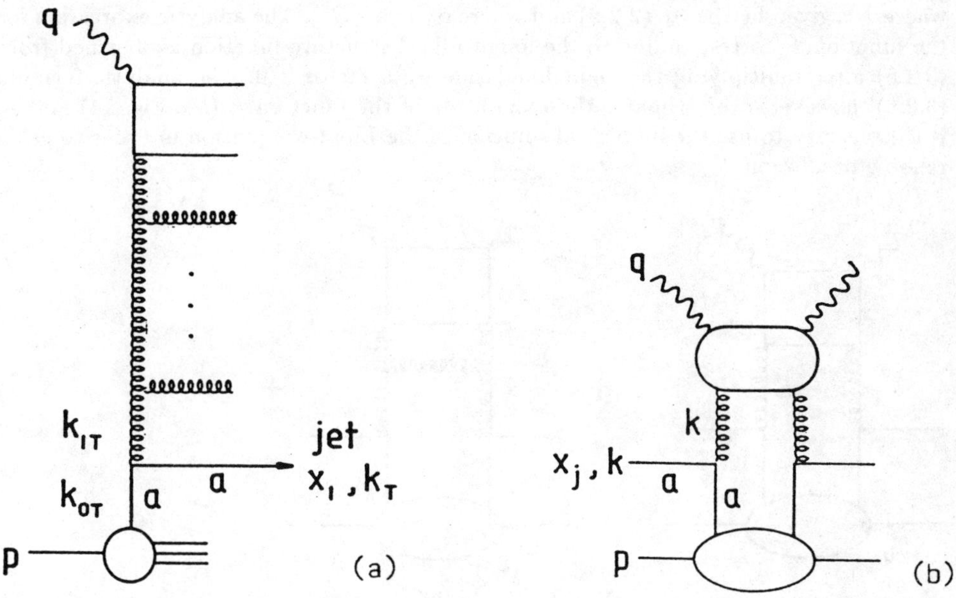

Fig. 11. (a) Diagrammatic representation of a deep-inelastic event which contains an identified jet with longitudinal and transverse momentum $x_j p$ and k_{1T}, respectively. x_j is chosen as large as experimentally feasible ($x_j \sim 0.1$) and so we assume strong ordering of the longitudinal, as well as transverse, momentum at the parton a - gluon vertex. Parton a can be either a quark or a gluon. (b) The diagram giving the cross section for deep inelastic scattering events containing an identified jet of longitudinal and transverse momentum $x_j p$ and k, respectively.

with

$$H(z,k^2,Q^2) = \frac{3\alpha_s(k^2)}{\pi} F(z,k^2,Q^2), \tag{3.2.4}$$

where the kernel K_L is defined by the eq. (2.2.2). The boundary condition for this equation at some $z = z_0$ ($z_0 < 1$) is given by the quark box (and crossed box) contribution $\bar{F}^{(0)}_{T,L}(k^2,Q^2)$ to the gluon+"γ" scattering (see Fig. 13). The functions $\bar{F}^{(0)}_{T,L}(k^2,Q^2)$ are defined in terms of $F^{(0)}_{T,L}(x',k^2,Q^2)$ given by the formulas (3.1.13) and (3.1.14)

$$\bar{F}^{(0)}_{T,L}(k^2,Q^2) = \int_0^1 \frac{dx'}{x'} F^{(0)}_{T,L}(x',k^2,Q^2). \tag{3.2.5}$$

It is important to notice that the function $F(z,k^2,Q^2)$ is, in principle, exactly calculable within perturbative QCD (including its normalisation). When the running coupling constant effects are neglected then the leading small z behaviour of $F(z,k^2,Q^2)$ can be obtained analytically [26,43,45-47]

$$F(z,k^2,Q^2) = \frac{9\pi^2}{512} \frac{2\sum e_q^2 \bar{\alpha}_s^{1/2}}{\sqrt{21\zeta(3)/2}} (k^2Q^2)^{1/2} \frac{z^{-\lambda}}{\sqrt{ln(1/z)}} \{1 + O[1/ln(1/z)]\}, \tag{3.2.6}$$

where λ is given by the eq. (2.2.9) and where $\bar{\alpha}_s = \alpha_s(Q^2)$. The analytic expression for the function F corresponding to the longitudinal structure function is obtained from (3.2.6) after multiplying the right hand side by a factor 2/9. The analytic formula (3.2.6), however, overestimates the magnitude of the function F (see Fig. 14) and so it is necessary to use the numerical solution of the Lipatov equation in order to get a reliable prediction.

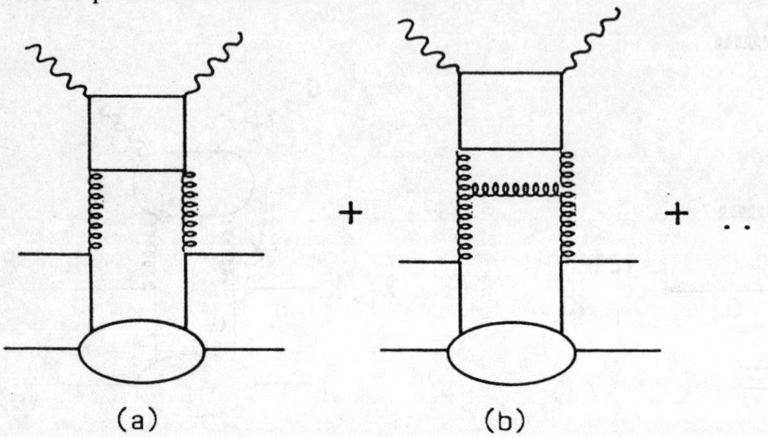

Fig. 12. the leading $\log(x_j/x)$ approximation to the process shown in Fig. 11.

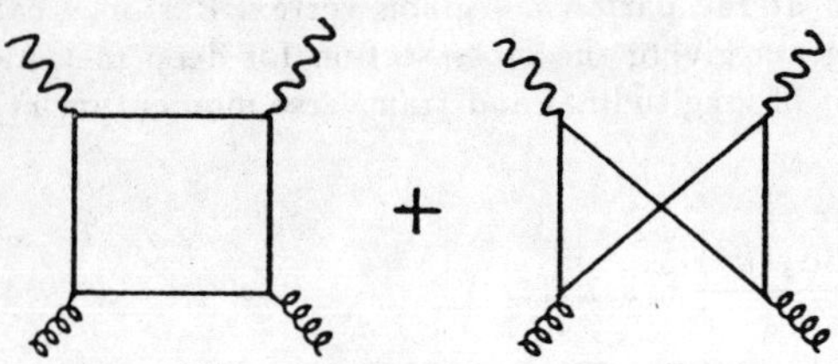

Fig. 13. The two diagrams embodied in the quark box diagram of Fig. 12.

Let us now discuss the feasibility of using the deep inelastic events which contain a measured jet to identify the singular $z^{-\lambda}$ type behaviour at HERA [27,45,46,48].

One practical limitation is that jets can only be measured if they are emitted at sufficiently large angles ($\theta_j > 5^0$) to the proton beam direction in the HERA laboratory frame. The relation between the kinematic variables describing the jet is illustrated by Fig. 15. We see that large x_j jets are only emitted at small θ_j; for a given θ_j we can reach larger x_j by observing jets with larger k_T^2 but with a depleted event rate. Fig. 16 shows the acceptance regions for these jets for various choices of the cuts on the jet variables x_j and θ_j (together with the constraints $\frac{1}{2}Q^2 < k_T^2 < 2Q^2$ and $z \equiv x/x_j < 0.1$).

In order to identify the Lipatov $z^{-\lambda}$ behaviour we need deep inelastic + jet events, with $k_T^2 \simeq Q^2$, over an interval of $z \equiv x/x_j$ which covers values of z as small as is experimentally possible. As a compromise we select the region $x_j > 0.05$ and $x < 2 * 10^{-3}$.

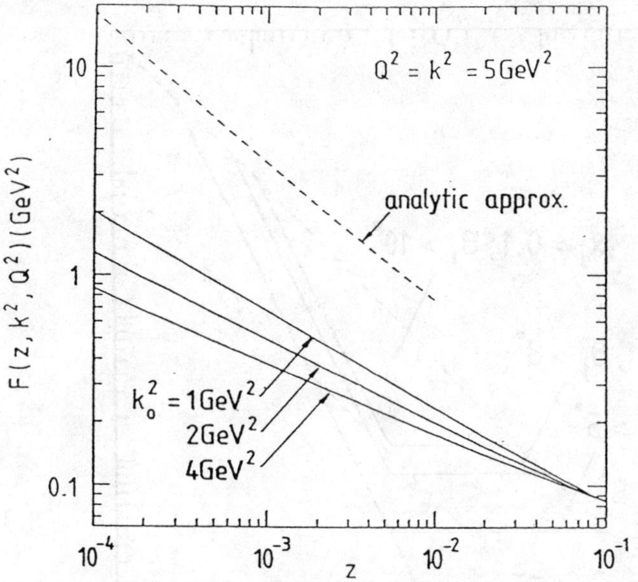

Fig. 14. The photon-gluon structure function factor $F(z, k^2, Q^2)$ corresponding to the transverse differential structure function for deep inelastic (x, Q^2) events with an identified jet (x_j, k^2), as a function of $z = x/x_j$ [26]. The continuous curves are calculated from the integro-differential form (3.2.3) of the Lipatov equation for three different choices of the transverse momentum cutoff, $k_0^2 = 1, 2$ and 4 GeV^2, respectively. The solution is matched to the quark box driving term at $z = z_0 = 0.1$. The dashed curve shows the analytic leading $\log(z_0/z)$ approximation, eq. (3.2.6).

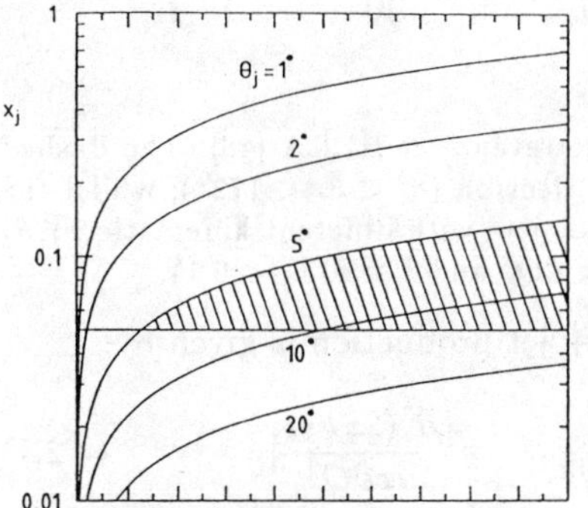

Fig. 15. The relation between the jet kinematic variables for deep inelastic + jet events with $x = 5 * 10^{-3}$ and $Q^2 = 20$ GeV^2 [45,46,48]. θ_j, the jet angle to the proton direction in the HERA laboratory frame, is not uniquely specified by $(x, Q^2; x_j, k_T^2)$. Varying the remaining azimuthal angle transforms the lines of constant θ_j into narrow bands in the x_j, k_T^2 plane. The lines shown are obtained by averaging over the azimuthal degree of freedom. The cross-sections shown in Fig. 17 correspond to jets lying in the shaded band, but subject to the additional constraint $\frac{1}{2}Q^2 < k_T^2 < 2Q^2$.

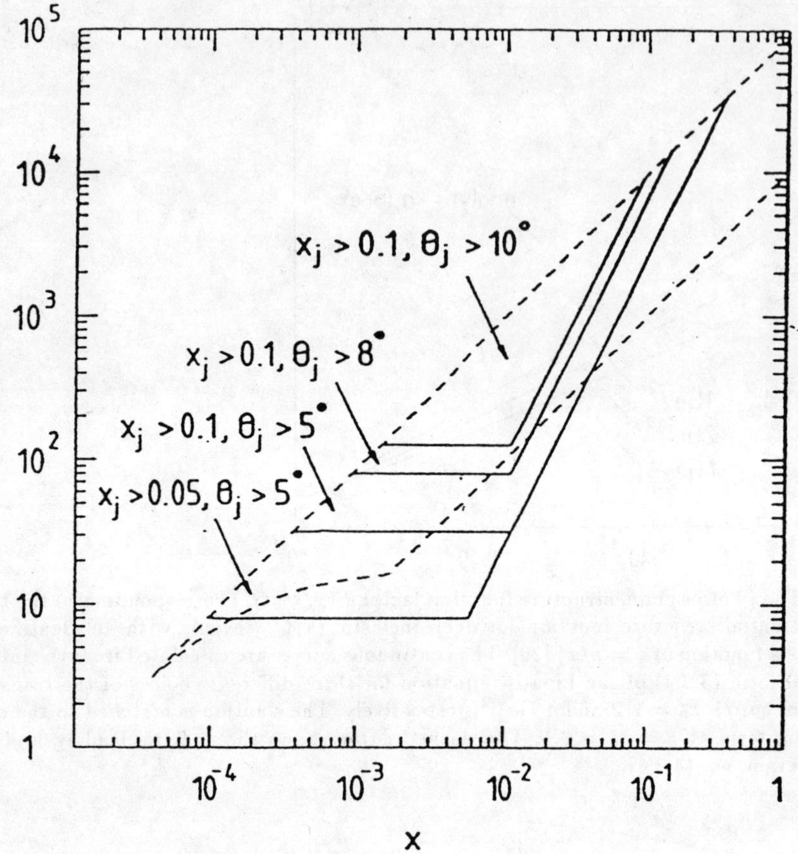

Fig. 16. The acceptance region for deep-inelastic + jet events at HERA [48]. The dashed curves outline the acceptance regions for the detected electron ($8^0 < \theta_e < 172^0$), whilst the solid curves outline the triangular acceptance region for jets with different kinematic x_j, θ_j cuts applied (together with the constraint $\frac{1}{2}Q^2 < k_T^2 < 2Q^2$ and $z \equiv x/x_j < 0.1$).

The differential cross section of deep inelastic + jet production is given by:

$$\frac{\partial^2 \sigma}{\partial x \partial Q^2} = \int dx_j \int dk_T^2 \frac{4\pi\alpha^2}{xQ^4}[(1-y)\frac{\partial^2 F_2}{\partial x \partial Q^2} + \frac{1}{2}y^2 \frac{\partial^2(2xF_1)}{\partial x \partial Q^2}] , \tag{3.2.7}$$

where the differential structure functions are each given by equations of the form of (3.2.1) with the boundary conditions, $\bar{F}^{(0)}$, for F appropriate to $F_2 \equiv F_T + F_L$ and $2xF_1 \equiv F_T$. As usual $y = Q^2/xs$ where $\sqrt{s}$ is the CM energy of the incoming electron-proton system. Fig. 17 shows the predicted x dependence of the deep inelastic+jet cross-section relevant for HERA [48]. For reasons given above, the regions of integrations in (3.2.7) are restricted to $x_j > 0.05$ and $\theta_j > 5^0$, but subject to the additional constraint $\frac{1}{2}Q^2 < k_T^2 < 2Q^2$. The continuous curves give the values of the cross-section when the Lipatov effects are included. These are to to be contrasted with the dashed curves which show the values when the Lipatov effect is neglected, that is when just the quark box approximation is used to evaluate $\partial^2 F_2/\partial x \partial Q^2$ and $\partial^2(2xF_1)/\partial x \partial Q^2$. The steep rise of the continuous curves with decreasing x (i.e. decreasing $z \equiv x/x_j$) reflects the $z^{-\lambda}$ Lipatov effect generated by the gluon radiation.

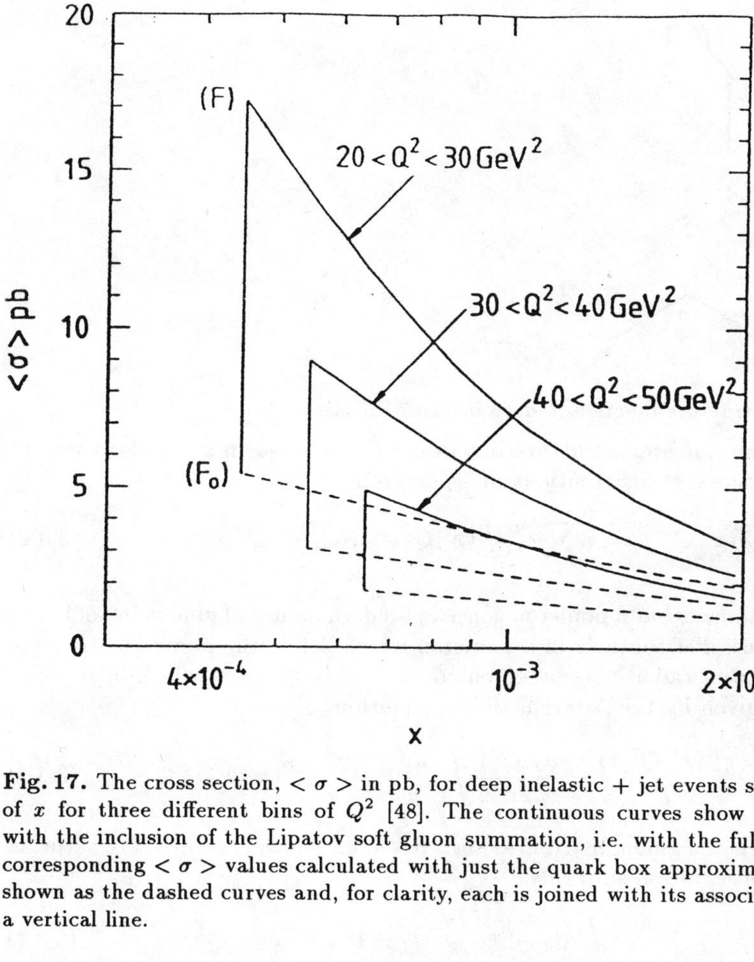

Fig. 17. The cross section, $<\sigma>$ in pb, for deep inelastic + jet events shown as a function of x for three different bins of Q^2 [48]. The continuous curves show $<\sigma>$ calculated with the inclusion of the Lipatov soft gluon summation, i.e. with the full $F(z, k^2, Q^2)$. The corresponding $<\sigma>$ values calculated with just the quark box approximation $F = \bar{F}^{(0)}$ are shown as the dashed curves and, for clarity, each is joined with its associated solid curve by a vertical line.

3.3 Deep Inelastic Diffraction

The deep inelastic diffraction is the process "γ"$(Q^2) + p \to X + p$ with a large rapidity gap between the recoiled proton and the hadronic system X. It is assumed to be dominated by the pomeron exchange as illustrated in Fig. 18. The differential diffractive structure function is then related to the Pomeron structure function F_2^P [49]:

$$\frac{\partial^2 F_2^{diff}}{\partial t \partial \xi} = \frac{\beta^2(t)}{16\pi\xi} F_2^P(\frac{x}{\xi}, Q^2, t), \tag{3.3.1}$$

where $\xi = \delta q/pq$ and $t = \delta^2$. Note that the variable $x' = x/\xi$ is the Bjorken scaling variable corresponding to deep inelastic "scattering" on a pomeron since $x' = Q^2/2\delta q$. The function $\beta(t)$ is the pomeron coupling to the nucleon and we assume that the pomeron has its intercept equal to 1.

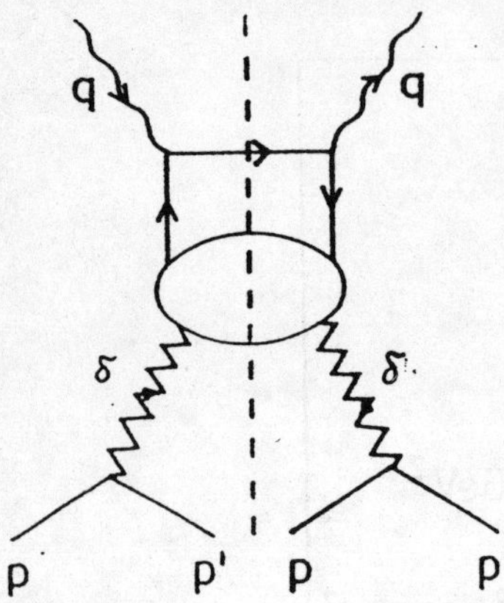

Fig. 18. Pomeron structure function in deep inelastic diffraction.

For large Q^2 the pomeron structure function F_2^P is related in a standard way to the quark (and antiquark) distributions in a pomeron

$$F_2^P(x', Q^2, t) = x \sum_i e_i^2 [q_i^P(x', Q^2, t) + \bar{q}_i^P(x', Q^2, t)]. \tag{3.3.2}$$

It is usually assumed that a pomeron consists predominantly of gluons [50-52]. The quark and antiquark distributions in a pomeron which define the pomeron structure function are generated radiatively by gluon conversion to $q\bar{q}$ pairs which in the first approximation is given by the Altarelli- Parisi equation:

$$Q^2 \frac{\partial q_i^P(x', Q^2, t)}{\partial Q^2} = \frac{\alpha_s(Q^2)}{2\pi} \int_{x'}^1 \frac{dy}{y} P_{gq}(\frac{x'}{y}) g^P(y, Q^2, t), \tag{3.3.3}$$

where $g^P(y, Q^2, t)$ is the gluon distribution in a pomeron which is assumed to saturate the momentum conservation sum rule:

$$\int_0^1 dy y g^P(y, Q^2, t) = 1. \tag{3.3.4}$$

In other models [49] it is assumed on the basis of the pomeron-photon analogy that the pomeron couples directly to quarks. In this approach the quark distributions are not constrained to satisfy the momentum conservation sum rules.

If the pomeron has a large gluon component and if its radius R_P is assumed to be small (i.e. of the order of 0.1 fm or so), then one may expect that the non-linear shadowing effects discussed in Sec. 2.3 can become very important [52]. This comes from the fact that the nonlinear shadowing terms are now proportional to $1/R_P^2$.

An interesting possibility was considered in refs. [53-57] that deep inelastic diffraction can be calculated almost entirely within perturbative QCD from triple gluonic ladder diagrams i.e. from the same diagrams which describe parton screening effects which we discussed in Sec. (2.3) (see Fig. 19). The use of perturbative QCD can be justified in the analysis of large k_T jet production of the diffractively produced system (see Fig. 20). The measurement of jet production in the diffractively produced system can also be a complementary test of the gluon distributions at small x. The simplest process in this case is the diffractive production of two jets and the cross-section of this process is [55]

$$\frac{\partial^3\sigma^D}{\partial t \partial k_T^2 \partial z} = [F_N^{2G}(t)]^2[x_j g(x_j, k_T^2)]^2 \sum_i e_i^2 \frac{4\pi^2\alpha_{em}}{3Q^2(k_T^2)^2}\alpha_s z^2(1-z)^2, \tag{3.3.5}$$

where $z = x/(x + x_j)$ and $[F_N^{2G}(t)]$ is the two-gluon-proton form factor. The variable x_j denotes the longitudinal momentum fraction of the jet i.e. its four momentum k is

$$k \simeq x_j p + \mathbf{k}_T. \tag{3.3.6}$$

It may be seen that the cross section for jet production is very sensitive to gluon distribution being proportional to $[x_j g(x_j, k_T^2)]^2$.

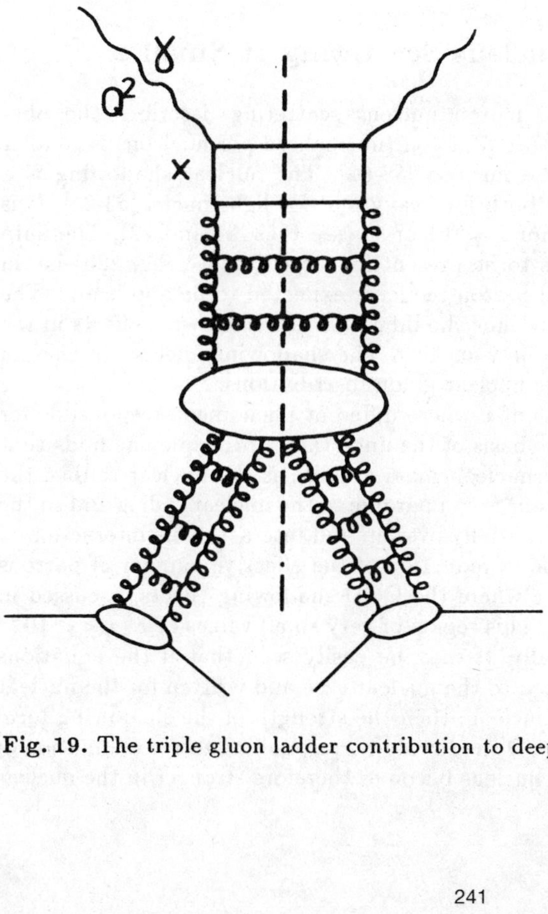

Fig. 19. The triple gluon ladder contribution to deep inelastic diffraction.

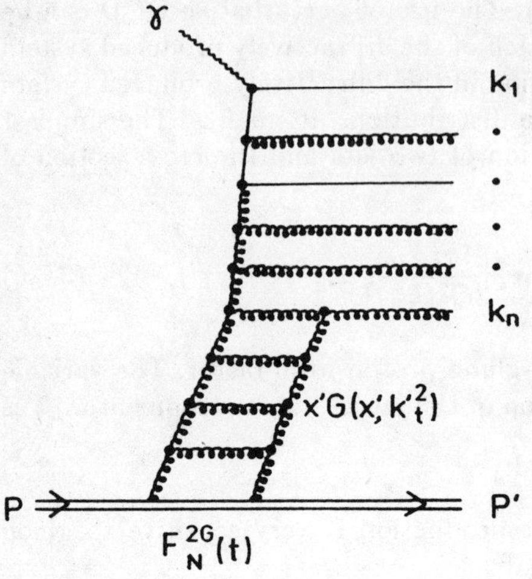

Fig. 20. The diagram representing the diffractive production of jets.

4 Shadowing in Lepton-Nucleus Scattering at Small x

The nuclear shadowing in inelastic lepton nucleus scattering describes the phenomenon in which the nuclear structure function (normalised per nucleon) is smaller than the structure function of a free nucleon [58-62]. The nuclear shadowing is a firmly established experimental fact both for heavy and for light nuclei [63-65]. It is a small x effect since it takes place for $x \leq 0.1$ or so (see Figs. 21 and 22). The data also suggest that shadowing persists to be present for large Q^2 (see Fig. 22b) i.e. in the region where the QCD improved parton model is expected to be applicable. The shadowing in deep inelastic scattering thus should reflect the shadowing effects in the quark (and antiquark) distributions in a nucleus. The shadowing effects are also, of course, expected to be present in the nuclear gluon distributions.

One can distinguish three regions of x where different phenomena responsible for nuclear shadowing can occur. On the basis of the uncertainty principle one finds that for $x < 1/(2MR_A)$, where M is the nucleon mass and R_A is the nuclear radius, the partons can occupy longitudinal distances comparable to the nuclear radius and so the partons from different nucleons can spatially overlap and interact. This interaction of partons from different nucleons leads to reduction of the effective number of partons in a nucleus. It is in this region of x where the QCD shadowing effects discussed in Sec. (2.3) become important [66,67]. This region of very small values of x ($x << 10^{-2}$ or so) is very interesting theoretically. It may be easily seen that if the equations discussed in the Sec. (2.3) are adapted to the nuclear case and written for the nuclear gluon distributions normalised per nucleon, then the strength of the shadowing term is enhanced by a factor $A^{1/3}$. The QCD shadowing corresponding to recombination of partons from different nucleons in a nucleus becomes therefore stronger in the nuclear environment.

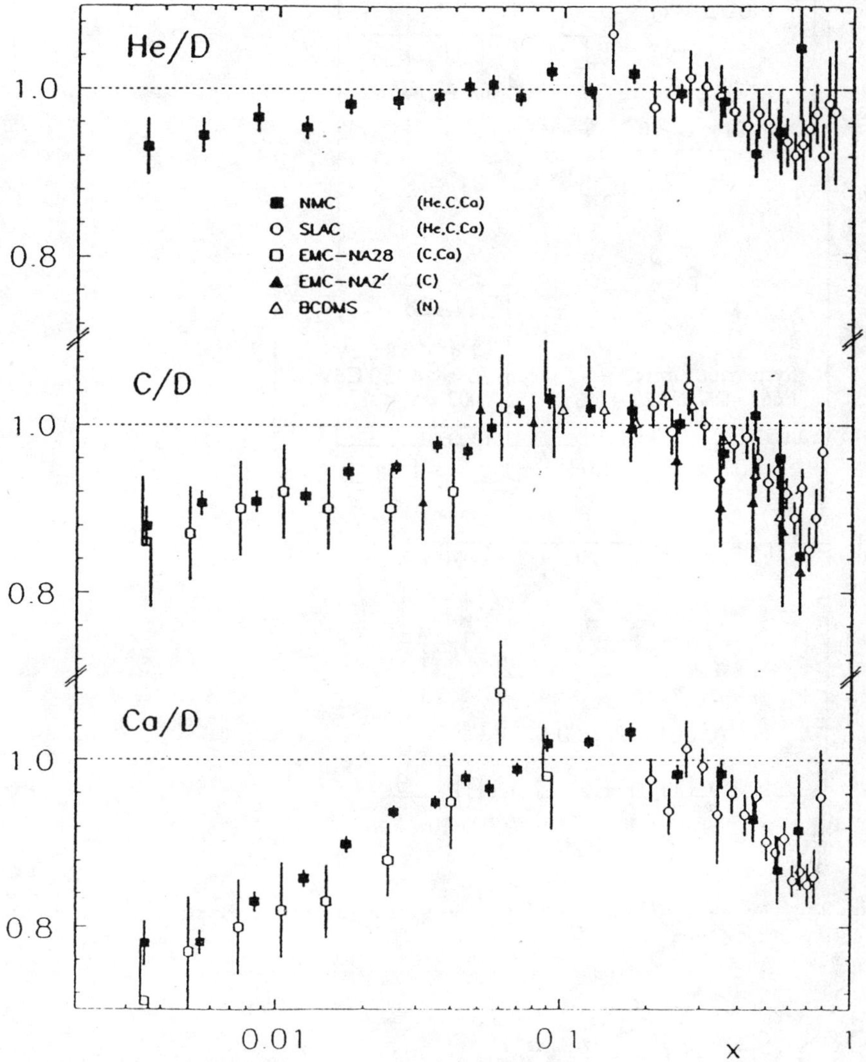

Fig. 21. Compilation of structure function ratios for different nuclei (from [62]).

In the region of moderately small x with $1/(2MR_A) < x < 1/(Mr)$, where r is the average distance between nucleons in a nucleus (i.e. $r \simeq 1fm$), one expects that the shadowing will decrease with increasing x. The x dependence of shadowing in this region is governed by the nuclear form factor. Finally for $x > 1/(Mr)$ the shadowing is negligible.

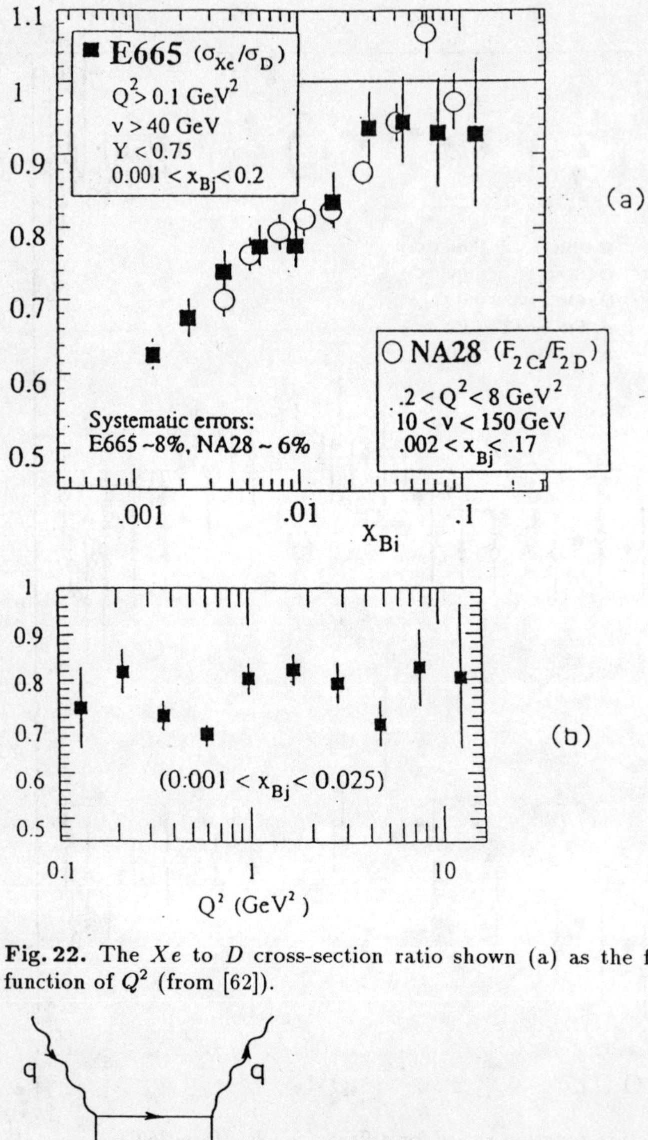

Fig. 22. The Xe to D cross-section ratio shown (a) as the function of x and (b) as the function of Q^2 (from [62]).

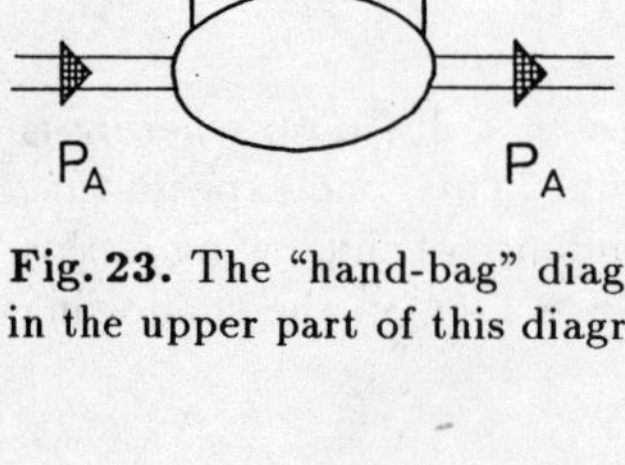

Fig. 23. The "hand-bag" diagram for the deep inelastic lepton-nucleus scattering. The lines in the upper part of this diagram denote quarks (or antiquarks).

The parton model is described by the "hand-bag" diagram for the virtual Compton scattering amplitude (Fig. 23). It is this hand-bag structure of the diagram which guarantees Bjorken scaling (modulo perturbative QCD corrections) independently of the structure of the lower part of this diagram. One can represent the diagram corresponding to the interaction of the virtual photon with the nucleus in a form of the multiple interaction series. Different terms in this series correspond to different numbers of nucleons participating in the interaction. The shadowing comes from those terms where at least two nucleons participate in the interaction. Various models of shadowing correspond to the various structure details of the corresponding diagrams.

At small x one can assume that the multiple interaction diagrams are dominated by the pomeron exchange shown in Fig. 24. The contribution of the double interaction diagram can then directly be related to the pomeron structure function discussed in the Sec. (3.1) i.e.

$$\Delta F_{2A} = -8\int_{\xi_0}^{1}\frac{d\xi}{\xi}\int d^2\mathbf{b}dz_1dz_2\Theta(z_1-z_2)n_A(\mathbf{b},z_1)n_A(\mathbf{b},z_2)cos[M\xi(z_1-z_2)]\frac{\partial^2F_2^{diff}}{\partial\xi\partial t}$$

$$\simeq -\frac{\pi}{2}\int_{x}^{1}\frac{d\xi}{\xi}\int d^2\mathbf{b}dz_1dz_2\Theta(z_1-z_2)n_A(\mathbf{b},z_1)n_A(\mathbf{b},z_2)cos[M\xi(z_1-z_2)]$$

$$\beta^2(0)F_2^P(\frac{x}{\xi},Q^2,t=0), \tag{4.1}$$

where

$$\xi_0 = x(1+M_{x0}^2) \tag{4.2}$$

and where M_{x0}^2 is the lowest mass squared of the diffractively produced system. The negative sign of the shadowing term is related to the fact that the amplitude of the diffractive scattering is imaginary. The definition of various variables and functions in the formula (4.1) is the same as in the Sec. (3.3) and the function $n_A(\mathbf{b},z)$ is the nucleon number density of the nucleus at the point $(\mathbf{b},z)$. It should be noted that the x' scaling of the pomeron structure function ($x'=x/\xi$) implies the Bjorken scaling of the shadowing term ΔF_{2A}.

The double interaction diagram neglects the interaction of the diffractively produced system within the nucleus and this approximation leads to the overestimate of nuclear shadowing for heavy nuclei. There exist several approximate model dependent ways to take into account this interaction [62,68-70].

For low Q^2 the parton model and so the partonic mechanism of shadowing may not be applicable. In this low Q^2 region the natural mechanism of shadowing is that related to the vector meson dominance i.e. to the multiple scattering of vector mesons which couple to virtual photons [58,59].

Let us finally consider, as an illustration, the shadowing in inelastic lepton deuteron scattering [71-73]. It is entirely given by the double interaction diagram (Fig. 25b) and so it is free of the model dependent assumptions concerning the multiple rescattering of the diffractively produced system.

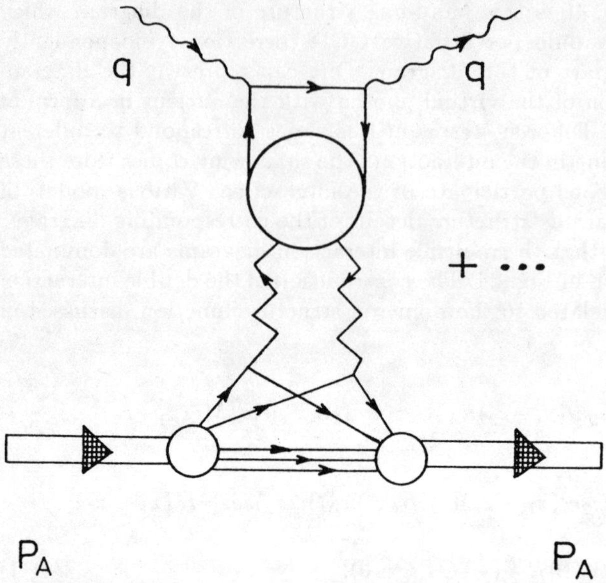

Fig. 24. The double Pomeron exchange contribution to shadowing in deep inelastic lepton-nucleus scattering. The dots represent other contributions to shadowing in which more than two nucleons participate in the interaction. Those terms correspond to multiple rescattering of the diffractively produced system within the nucleus.

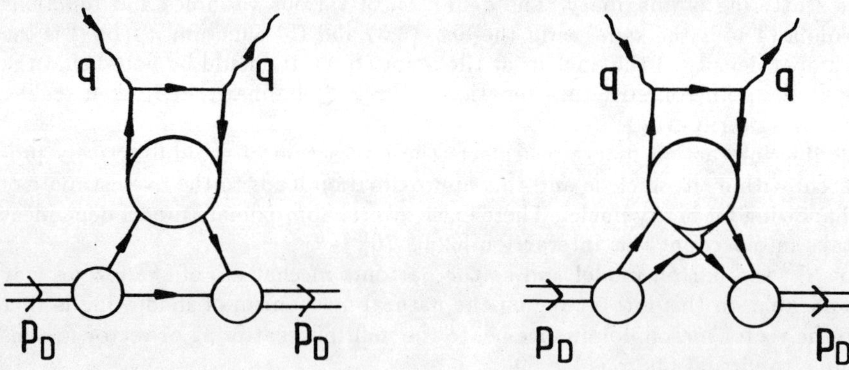

Fig. 25. Hand-bag diagrams for single (a) and double (b) interaction for the virtual Compton scattering on a deuteron. The lines in the lower part of the diagrams denote nucleons while those in the upper part denote quarks (antiquarks).

The deuteron structure function $F_{2D}(x,Q^2)$ is related in the following way to the proton and neutron structure functions $F_{2p}(x,Q^2)$ and $F_{2n}(x,Q^2)$ and to the shadowing term $\delta F_2(x,Q^2)$ normalised per nucleon:

$$F_{2D}(x,Q^2) = F_{2p}(x,Q^2) + F_{2n}(x,Q^2) - 2\delta F_2(x,Q^2) \tag{4.3}$$

with $\delta F_2(x,Q^2) \geq 0$.

A quantitative estimate of the shadowing term in inelastic lepton-deuteron scattering is important and interesting at least for the following two reasons:

(a) The inelastic lepton-deuteron scattering serves as the unique tool for determining the neutron structure function F_{2n}. The recent measurements of the structure function ratios F_{2D}/F_{2p} performed by the New Muon Collaboration [74,75] are very precise and so even the relatively small shadowing term can in principle affect extraction of the neutron structure function from the experimental data. This can in turn affect determination of other quantities like for instance the Gottfried sum I_G:

$$I_G = \int_0^1 \frac{dx}{x}[F_{2p}(x,Q^2) - F_{2n}(x,Q^2)]. \tag{4.4}$$

(b) In the region of very small values of $x < 10^{-2}$ and for large Q^2, which may become accessible in HERA measurements, the shadowing should reveal various QCD effects including, in particular, parton screening from two different nucleons in the deuteron.

As discussed above the double interaction term relates the shadowing to the deep inelastic diffractive production. This relation takes the following form for the deuteron:

$$\delta F_2(x,Q^2) = \int d^2\boldsymbol{\delta}_T \int_{\xi_0}^1 d\xi S_D(\boldsymbol{\delta}^2)\frac{\partial^2 F_2^{diff}}{\partial\xi\partial t} . \tag{4.5}$$

In this formula $S(\boldsymbol{\delta}^2)$ is the deuteron form factor and $\frac{\partial^2 F_2^{diff}}{\partial\xi\partial t}$ is the diffractive structure function (see Sec. 3.3). We define $\xi = 2\delta q/p_D q$ where δ is the four momentum corresponding to a pomeron. We also have

$$t \simeq -\boldsymbol{\delta}^2 \tag{4.6}$$

and

$$\boldsymbol{\delta}^2 = \boldsymbol{\delta}_T^2 + M^2\xi^2. \tag{4.7}$$

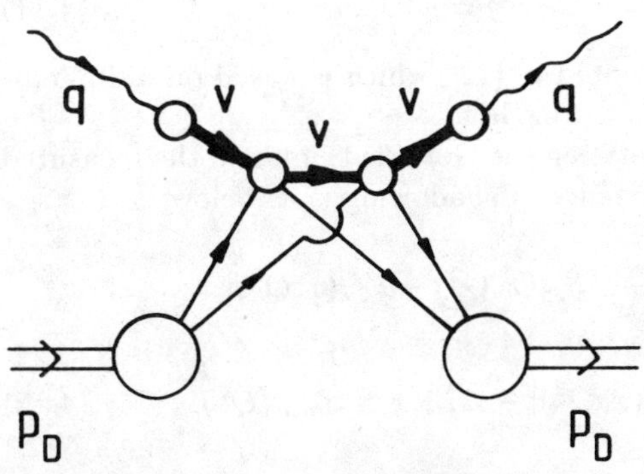

Fig. 26. Double scattering of vector mesons.

The region of low M_x^2 where M_x is the mass of the diffractively produced system, is dominated by the diffractive production of low mass vector mesons. It is assumed to be described by the Vector-Meson-Dominance model. In this model the contribution of the low-mass vector mesons to nuclear shadowing corresponds to double scattering of vector mesons which couple to virtual photons (Fig. 26). The corresponding shadowing term $\delta F_{2D}^{(v)}$ is:

$$\delta F_{2D}^{(v)} = \frac{Q^2}{4\pi} \sum_v \frac{M_v^4 \delta\sigma_{vD}}{\gamma_v^2 (Q^2 + M_v^2)^2}, \tag{4.8}$$

where M_v is the mass of the vector meson v and the constants γ_v^2 can be calculated from the leptonic width of the vector meson v [58,76]. The cross-section $\delta\sigma_{vD}$ is given by:

$$\delta\sigma_{vD} = \frac{1}{16\pi^2} \sigma_{vN}^2 \int d^2\boldsymbol{\delta}_T S_D(\boldsymbol{\delta}^2), \tag{4.9}$$

where now $\boldsymbol{\delta}^2 = \boldsymbol{\delta}_T^2 + M^2 x^2 (M_v^2/Q^2 + 1)^2$ and σ_{vN} is the total vector meson nucleon total cross-section. It can be seen from (4.8) that the double scattering of vector mesons gives vanishing contributions to shadowing in the large Q^2 limit.

In Fig. 27 we plot the results for $\delta F_{2D} = \delta F_{2D}^v + \delta F_{2D}^p$ as functions of x for fixed Q^2. We also plot separately the vector meson δF_{2D}^v and parton mechanism δF_{2D}^p contributions. The partonic contribution to shadowing is defined by the eq. (4.5) and (3.3.1) and the parametrisation of the pomeron structure function is the same as in [71]. The pattern of x dependence changes weakly with Q^2. The x dependence is determined by the "$1/\xi$" diffractive spectrum (see eq. (3.3.1)) for small values of x and by the deuteron form factor for higher values of x. The latter stems from the fact that the form-factor argument, $\boldsymbol{\delta}^2$, depends upon ξ. It can also be seen that the shadowing in deuteron is a weak (i.e. 2-3%) effect. Using the results presented in Fig. 27 we may estimate the contribution of shadowing to the Gottfried sum $\Delta I_G(x > x_{min}, Q^2)$ defined as:

$$\Delta I_G(x > x_{min}, Q^2) = -2 \int_{x_{min}}^{0.1} \frac{dx}{x} \delta F_{2D}(x, Q^2). \tag{4.10}$$

For $Q^2 = 4$ GeV^2 and for $x_{min} = 0.004$ (i.e. for the values corresponding to experimental measurements [75]) we find

$$\Delta I_G(x > x_{min}, Q^2) = 4 \text{ GeV}^2) = -0.025. \tag{4.11}$$

The independent estimate of ΔI_G presented in [72], which is based on a different model of diffractive production, gives $\Delta I_G = -0.043$.

It should be noticed that the relation between the Gottfried sum and the measured structure functions F_{2D} and F_{2p} in the presence of shadowing is as below:

$$I_G(x_{min}, x_{max}, Q^2) \equiv \int_{x_{min}}^{x_{max}} \frac{dx}{x} [F_{2p}(x, Q^2) - F_{2n}(x, Q^2)] =$$

$$\int_{x_{min}}^{x_{max}} \frac{dx}{x} [2F_{2p}(x, Q^2) - F_{2D}(x, Q^2)] + \Delta I_G(x > x_{min}, Q^2). \tag{4.12}$$

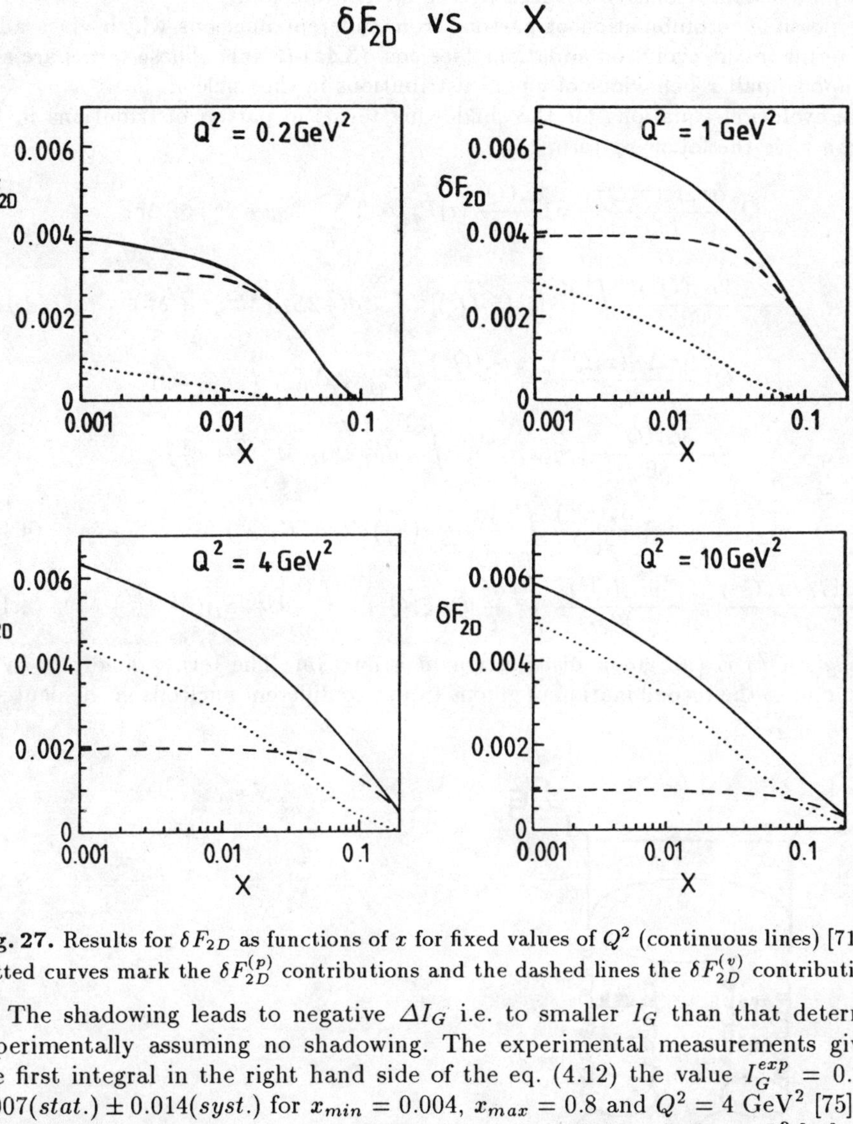

Fig. 27. Results for δF_{2D} as functions of x for fixed values of Q^2 (continuous lines) [71]. The dotted curves mark the $\delta F_{2D}^{(p)}$ contributions and the dashed lines the $\delta F_{2D}^{(v)}$ contributions.

The shadowing leads to negative ΔI_G i.e. to smaller I_G than that determined experimentally assuming no shadowing. The experimental measurements give for the first integral in the right hand side of the eq. (4.12) the value $I_G^{exp} = 0.227 \pm 0.007(stat.) \pm 0.014(syst.)$ for $x_{min} = 0.004$, $x_{max} = 0.8$ and $Q^2 = 4\ \text{GeV}^2$ [75].

Let us now discuss the possible QCD effects on shadowing for large Q^2 [71]. In the large Q^2 region the shadowing contribution $\Delta F_{2D} \equiv -2\delta F_{2D}$ is related to the shadowing terms Δq_i and $\Delta \bar{q}_i$ in the quark and antiquark distributions in the deuteron:

$$\Delta F_{2D}(x, Q^2) = x \sum_i e_i^2 [\Delta q_i(x, Q^2) + \Delta \bar{q}_i(x, Q^2). \tag{4.13}$$

The shadowing terms Δg is also present in the gluon distribution in the deuteron.

There are two possible effects which can modify shadowing contributions to parton distributions:

(1) mild scaling violation induced by the QCD evolution,

(2) possible recombination of partons from different nucleons which gives additional terms in the evolution equations (see eqs. (3.4a)-(3.4c)). Those terms are sensitive upon small x behaviour of gluon distributions in the nucleon.

The evolution equations for the shadowing terms in parton distributions in the deuteron have the following form:

$$Q^2\frac{\partial x\Delta g(x,Q^2)}{\partial Q^2} = \frac{\alpha_s(Q^2)}{2\pi}x[P_{gq}\otimes 2\sum_i \Delta q_i + P_{gg}\otimes \Delta g]$$

$$-\frac{9\alpha_s(Q^2)^2}{8Q^2}\int_x^1 \frac{d\xi}{\xi}[\xi g(\xi,Q^2)]^2\int_0^\infty d\delta_T^2 2S_D(M^2\xi^2+\delta_T^2) \qquad (4.14a)$$

$$Q^2\frac{\partial x\Delta q_i(x,Q^2)}{\partial Q^2} = \frac{\alpha_s(Q^2)}{2\pi}x[P_{qq}\otimes \Delta q_i + P_{qg}\otimes \Delta g]$$

$$-\frac{3\alpha_s(Q^2)^2}{80Q^2}[xg(x,Q^2)]^2\int_0^\infty d\delta_T^2 2S_D(M^2x^2+\delta_T^2)$$

$$+\frac{\alpha_s(Q^2)}{\pi Q^2}\int_x^1 \frac{dx'}{x'}\frac{x}{x'}\gamma(\frac{x}{x'})x'G_H(x',Q^2) \qquad (4.14b)$$

$$Q^2\frac{\partial xG_H(x,Q^2)}{\partial Q^2} = -\frac{9\alpha_s(Q^2)^2}{8Q^2}\int_x^1 \frac{d\xi}{\xi}[\xi g(\xi,Q^2)]^2\int_0^\infty d\delta_T^2 2S_D(M^2\xi^2+\delta_T^2), \qquad (4.14c)$$

where $g(x,Q^2)$ is the gluon distribution in a nucleon. The terms quadratic in xg correspond to the recombination of gluons from two different nucleons in the deuteron (see Fig. 28).

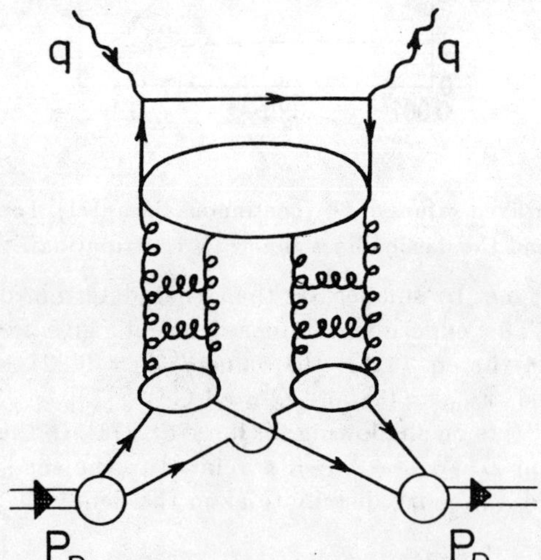

Fig. 28. The two gluon ladder exchange for the double interaction diagram. The upper blob denotes different perturbative QCD couplings of the four gluons to the $q\bar{q}$ pair.

The QCD corrections to shadowing become important in the region of very small values of x and for moderately large values of Q^2, that is in the kinematical range of the possible HERA measurements. They are displayed in Fig. 29 and compared with the prediction of the partonic mechanism without the QCD evolution effects. Two points are worthwhile to be noticed:

(1)The large amount of gluons in a pomeron (see Sec. 3.3) implies a relatively large shadowing term Δg in the gluon distributions in the deuteron. This term enhances the shadowing effects in the quark and antiquark distributions (and so in F_{2D}) through the $q\bar{q}$ radiation from the gluons which is generated by the QCD evolution.

(2) The parton recombination from two different nucleons is very sensitive to the gluon distributions in the nucleon at low x.

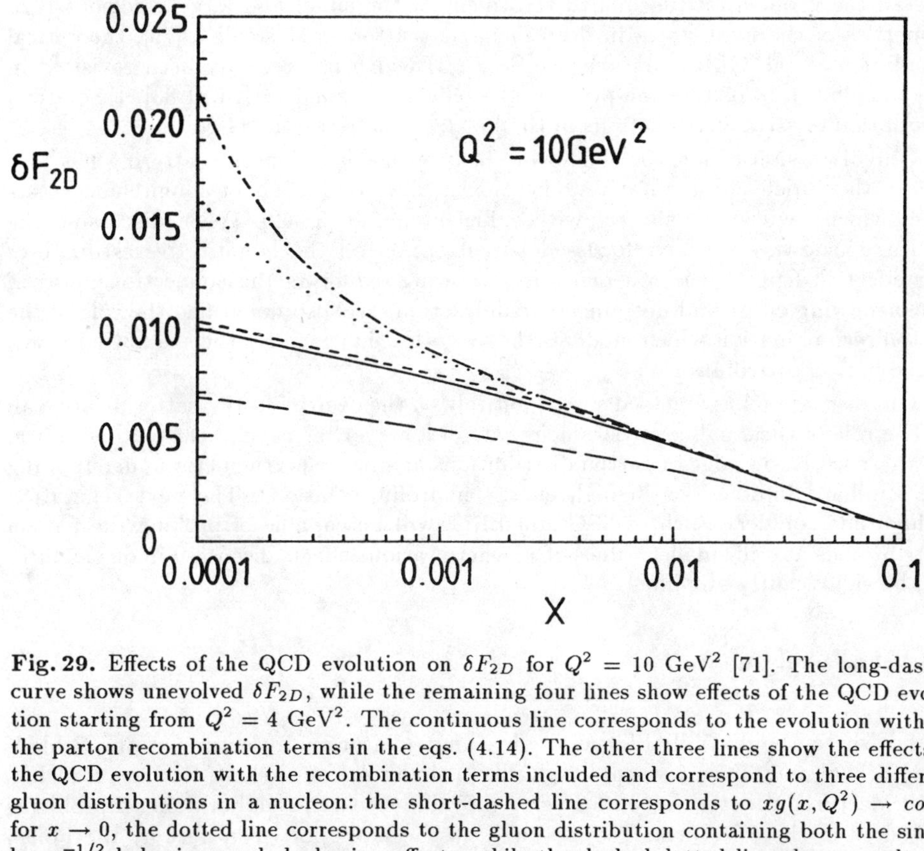

Fig. 29. Effects of the QCD evolution on δF_{2D} for $Q^2 = 10$ GeV^2 [71]. The long-dashed curve shows unevolved δF_{2D}, while the remaining four lines show effects of the QCD evolution starting from $Q^2 = 4$ GeV^2. The continuous line corresponds to the evolution without the parton recombination terms in the eqs. (4.14). The other three lines show the effects of the QCD evolution with the recombination terms included and correspond to three different gluon distributions in a nucleon: the short-dashed line corresponds to $xg(x, Q^2) \rightarrow$ *const* for $x \rightarrow 0$, the dotted line corresponds to the gluon distribution containing both the singular $x^{-1/2}$ behaviour and shadowing effects, while the dashed-dotted line shows results for $xg(x, Q^2) \rightarrow x^{-1/2}$.

5 Summary and Outlook

In these lectures we have discussed the elements of the QCD expectations concerning the small x behaviour of parton distributions and possible phenomenological implications of those expectations for deep inelastic lepton - hadron scattering. When discussing the phenomenological implications we have, in particular, emphasised the role of the jet production in deep inelastic scattering which will be the "landmark" measurement of the small x physics. We have pointed out that HERA can in principle measure this process although this measurement would certainly be very difficult.

There have been several theoretical developments concerning the small x physics which we have not discussed in sufficient detail but which are certainly very topical and important. They are, in particular: the small x behaviour of the parton distributions in a photon [77], theoretical issues related to the factorisation theorem at small x [41,78], the problem of the unified treatment of the small and large x regions [22], properties of the final states in deep inelastic scattering at small x [79], theoretical foundation of the GLR equation (see Sec. 2.3) which has recently been revisited in [80], possible role of the non-perturbative effects at small x [24,81-83], theoretical description of structure functions in the low Q^2, low x region [84-86].

Our discussion of nuclear shadowing in deep inelastic lepton scattering has also been rather brief and qualitative. There exist, however, excellent comprehensive articles which review this subject with sufficient detail [58-62]. When discussing the nuclear shadowing we have focussed our attention on the quantitative estimate of this effect in lepton-deuteron inelastic scattering exploring the connection between the shadowing effect and deep inelastic diffraction. We also quantified the role of the parton recombination which modifies the evolution equation for the shadowing terms in the parton distributions.

Our discussion has focussed almost entirely on the deep inelastic scattering at small x. The role of the small x physics does extend however far beyond the lepton-hadron interactions. Knowledge of parton distributions at small x is crucial for understanding the semihard interactions in high energy hadronic collisions. The next generation of hadronic colliders, such as SSC and LHC, will be capable of probing the parton distributions at still smaller values of x, where various effects discussed in our lectures will be significantly amplified.

Acknowledgments

I wish to thank Willi Plessas for his kind invitation to lecture at the Schladming School and for very warm hospitality. I thank Adrian Askew, Barbara Badelek, Alan Martin and Peter Sutton for very enjoyable research collaboration on problems presented in these lectures. This research has been supported in part by the Polish Committee for Scientific Research, Grant No. 2 0198 91 01.

References

[1] R.G. Roberts, "The Structure of the Proton. Deep Inelastic Scattering", Cambridge University Press, Cambridge, New York, Port Chester, Melbourne, Sydney, 1990.

[2] W.J. Stirling, These Proceedings.

[3] G. Altarelli, Phys. Rep. **81** (1982) 1.

[4] E. Reya, Phys. Rep. **69** (1981) 195.

[5] Yu.L. Dokshitzer, D.I. Dyakonov and S.I. Troyan, Phys. Rep. **58** (1980) 265.

[6] L.V. Gribov, E.M. Levin and M.G. Ryskin, Phys. Rep. **100** (1983) 1.

[7] E.M. Levin, "Parton Density at Small x_B", Review talk at "QCD - 20 Years Later", Aachen, June 9-13, 1992, DESY preprint DESY 92-122.

[8] A.D. Martin, Acta Phys. Polon. **B22** (1991) 1093; Talk at 22nd International Symposium on Multiparticle Dynamics, Santiago de Compostela, Spain, July 1992.

[9] B. Badełek, K. Charchula, M. Krawczyk and J. Kwieciński, Rev. Mod. Phys. **64** (1992) 927.

[10] Proceedings of the Topical Workshop on the Small x Behaviour of Deep Inelastic Scattering Structure Functions in QCD, DESY, Hamburg, Germany, May 1990, Nucl. Phys. B (Proc. Suppl.) **18C** (1990) 172.

[11] Proceedings of the Workshop "Physics at HERA", DESY, Hamburg,Germany, October 1992, edited by W. Buchmüller and G. Ingelman.

[12] G. Wolf, These Proceedings.

[13] J. Feltesse, "HERA, The New Frontier" (Presented at SLAC Summer Institute, Stanford, CA, USA, August 1991) Saclay preprint DAPNIA-SPP-92-01.

[14] G. Altarelli and G. Parisi, Nucl. Phys. **B126** (1977) 298.

[15] E.A. Kuraev, L.N. Lipatov and V. Fadin, Zh. Eksp. Teor. Fiz. **72** (1977) 373 [Sov. Phys. JETP **45** (1977) 199].

[16] Ya. Ya. Balitskij and L.N. Lipatov, Yad. Fiz. **28** (1978) 1597 [Sov. J. Nucl. Phys. **28** (1978) 822]

[17] L.N. Lipatov, in "Perturbative QCD", edited by A.H. Mueller, (World Scientific, Singapore, 1989), p.411.

[18] J.B. Bronzan and R.L. Sugar, Phys. Rev. **D17** (1978) 585.

[19] T. Jaroszewicz, Acta Phys. Polon. **B11** (1980) 965.

[20] M. Ciafaloni, Nucl. Phys. **B296** (1988) 49

[21] S. Catani, F. Fiorani and G. Marchesini, Phys. Lett. **B234** (1990) 339; Nucl. Phys. **B336** (1990) 18.

[22] G. Marchesini, Proceedings of the Workshop "QCD at 200 TeV", Erice, Italy, June 1990, edited by L. Cifarelli and Yu.L. Dokshitzer, Plenum Press, New York, 1992, p. 183.

[23] T. Jaroszewicz, Phys. Lett. **B116** (1982) 291.

[24] R.E. Hancock and D.A. Ross, Nucl. Phys. **B383** (1992) 575.

[25] J. Kwieciński, A.D. Martin and P.J. Sutton, Phys. Rev. **D44** (1991) 260.

[26] J. Kwieciński, A.D. Martin and P.J. Sutton, Phys. Rev. **D46** (1992) 921.

[27] A.D. Martin, J. Kwieciński and P.J.Sutton, Nucl. Phys. B (Proc. Suppl.) **29A** (1992) 67.

[28] D. Strózik-Kotlorz, Z. Phys. **C53** (1992) 93.

[29] A.H. Mueller, Nucl. Phys. **B335** (1990) 115.
[30] E.M. Levin and M.G. Ryskin, Phys. Rep. **189** (1990) 267.
[31] A.H. Mueller, Nucl. Phys. B (Proc. Suppl.) **18C**(1990) 125.
[32] J. Kwieciński, A.D. Martin and P.J. Sutton, Phys. Lett. **B264** (1991) 199.
[33] A.H. Mueller and J. Qiu, Nucl. Phys. **B268** (1986) 427.
[34] J. Bartels and E.M. Levin, Nucl. Phys. **B387** (1992) 617.
[35] J. Bartels, J. Blümlein and G. Schuler, Z. Phys. **C50** (1991) 91.
[36] J.C. Collins and J. Kwieciński, Nucl. Phys. **B335** (1990) 89.
[37] E.M. Levin, G. Marchesini, M.G. Ryskin and B.R. Webber, Nucl. Phys. **B357** (1991) 167.
[38] J. Kwieciński, A.D. Martin, R.G. Roberts and W.J. Stirling, Phys. Rev. **D42** (1990) 3645.
[39] A.D. Martin, R.G. Roberts and W.J. Stirling, "New Information on Parton Distributions", Durham Univ. preprint DTP/92/16 (RAL-92-021).
[40] A.J. Askew, J. Kwieciński, A.D. Martin and P.J. Sutton, "QCD Predictions for Deep Inelastic Structure Functions at HERA", Durham Univ. preprint DTP/ 92/78.
[41] S. Catani, M. Ciafaloni and F. Hautman, Phys. Lett. **B242** (1990) 91; Nucl. Phys. **B366** (1991) 135.
[42] E.M. Levin and M.G. Ryskin, Yad. Fiz. **53** (1991) 1052.
[43] S. Catani, M. Ciafaloni and F. Hautman, Proceedings of the Workshop "Physics at HERA", DESY, Hamburg, Germany, October 1992, edited by W. Buchmüller and G. Ingelman.
[44] A.H. Mueller, J.Phys. **G17** (1991) 1443.
[45] J. Bartels, M. Besancon, A. De Roeck and J. Kurzhoefer, "Measurements of Hot Spots at HERA", Proceedings of the Workshop "Physics at HERA", DESY, Hamburg, Germany, October 1992, edited by W. Buchmüller and G. Ingelman.
[46] J. Bartels, M. Loewe and A. De Roeck, Z. Phys. **C54** (1992) 635; A. de Roeck Nucl. Phys. B (Proc. Suppl.) **29A** (1992) 61;
[47] W.K. Tang, Phys. Lett. **B278** (1992) 363.
[48] J. Kwieciński, A.D. Martin and P.J. Sutton, Phys. Lett. **B287** (1992) 254.
[49] A. Donnachie and P.V. Landshoff, Nucl. Phys. **B244** (1984) 322; Phys. Lett. **B191** (1987) 309; Nucl. Phys. **B303** (1988) 634.
[50] G. Ingelman and P. Schlein, Phys. Lett. **B152** (1985) 256.
[51] E.L. Berger, J.C. Collins, D.E. Soper and G. Sterman, Nucl. Phys. **B286** (1987) 704.
[52] G. Ingelman and K. Janson-Prytz, Proceedings of the Workshop "Physics at HERA", DESY, Hamburg, Germany, October 1992, edited by W. Buchmüller and G. Ingelman; Nucl. Phys. B (Proc. Suppl.) **29A** (1992) 97.
[53] J. Bartels and G. Ingelman Phys. Lett. **B235** (1990) 175; G. Ingelman, Nucl. Phys. B(Proc. Suppl.) **18C** (1990) 172.
[54] M.G. Ryskin, Sov. J. Nucl. Phys., **52** (1990) 529.
[55] M.G. Ryskin, Proceedings of the Workshop "Physics at HERA", DESY, Hamburg, Germany, October 1992, edited by W. Buchmüller and G. Ingelman.
[56] N.N. Nikolaev and B.G. Zakharov, Z. Phys. **C53** (1992) 331.
[57] E.M. Levin and W. Wüsthoff, "Photon Diffractive Dissociation in Deep Inelastic Scattering", DESY preprint DESY 92-166.

[58] T.H. Bauer et al., Rev. Mod. Phys. **50** (1978) 261.
[59] G. Grammer Jr. and J.D. Sullivan, in Electromagnetic Interactions of Hadrons, ed. A. Donnachie and G. Shaw, (Plenum, New York, 1978).
[60] L. Frankfurt and M. Strikman, Phys. Rep. **160** (1988) 235.
[61] R.J.M. Covolan and E. Predazzi, Lectures at the Folgaria Winter School, February 1990, Univ. of Torino preprint DFTT 21/90.
[62] M. Arneodo, "Nuclear Effects in Structure Functions", CERN preprint CERN-PPE/92-113.
[63] EMC, M. Arneodo et al., Nucl. Phys. **B333** (1990) 1;
[64] NMC, P. Amaudruz et al., Z. Phys. **C51** (1991) 387.
[65] E-665, M.R. Adams et al., Fermilab-Pub-92-51-E.
[66] E.M. Levin and M.G. Ryskin, Yad. Fiz. **41** (1985) 472.
[67] J. Qiu, Nucl. Phys. **B291** (1987) 746.
[68] L.L. Frankfurt and M.I. Strikman, Nucl. Phys. **B316** (1989) 340.
[69] S.J. Brodsky and H.J. Lu, Phys. Rev. Lett. **64** (1990) 1342.
[70] J. Kwieciński, Z. Phys. **C45** (1990) 461.
[71] B. Badełek and J. Kwieciński, Nucl. Phys. **B370** (1992) 278; Nucl. Phys. B (Proc. Suppl.) **29A** (1992) 30.
[72] V.R. Zoller, Phys. Lett. **B279** (1992) 145; Z. Phys. **C54** (1992) 425.
[73] W. Melnitchouk and A.W. Thomas, "Shadowing in Deuterium", Univ. of Adelaide preprint ADP-92-192/T120.
[74] NMC, D. Allasia et al., Phys. Lett. **B249** (1990) 366.
[75] NMC, P. Amaudruz et al., Phys. Rev. Lett. **66** (1991) 387.
[76] J. Kwieciński and B. Badełek, Z. Phys., **C43** (1989) 251.
[77] J.R. Forshaw and P.N. Harriman, Phys. Rev. **D46** (1992) 3778.
[78] J.C. Collins and R.K. Ellis, Nucl. Phys. **B360** (1991) 3.
[79] B.R. Webber, Proceedings of the Workshop "Physics at HERA", DESY, Hamburg, Germany, October 1992, edited by W. Buchmüller and G. Ingelman.
[80] J. Bartels, "Unitarity Corrections to the Lipatov Pomeron and the Small x Region in Deep Inelastic Scattering", DESY preprint DESY-92-114.
[81] P.V. Landshoff and O. Nachtmann, Z. Phys. **C35** (1987) 405.
[82] J.R. Cudell, A. Donnachie and P.V. Landshoff, Nucl. Phys. **B322** (1989) 55.
[83] A. Donnachie and P.. Landshoff, Nucl. Phys. **B311** (1989) 509.
[84] B. Badełek and J. Kwieciński, Phys. Lett. **B295** (1992) 263.
[85] H. Abramowicz et al., Phys. Lett. **B269** (1991) 465.
[86] M. Glück, E. Reya and A. Vogt, Z. Phys. **C48** (1990) 471; **C53** (1992) 127.

Nucleon Structure Functions

A. Milsztajn and M. Virchaux

DAPNIA/SPP, C.E. Saclay, F-91191 Gif-sur-Yvette, France

1 Introduction

This presentation is a review of the present status of the measurement of the nucleon structure functions with emphasis on what they have brought to our understanding of quantum chromodynamics (QCD). The first part consists of an overview of the last decade of structure functions measurements from deep inelastic scattering of lepton beams on fixed targets. The experimental situation was considerably clarified during the last year. The coincidence in time between this clarification and the first collisions at HERA makes this review timely. The second part is devoted to the QCD analyses of these data and to the various measurements of the strong coupling constant α_S that they yield. In the third and last part, a few remarks are made on what will be (or could be) the future of this field of experimental physics.

A general introduction to QCD can be found in these proceedings [0], where a discussion of the existing parton distribution parametrisations is also given. The polarised structure functions [1] as well as HERA [2] and the very low x regime [3] are also covered by other speakers.

2 Present Status of Structure Function Measurements

2.1 Definition of Structure Functions

2.1.1 Deep Inelastic Scattering Differential Cross Sections

The various structure functions of the nucleon are usually measured through deep inelastic scattering of leptons (electrons, muons or neutrinos) acting as elementary probes on target nucleons. In the Born approximation, deep inelastic scattering (DIS) is represented, as in Fig. 1, by the exchange of one boson. This boson is a virtual photon in the case of electromagnetic scattering (electron or muon beams), a Z^0 or a charged W in the case of weak scattering (neutrino beams). In the latter case, Z^0's correspond to neutral currents - neutrino in, neutrino out - and W^+'s or W^-'s to charged currents - neutrino in, muon (or electron) out.

The square of the four-momentum of the exchanged boson (four-momentum transfer) $-Q^2$ is related to the energy E of the incident lepton, to the energy E' of the scattered one and to the scattering angle θ through the simple relation

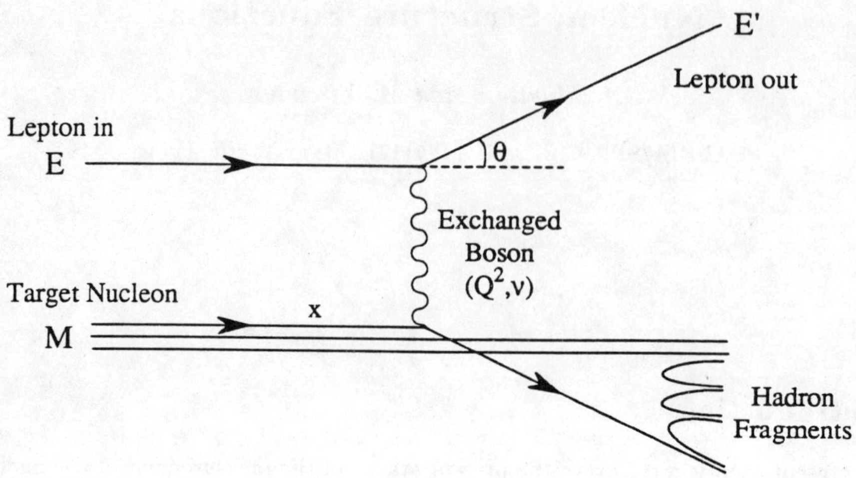

Fig. 1. Deep inelastic scattering of a lepton on a target nucleon.

$Q^2 = 2EE'(1-\cos\theta)$ valid up to terms of order m_μ^2 or m_e^2. If we call ν the transferred energy (energy of the boson in the laboratory frame, equal to $E-E'$) and M the mass of the target nucleon, then the two scaling variables x and y are defined as $x = Q^2/2M\nu$ and $y = \nu/E$. For unpolarised nucleons, there are three independent structure functions, e.g. $F_2(x,Q^2)$, $xF_3(x,Q^2)$, and $R(x,Q^2)$, the last one (R) corresponding to the ratio of longitudinally to transversely polarised virtual photon absorption cross sections. Instead of R the structure functions F_1 or F_L are sometimes used ($F_L = F_2 - 2xF_1$ and $R \simeq F_L/F_2$). In the one-boson-exchange approximation, these structure functions are related to the DIS differential cross sections through the following relation

$$\frac{\mathrm{d}^2\sigma(x,Q^2)}{\mathrm{d}x\mathrm{d}Q^2} = 4\pi\alpha^2\frac{M\nu}{Q^6E^2}\left(2EE' - \frac{Q^2}{2} + \frac{\nu^2+Q^2}{1+R(x,Q^2)}\right)F_2(x,Q^2) \tag{1}$$

for charged lepton scattering, and through a slightly more complicated one

$$\begin{aligned}\frac{\mathrm{d}^2\sigma(x,Q^2)}{\mathrm{d}x\mathrm{d}Q^2} = \frac{G^2}{2\pi}\frac{M\nu}{Q^2E^2}\Bigg[&\left(2EE' - \frac{Q^2}{2} + \frac{\nu^2+Q^2}{1+R(x,Q^2)}\right)F_2(x,Q^2)\\ &\pm(2E\nu-\nu^2)\,xF_3(x,Q^2)\Bigg]\end{aligned} \tag{2}$$

for neutrino scattering, where the $\pm$ sign corresponds to neutrino (+) or antineutrino (−) scattering.

For charged lepton scattering, the differential cross section is directly proportional to F_2 and depends only weakly on R. So, measurements of this cross section yield primarily measurements of F_2 while R can only be estimated with less precision from measurements of the cross section at the same (x,Q^2) points for several different beam energies (E). Usually, one assumes the function $R(x,Q^2)$ to be equal to some parametrisation compatible with existing measurements and with the QCD prediction (see 2.1.3) so that F_2 can be directly extracted from the measurements of the differential cross section.

In neutrino scattering, the function F_2 can be obtained in much the same way by using the sum of the differential cross sections of neutrino and antineutrino scattering (the term proportional to xF_3 then vanishes) while the difference of these cross sections yields a measurement of xF_3.

At present fixed-target energies, the magnitude of the cross sections is of course much higher in charged lepton scattering, which involves electromagnetic interactions, than in neutrino scattering, which occurs only through weak interactions. That is why it is much easier to gather the high statistics of experimental DIS events needed for precise measurements of the differential cross sections with charged leptons than with neutrinos. In particular, all significant structure functions measurements with neutrino beams have been made, up to now, on dense targets (most of the time iron). Measurements on "light" targets, hydrogen or deuterium, which allow the determination of the structure function of free or quasi-free nucleons, only comes from experiments using electron or muon beams. On the other hand, the neutrino cross sections do not have the strong Q^2 dependence of the charged lepton cross sections that, due to the photon propagator term, vary like $1/Q^4$. This strong dependence on a kinematical variable makes the measurements more sensitive to any uncertainty in the knowledge of the experimental resolution. So, in general, charged lepton measurements are dominated by systematic uncertainties (it actually tends to become so also for the latest neutrino results).

2.1.2 Quark-Parton Model

In the quark-parton model, the variable x is equal to the fraction of the four-momentum of the nucleon carried by the parton on which the scattering has occurred. The structure functions are then connected to distributions along x of the quark constituents of the nucleon: $q_f(x)$ for quarks of flavour f. More precisely, $q_f(x)\mathrm{d}x$ represents the probability of finding a quark of flavour f carrying a fraction of the four-momentum of the nucleon between x and $x+\mathrm{d}x$. As an example, the structure function F_2 measured in charged lepton scattering on a nucleon N is connected to the quark distributions in the following way

$$F_2^{l^\pm N}(x) = \sum_f e_f^2\, x\,(q_f(x) + \bar{q}_f(x)) , \tag{3}$$

where f is the flavour index ($f = 1, 2, 3, ... \Leftrightarrow$ quark $= u, d, s, ...$), e_f is the electrical charge of the quark of flavour f (in units of the electron charge) and $\bar{q}$ represents the distributions of anti-quarks. For structure functions measured in neutrino scattering, the relations are

$$F_2^{\nu N}(x) = \sum_f x\,(q_f(x) + \bar{q}_f(x)) \tag{4}$$

$$xF_3^{\nu N}(x) = \sum_f x\,(q_f(x) - \bar{q}_f(x)) . \tag{5}$$

In this case, F_2 is proportional to the sum of all quark distributions, which, being invariant under any flavour permutation, is called the (flavour-) singlet quark distribution $q^{SI}(x)$. The function xF_3 is, on the contrary, a purely non-singlet quark distribution equal to the distribution of valence quarks. A non-singlet distribution is defined as a linear combination of quark distributions orthogonal to the singlet quark distribution q^{SI} in the $2N_f$ dimensional space of distributions of the different flavours (N_f being the number of flavours and the factor 2 taking into account separately quarks and anti-quarks). So, in this space, the non-singlet distributions form a hyper-plane of dimension $2N_f - 1$ and can always be decomposed into linear combinations of differences of quark distributions $q_i(x) - q_j(x)$, where for example: $i, j = 1, 2, 3, ..., 2N_f \Leftrightarrow$ quark $= u, \bar{u}, d, \bar{d}, ..., \bar{t}$. The basic interest of this decomposition of structure functions into singlet (SI) and non-singlet (NS) parts will appear in the discussion of QCD analyses (Section 2). In this naive quark-parton model, quark distributions, and consequently structure functions, only depend on x. When strong interactions between quarks are taken into account, i.e. in the framework of QCD, an additional dependence of the structure functions on the energy scale (proportional to Q^2) is introduced and "effective" quark distributions $q(x, Q^2)$ are defined (see 2.1.1).

To end this brief introduction with a few more remarks, one can notice that, in charged lepton scattering, the average of F_2 measured on neutron and proton, i.e. $(F_2^{l^\pm n} + F_2^{l^\pm p})/2$, is almost exactly proportional to the singlet quark distribution (for details see section 1.3.2.). This average corresponds to the structure function F_2 (normalised per nucleon) measured on an isoscalar target, i.e. on a target containing equal numbers of protons and neutrons. The lightest example of such a target is of course deuterium. Moreover, as the measurement of F_2 on free protons $F_2^{l^\pm p}$ can be obtained using hydrogen targets, F_2 of neutrons $F_2^{l^\pm n}$ can be extracted from F_2 measured on deuteron $F_2^{l^\pm d}$ (or more generally on any isoscalar nucleus, but the nuclear effects, which blur the extraction of $F_2^{l^\pm n}$, are likely to be smallest in deuteron). This gives access to the structure function $F_2^{l^\pm p} - F_2^{l^\pm n} \simeq 2\,(F_2^{l^\pm p} - F_2^{l^\pm d})$ which, like $xF_3^{\nu N}$, is a pure flavour non-singlet. Neutrino scattering nevertheless keeps an advantage on charged lepton scattering for what concerns flavour discrimination. Indeed, a neutrino can only interact with, for example, $\bar{u}$ and d quarks while an antineutrino interacts with u and $\bar{d}$ quarks. And as, due to isospin symmetry, the distribution of u (resp. $\bar{u}$, d and $\bar{d}$) quarks in the proton is equal to the distribution of d (resp. $\bar{d}$, u and $\bar{u}$) quarks in the neutron, by combining measurements from neutrino and antineutrino scattering on proton and neutron (or deuteron) one should in principle be able to determine the individual distributions $u(x), \bar{u}(x), d(x)$ and $\bar{d}(x)$.

2.2 Experimental Results

The very first experiments of deep inelastic scattering have been performed with the 20 GeV electron beam delivered by the Stanford linear accelerator in the end of the 60's. They have led to the fundamental discovery of the existence of a substructure in the nucleon, i.e. the quark-partons [4]. Slightly later, together with the first Fermilab deep inelastic experiment, they led to a second fundamental discovery: the existence of the scaling violations [5], i.e. the non-zero Q^2-variation of structure functions. Since that discovery era, many other fixed target experiments have been studying deep inelastic scattering with the neutrino and muon beams that became available at Fermilab and at CERN in the 70's. Almost all these experiments have by now published their final results and it seems that the second period of the "golden age" of deep inelastic scattering, the one of general study and global measurements,

Table 1 : High statistics measurements of structure functions

Beam	Target	Experiment	Final pub.	Remarks
	Heavy			
ν	$CaCO_3$	CHARM	1984	
ν	Fe	CDHS	1984	
μ	Fe	BFP	1986	
μ	Fe	EMC-NA2	1986	
μ	C	BCDMS	1987	$x > 0.25$ $Q^2 > 25\,\text{GeV}^2$
ν	Fe	CDHSW	1989	
ν	Fe	CCFR	1992	Prelim.
	Light			
μ	H_2, D_2	CHIO	1979	
ν	D_2	BEBC WA25	1985	
μ	H_2, D_2	EMC-NA2	1985, 87	$5 < Q^2 < 200\,\text{GeV}^2$
μ	H_2, D_2	BCDMS	1989, 90	$8 < Q^2 < 260\,\text{GeV}^2$
ν	H_2	BEBC WA21	1989	
μ	D_2	EMC-NA28	1989	$x < 0.14$
e	H_2, D_2	SLAC	1990	All 1970-85 data $0.5 < Q^2 < 20\,\text{GeV}^2$
μ	H_2, D_2	NMC	1992	Prelim. $x > 0.008$

is over. A third period is now starting that aims at three more specific domains: first, the very low x and high Q^2 region that is going to be extensively studied at the HERA collider, second, the polarised structure functions and spin physics that are presently investigated at CERN by the SMC experiment and at SLAC by the E142 and E143 experiments, and third (maybe) ultra-precise measurements over a wide x range (from 0.01 up to 1.) that can be performed with existing muon or neutrino beams by new-generation experiments which up to now only exist as projects (see Section 3). So, it appears timely to review the results obtained up to now.

All deep inelastic experiments that have produced statistically significant measurements of the structure functions R, F_2, and xF_3 are listed in Table 1 in chronological order of appearance of their final publication. They are arbitrarily split in two parts, one for the results obtained on "light" targets (i.e. hydrogen or deuterium) and the other one for the "heavy" targets (all other nuclei). A few experiments have produced measurements on both types of targets. In what follows, we will concentrate on the most significant results, so that we will only mention the principal oldest results, the statistical and systematic accuracies of which have been greatly superseded by the latest ones. In any case, the limited precision of these early results makes them compatible with all the more recent and more accurate data even when the latter significantly differ between themselves. Among the large number of these somewhat obsolete results, we have mentioned in Table 1 the ones from the following collaborations:

• The CHARM experiment (WA18), which from 1978 to 1984 was using the CERN neutrino beam with a marble target ($CaCO_3$) to measure the structure functions F_L, F_2, and xF_3 [6]. This experiment has subsequently been upgraded (CHARM-2) and has been running almost up to now, but essentially for electroweak tests and measurement of $\sin^2\theta_W$ [7].

• The CDHS experiment (WA1), which was using the same CERN neutrino beam (from 1976 up to 1980) with an iron target, to get measurements of F_2, xF_3, and R [8]. An upgraded version of this experiment, CDHSW, up to 1984 used the same beam in its "wide band" mode in order to get a much more statistically precise measurement (see below).

• The BFP experiment, which was using the Fermilab muon beam and an iron target from 1978 up to 1982 to get measurements of F_2 [9].

• The CHIO experiment, completed in 1976, which ran at the Fermilab muon beam and which has been the first high-energy muon experiment to produce results with light targets [10].

• The BEBC – WA21 and BEBC – WA25 experiments, run at the CERN neutrino beam. These are the only bubble chamber experiments we quote here because they brought the highest statistics neutrino measurements on light targets, hydrogen and deuterium, respectively [11,12]; total statistics is, of course, rather low (20K events each).

The other results listed in Table 1 are more recent and/or more precise and we will now discuss and compare them in more detail. They come from:

• The EMC experiment, one of the main deep inelastic experiments of the 80's, which started using the CERN muon beam in 1978 and had many extensions and upgrades, the final one, called the SMC experiment being still running for a study of polarised structure functions. The EMC apparatus [13] is based on a big air-core dipole magnet located behind a target up to 6 m long. Tracks from the particles resulting from a deep inelastic interaction are recorded with an array of drift and proportional wire chambers. The original EMC experiment (NA2), using beam energies from 120 to 280 GeV, has produced measurements of the structure function F_2 , both on a heavy target (iron) [14] and on light ones (hydrogen and deuterium) [15,16] over a rather wide kinematical range ($0.03 < x < 0.8$, $3\,\mathrm{GeV}^2 < Q^2 < 250\,\mathrm{GeV}^2$). The light target results were the first high statistics measurements of the proton and deuteron F_2. It is to be noted that the EMC data on light targets have recently been re-analysed by ex-members of the EMC collaboration [17].

• The EMC′ experiment (NA28), one of the many extensions of the original EMC, dedicated in part to a low x measurement of F_2. This measurement [18], albeit of limited accuracy, improved on the CHIO results and covered the very low x region

($0.002 < x \leq 0.14$); it is now superseded by the recent and much more precise results from NMC (see below).

• The BCDMS experiment (NA4), another important experiment of the last decade, which was run in parallel with EMC at the CERN muon beam from 1978 up to 1985. The apparatus of this experiment [19] was completely different from the one of EMC. It consisted of a long (50 m) toroid of magnetised iron surrounding a target up to 40 m long. Particles in the hadronic shower resulting from a deep inelastic interaction were quickly absorbed in the iron and only the scattered muon survived. Its trajectory, bent by the magnetic field, was recorded with 80 planes of proportional wire chambers interspersed along the toroid. Due to its high luminosity this experiment has produced the most precise measurement of F_2 at high Q^2 both on a heavy target (carbon) and on light ones (hydrogen and deuterium). The carbon result [20] comes from a dedicated high-Q^2 high-x measurement ($0.25 \leq x \leq 0.75$, $25\,\mathrm{GeV}^2 < Q^2 < 260\,\mathrm{GeV}^2$) while the light target results [21,22] correspond to a wider kinematical range ($0.07 \leq x \leq 0.75$, $7\,\mathrm{GeV}^2 < Q^2 < 260\,\mathrm{GeV}^2$).

• The CDHSW experiment (WA1/2), one of the main neutrino experiments to this day, which used an upgraded version of the CDHS apparatus and was run in the CERN neutrino beam (in parallel with CHARM) from 1982 up to 1984. Its apparatus [23] consisted of a big iron toroid (20 m long, 4 m in diameter), interspersed with scintillator to act as a calorimeter (to measure the energy ν of the hadronic shower), magnetised and equipped with drift chambers to measure the energy of the scattered muon in charged current interactions. This experiment used the CERN neutrino beam in the "wide band" mode which allowed a considerable increase in neutrino intensity and thus in statistics. The drawback of such a mode is that, on an event-by-event basis, the energy of the incident neutrino is less constrained, making the measurements more sensitive to systematics. Compared to the earlier results from CDHS, the measurements of F_2 and xF_3 from CDHSW [24] cover a similar kinematical range ($0.015 \leq x \leq 0.65$, $0.5\,\mathrm{GeV}^2 < Q^2 < 200\,\mathrm{GeV}^2$) but are much more precise. The systematic differences, concentrated at low x, that exist between the two results have been traced back to the use of different assumptions and corrections in the extraction of the structure functions (concerning mainly the radiative corrections but also the strange sea corrections, the total cross section normalisation and the assumption on R - see Sections 1.3.3 and 1.3.4).

• The CCFR experiment, the other big neutrino scattering experiment of the last decade, was run at the Fermilab neutrino beam from 1979 (at that early time it was the CCFRR experiment) until 1988. Its apparatus [25] is similar to the one of CDHS with an important difference; the calorimetry is spatially separated from the muon spectrometer, i.e. the big iron target interspersed with scintillator ($3 \times 3 \times 17.7\,\mathrm{m}^3$) is non-magnetic and is followed by a magnetised iron toroid (3.6 m in diameter, 16 m long) equipped with drift chambers to measure the energy of scattered muons coming out of the target-calorimeter. The first results from this experiment, published in 1984 [26], were based on lower statistics (especially for antineutrinos) but, from 1985 on, the CCFR experiment was able to take advantage of the high intensity high energy (up to 600 GeV) neutrino beam from the Fermilab Tevatron. Final results from these high energy CCFR data have appeared recently [27]. Their accuracy is very impressive and they probably will remain the most precise neutrino data for quite some time.

• The SLAC experiments. There was a large number of experiments at the Stanford linear electron accelerator spanning the time period 1970 to 1985. In 1990 has appeared a coherent global reanalysis of electron scattering data on hydrogen and deuterium targets from the most important of these experiments, leading to a precise determination of F_2 [28]. The main improvements compared to original publications of the individual experiments are a better determination of $R(x, Q^2)$ [29] and a more correct treatment of radiative corrections. This allowed the F_2 measurements to cover with good precision a wide kinematical domain in x ($0.07 \leq x \leq 0.85$) and also in Q^2 although the maximum energy of the SLAC electron beam (21 GeV) limited it to $\sim 25\,\mathrm{GeV}^2$.

• The NMC experiment (NA37), one of the last extensions of the original EMC, that was running an upgraded version of the EMC apparatus in the CERN muon beam from 1986 to 1989 with the goal, among others, to get precise measurements of F_2 on hydrogen and deuterium targets down to very low values of x (and an even more precise measurement of the ratio of the neutron to proton F_2's, F_2^n/F_2^p). Compared to the EMC apparatus originally used for the structure function measurements, the main improvement consisted in the addition of new proportional wire chambers behind the magnet to help in the pattern recognition and momentum measurement of the scattered muon. Furthermore, measurements on proton and deuteron have been undertaken simultaneously by using two pairs of 3 m long targets (H_2-D_2 and D_2-H_2) exposed alternatively to the beam. The first F_2 measurements of NMC have appeared recently [30]. The bulk of the data is at low x, the measured x range extending down to 0.008 (and up to 0.5 but with smaller statistics than SLAC or BCDMS data for $x > 0.2$). The Q^2 range overlaps with both the low energy measurements of SLAC and the high energy ones of EMC/BCDMS, filling the gap that exists along Q^2 between these two sets of measurements for $0.07 < x < 0.3$.

From approximately 1987 on, it has become more and more widely known that the high statistics F_2 data of the two CERN muon experiments, EMC and BCDMS, were in marginal agreement, to say the least [31]. The first hint of this problem appeared with the results on heavy targets (EMC iron vs. BCDMS carbon). The x dependence of F_2 was somewhat steeper in the BCDMS data but the significance of this difference was limited, due to the restricted x range ($x \geq 0.275$) of the BCDMS data. Later, with the results on light targets, extending down to lower x, the problem was confirmed and the differences between the two results reached 15% at low x ($x = 0.07$). The only other high statistics data on light targets, the ones from SLAC, although favouring the steeper x dependence of the BCDMS data (see [32]) could not unambiguously resolve the discrepancy because they cover an almost completely disjoined Q^2 range. The CDHSW data on iron were also in better agreement with BCDMS carbon results than with EMC iron's under consistent assumptions on R (see [33]).

This long-standing controversy had cast some doubts on the validity of the claimed accuracy of these high statistics data but, during the last year, two new high statistics measurements appeared (from NMC and CCFR) that definitively resolved the problem, as we shall see in the next section.

2.3 Comparison of High Statistics Measurements of F_2

2.3.1 Definition of a Comparable Set of Measurements

As far as the decomposition of structure functions in the quark flavour space (see 1.1.2) is concerned, the existing F_2 measurements can be separated in two main classes, first, the results obtained with isoscalar targets that correspond to a pure or almost pure (see next subsection) singlet structure function and, second, the ones obtained with hydrogen targets that correspond to a linear combination of a singlet and a non-singlet part. All the heavy targets used in high statistics deep inelastic studies (i.e. iron and carbon) and half of the light targets (i.e. deuterium) are (approximately or exactly) isoscalar. So, the isoscalar F_2 results are more numerous and we shall now go through a comparison of those, which are listed below (in any case, a comparison of the hydrogen results leads to similar conclusions).

- The EMC F_2 result on iron, $F_{2\,EMC}^{Fe}$ [14],
- the EMC F_2 result on deuterium, $F_{2\,EMC}^{D}$ [16],
- the EMC F_2 result on reanalysed, $F_{2\,EMC\ Rean.}^{D}$ [17],
- the SLAC F_2 result on deuterium, $F_{2\,SLAC}^{D}$ [28],
- the BCDMS F_2 result on carbon, $F_{2\,BCDMS}^{C}$ [20],
- the BCDMS F_2 result on deuterium,$F_{2\,BCDMS}^{D}$ [22],
- the CDHSW F_2 result on iron,$F_{2\,CDHSW}^{Fe}$ [24],
- the NMC F_2 result on deuterium, $F_{2\,NMC}^{D}$ [30],
- the CCFR F_2 result on iron, $F_{2\,CCFR}^{Fe}$ [27].

All these results (except $F_{2\,EMC}^{D}$) correspond to more than half a million recorded deep inelastic events; we take this as our definition of *high statistics* F_2 measurements.

2.3.2 "Mean Square Charge" and "Nuclear Effects" Corrections

As we have seen in Eq. 4, the structure function F_2 measured in neutrino scattering corresponds to a pure singlet structure function $F_2^{\nu} = xq^{SI}(x)$. From Eq. 3, one can derive F_2 measured in charged lepton scattering on an isoscalar target I (average of proton and neutron),

$$F_2^{l^\pm I} = \frac{5}{18}x\left[u+\bar{u}+d+\bar{d}+s+\bar{s}+c+\bar{c}-\frac{3}{5}(s+\bar{s}-c-\bar{c})\right], \quad (6)$$

where u, $\bar{u}$, d,... stand for the distributions along x, $u(x)$, $\bar{u}(x)$, $d(x)$,... of the quarks of the respective flavours. (At fixed-target experiment energies, one can safely neglect b and t quark distributions.) As the sum of the distributions of all flavours is the singlet quark distribution q^{SI}, one has

$$F_2^{l^\pm I} = \frac{5}{18}x\left[q^{SI}-\frac{6}{5}(s-c)\right]. \quad (7)$$

So, the ratio between isoscalar F_2's measured in charged lepton and neutrino scattering is

$$\frac{F_2^{l^\pm I}}{F_2^{\nu I}} = \frac{5}{18}\left[1-\frac{6}{5}\frac{s-c}{q^{SI}}\right], \quad (8)$$

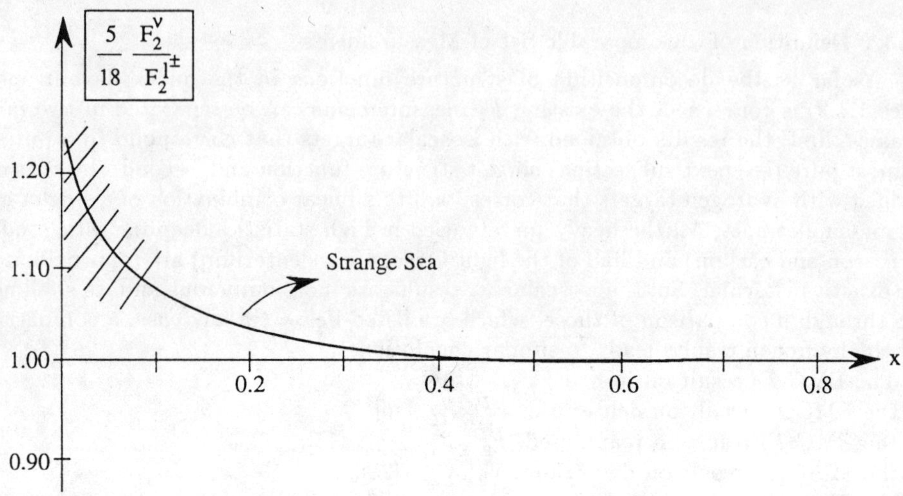

Fig.2. "Mean square charge" correction (see text).

i.e. almost constant along x, but for the small correction proportional to the difference of the distributions of strange and charm sea quarks that is non-negligible only at small x.

So, for a consistent comparison of isoscalar F_2's, one has, beforehand, either to multiply the neutrino results or to divide the charged lepton ones by the ratio given in Eq. 8. This correction is often called "mean square charge correction". In what follows, we have arbitrarily chosen to correct the neutrino results to make them correspond to the (more numerous) charged lepton ones. The correction ratio used is displayed in Fig. 2 as a function of x. It is almost constant, equal to 5/18, except at small x where the $s - c$ term becomes non-negligible. This term is obtained from a parametrisation of the distribution of strange (and charmed) sea quarks estimated from neutrino data for $\sim 0.1 < x <\sim 0.2$ and extrapolated to lower x values. The hatched area indicates the region where uncertainties in this correction are high (bigger than 5% in absolute). We have neglected the Q^2 variation of the $s - c$ term, which is anyway predicted by perturbative QCD to be small (at least at low x).

Another correction is needed, for a consistent comparison of isoscalar F_2's, to take into account the nuclear effects discovered by the EMC experiment [34] (and thus often generically called the "EMC effect"). Due, possibly, to the different binding energies of nucleons inside different nuclei, the x dependence of F_2 (per nucleon) measured on heavy isoscalar nuclei is different from the one of F_2 measured on the lightest isoscalar nucleus, i.e. deuterium. Since this discovery (1983), the ratios F_2^A/F_2^D have been measured by several deep inelastic experiments (see, e.g. [35]), often with a better precision than the structure functions themselves. The results of these measurements are in good agreement (but for the original measurement of EMC which had some problems at low x) and there is, by now, a well established shape for the x dependence of the ratio of F_2 measured, e.g. on iron and deuterium, F_2^{Fe}/F_2^D . Up to now,

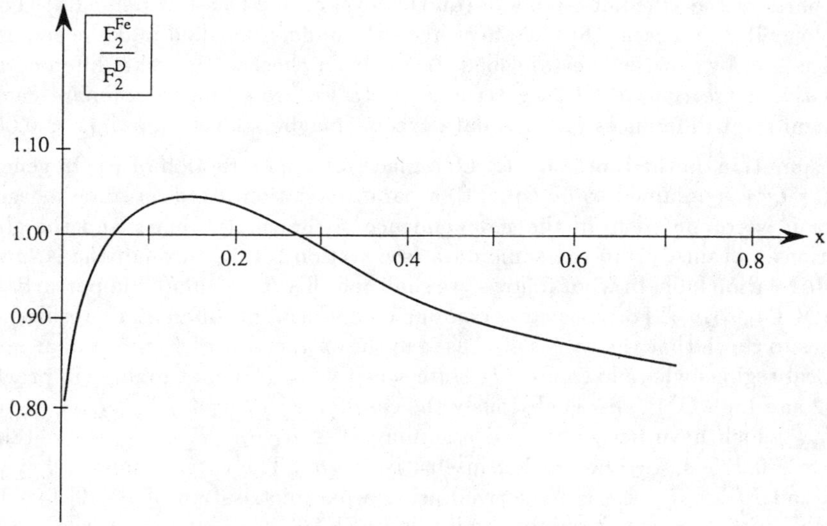

Fig. 3. "Nuclear effects" correction (see text).

the measurements of such ratios have not shown any significant Q^2 dependence so that, within the present accuracy of F_2 data, it is sufficient to apply an x dependent correction to any F_2 measurement on a heavy target to compare it to deuterium results. So, we have parametrised along x all the available data on the ratio F_2^{Fe}/F_2^D, got the result displayed in Fig. 3, and used this parametrisation to correct F_2 results obtained on iron to compare them to deuterium data. We have chosen to correct the F_2 result obtained on carbon (from BCDMS) with the same parametrisation because nuclear effects measurements show only small differences between F_2^C/F_2^D and F_2^{Fe}/F_2^D.

2.3.3 Consistency of Corrections and Assumptions in Published Data

By applying adequately the two x dependent corrections displayed in Figs. 2 and 3, one can make all F_2 measurements on isoscalar targets correspond to measurements on a deuterium target with charged lepton beams, i.e. to $F_2^{l^\pm D}$, so that they can, in principle, be directly compared. Yet, a few more precautions are needed. Indeed, the original published F_2 measurements, that are derived from the measured differential cross sections, already contain several corrections and assumptions that may, or may not, be consistent from one result to the other. These corrections and assumptions are listed below, starting with the ones that concern *both* charged lepton and neutrino results:

• The radiative corrections, that take into account higher orders than the leading one (exchange of one boson - see Fig. 1) in the lepton-quark scattering (in neutrino scattering, these corrections also include a term due to the W or Z propagator). In the case of charged leptons, these radiative corrections have been computed from two different approaches ([36] and [37]), and the two computed corrections exhibit differences in some parts of the kinematic domain (on the level of one to a few percents). These generally small differences [38] have been recently understood [39] and the radiative corrections are, by now, well established. It has been checked that the differences in the radiative corrections used to get the F_2 data we are going to compare cannot induce significant differences in these data except, maybe, at very low x ($x < 0.05$).

• The assumption on the function $R(x, Q^2)$, made for the extraction of F_2. In general, either $R(x, Q^2)$ is assumed to be equal to a parametrisation fitted on some measurements, or it is computed from the x dependence of the singlet quark and the gluon distributions, obtained from the same data (see section 2.1.3), through the *Altarelli-Martinelli* relation [40] . In what follows, we shall call this R, calculated in perturbative QCD (pQCD), R_{QCD}. For consistency in our comparison of different F_2 results, one simply has to check that the values of R used in the extractions of F_2 are similar in the kinematical regions where extracted F_2's are sensitive to R, i.e. at high y (in practice $x <\sim 0.2$ and high Q^2). This is obviously the case for $F^D_{2\,BCDMS}$, $F^D_{2\,EMC\,Rean.}$ and $F^C_{2\,BCDMS}$ which have been obtained assuming $R \equiv R_{QCD}$ ($F^C_{2\,BCDMS}$, obtained only for $x \geq 0.275$ is, anyhow, rather insensitive to R). The three results, $F^D_{2\,SLAC}$, $F^D_{2\,NMC}$ and $F^{Fe}_{2\,CCFR}$ have been derived using a parametrisation of R [29] fitted on all available measurements (mainly coming from SLAC, see Fig. 15). It was checked that this parametrisation is compatible with R_{QCD} at large Q^2. The parametrisation of R used to derive $F^{Fe}_{2\,CDHSW}$, fitted on CDHSW measurements, is also (roughly) compatible with R_{QCD} at large Q^2. The two remaining measurements, $F^{Fe}_{2\,EMC}$ and $F^D_{2\,EMC}$, have been published assuming $R \equiv 0$ so that, for the comparison, we had to correct them to make them correspond to R_{QCD}. This correction is easy to compute for $F^D_{2\,EMC}$ which was measured using a single beam energy (280 GeV) but a little more complicated for $F^{Fe}_{2\,EMC}$ that corresponds to an average of measurements performed at four different beam energies (in that case we have employed the method described in [32]).

• The "non-isoscalarity" correction that has to be applied when the target used does not have an exactly equal number of protons and neutrons, e.g. for iron targets where, in average over the iron isotopes, $Z = 26$ and $N = 30$. As long as the proton F_2 structure function is roughly known in the same kinematical domain as the isoscalar F_2 measurement (which is indeed the case), this correction is straightforward. All the heavy target F_2 measurements involved in this comparison have been adequately corrected for non-isoscalarity.

2.3.4 Corrections and Assumptions Specific to Neutrino Data

We shall now list the various corrections and assumptions used specifically in the F_2 extraction from *neutrino* scattering:

• The "strange sea" correction. The structure function proportional to the sum of the differential cross sections of neutrino and antineutrino scattering - see Section 1.1, Eq. 2 - is, actually, not exactly equal to the purely singlet structure function xq^{SI} . With the simplified notation σ and $\bar{\sigma}$ for the differential cross sections of neutrinos and antineutrinos, the purely singlet structure function (the one usually published by neutrino experiments) is $F_2 = xq^{SI} = K\,(\sigma + \bar{\sigma}) + K'\,(s - c)$, where K and K' are kinematical factors depending on E, x and Q^2. So, in order to get the purely singlet

F_2 structure function, a small correction proportional to $s - c$ (i.e. mostly to s), is applied to the cross section measurements. This correction has been applied in the two neutrino results we consider in our comparison, i.e. $F_{2\,CDHSW}^{Fe}$ and $F_{2\,CCFR}^{Fe}$.

• The "charm threshold" correction, usually done according to the "slow rescaling" prescription [41]. In the quark-parton model, when, for example, a neutrino interacts with a strange quark through a charged current interaction, the outgoing particles are a μ^- and a c quark. As the mass of the c quark, contrary to the ones of u, d and s quarks, is non-negligible, this type of interaction must have a kind of energy threshold. The values of $F_{2\,CCFR}^{Fe}$ have been corrected with the standard slow rescaling procedure but published values of $F_{2\,CDHSW}^{Fe}$ are *not* corrected for the c production threshold. So, in what follows, we have used specific F_2 values from CDHSW [42], identical to the published ones but for the slow rescaling correction, applied in the same way as for CCFR results.

• The assumption on the values of the neutrino (and antineutrino) total cross section energy slopes. In order to normalise the F_2 measurements, one has to use some values for the slopes of the linear dependence upon the energy of the total neutrino and antineutrino cross sections, σ_{tot}/E and $\bar{\sigma}_{tot}/E$. Usually, the values of these total cross section slopes, used in the F_2 extraction, come from measurements from the same experiment or from an average of all available total cross section determinations. The values of σ_{tot}/E and $\bar{\sigma}_{tot}/E$ assumed in the two neutrino F_2 measurements involved in the present comparison are somewhat different, they are $0.703 \cdot 10^{-42}$ m^2/GeV and $0.331 \cdot 10^{-42}$ m^2/GeV for the CDHSW results and $0.676 \cdot 10^{-42}$ m^2/GeV and $\sim 0.350 \cdot 10^{-42}$ m^2/GeV for the CCFR results. The value of $\bar{\sigma}_{tot}/E$ for CCFR is given here only approximately, because a non-negligible E dependence of this value was actually assumed by CCFR. As these total cross section slopes are, in some sense, part of the measurements of the individual experiments, we have chosen *not* to correct the F_2 results to make them correspond to the same total cross sections. It is then possible that some differences between the two results can be, at least partly, explained by these different assumptions on the total cross sections (see later)(*). One can remark that the values assumed by CCFR are closer to the present world average than the ones assumed by CDHSW.

2.3.5 Comparison of Corrected F_2 Results

After having checked the consistency of all assumptions and corrections used in the extraction of the isoscalar F_2 measurements we are considering, and having, when necessary, corrected the published F_2 values, we can now apply the "mean square charge" correction (Section 1.3.2, Eq. 8, Fig. 2) to neutrino results (i.e. $F_{2\,CDHSW}^{Fe}$ and $F_{2\,CCFR}^{Fe}$) and the "nuclear effects" correction (Section 1.3.2, Fig. 3) to heavy target results (i.e. $F_{2\,EMC}^{Fe}$, $F_{2\,BCDMS}^{C}$, $F_{2\,CDHSW}^{Fe}$, and $F_{2\,CCFR}^{Fe}$) to make them correspond to $F_2^{l^\pm D}$. We can then proceed to a consistent and significant comparison of these results.

In order to avoid too many figures, we have decided to compare each F_2 result, coherently corrected to correspond to $F_2^{l^\pm D}$, to a single reference. For this reference,

(*) This remark also applies to the bubble chamber neutrino measurements, WA25 and WA21.

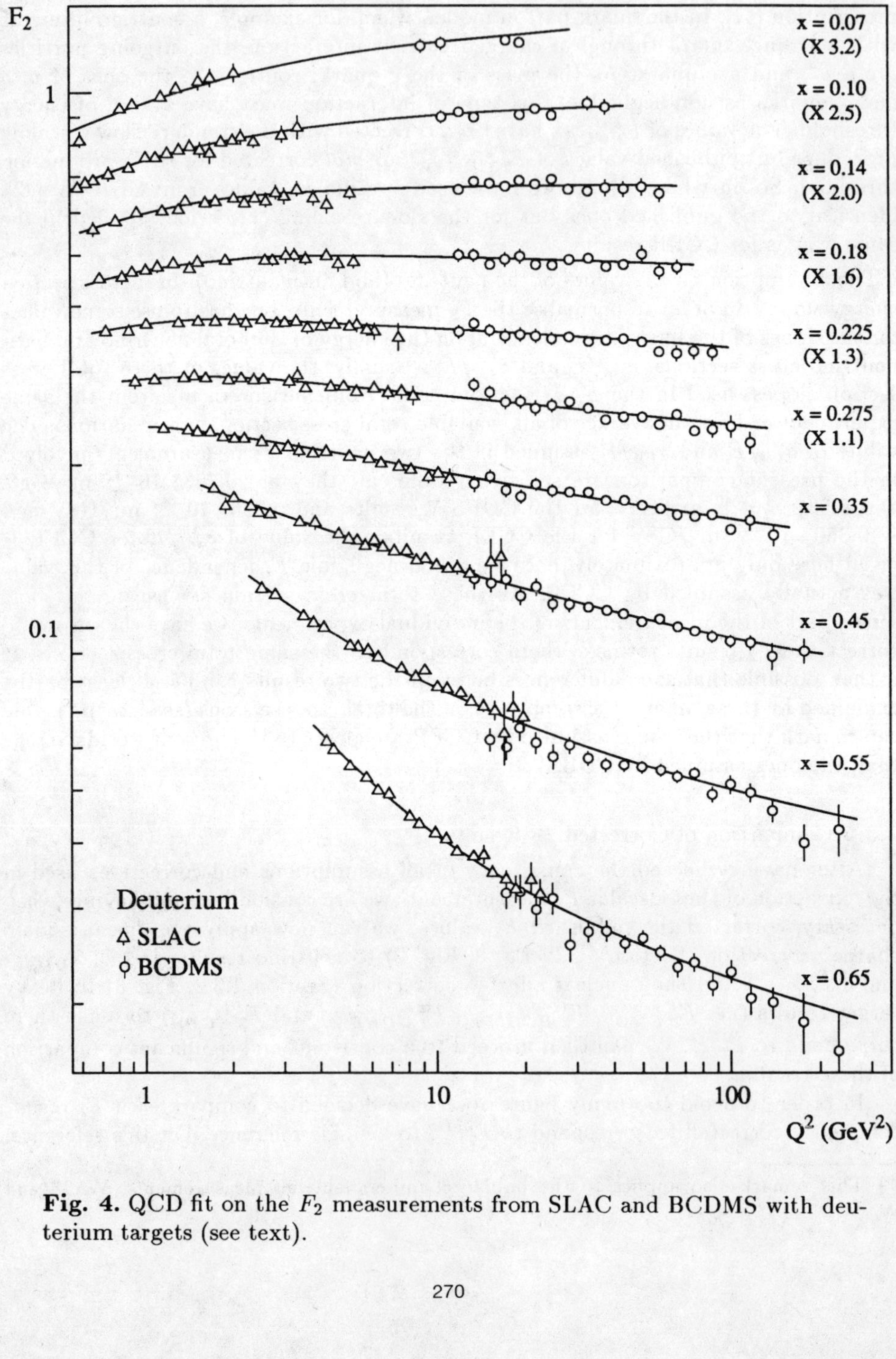

Fig. 4. QCD fit on the F_2 measurements from SLAC and BCDMS with deuterium targets (see text).

we have chosen the result of a global QCD fit performed on the F_2 data from SLAC and BCDMS on hydrogen and deuterium [43]. We believe that this result represents the best knowledge of F_2 we had one year ago, before the appearance of the two measurements from NMC and CCFR. The result of this fit, concerning $F_2^{l^\pm D}(x,Q^2)$, is represented in Fig. 4 by a solid line, as a function of Q^2 for the different x-bins, together with the data on which it has been fitted i.e. $F^D_{2\,SLAC}$ and $F^D_{2\,BCDMS}$. The error bars on these data represent total errors, i.e. statistical and systematic ones combined in quadrature, but do not include overall normalisation errors which amount to 2% for SLAC and 3% for BCDMS. The agreement between the fit and the data of both SLAC and BCDMS is good. In this figure, according to the result of the fit and as discussed in [43], the relative normalisation of the two data sets is shifted by 0.4% (0.1 standard deviation) and the BCDMS data correspond to a value of their main systematic uncertainty 1.2 standard deviation off their central value (for $x < 0.55$, at the scale of the figure, both effects are in any case hardly visible). The x binning is the one of the original data and all x-bins are represented but the last one ($x = 0.75$) for BCDMS and the last two ($x = 0.75, 0.85$) for SLAC. These large x bins are omitted here for clarity of the figures, and because many results do not extend to such high x values. When needed, the other results that we are going to compare to this "SLAC + BCDMS reference" will be re-binned along x to correspond to the same binning. Any sensible re-binning procedure leads to very similar results and identical conclusions. A few results include significant F_2 measurements down to lower x values than the lowest one of SLAC and BCDMS data ($x = 0.07$). These low x results will be compared later (Section 1.3.7) and we shall now go on with the comparison in the x range $0.07 \leq x \leq 0.65$.

The result $F^D_{2\,EMC}$ is represented in Fig. 5 together with the SLAC + BCDMS reference fit. Here also, the error bars represent total errors but for the normalisation uncertainty which amounts to 5%. The only correction compared to published data is the one that makes the measurement correspond to $R \equiv R_{QCD}$ (instead of $R \equiv 0$), in particular there is no global renormalisation of the data. It appears clearly that these EMC data exhibit an x dependent discrepancy with the reference fit; at large x ($x > 0.30$), $F^D_{2\,EMC}$ is compatible with the reference while for lower x values it is more and more significantly below the reference as x decreases. The discrepancy reaches 15% at $x = 0.07$.

In Fig. 6 $F^D_{2\,EMC\,Rean.}$ is displayed, still with total errors (not including the normalisation uncertainty of 5%) and compared to the same reference fit. Here, no corrections to the published data (assuming $R = R_{QCD}$) were needed. Compared to the previous result $F^D_{2\,EMC}$, some additional raw data at low Q^2 have been used in the extraction of F_2 and the point to point fluctuations are reduced, especially at large x. Nevertheless, the difference with the reference fit remains: it is slightly reduced at high Q^2 low x, but is even enhanced for the additional low Q^2 data which, in the first x bin, are almost 20% below the reference (at higher x, the situation does not change much). Although it can reduce the overall mismatch, a normalisation error cannot account for this x dependent discrepancy. The Q^2 evolution also seems somewhat different; flatter at large x and steeper at low x.

The result from EMC on heavy target, i.e. $F^{Fe}_{2\,EMC}$, is represented in Fig. 7 still with total errors and compared to the same reference fit. Here, the normalisation uncertainty is 5% and, compared to published data, two corrections have been made,

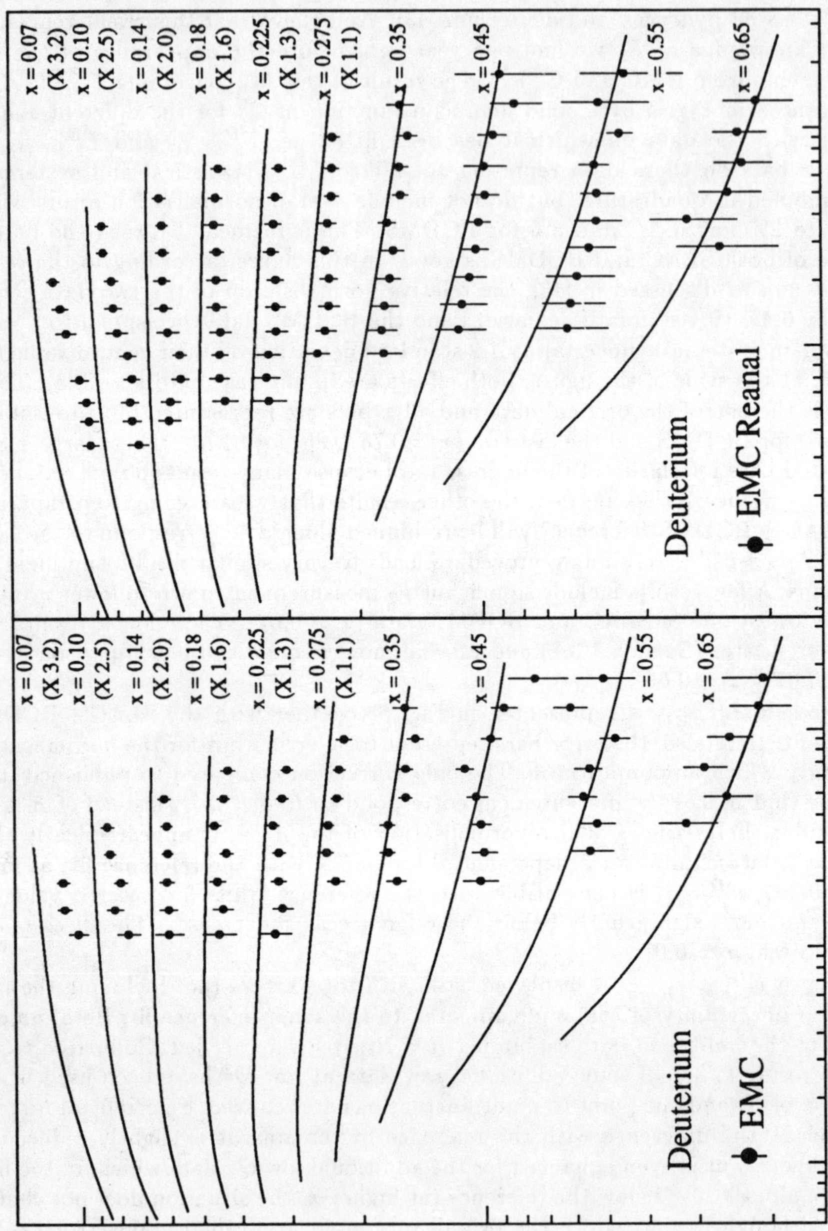

Fig. 5. F_2 measurements on deuterium from EMC compared to the SLAC+ BCDMS reference fit.

Fig. 6. The same measurements reanalysed (see text).

the one that makes the measurement correspond to $R \equiv R_{QCD}$ (instead of $R \equiv 0$) and the "nuclear effects" correction (Section 1.3.2). These data compare to the reference in, qualitatively, the same way as the other EMC results, i.e. at large x (here, $x > 0.4$) there is a good agreement but at smaller x the F_2 points are significantly below the reference. The magnitude of the difference, however, is smaller than for the light targets EMC results. The Q^2 evolution of the data is similar to the one of the reference, but for the lowest Q^2 points they exhibit a tendency to drop.

In Fig. 8, $F_{2\,CDHSW}^{Fe}$ is displayed with total errors and again compared to the same reference fit. Here, to make this neutrino measurement on iron comparable to the reference, three corrections to the published data were needed, the "mean square charge" and "nuclear effects" ones (Section 1.3.2) and the "slow rescaling" one (Section 1.3.4). If one divides these CDHSW data in two parts along Q^2 in each x bin, the high Q^2 results agree very well with the reference while the low Q^2 measurements (typically the first four to five points in each x bin) are higher than the reference at high x and lower at low x. In this low Q^2 region, the measurement errors are dominated by systematics and are highly point to point correlated so that the observed differences with the reference fit may actually correspond to only $<\sim 2$ standard deviations of one of the main systematic error, like the calorimeter calibration uncertainty. So, taking into account the systematic errors and their correlations, the agreement between $F_{2\,CDHSW}^{Fe}$ and the reference is satisfactory. (Were it not for the recent CCFR results, these different x behaviours at low Q^2 could be ascribed to different higher twist terms - Section 2.2.4 - in neutral and charged lepton scattering.)

The very high luminosity heavy target result of BCDMS, $F_{2\,BCDMS}^{C}$, is displayed in Fig. 9 together with the reference fit. Once again, the error bars represent total errors but for the normalisation uncertainty of 3%. The only correction we had to apply to the published results was the "nuclear effects" one. The data, available only for $x \geq 0.275$ and $Q^2 \geq 26\,\mathrm{GeV}^2$, exhibit a Q^2 evolution very similar to the one of the reference but are systematically above it, at least for $x < 0.5$. This is within a $\sim 3\%$ normalisation offset (~ 1 standard deviation). However, in the last x-bin ($x = 0.65$), the data points lie on top of the reference so that a 3% downward renormalisation will bring them below it. From the published correlated systematic errors [44], one can find out that this remaining x-dependent difference is compatible with $\sim$1.2 standard deviations of the calibration uncertainty of the BCDMS data (as in the case of $F_{2\,BCDMS}^{D}$ corresponding to the reference fit). So, within 1 standard deviation of normalisation uncertainty and 1.2 standard deviations of energy calibration systematics, $F_{2\,BCDMS}^{C}$ is compatible with the reference fit.

At that point, we have gone through the comparison of the high statistics isoscalar F_2 data available more than one year ago. The situation then could be summarised as follows. Light target data from SLAC and BCDMS are compatible and well described by a common QCD fit. Data from EMC and BCDMS disagree, the discrepancy being more pronounced for light targets than for heavy targets. Data from SLAC and EMC, however, are not incompatible (common fits of these two data sets are acceptable). And finally, CDHSW data are in better (although not very good) agreement with the BCDMS ones. So, the situation was not particularly clear until the results from NMC and CCFR.

$F_{2\,NMC}^{D}$ is represented in Fig. 10 together with the reference fit. The NMC result is split in two data sets, each one corresponding to a given muon beam energy, 90

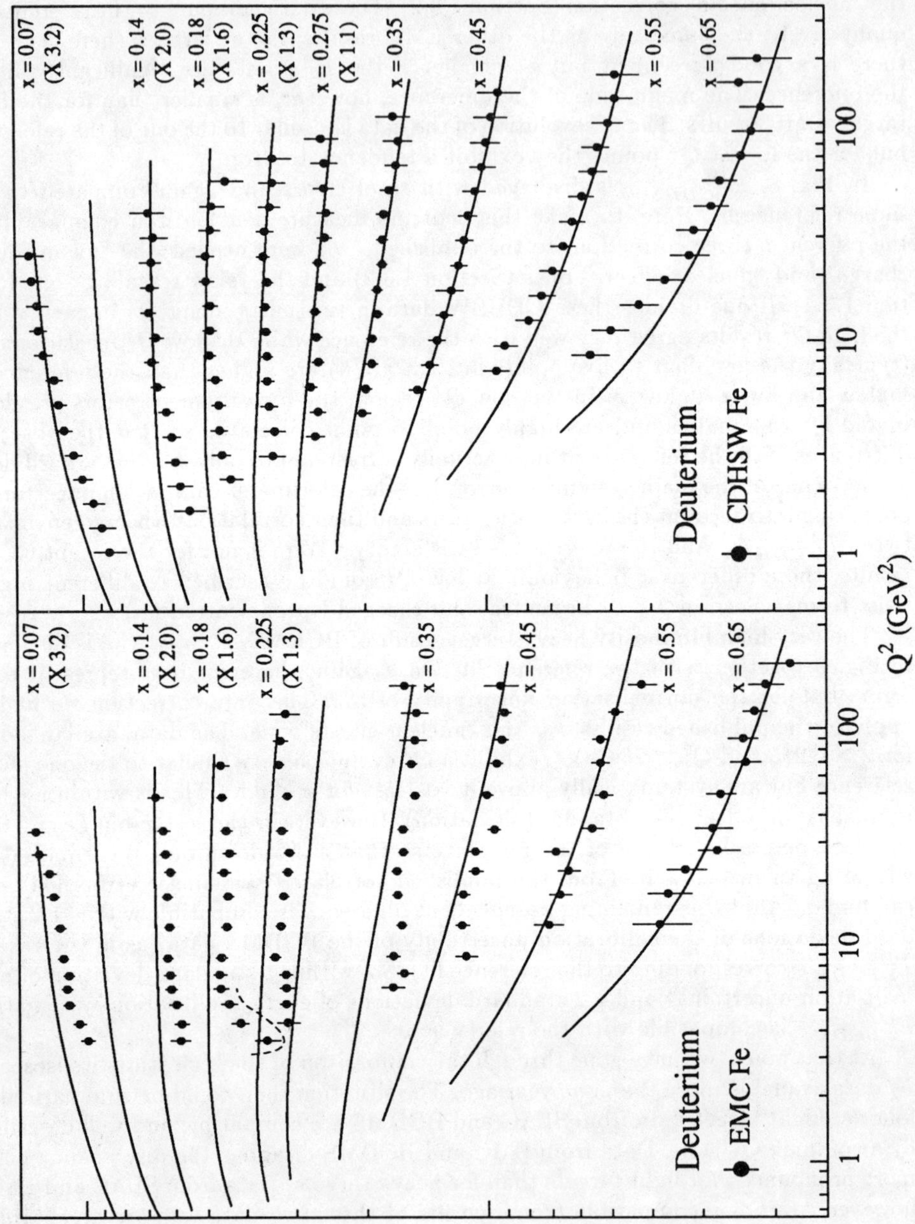

Fig. 7. F_2 measurements on iron from EMC compared to the reference fit.

Fig. 8. F_2 measurements on iron from CDHSW compared to the reference fit (see text).

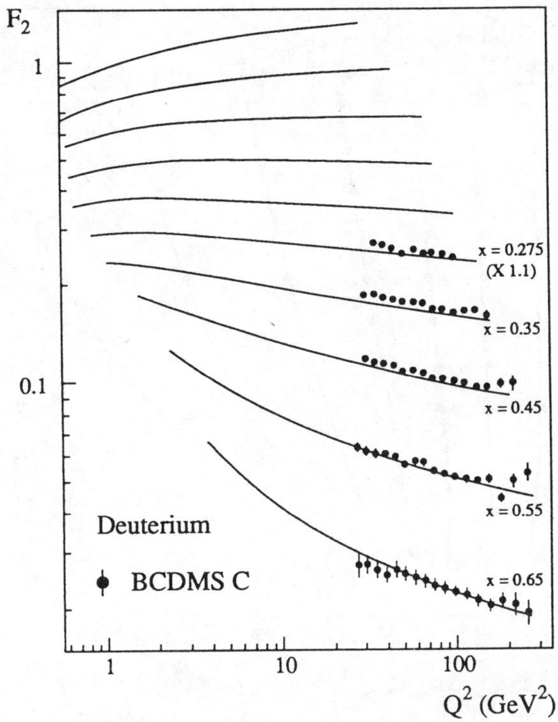

Fig. 9. F_2 measurements on carbon from BCDMS compared to the reference fit.

and 280 GeV. The error bars correspond to total errors but for the normalisation uncertainties, which amount to 1.6% for the 90 GeV data set and 2.6% for the 280 GeV one. An important feature of the NMC data is that they extend to very low x ($x = 0.008$) so that only about half of them are represented in Fig. 10. In the region of overlap with the reference ($0.07 \leq x \leq 0.5$), these data are in good agreement with it; this is obvious for the 90 GeV data while the 280 GeV ones are globally 1.5% below the fit (0.6 standard deviations of normalisation uncertainty). A similar slight renormalisation of the 280 GeV data is also favored by QCD fits to the NMC data alone (see 2.3.4). The overall agreement is very good; in particular the x and Q^2 dependencies of the data perfectly agree with the ones of the reference.

As can then be expected, the NMC F_2 data disagree with the EMC results. This is particularly interesting in view of the strong similarity of the NMC apparatus with the EMC one. One of the main improvements between the NMC and the EMC – NA2 apparatus consisted in the addition of proportional wire chambers to allow the reconstruction of the scattered muons down to lower values of x. These new chambers partly overlap with some of the big drift chambers used up to then for the scattered muon reconstruction. So, they have allowed to check the efficiency of the scattered muon pattern recognition using (only) the drift chambers. Important inefficiencies

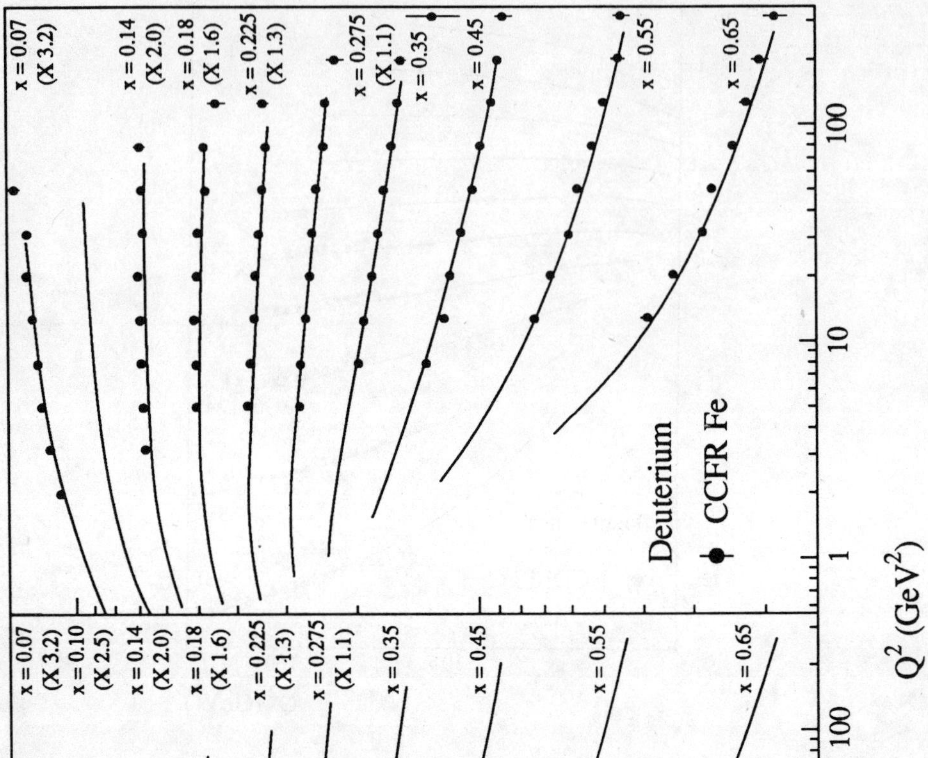

Fig. 10. The recent F_2 measurements on deuterium from NMC compared to the reference fit.

Fig. 11. The recent F_2 measurements on iron from CCFR compared to the reference fit (see text). Error bars are statistical only.

were unveiled, strongly correlated with the charged track multiplicities, themselves functions of the energy ν of the hadronic shower. Careful Monte Carlo studies showed that such a ν dependent reconstruction inefficiency was also present, although much smaller, when using the proportional chambers. As the effect in the drift chambers was so important and could not be adequately simulated by Monte Carlo, the NMC collaboration has used *only* events reconstructed in the new proportional chambers for their recent publication [30]. This explains why the Q^2 range covered by each single beam energy data set is rather small. In view of the Monte Carlo studies, an additional ν dependent correction for reconstruction inefficiencies (reaching 8% at the highest ν) was applied to the F_2 measurement. This correction mainly affects the low x results. It seems likely that the differences between the F_2 results of EMC – NA2 and BCDMS arose from such ν dependent reconstruction inefficiencies that can describe well the x dependent pattern of the difference. (In the EMC analysis, reconstruction inefficiencies were considered to be independent of the kinematics.) In any case, the NMC results have resolved the long standing BCDMS/EMC discrepancy (in favour of BCDMS).

This clarification of the experimental situation was still strengthened by the recent CCFR results. The F_2 measurement from CCFR, $F_{2\ CCFR}^{Fe}$, is displayed in Fig. 11, once again with the same reference fit. Contrary to other F_2 figures here, the error bars in this figure are *only statistical* because the estimate of the point to point systematic errors is not available yet. (These errors are expected to be small.) Two corrections were needed to make the CCFR measurement comparable to the reference; the "mean square charge" and "nuclear effects" ones (see Section 1.3.2 and Figs. 2 and 3). The data exhibit a global tendency to be slightly higher than the reference but this is compatible with a $\sim 2\%$ normalisation offset which is certainly within the normalisation uncertainties of both the data and the reference fit. Besides that, and in view of the remarkable statistical accuracy of the data, the agreement between the CCFR result and the reference is really impressive (the extremely slight residual x dependent difference around $x = 0.45$ can easily be attributed to imperfect "nuclear effects" correction). In particular, in each x bin, the Q^2 evolution of $F_{2\ CCFR}^{Fe}$ is, over a wide range, in striking agreement with the one of the reference fit. It is the first time, in the history of structure function measurements, that completely independent results agree to this level of accuracy. As a consequence, the CCFR data are in disagreement with the EMC ones. On top of that, the CCFR collaboration also claims a strong disagreement with the CDHSW results [27]. We think that, although the agreement between $F_{2\ CCFR}^{Fe}$ and $F_{2\ CDHSW}^{Fe}$ is not completely satisfactory, especially at low x (see also Fig. 8), the CCFR collaboration overemphasizes the discrepancy for two reasons; first, they compare both data sets without any systematic errors (those are important, at least for the CDHSW result) and second, they do not apply the "slow rescaling" correction to the published CDHSW data (or remove this correction from their data), thus enhancing the difference at low x and low Q^2.

2.3.6 Conclusion of the Comparisons

The experimental situation of structure function measurements, which during a few years suffered from the BCDMS/EMC discrepancy, has been definitively clarified by the recent results from NMC and CCFR. This was shown here for the isoscalar

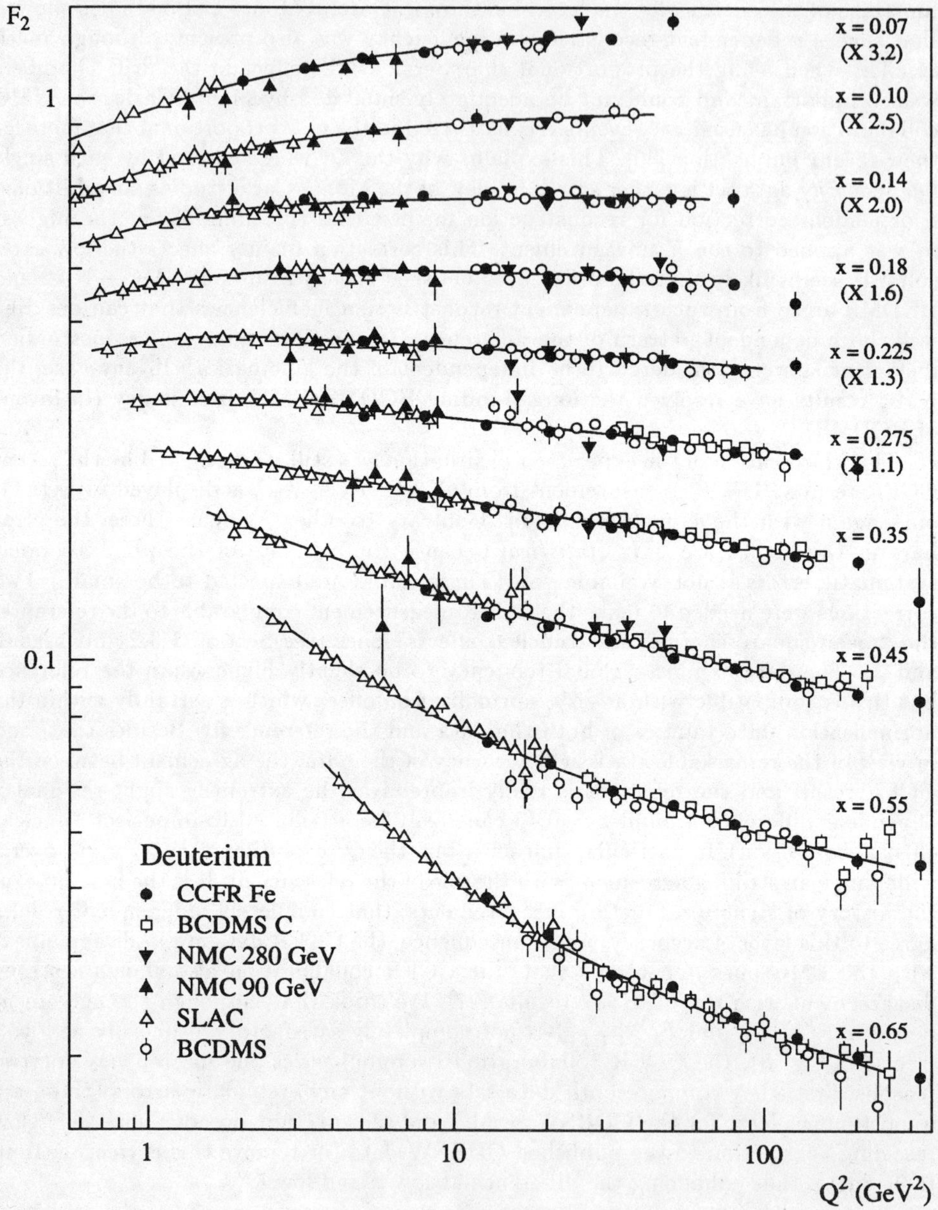

Fig. 12. The consistent set of high accuracy F_2 measurements (slightly rescaled - see text) from SLAC, BCDMS, NMC, and CCFR with the reference fit.

F_2 results, but identical conclusions can be reached with the F_2 measurements on hydrogen from SLAC, BCDMS, EMC, and NMC. Not only do the NMC and CCFR results agree with BCDMS and SLAC data but, at least as important, the NMC data analysis has unveiled a likely explanation for the x dependent bias of the EMC F_2 results.

So, there is now a set of high accuracy structure function measurements from SLAC, BCDMS, NMC, and CCFR in very good agreement. To illustrate this, we have plotted in Fig. 12 all isoscalar F_2 results from these experiments corrected for the small offsets in their main systematic uncertainties (mainly normalisation) that optimise the mutual agreement. These results are:

- $F_{2\,SLAC}^{D}$ taken arbitrarily as the normalisation reference,
- $F_{2\,BCDMS}^{D}$ with $+0.1\,\sigma$ of normalisation (+0.4%) and $+1.2\,\sigma$ of calibration,
- $F_{2\,NMC}^{D}$ with $+0.6\,\sigma$ of normalisation (+1.5%) in the 280 GeV data set,
- $F_{2\,BCDMS}^{C}$ with $-1.0\,\sigma$ of normalisation (−3.0%) and $+1.2\,\sigma$ of calibration,
- $F_{2\,CCFR}^{Fe}$ with $< 1.0\,\sigma$ of normalisation (−2.0%).

The other high statistics results are the ones from EMC and CDHSW. While the EMC data are in disagreement with this set of consistent results, the CDHSW measurements may be considered in agreement, although marginal, with it. The difference observed at low Q^2 between the CDHSW measurements (Fig. 8) and the coherent set of data of Fig. 12 can be attributed to $<\sim 2$ standard deviations of the low ν (i.e. low Q^2) dominant systematic error (hadronic calorimeter calibration), which is, as stated in the CDHSW publication, strongly point-to-point correlated.

2.3.7 Comparison of Low x Results

As already mentioned, some of the F_2 measurements extend to a lower x region than displayed in Figs. 4 to 12 ($x \geq 0.07$). This is the case for the recent NMC results, which are the first precise light target F_2 measurements to cover such a low x domain ($0.008 \leq x \leq 0.05$). These NMC low x data are displayed in Fig. 13 (open symbols). Their error bars correspond to total errors and for consistency with Fig.12 (see above) the 280 GeV data set is renormalised by +1.5%. Compared to what is shown in Figs. 5 to 7, the original EMC results included F_2 measurements in two more low x bins ($x = 0.03$ and 0.05); as expected in view of the systematic problem discussed above, these are (much) below the NMC results. They are not shown in Fig.13. Neither is the dedicated low x measurement of EMC-NA28, because of its lower statistics. In any case, within its large point to point errors and its normalisation uncertainty (7%), this measurement is in agreement with both EMC and NMC results. Results from CDHSW and CCFR also include measurements in two low x bins, $x = 0.015$ and 0.045. These measurements are also displayed in Fig. 13 (closed symbols), after having been re-binned along x ($0.015 \to 0.0175$ and $0.045 \to 0.05$) to allow a comparison with the NMC results. Of course these neutrino data are corrected for "mean square charge" and "nuclear effects" (CDHSW data also for "slow rescaling") and, still for consistency with Fig. 12, CCFR results are renormalised by −2%. In the bin $x = 0.05$, the CDHSW results are, within their larger errors, in agreement with the NMC data while the precise CCFR measurements are slightly above (by $\sim 3\%$). At $x = 0.0175$, CDHSW data are clearly off the NMC results (more than 10% above) while the CCFR data are only a few percent above. As a conclusion, and in view of the missing

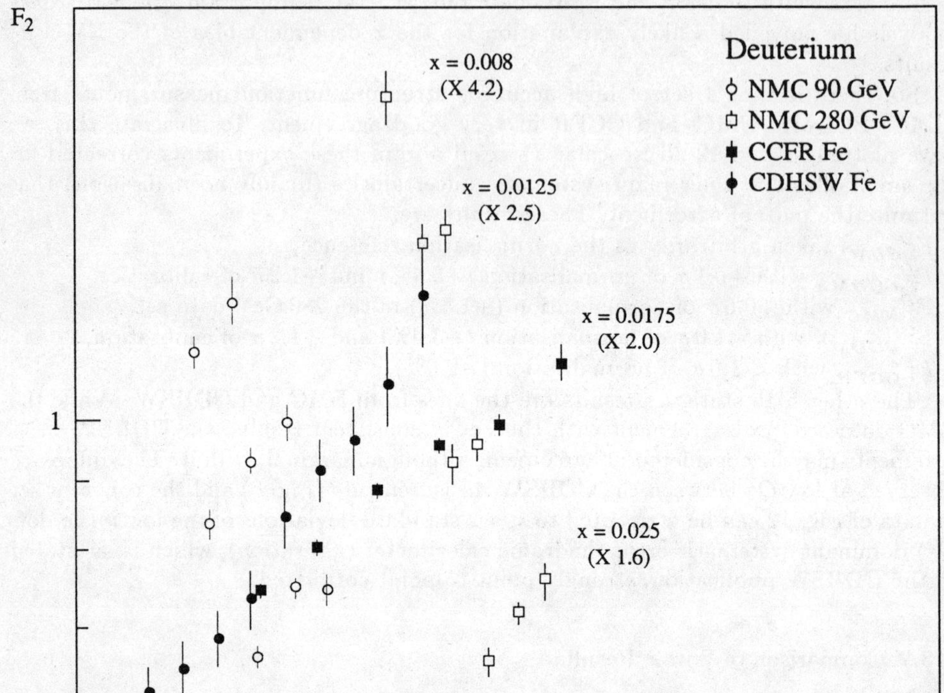

Fig. 13. Low x measurements of F_2 from NMC, CDHSW and CCFR (see text). CCFR errors are statistical only.

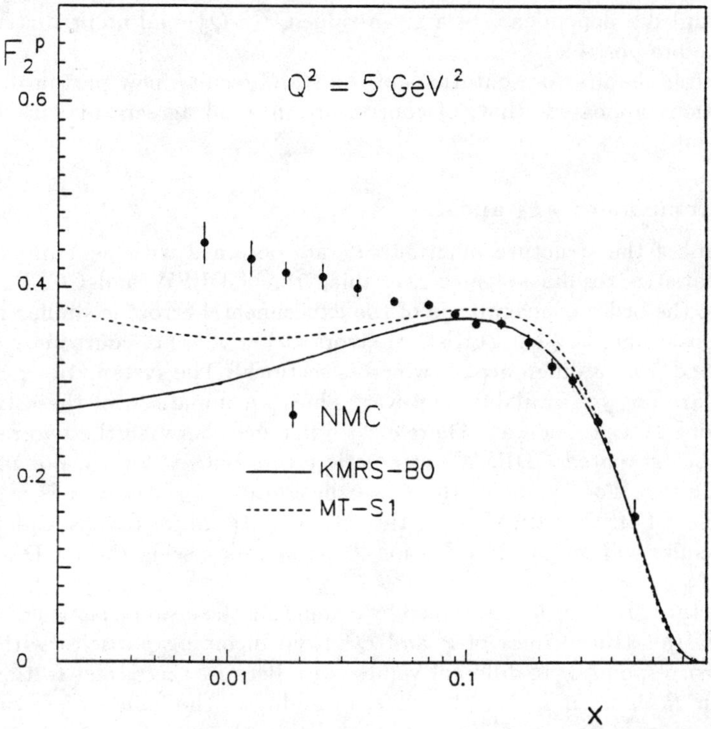

Fig. 14. The x dependence of the recent F_2 measurements from NMC compared to results from pre-existing parton distribution parametrisations.

systematic errors on CCFR results and of the *large* uncertainties that exist at very low x in the corrections applied here to neutrino data (concerning mainly the $s - c$ contribution - see Fig. 2 - but also $R(x, Q^2)$ and the radiative corrections), one can say that, except maybe for the high Q^2 CDHSW data in the lowest x bin, the agreement between NMC data and results from the two neutrino experiments is satisfactory at low x.

2.3.8 Low x Parton Distributions

As we have already remarked, before the appearance of the NMC results, no precise light target F_2 measurements were available at low x ($x < 0.07$). So, all the parametrisations of the parton distributions inside the nucleon were fitted on higher x data and it is interesting to remark that most of them failed in their low x extrapolations. This is illustrated in Fig. 14, where the NMC F_2 measurement on hydrogen is displayed as a function of x for $Q^2 = 5\,\mathrm{GeV}^2$ together with two curves corresponding to F_2^H at the same Q^2 as obtained from two parton distribution parametrisations published before the NMC results. This uncertainty in the extrapolation of the parton distributions along x is not surprising because, within the present framework of perturbative QCD, only the Q^2 evolution of the parton distributions can be computed (from an assumed x dependence at a given value $Q^2 = Q_0^2$) and no predictions on the x dependence are possible.

Shortly after the first presentations of the NMC results, new parton distribution parametrisations appeared, that, of course, are in good agreement with the low x NMC data [45].

2.4 Measurements of xF_3 and R

Measurements of the structure function xF_3 are obtained with neutrino scattering. Two high statistics results are now available, from CDHSW and CCFR. For both these results, the order of magnitude of the experimental errors is similar to the one of the F_2 measurements. (The statistical errors at low x are of course larger because xF_3 is obtained from a difference of two cross sections.) The systematic errors for the CCFR data are not yet available, however. Thus, a comparison of these two results cannot be done in a proper way. There exist differences between the two results that go beyond the estimated CDHSW systematic errors but, as for F_2, not much more than a factor two. For example, the Q^2 evolution of xF_3 at low x is significantly steeper in the CDHSW data than in the CCFR ones. More details and figures on these xF_3 results will be given in Section 2.3, when discussing the QCD analyses of these data.

The structure function R is obtained by comparing the deep inelastic cross-sections measured for the same values of x and Q^2 from incoming particles with different energies E (corresponding to different values of y). Because these cross sections depend much less on R than on F_2 and because, in addition, the values of R turn out to be small, R is measured with much less precision than F_2 (see 1.1.1, Eqs. 1 and 2). Presently, the most precise measurements at low Q^2 have been obtained by the SLAC – E140 experiment [38] and in the reanalysis of all SLAC experiments that it triggered [29]. At higher Q^2, precise results come from CDHSW [24] and from BCDMS [20,21,22]. Other determinations were obtained by EMC [14,15] and CHARM [6]. (More data are expected soon from the analyses of the CCFR and NMC data, and from an extension of the E140 experiment, run in 1991.)

Most of the published results are shown in Fig. 15, taken from [29]. The high Q^2 measurements of R all agree with the QCD prediction discussed in Section 2.1.3; in this part of the kinematic domain, the contribution from target mass effects [56] is small. In [29], the SLAC measurements have been compared to the QCD prediction, both with and without target mass effects. The stated conclusion is that the SLAC data on R lie systematically above the predictions; this may be explained by a substantial dynamical higher twist contribution to R, on top of target mass effects. This conclusion owes much to the E140 results, which have a systematic accuracy five to ten times better than all other determinations.

In the reanalysis of all SLAC data, it was also concluded that, within errors, R is the same for the proton and deuteron, $R^d - R^p = -0.001 \pm 0.009$ (stat.) ± 0.009 (syst.), for $x > 0.10$ at a typical Q^2 of 5 GeV^2. A recent analysis of the NMC data at lower x led to the same conclusion with a comparable precision, $R^d - R^p = 0.031 \pm 0.016$ (stat.) ± 0.011 (syst.) [72].

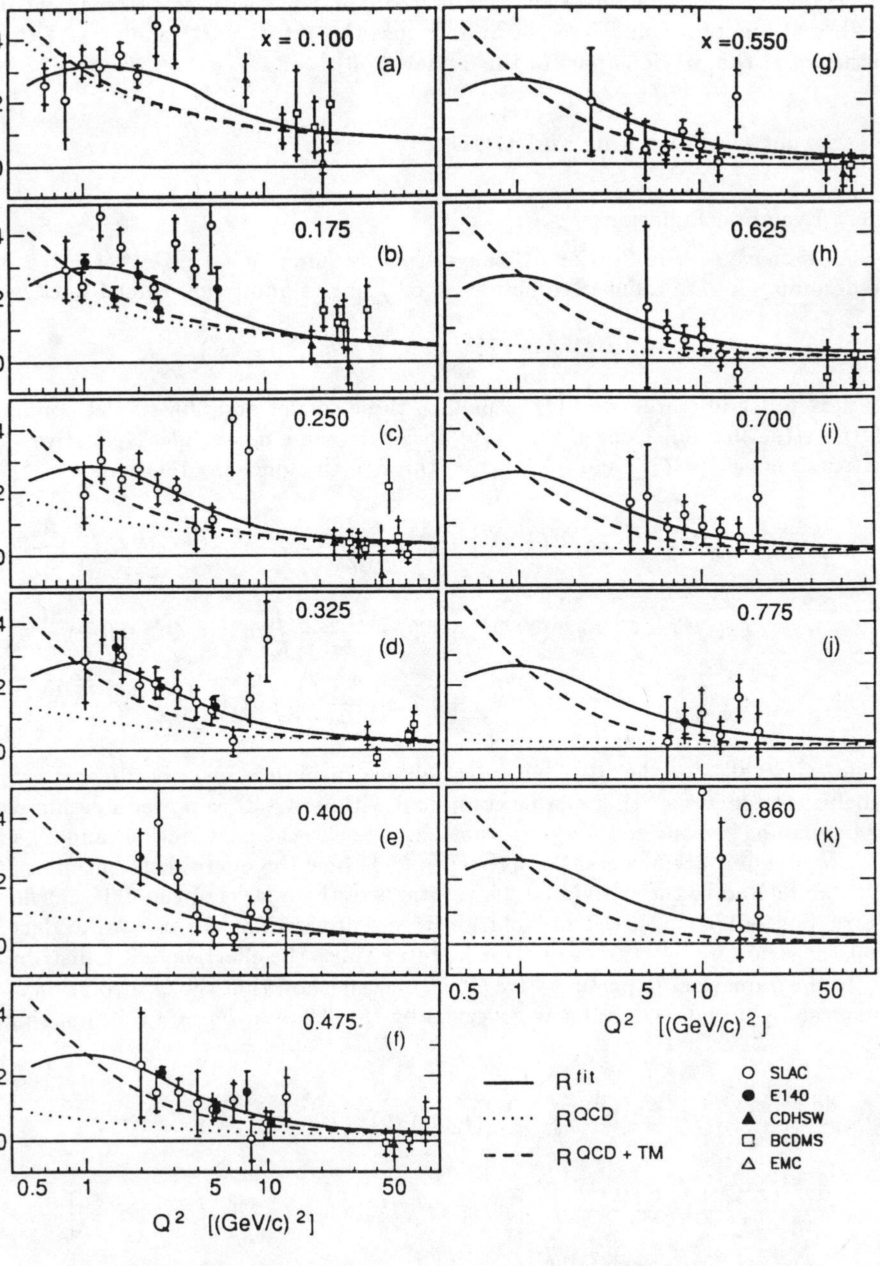

Fig. 15. Measurements of $R(x, Q^2)$ from SLAC, CDHSW, BCDMS, and EMC.

3 QCD Analyses of Structure Function Data

In this section, we will first discuss the principles of QCD fits and the underlying theory (2.1), then how these fits are done in practice with some technical details (2.2), and finally we shall present and discuss the actual QCD analyses of the data discussed in the previous part of this review (2.3).

3.1 Structure Functions and QCD

3.1.1 Evolution Equations

As we have seen in Section 1.1, any structure function $F_i(x,Q^2)$ $(i=1,2,3)$ can be decomposed into a linear combination of a singlet and a non-singlet part,

$$F_i(x,Q^2) = a\,F_i^{NS}(x,Q^2) + b\,F_i^{SI}(x,Q^2)\,. \tag{9}$$

At next to leading order (NLO) of pQCD, these singlet and non-singlet components of the structure functions are related to singlet and non-singlet "effective" quark distributions $q^{SI}(x,Q^2)$ and $q^{NS}(x,Q^2)$ through the following relations,

$$F_i^{NS}(x,Q^2) = C_i\,x\,q^{NS}(x,Q^2) + C_i\,x\,\frac{\alpha_S(Q^2)}{2\pi}\int_x^1 q^{NS}(t,Q^2)\,f_{iq}(\frac{x}{t})\frac{\mathrm{d}t}{t} \tag{10}$$

$$\begin{aligned} F_i^{SI}(x,Q^2) = {} & C_i'\,x\,q^{SI}(x,Q^2) + C_i'\,x\,\frac{\alpha_S(Q^2)}{2\pi}\int_x^1 q^{SI}(t,Q^2)\,f_{iq}(\frac{x}{t})\frac{\mathrm{d}t}{t} \\ & + C_i''\,x\,\frac{\alpha_S(Q^2)}{2\pi}\int_x^1 G(t,Q^2)\,f_{iG}(\frac{x}{t})\frac{\mathrm{d}t}{t}\,, \end{aligned} \tag{11}$$

where G stands for the distribution of gluons inside the nucleon, f_{iq} and f_{iG} are "coefficient functions" that can be computed within pQCD as power series in $\alpha_S(Q^2)$ (the "running" strong coupling constant, discussed in the next section) and C_i, C_i' and C_i'' are constant coefficients. We have seen (1.1) how the quark distributions q^{SI} and q^{NS} are related to the individual distributions of the quarks of the different flavours. At leading order (LO), the convolution terms proportional to α_S are neglected, so that the structure functions are identified to x times the effective quark distributions.

In the framework of perturbative QCD, one can show that the Q^2 evolution of these quark and gluon distributions is governed by the *Altarelli-Parisi* evolution equations [46],

$$\frac{\mathrm{d}q^{NS}(x,Q^2)}{\mathrm{d}\ln Q^2} = \frac{\alpha_S(Q^2)}{2\pi}\int_x^1 q^{NS}(t,Q^2)\,P^{NS}(\frac{x}{t})\,\frac{\mathrm{d}t}{t} \tag{12}$$

$$\frac{\mathrm{d}q^{SI}(x,Q^2)}{\mathrm{d}\ln Q^2} = \frac{\alpha_S(Q^2)}{2\pi}\int_x^1 \left[q^{SI}(t,Q^2)\,P_{qq}(\frac{x}{t}) + C_q\,G(t,Q^2)\,P_{qG}(\frac{x}{t})\right]\frac{\mathrm{d}t}{t} \tag{13}$$

$$\frac{\mathrm{d}G(x,Q^2)}{\mathrm{d}\ln Q^2} = \frac{\alpha_S(Q^2)}{2\pi}\int_x^1 \left[G(t,Q^2)\,P_{GG}(\frac{x}{t}) + C_G\,q^{SI}(t,Q^2)\,P_{Gq}(\frac{x}{t})\right]\frac{\mathrm{d}t}{t}\,, \tag{14}$$

where P^{NS}, P_{qq},... P_{Gq} are "splitting functions" that can be computed within pQCD as power series in $\alpha_S(Q^2)$. So, for a non-singlet structure function, as soon as its x-dependence is known at any given Q^2 value (Q_0^2), its Q^2 evolution is unambiguously determined, over the whole Q^2 range where pQCD is valid, through a single integro-differential equation (Eq. 12). The case of the singlet structure function is slightly more complicated because its Q^2 evolution is governed, together with the one of the gluon distribution, by a system of two coupled integro-differential equations (Eqs. 13 and 14). So, in principle, to get the predicted Q^2 evolution of the singlet structure function (i.e. to solve the system of coupled equations), one needs to know the x dependence - at any given Q_0^2 - of both the singlet quark distribution q^{SI} *and* the gluon distribution G. In practice, however, the convolution term $G \bullet P_{qG}$ in Eq. 13 becomes usually negligible for $x >\sim 0.3$ so that, at large x, the Q^2 evolution of the singlet structure function is governed by a single integro-differential equation almost identical to the one governing the non-singlet evolution. Indeed the two splitting functions P^{NS} and P_{qq} differ only at very low x ($x <\sim 0.01$). There is, actually, at NLO, two different non-singlet splitting functions P_+^{NS} and P_-^{NS} corresponding to different types of non-singlet structure functions, but they also only differ at very low x.

Of course, as any quantity perturbatively computed in a renormalisable field theory, the coefficient functions and the splitting functions depend, at NLO, on the renormalisation scheme. Moreover, these functions also depend on an other arbitrary choice, the "factorisation scheme". Indeed, in the derivation of the evolution equations at NLO, terms of order α_S can be taken into account either in the coefficient functions f_i or in the splitting functions P (e.g. in Eqs. 10 or 12). So, an ambiguity appears in the definition of the effective quark distributions linked to the arbitrary choice of the coefficient functions f_{iq}. Any choice for one of these functions (e.g. f_{2q}) is possible but, once this choice is made, all the other coefficient functions and all the splitting functions are unambiguously defined (see, e.g. [47]). Of course, the differences, in any observable quantity, induced by different choices of renormalisation or factorisation schemes are one order of α_S higher than the one of the perturbative computation (here NLO). So, in the limit of infinite number of terms in the perturbative expansions, any renormalisation or factorisation scheme yields identical observable results. In practice, the renormalisation and factorisation schemes most often used are the so-called $\overline{\text{MS}}$ ones. The $\overline{\text{MS}}$ factorisation scheme is a "natural" choice in the partonic view because it corresponds to a definition of effective quark distributions such that the basic partonic sum rules - as the "momentum sum rule" (see below) - are conserved for any value of Q^2. At present, full computations of the various splitting and coefficient functions are available at NLO in the $\overline{\text{MS}}$ schemes [48] and the computation of the coefficient functions at NNLO was performed recently [49].

3.1.2 The "Running" Coupling Constant α_S and Λ_{QCD}

The evolution of the effective coupling constant of QCD with the energy scale μ is governed by the renormalisation group equation (RGE - see, e.g. [50]),

$$\mu \frac{d\alpha_S(\mu^2)}{d\mu} = \beta(\alpha_S) = -\beta_0' \alpha_S^2(\mu^2) - \beta_1' \alpha_S^3(\mu^2) - \beta_2' \alpha_S^4(\mu^2) - \cdots \quad , \tag{15}$$

where the coefficients β_i' of the perturbative expansion of the "beta" function depend on the number N_f of flavours active at the energy scale μ. Up to now, we have implicitly assumed, as is often done, that the energy scale μ is $\sqrt{Q^2}$. The consequences of this choice will be discussed in the Section 2.2.5 devoted to the "energy scale uncertainties".

At NLO, a solution of the RGE (Eq. 15) often used is the following one

$$\alpha_S(Q^2) = \frac{4\pi}{\beta_0 \ln(Q^2/\Lambda^2)} \left[1 - \frac{\beta_1}{\beta_0^2} \frac{\ln\ln(Q^2/\Lambda^2)}{\ln(Q^2/\Lambda^2)}\right] , \tag{16}$$

where, in the $\overline{\mathrm{MS}}$ renormalisation scheme,

$$\beta_0 = 11 - \frac{2}{3} N_f \ , \quad \beta_1 = 102 - \frac{38}{3} N_f \ , \tag{17}$$

and Λ is an integration constant, widely known as the QCD mass scale parameter. One has to keep in mind, however, that a relation, such as Eq. 16, between Λ and α_S depends on the renormalisation scheme and, even within the same scheme and at a given order (like NLO), is not unique; different relations at the same order can differ by higher order terms. Moreover, the variation of α_S with the energy scale is of course continuous; as, at each flavour threshold, the number N_f of active flavours and thus the coefficients β_i change, the value of the integrating constant Λ has to be changed too (see, e.g. [51]). So, when quoting a value of Λ, one has to specify to what renormalisation scheme and to how many flavours it corresponds, e.g. $\Lambda^{(4)}_{\overline{\mathrm{MS}}}$ for the $\overline{\mathrm{MS}}$ scheme and 4 flavours. For completeness, one should also specify the exact form of the relation between Λ and α_S that was assumed (this is rarely done).

It used to be popular to give, as a result of some experimental measurement of the "strength" of the QCD coupling, a value of $\Lambda^{(i)}_{\overline{\mathrm{MS}}}$ instead of a value of α_S at a given energy scale. At present however, in view of the many assumptions entering the definition of Λ, the tendency is to give as a result the value of α_S at a well specified energy scale μ_0 (e.g. the Z^0 mass M_Z). Then, α_S can be obtained at any Q^2 value from $\alpha_S(\mu_0^2)$ directly through the RGE, possibly changing the coefficients β_i' at each flavour threshold. As the perturbative expansion of the beta function is available up to NNLO, the precision on the computed Q^2 evolution of α_S can be far beyond any present experimental accuracy.

3.1.3 The QCD "Prediction" for R

As we have mentioned in Section 1.1.1, instead of R one can use the longitudinal structure function F_L. The relation between these two functions is

$$R(x, Q^2) = \frac{F_L(x,Q^2) + (4M^2x^2/Q^2)F_2(x,Q^2)}{F_2(x,Q^2) - F_L(x,Q^2)} \simeq \frac{F_L(x,Q^2)}{F_2(x,Q^2)} , \tag{18}$$

and, within pQCD, the structure function F_L is linked to the singlet quark distribution q^{SI} and the gluon distribution G through the *Altarelli-Martinelli* relation [40],

$$F_L(x,Q^2) = \frac{\alpha_S(Q^2)}{2\pi} x^2 \int_x^1 \left[\frac{8}{3} F_2(t,Q^2) + 4 N_f C_2 (1 - \frac{x}{t}) t G(t,Q^2)\right] \frac{dt}{t^3} , \tag{19}$$

where $F_2 \simeq C_2\, x\, q^{SI}$, i.e. $C_2 = 1$ for neutrino structure functions and $C_2 = \sum_f e_f^2 \,/\, N_f$ for charged lepton ones.

This constitutes an absolute prediction of pQCD for the ratio R, from the x dependence of the structure function F_2 (easily measured) and of the gluon distribution G (obtained indirectly from scaling violations at low x). In Eq. 19, the glue contribution is numerically important at low x only.

3.1.4 QCD sum rules

Here, we shall only list three of the main sum rules on the quark and gluon distributions, that can be derived within the parton model and pQCD.

• One of the most fundamental sum rules is certainly the "momentum" sum rule which simply ensures that the four-momentum carried by all the quarks and gluons is equal to the one of the nucleon,

$$\int_0^1 \left(x\, q^{SI}(x) + x\, G(x)\right) \mathrm{d}x = 1 \quad . \tag{20}$$

This sum rule is always valid at LO, where q^{SI} is defined as proportional to F_2^{SI}/x. At NLO, we have seen that the definition of q^{SI} depends on the factorisation scheme. Only the effective quark and gluon distributions $q^{SI}(x, Q^2)$ and $G(x, Q^2)$ defined in the $\overline{\mathrm{MS}}$ factorisation scheme verify the momentum sum rule at any Q^2 value.

• Another important sum rule is the *Gross–Llewellyn-Smith* sum rule [52], dealing with the non-singlet structure function xF_3 measured in neutrino scattering, which reflects the fact that the number of valence quarks in the nucleon is 3,

$$\int_0^1 x\, F_3(x)\, \frac{\mathrm{d}x}{x} = \int_0^1 \sum_f (q_f(x) - \bar{q}_f(x))\, \mathrm{d}x = 3\,(1 - \frac{\alpha_S}{\pi} + \cdots) \quad . \tag{21}$$

The α_S/π correction takes into account the definition of the effective quark distributions in the $\overline{\mathrm{MS}}$ factorisation scheme.

• The last sum rule we shall mention here is the *Gottfried* sum rule [53], dealing with the non-singlet structure function $F_2^p - F_2^n$, the difference of F_2 measured on proton and on neutron in charged lepton scattering. This sum rule expresses the fact that there is one more u valence quark than d valence quark in the proton and is only valid under the assumption that the seas of u and d quarks in the proton are equal ($\bar{u} = \bar{d}$). It is written as

$$\int_0^1 (F_2^p(x) - F_2^n(x))\, \frac{\mathrm{d}x}{x} = \frac{1}{3} \int_0^1 \Big(u_v(x) - d_v(x) + 2\,(\bar{u}(x) - \bar{d}(x))\Big)\, \mathrm{d}x = \frac{1}{3} \ , \tag{22}$$

where u_v and d_v stand for the distributions of valence u and d quarks.

3.2 QCD Fits in Practice

In this section we shall deal with technical items that can play an important role when analysing, in the framework of QCD, structure function measurements.

3.2.1 QCD Fits on Structure Function Data

Performing a "QCD fit" on a set of structure function measurements consists, first, of choosing (somehow arbitrarily) two fixed values of Q^2, Q_0^2 and Q_1^2, within the measured Q^2 range and often taken to be equal, and then, in finding through an iterative procedure both

- the values of the coefficients of three parametrisations $\tilde{q}^{SI}(x)$, $\tilde{q}^{NS}(x)$ and $\tilde{G}(x)$ describing the x dependence of the effective quark and gluon distributions $q^{SI}(x,Q_0^2)$, $q^{NS}(x,Q_0^2)$ and $G(x,Q_0^2)$, at $Q^2 = Q_0^2$, and
- the value of α_S, $\alpha_S(Q_1^2)$, at $Q^2 = Q_1^2$,

that optimise the agreement, over the whole x and Q^2 domain, between the measured structure functions and the ones that can be computed from the parametrisations, $\tilde{q}^{SI}(x)$, $\tilde{q}^{NS}(x)$, $\tilde{G}(x)$ and the value $\alpha_S(Q_1^2)$ using Eqs. 9 to 15. More precisely, one obtains these "computed" structure functions as follows. First, the Q^2 evolution of α_S (from $\alpha_S(Q_1^2)$) is derived through the renormalisation group equation (Eq. 15). Then, from $\alpha_S(Q^2)$ and the parametrised x dependencies of the parton distributions at Q_0^2, the Q^2 evolution of these parton distributions is obtained through the evolution equations (Eqs. 12 to 14) in order to get two-dimensional parton distributions. Finally, the values of the computed structure functions are obtained, at each measured (x, Q^2) point, from $\alpha_S(Q^2)$ and the two-dimensional parton distributions through the Eqs. 9 to 11.

In such QCD fits, the x dependence of the quark distributions (i.e. $\tilde{q}^{SI}(x)$ and $\tilde{q}^{NS}(x)$) is mainly constrained by the x dependence of the measured structure functions, while α_S and the x dependence of the gluon distribution (i.e. $\tilde{G}(x)$) are constrained *only* by the Q^2 evolution of the measurements.

In the case of QCD fits to purely non-singlet structure functions, the gluon distribution plays no role so that the *only* free parameter having an influence on the Q^2 evolution is α_S. It is actually so also for singlet structure functions at large x. Indeed, we have mentioned in Section 2.1.1 that, in the evolution equation of the singlet quark distribution (Eq. 13), the gluon convolution term becomes negligible for $x >\sim 0.3$ so that the equation governing the Q^2 evolution of q^{SI} becomes almost identical to the one governing the evolution of q^{NS} (Eq. 12). So, through QCD fits, the scaling violations of any structure function at large x ($x >\sim 0.3$) or of a non-singlet structure function over the whole x range, provide a measurement of α_S, while the scaling violations of a singlet structure function at small x yield more a determination of the gluon distribution in the nucleon.

3.2.2 Testing the Significance of QCD Fits

Such QCD fits on structure functions data may appear rather "straightforward". However, to get a meaningful result (e.g. on a determination of α_S), many precautions are in order. In particular, there are no predictions at all from pQCD on the x dependence of the quark distributions; in other terms, Eqs. 9 to 15 just link the

Q^2 evolution of structure functions to their x dependencies but do not constrain at all these x dependencies. So, it is important, when performing QCD fits, to use parametrisations $\tilde{q}^{SI}(x)$, $\tilde{q}^{NS}(x)$ (and, when needed, $\tilde{G}(x)$) as versatile as possible, i.e. in practice containing enough free parameters (ultimately, as many as the number of x bins of the fitted data), so that they can reproduce with good precision any measured x dependence of structure functions (at any Q^2 value). If this condition is not fulfilled, it may happen that, for example, because of the convolution terms in Eqs. 10 and 11 (factors of $\alpha_S(Q^2)$), there exists a specific value of $\alpha_S(Q_1^2)$ for which the (rigid) assumed analytical shape of the parametrisations $\tilde{q}^{SI}(x)$ and $\tilde{q}^{NS}(x)$ corresponds better to the actual measured x dependence of the data. This specific value of $\alpha_S(Q_1^2)$ will then be favoured in the fit, leading to a biased result. One can test that the chosen parametrisations are versatile enough by checking that the results of the fit are independent of the chosen value of Q_0^2, as indeed they must be.

In any case, these QCD fits must contain many free parameters (typically between 10 and 30). As most of these parameters are independent of perturbative QCD, it is important to check whether, in the result of a QCD fit on some measured structure function F, the features most specific to pQCD are in good agreement with the data. These features of course deal with the Q^2 evolution and there is a simple way to quantify this Q^2 evolution and to compare it to the QCD predictions. Indeed, in the non-singlet evolution equation Eq. 12 (or in the singlet one for large x - see above), the convolution term $q^{NS} \bullet P^{NS}$ (or $q^{SI} \bullet P_{qq}$) is numerically nearly proportional to $q^{NS}(x)$ (or $q^{SI}(x)$). So, the logarithmic derivative $\mathrm{d}\ln q/\mathrm{d}\ln Q^2$ ($q \equiv q^{NS}$ or q^{SI}) of the quark distribution and, consequently, the logarithmic derivative $\mathrm{d}\ln F/\mathrm{d}\ln Q^2$ of the structure function are nearly proportional to $\alpha_S(Q^2)$ with an x dependent proportionality coefficient that depends only slightly on the x dependence of the measured structure function F. In other terms, in each measured x bin x_i, one has,

$$\frac{\mathrm{d}\ln F(x_i, Q^2)}{\mathrm{d}\ln Q^2} \simeq \alpha_S(Q^2)\, K(x_i) \quad , \tag{23}$$

where the coefficient $K(x_i)$ is essentially given by pQCD with just a slight dependence on the shape along x of $F(x, Q_0^2)$. The test of perturbative QCD then consists in checking whether a single value of $\alpha_S(Q_1^2)$ is able to describe all the measured logarithmic derivatives, over the whole x range for a non-singlet structure function or at large x for any other one.

In practice one computes, in each measured x bin, the logarithmic derivative $\mathrm{d}\ln F/\mathrm{d}\ln Q^2$, averaged over Q^2, for the data as well as for the "computed" structure function corresponding to the result of the QCD fit. One may restrict, in each x bin, the Q^2 range over which the average logarithmic derivatives are computed by a cut $Q^2 > Q^2_{min}$ to get rid of a kinematical region where non-logarithmic effects (like higher-twists - see below) which do not obey the *Altarelli-Parisi* equations may blur the validity of pQCD. These measured and "predicted" averaged logarithmic derivatives may then be compared along x. If a good agreement is observed over the whole x range, in the case of a non-singlet structure function, or at least for $x >\sim 0.3$, in the case of any other one, it provides both a proof that the value of $\alpha_S(Q_1^2)$ obtained in the fit is meaningful (the log. derivatives are proportional to α_S) and a significant test of perturbative QCD (all the log. derivatives correspond to the same α_S). One can remark here that this test, contrary to many other QCD tests, is free of any uncertainty in the description of hadronisation.

For singlet structure functions, the averaged logarithmic derivatives at small x are dependent on the gluon distribution as well as on α_S. One can show that, numerically, the differences between the measured derivative at $x = x_i$ and the one expected from pQCD when assuming no glue at all ($G \equiv 0$) are linked to the integral of $xG(x)$ from x_i to 1. The agreement between measured and predicted logarithmic derivatives at small x proves essentially that the gluon distribution obtained through the fit is meaningful.

These general considerations on QCD fits are schematically summarised in Fig. 16. Points representing a hypothetical measurement of a structure function F are displayed in the $\ln F$ vs. $\ln Q^2$ plane for different x bins x_i. Dashed curves represent the "computed" F resulting from a QCD fit on these data. After Q^2 cuts, that may not be the same in all x bins, the Q^2 averaged logarithmic derivatives are computed from both the measured and computed F values. These average derivatives are symbolized by solid straight lines fitted to the measured points or to the dashed curves. Their values are plotted along x at the bottom of the figure distinguishing between the two cases of a non-singlet or a singlet structure function F. The measured derivatives (points) agree with the predicted ones (solid curves) for the value of α_S resulting from the fit. To visualize the sensitivity of the measurements to α_S, the derivatives that would be predicted for other values of α_S are indicated by dashed curves.

3.2.3 QCD Fits and the Momentum Sum Rule

When one computes a structure function at a given point (x_0, Q_0^2) from some assumed parton distributions through Eqs. 9 to 11, only the values of these parton distributions for $x \geq x_0$ are needed. So, a QCD fit on some structure function measurement, for which the lowest x bin is x_{min}, cannot constrain in any way the parton distributions for $x < x_{min}$. It is then clear that, when performing a QCD fit on structure function data - containing singlet contributions - for which x_{min} is not "very low" (in practice, larger than a few percents), it may be "dangerous" to force the fitted parton distribution parametrisations $\tilde{q}^{SI}(x)$ and $\tilde{G}(x)$ to verify the momentum sum rule. Indeed, if one does so, one implicitly assumes that the x behaviour of these parton distributions between 0 and x_{min} is the one extrapolated from the parametrisations fitted on the data for $x \geq x_{min}$ with the *a-priori* chosen analytical form of these parametrisations. But, the actual behaviour of these quark and gluon distributions for $x < x_{min}$ may be quite different from the assumed simple analytical ones, especially in view of the divergences that are likely to occur when $x \to 0$. So, the fraction of momentum carried by the partons for $x < x_{min}$ may actually be different (larger or smaller) than the one given by the integral of $x\tilde{q}^{SI}(x) + x\tilde{G}(x)$ between 0 and x_{min}. If this contribution of the "unseen" region ($0 < x < x_{min}$) to the total momentum is non-negligible (i.e. in practice if $x_{min} > \sim 0.02$), it is clear that imposing the momentum sum rule on the fitted parametrisations may induce a bias in the result of the fit. In principle, one should also worry about the "unseen" large x region $x > x_{max}$ as this region generally covers a wider x range than the low x one. In practice, however, taking into account the fact that the parton distributions are expected to decrease monotonously and regularly (roughly as $(1-x)^\eta$) at large x, the

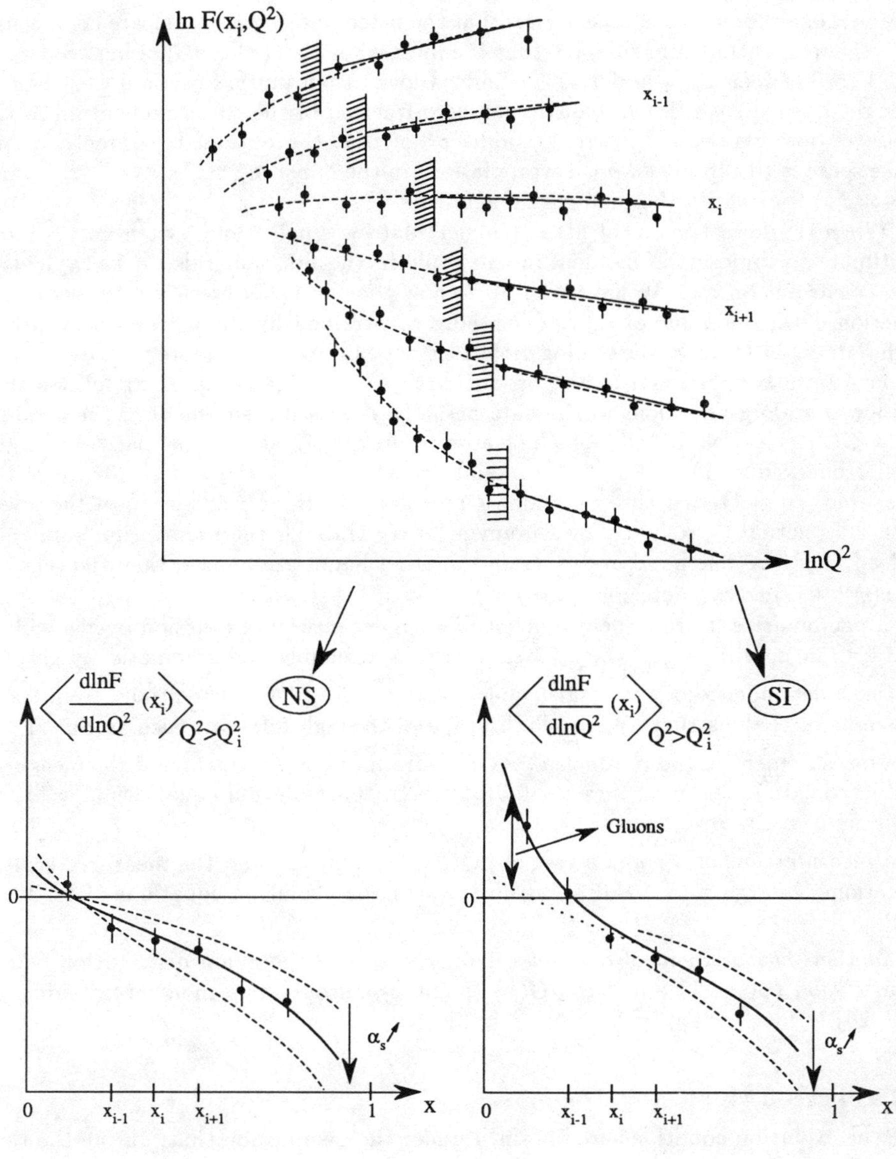

Fig. 16. Principles of QCD tests and measurements from structure function data.

fraction of the total momentum carried by the parton distributions for $x > x_{max}$ is usually smaller and better known than the one corresponding to $x < x_{min}$.

So, it appears "safer" to perform QCD fits on (singlet) structure function data corresponding to x_{min} larger than a few percents *without* imposing the momentum sum rule on the parton distribution parametrisations to be fitted. In these cases, however, one should check *a posteriori* that the fitted parametrisations are compatible with the momentum sum rule, i.e. that the integral of $x\tilde{q}^{SI}(x) + x\tilde{G}(x)$ between x_{min} and 1 (or between x_{min} and x_{max}) is indeed lower than unity. It should even be low enough to leave room for a realistic non-zero fraction of the total momentum in the "unseen" low x region. In fact, a simple proof that the obtained parametrisations make sense is to exhibit *ad-hoc* extrapolations in the "unseen" regions ($x < x_{min}$ and $x > x_{max}$) that verify the momentum sum rule.

When the lowest x bin of fitted (singlet) data is small enough, it becomes more legitimate to impose the momentum sum rule in the fits, and this, in fact, yields a new constraint on α_S. We have seen up to now that, in QCD fits of singlet structure function data, the value of α_S is essentially constrained by the Q^2 evolution of the high x data. At lower x, the scaling violations depend, more and more with decreasing x, on a term proportional to the product $\alpha_S \times G \bullet P_{qG}$ (Eq. 13). It follows that the low x scaling violations of the data actually yield a measurement of the product $\alpha_S \times G(x)$ (i.e. not directly of $G(x)$ as was previously stated for clarity). As the singlet quark distribution $q^{SI}(x)$ is mainly constrained by the x dependence of the measured (singlet) structure function, it appears that the Q^2 evolution of the low x data does actually provide a measurement of α_S through the momentum sum rule. Indeed, if q^{SI} on one hand and $\alpha_S G$ on the other hand, are known, then the relation $\int_0^1 x(q^{SI} + G)\mathrm{d}x = 1$ determines α_S.

To summarize, from a measurement of a singlet structure function over a wide x and Q^2 range, with, e.g. $x_{min} < 0.01$, one can in principle and schematically get,

1) the x dependence of the singlet quark distribution $q^{SI}(x)$, essentially from the x dependence of the structure function itself, and through relations such as Eq. 11,

2) a measurement of the product $\alpha_S \times G(x)$, from $q^{SI}(x)$ (point 1) and the measured scaling violations at small x ($x <\sim 0.3$), through the evolution equations (Eqs. 12 to 14),

3) a measurement of α_S *and* a test of pQCD, from $q^{SI}(x)$ and the measured scaling violations at large x ($x >\sim 0.3$), again through the evolution equations (Eqs. 12 to 14),

4) another measurement of α_S *and* a determination of the gluon distribution $G(x)$, from $q^{SI}(x)$ (point 1) and $\alpha_S \times G(x)$ (point 2), through the momentum sum rule (Eq. 20).

3.2.4 Higher Twists

The evolution equations are obtained under the assumption that, during the time of the interaction, the quark on which the lepton scatters does not interact with the other partons inside the nucleon ("spectator" partons). It is not possible to compute perturbatively in QCD the exact correction terms to the evolution equations induced by the parton-parton interactions. It is known, however, (see, e.g. [54]) that these terms, called "higher twists", vary like a power series in $1/Q^2$ and are thus expected to become negligible at "high enough" Q^2, i.e. above certain Q^2 thresholds that may depend on x. These thresholds cannot be quantified within pQCD but they are usually considered to be between a few and a few tens of GeV^2; experimental results have actually confirmed these estimates (see 2.3.2).

Measurements of structure functions with fixed-target experiments often extend down to such values of Q^2. It is then wise to take into account the effect of possible higher twist terms in the QCD analyses of these data. In most cases, this is done by applying a low Q^2 cut on the data, and in studying the variation of the results of QCD fits for different Q^2 cuts. In some cases, however, the effect of the first terms of the higher twist series ("twist-4") are explicitly taken into account with coefficients that are fitted on the data themselves. More precisely, the function that is fitted to the data is,

$$F^{HT}(x,Q^2) = F^{LT}(x,Q^2)\left(1+\frac{C(x)}{Q^2}\right) , \tag{24}$$

where F^{LT} obeys the pQCD Q^2 evolution according to the *Altarelli-Parisi* equations and $C(x)$ is a parametrisation of the x dependence of the twist-4 coefficient that is fitted to the data (together with $\tilde{q}^{SI}(x)$, $\tilde{q}^{NS}(x)$, $\tilde{G}(x)$, and α_S). In order not to bias the result with any *a-priori* chosen analytical form for $C(x)$, one may choose to use one free parameter C_i for *each* measured x bin, x_i. Of course, this is possible only if the accuracy of the data is high enough and the covered Q^2 range large enough to allow a significant determination of as many higher twist coefficients as there are x bins.

3.2.5 Energy Scale Uncertainties

As mentioned in Section 2.1.2, the energy scale of the effective coupling constant that appears in the evolution equations (in front of the convolution integrals in Eq. 12 to 14) is not well determined. It is related to Q^2 but may *a-priori* be anything like $\mu^2 = kQ^2$. This induces a "theoretical uncertainty", which is actually due to the limited number of terms in the perturbative expansion (NLO) of the splitting functions $P(t) = P_0(t) + \alpha_S(\mu^2)P_1(t) + \cdots$ (where $P \equiv P^{NS}$, $P_{qq}, \ldots$) and of the coefficient functions $f(t)$. Indeed, the complete perturbative expansion (i.e. with an infinite number of terms) does not depend on such a scale (i.e. on k). Another scale uncertainty arises from the limited number of terms in the factorisation-scheme dependent relation that connects the measured structure function to the effective parton distributions (Eq. 10 and 11).

A way to study such uncertainties is to perform, on the same data set, several QCD fits for different values of the two parameters k_1 and k_2 entering the equations used in the fits, as indicated below for the simpler case of a non-singlet evolution. (The coefficient k_1 is connected to the factorisation scale while k_2 is connected to the renormalisation scale.)

$$F(x,Q^2) = C\int_x^1 q(t,k_1Q^2)\left[\delta(\frac{x}{t}-1)+\frac{\alpha_S(k_2Q^2)}{2\pi}\left(f(\frac{x}{t})-P_0(\frac{x}{t})\ln k_1\right)\right]\frac{\mathrm{d}t}{t} , \tag{25}$$

$$\frac{\mathrm{d}q(x,M^2)}{\mathrm{d}\ln M^2} = \frac{\alpha_S(k_2M^2)}{2\pi}\int_x^1 q(z,M^2)\left[P_0(\frac{x}{z})+ \right.$$
$$\left. +\alpha_S(k_2M^2)\left(P_1(\frac{x}{z})+\frac{\beta_0}{4\pi}P_0(\frac{x}{z})\ln k_2\right)\right]\frac{\mathrm{d}z}{z}\,. \tag{26}$$

From the spread of the resulting values of α_S (at any chosen energy scale) when k_1 and k_2 are varied around unity (without degrading too much the quality of the fit), a theoretical scale uncertainty on the α_S determination can be estimated.

3.2.6 Possible Refinements of QCD Fits

Here, we shall pragmatically go through the list of technical items that may have some influence on the result of QCD fits (and on their validity). These items can be distributed in three classes according to whether they deal with experimental, phenomenological or theoretical matters.

On the *experimental* side, the only important technicality concerns
• the systematic errors and their point-to-point correlations. As already mentioned, uncertainties in structure function measurements are often dominated by systematic errors. These errors may arise from a number of different sources, but in general it is possible to estimate a pattern of point-to-point totally correlated errors corresponding to each source. It is then important to take into account in the QCD fits these correlations, or at least the ones of the dominant systematic errors. Indeed, the wide-spread procedure which attempts to take into account systematics by using in the fits total errors (i.e. statistical and systematic added in quadrature, assumed to be point-to-point uncorrelated) in place of the statistical ones often leads to biased results. These biases concern not only the errors on the measured parameters (like α_S) but often also the *central values* of these parameters.

On the *phenomenological* side, many of the most important technicalities have already been discussed. They concern
• the versatility of the parton distribution parametrisations (see 2.2.2),
• the higher twist terms (see 2.2.4),
• the constraint of the momentum sum rule, imposed or not on the parton distributions to be fitted (this refers to singlet structure functions only - see 2.2.3),
• and the "Fermi motion" and/or nuclear effects corrections that may be applied to heavy target data. In principle, such corrections should *not* be applied for QCD fits. One can easily show that, as long as nucleons retain their integrity inside a nucleus (no "delocalised" partons), then, whatever their motion in the nucleus, the measured structure functions of this nucleus still obey the evolution equations [55]. In fact, the evolution equations to be used should be slightly modified; the upper boundary of the integrals in Eqs . 12 to 14 should be moved from 1 to A (A being the number of nucleons in the nucleus). However, while it is true that measured values of x can in principle reach A, the measured structure functions are so small for $1 < x \leq A$ (see, e.g. [24]) that the contributions to the integrals from this $x > 1$ region are negligible. In any case, we have seen that the predicted Q^2 evolution of a structure function depends only slightly on its actual x shape. So, any nuclear effect correction will hardly modify the x dependence of a structure function enough to make it correspond

to a significantly different Q^2 evolution. In practice, nuclear effect corrections are essentially irrelevant for QCD fits.

On the *theoretical* side finally, several technical points have also to be considered.

• The order of the pQCD computations of the splitting and coefficient functions used in the fits. One has to realize that, as far as the α_S determination is concerned, results obtained at LO are in principle meaningless because they cannot be compared to results obtained from other processes. Indeed, at LO, neglected terms are of the same order as the measured ones and are different from one process to the other; in other words, the energy scale uncertainties (see, 2.2.5) are as large as the measured quantities. By using computations at NLO (e.g [48]), these scale uncertainties are reduced by one order of magnitude. Given the recent improvement in the experimental accuracies, they nevertheless remain the dominant source of uncertainty in present α_S determinations from DIS. So, computations at NNLO would indeed be welcome; in view of the present level of experimental errors, such computations would allow to reduce the present error on the α_S determination by a factor ~ 2 in one go (!). Recently the impressive computations of the various coefficient functions at NNLO were published [49]; unfortunately, for a consistent use of these, the very arduous NNLO computations of the splitting functions are still missing.

• In the case of QCD fits to the structure function F_2, one has to check the consistency that must exist (at least at large Q^2 - see 1.3.3) between the assumption on the function $R(x,Q^2)$ made for the extraction of F_2 and the QCD prediction, $R_{QCD}(x,Q^2)$ (see 2.1.3), that can be computed from the parton distributions and the value of α_S resulting from the fit. Of course, the most consistent treatment is to use, at each step of the fitting procedure, F_2 values corresponding to R_{QCD} computed from the current parton distributions and α_S value.

• The "target mass corrections" (TMC). These corrections, that take into account the non-zero value of the mass of the target nucleon, are analogous to the higher twist effects (see 2.2.4), in particular they vary like power series in $1/Q^2$. The origin of these corrections being purely kinematical, however, they can be computed within pQCD from the measured structure functions F_1 and F_2 [56], unlike higher twists. In QCD fits one can *either* apply Q^2 cuts on the data, similar to the ones discussed in the higher twist section, and forget about TMC's *or* compute these TMC's and correct the data. In the latter case, one is probably using low Q^2 data and should also incorporate higher twist terms; the main effect of the TMC's is then to reduce the magnitude of the fitted higher twist coefficient $C(x)$.

• The way the Q^2 evolution of α_S is computed. For quite some time, the mass scale parameter $\Lambda^{(i)}_{\overline{\rm MS}}$ was the quantity actually fitted in QCD fits, and the Q^2 evolution of α_S was obtained through a relation of the type of Eq. 16 (no flavour thresholds were considered). At present, the tendency is to fit directly an α_S value at a given energy scale (Q_1^2) and to compute the Q^2 evolution of α_S from the RGE (Eq. 15), which allows an easier treatment of the possible flavour thresholds (see 2.1.2). The effect of this refinement in QCD fits is presently negligible, however.

• The charm threshold in the splitting function P_{qG}. This splitting function is related to the creation of a $q\bar{q}$ pair from a gluon. The mass of the c quark being non-negligible, it induces a threshold effect that will cause the splitting function P_{cG} to be different from the one (P_{qG}) corresponding to quasi-massless quarks ($q = u$, d or s). In

particular, P_{cG}, unlike P_{qG}, should depend on Q^2. The expression of $P_{cG}(x, Q^2)$ has been computed [57] and it can be adequately used in QCD fits. The effect of this refinement on the results of the fits is usually small but for the gluon distribution at small x ($x < 0.1$).
• The last possible refinement of QCD fits we shall mention here concerns the Q^2 evolution of the small $s - c$ component of some isoscalar F_2's. It is usually assumed in QCD fits that the Q^2 evolution of isoscalar F_2 structure functions is that of a pure flavour singlet. But we have seen that isoscalar F_2's measured with charged lepton scattering contain a small non-singlet component, proportional to $s-c$ (Section 1.3.2, Eq. 7). One can specifically take into account in the QCD fits the non-singlet Q^2 evolution of this small contribution to an otherwise essentially singlet structure function. As could be expected, the effect of this refinement was proven to be negligible [63].

3.3 Results of QCD Analyses of Structure Function Data

We shall now go through the results of the QCD analyses of the most significant structure function measurements, i.e. of the high statistics results discussed in Section 1.3.

3.3.1 List of the Most Significant QCD Analyses

These analyses, all performed at NLO unless otherwise stated, correspond to:
• The QCD fits performed by the BCDMS collaboration on their heavy target result $F^C_{2\,BCDMS}$ [58]. Here, these high x data ($x \geq 0.275$) were first analysed in the non-singlet approximation (i.e neglecting the gluon distribution influence) and then with the correct singlet treatment (although the gluon distribution was of course ill-constrained). A good agreement of the data with the specific predictions of pQCD was exhibited.
* The QCD fits performed by the BCDMS collaboration on their measurements on hydrogen $F^H_{2\,BCDMS}$ [59] and on deuterium $F^D_{2\,BCDMS}$ [22]. For essentially historical reasons, high x subsets of these data ($x \geq 0.275$) were first analysed in the non-singlet approximation, before the analyses of the whole data sets with the full singlet treatment (+ non-singlet in the case of hydrogen). A QCD analysis of the purely non-singlet structure function $F^p_2 - F^n_2$ obtained from these light target results was also published [60]. A good agreement between data and the pQCD predictions was observed for all results.
• The global QCD fit performed simultaneously on SLAC and BCDMS hydrogen and deuterium F_2 data [43]. This analysis, more refined and more complete than the previous four, has demonstrated that the Q^2 evolution of this precise set of data can be well described over a wide Q^2 range by pQCD with relatively small higher twist terms. It has, in our view, superseded the BCDMS analyses on their individual light target results, that we shall consequently no longer discuss.
* The QCD fits performed by the EMC collaboration on their heavy and light targets F_2 results [14,15]. The heavy target result $F^{Fe}_{2\,EMC}$ was actually not in good agreement with the pQCD predictions (as pointed out in [58]) but the analyses of the light target results, especially the one of the re-analysed hydrogen data [61], showed that the Q^2

evolution of these data can be satisfactorily described within pQCD. However, in view of the problem affecting the EMC data (see 1.3.5) and of the relatively large errors on the results of these fits, e.g. $\alpha_S(M_Z^2) = 0.109 \pm 0.009$ (exp.), we shall no longer discuss them here.

• The QCD fit performed by the CDHSW collaboration on their measurement of the purely non-singlet structure function xF_3 [24]. (Global fits of the differential cross sections, including singlet + non-singlet treatments were also performed, but at LO only.) This analysis does not yield a satisfactory result (an apparently marginal agreement with the specific QCD predictions) because of an, in our view, important defect: the correlations of the point-to-point systematic errors were *not* taken into account in the fits (but only in subsequent estimates of the uncertainties on the fit results). A parallel unpublished analysis performed by a few members of the CDHSW collaboration [62], that explicitly takes into account the correlations of the dominant systematic errors, has yielded much more convincing results.

• The *preliminary* results of QCD fits performed by the NMC collaboration on their recent light target results [63]. Hydrogen and deuterium results have been simultaneously fitted with a complete non-singlet + singlet treatment. These data, extending down to $x = 0.008$, allow the most precise determination of the gluon distribution at low x.

• The non-singlet QCD fits performed by the CCFR collaboration on their xF_3 measurement (only preliminary and partial results on singlet QCD fits performed on F_2 are available up to now) [27]. The observed scaling violations are in good agreement with the pQCD predictions, much more clearly than in the case of CDHSW results. The accuracy on the α_S determination is boosted by substituting F_2 for xF_3 at large x.

* Finally, the global QCD fit performed simultaneously on several structure function data (BCDMS, CDHSW, prelim. NMC, prelim. CCFR,...) and on results from prompt photon and Drell-Yan experiments (WA70, E605) [45,64]. This global fit is of prime interest for its detailed results concerning a set of parametrisations of the various parton distributions. However, as far as QCD tests and the determination of α_S are concerned, this analysis does not incorporate all the refinements we have discussed in Section 2.2. Furthermore, the data used on top of structure function ones (e.g. prompt photon results) essentially provide an additional constraint on the gluon distribution at high x, i.e. a confirmation that the glue is small at large x; as a consequence, they play little role for what concerns QCD tests and α_S measurements. So, about these particular matters, this analysis does not shed more light than the analyses of the individual structure function data.

We shall now detail the results of the QCD analyses tagged with a "•".

3.3.2 Global QCD Fit on SLAC + BCDMS Hydrogen + Deuterium F_2 Data

This analysis [43] has used the two highest statistics F_2 measurements on free or quasi-free nucleons available at the time (one year ago). This allowed a wide kinematical domain ($0.07 \leq x \leq 0.85$ and $0.5 \leq Q^2 \leq 260\,\mathrm{GeV}^2$) to be covered. The computer program used performs vectorized fully numerical computations of the convolution integrals involved in the evolution equations. It has been compared to other programs (see, e.g. [58]) and extensively checked to give reliable results (for many years and within several collaborations). The QCD fit was performed on all data simultaneously with

- full NLO singlet + non-singlet treatment;
- correlations of the main systematic errors included in the fit;
- as versatile as possible quark distribution parametrisations $\tilde{q}^{SI}(x)$ and $\tilde{q}^{NS}(x)$ (see 2.2.2), i.e. each, schematically containing, as many parameters as the number of x bins (for the less constrained gluon distribution parametrisation, two parameters proved to be enough); these parametrisations correspond to $Q_0^2 = 20\,\mathrm{GeV}^2$;
- inclusion of twist-4 terms without any assumed x dependence, i.e. one fitted parameter $C(x_i)$ per x bin and target material (see 2.2.4);
- target mass corrections (TMC), the only effect of these being to reduce the magnitude of the fitted higher twist coefficients (see 2.2.6) without modifying the quality of the fit (the χ^2 value);
- no constraint from the momentum sum rule imposed, because in the covered x region partons carry only $\sim 60\%$ of the nucleon momentum, i.e. $\int_0^{0.07} x(q^{SI} + G)\mathrm{d}x \simeq 0.40$ (see 2.2.3);
- consistent assumptions on $R(x, Q^2)$, taken to be equal to R_{QCD} at large Q^2 ;
- inclusion of the charm threshold in the splitting functions (P_{cG}) and in the Q^2 evolution of α_S, directly computed from the RGE (see 2.2.6).

The main result is that the data can be very well described over the whole kinematical region by this QCD fit, as shown in Fig. 17. In particular, the measured x dependence of the scaling violations at large Q^2 is in good agreement with the specific one predicted by pQCD, as shown in Fig. 18. This constitutes a significant test of the validity of pQCD (see 2.2.2). More specifically, the results can be summarised as follows.

On the *experimental* side, the systematic offsets in the data preferred by the fit are compatible with the published systematic errors. They correspond to 0.1 to 0.3 standard deviations of relative normalisation uncertainty between SLAC and BCDMS data and to 1.2 to 1.4 standard deviations of BCDMS spectrometer calibration uncertainty. These systematic offsets are also the ones that optimise the agreement between SLAC and BCDMS measurements, without *any* theoretical assumptions.

On the *phenomenological* side, first, the fitted gluon distribution is rather unprecise, $x\tilde{G}(x) = A(1+\eta)(1-x)^\eta$ with $A = 0.40 \pm 0.07$ and $\eta = 5.5 \pm 1.5$ (at $Q_0^2 = 20\,\mathrm{GeV}^2$). Second, the fitted higher twist coefficients, shown in Fig. 19, exhibit a similar x dependence for hydrogen and deuterium; there is a general indication that these terms are smaller in the proton F_2 than in the deuteron. This tendency was actually established by an independent analysis of several F_2^n/F_2^p data sets [65]. At low x ($x < 0.20$), these higher twist coefficients are strongly correlated with the gluon distribution. In Fig. 19, they correspond to the central gluon distribution ($A = 0.40$, $\eta = 5.5$) but there exists a shape for this distribution, still compatible with the data ($A = 0.48$, $\eta = 7$), that leads to null values for the fitted higher twist coefficients at $x < 0.20$ (in that case, the dashed and solid lines in Fig. 17 become indistinguishable). Finally, as a proof of the compatibility of the data with the momentum sum rule, it is worth noticing that the fitted singlet quark distribution corresponds to $\int_0^1 x\tilde{q}^{SI}\mathrm{d}x = 0.52 \pm 0.03$ so that, combined to $\int_0^1 x\tilde{G}\mathrm{d}x = A = 0.40 \pm 0.07$, it leads to a value of the momentum sum of 0.92 ± 0.08 which is satisfactory (see 2.2.3).

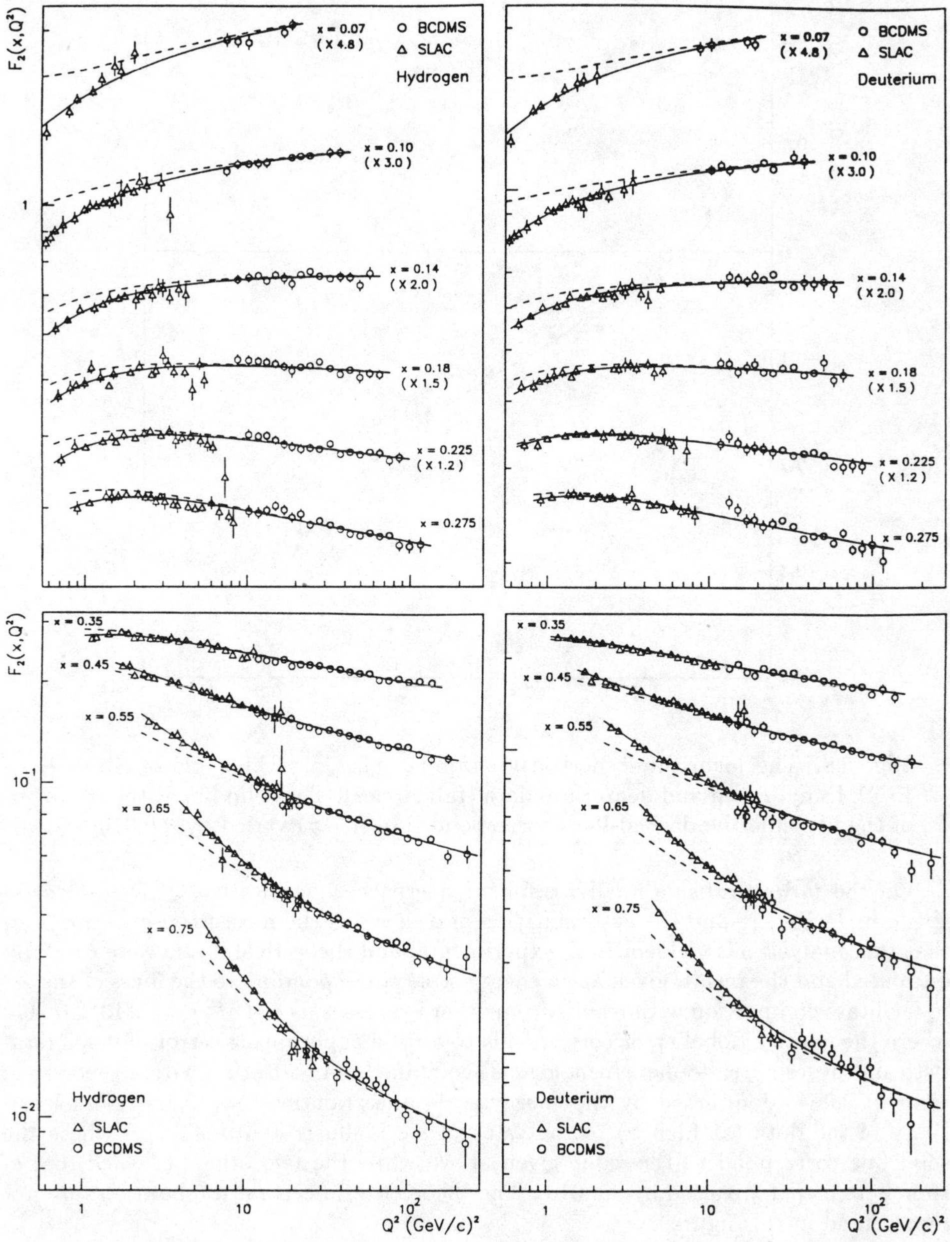

Fig. 17. QCD fit to SLAC and BCDMS hydrogen and deuterium F_2 data. The solid line is the result of the fit; the dashed line visualizes the Q^2 evolution without the higher twist effects (leading twist + TMC only).

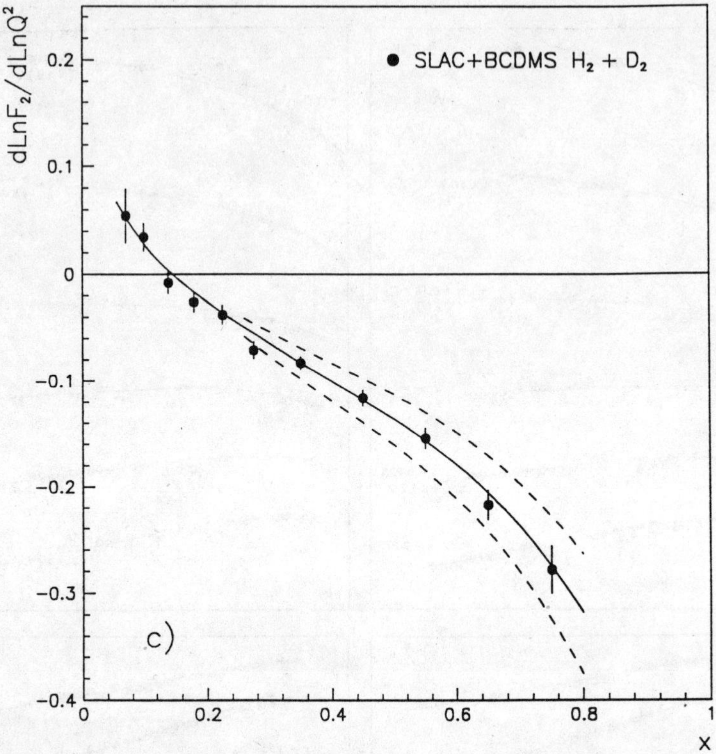

Fig. 18. The logarithmic derivatives $\mathrm{d}\ln F_2/\mathrm{d}\ln Q^2$ at high Q^2 of SLAC+BCDMS hydrogen and deuterium data (full circles). The solid line is the result of the fit while the dashed lines correspond to $\alpha_S(M_Z^2)$ varied by ± 0.010.

On the *theoretical* side finally, results concern the quantitative QCD test corresponding to Fig. 18 and the determination of α_S (one of the most precise to this day) that this analysis has yielded. Both experimental and theoretical errors were carefully estimated and the result, given at an energy scale corresponding to the mass of the Z^0 to facilitate comparison with results from other processes, is $\alpha_S(M_Z^2) = 0.113 \pm 0.005$, where the quoted global error corresponds to a total experimental error of 0.003 (statistical + systematic + phenomenological) combined in quadrature with a theoretical error of 0.004 (dominated by the energy scale uncertainties - see 2.2.5). The sensitivity of the data (at high x) to the value of α_S is illustrated in Fig. 18 where the solid line correspond to the value given above while the two others (dashed) corresponds to $\alpha_S(M_Z^2)$ varied by ± 0.010 . The theoretical uncertainties, however, are not represented in this figure.

3.3.3 QCD Fit on BCDMS Carbon F_2 Data

The $F_{2\ BCDMS}^{C}$ result is restricted to a large x large Q^2 region ($x \geq 0.275$, $Q^2 \geq 26\,\mathrm{GeV}^2$) but still constitutes one of the most precise F_2 measurement at large Q^2. The

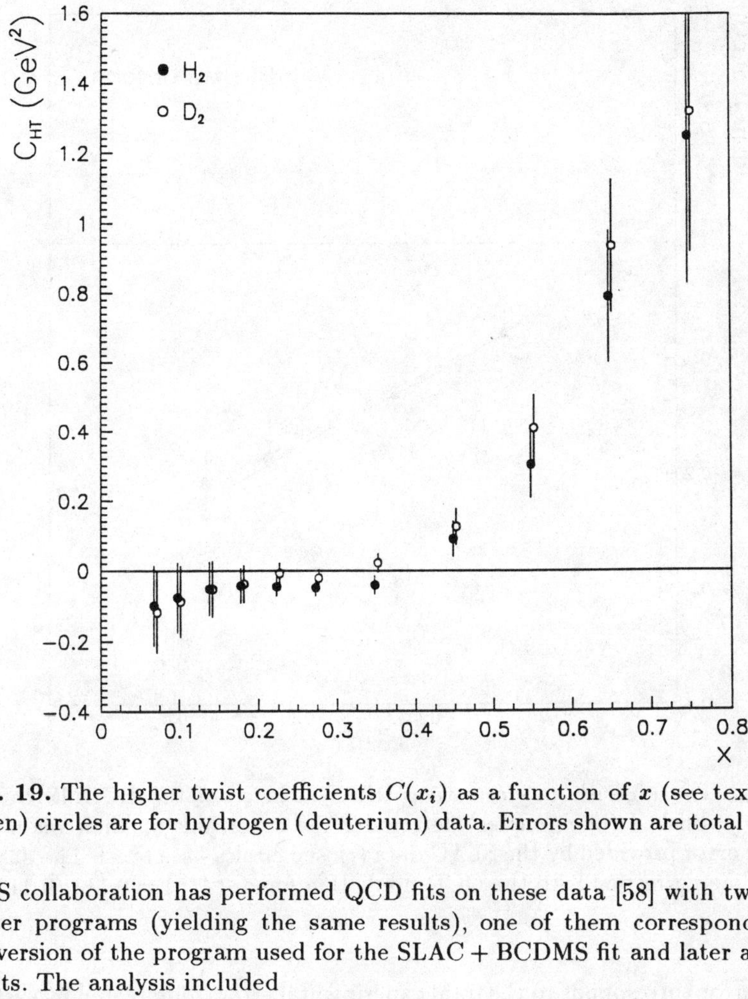

Fig. 19. The higher twist coefficients $C(x_i)$ as a function of x (see text). Full (open) circles are for hydrogen (deuterium) data. Errors shown are total errors.

BCDMS collaboration has performed QCD fits on these data [58] with two different computer programs (yielding the same results), one of them corresponding to an earlier version of the program used for the SLAC + BCDMS fit and later also for the NMC fits. The analysis included

- full NLO non-singlet or singlet treatment, the data ($x_{min} = 0.275$) being almost insensitive to the gluon distribution;
- correlations of the main systematic errors included in the fit;
- as versatile as possible quark distributions parametrisations (see previous section);
- no inclusion of higher twist terms, legitimate because of the high effective Q^2 cuts;
- no target mass corrections, legitimate for the same reason;
- of course, no constraint from the momentum sum rule imposed, because of the large x_{min};
- $R(x, Q^2)$ taken to be R_{QCD};
- inclusion of the charm threshold in the splitting functions (negligible effect).

The Q^2 evolution of the data is well described by pQCD and the main result of this analysis, $\Lambda^{(4)}_{\overline{\rm MS}} = 230 \pm 63\,\mathrm{MeV}$, translates into $\alpha_S(M_Z^2) = 0.111 \pm 0.005$,

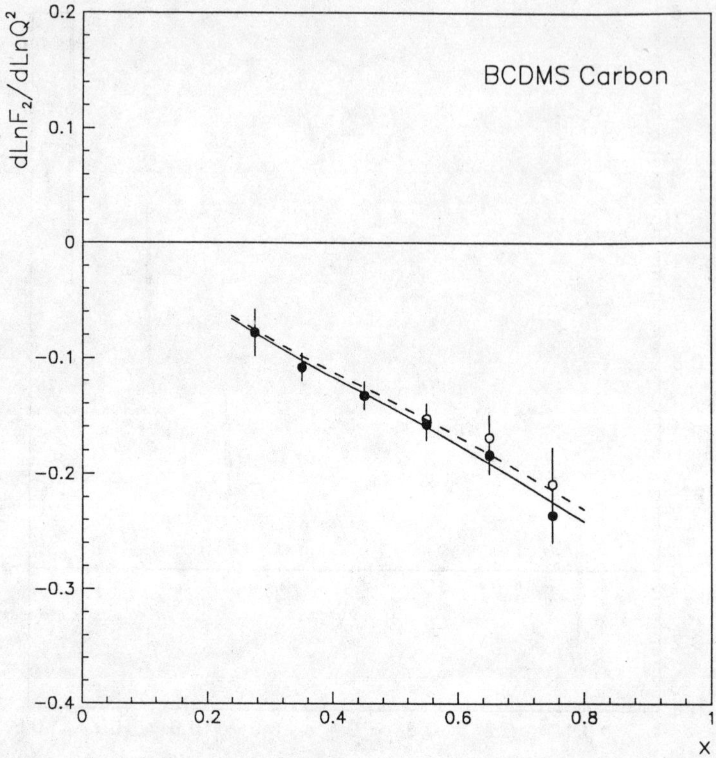

Fig. 20. The logarithmic derivatives $\mathrm{d}\ln F_2/\mathrm{d}\ln Q^2$ of the BCDMS carbon data as published (open circles) and under the constraint on their main systematic error provided by the SLAC data (closed circles - see text). The dashed (solid) line corresponds to the QCD prediction for $\alpha_S(M_Z^2) = 0.111$ (0.113).

where the error corresponds to the total experimental errors, dominated by systematic uncertainties (no theoretical uncertainties were considered).

One may wonder, however, what would happen to these BCDMS data if one constrained their main systematic uncertainty (spectrometer calibration) in the same way this error is constrained by the SLAC data for the light target results (see 1.3.5 and 2.3.2). This question makes sense because the spectrometer calibration error is likely to be the same for all data recorded in the BCDMS experiment. The answer is that, under this constraint (corresponding in average to +1.3 standard deviation of the calibration uncertainty), the agreement of the BCDMS carbon data with the QCD predictions becomes even better with a central value for $\Lambda^{(4)}_{\overline{\mathrm{MS}}}$ shifted to 260 MeV . This is illustrated in Fig. 20 where the x dependence of the measured scaling violations is compared to pQCD predictions. When using this constraint on the main systematic error of the BCDMS data, the total systematic error must be reduced accordingly, roughly from 60 to 50 MeV. It follows that the result from the BCDMS carbon F_2

data, under the constraint on their main systematic error provided by light target data, becomes $\alpha_S(M_Z^2) = 0.113 \pm 0.004$, where the error is experimental only.

3.3.4 Preliminary QCD Fits on NMC light target F_2 data

This *preliminary* QCD analysis has been performed by the NMC collaboration on their recent F_2 measurements [63]. The covered kinematical range includes, for the first time, precise measurements in the very low x region ($0.008 \leq x \leq 0.50$ and $0.7 \leq Q^2 \leq 48\text{GeV}^2$), making this result well suited for a precise determination of the gluon distribution. The computer program used is an evolved version of the one used in the SLAC + BCDMS analysis (2.3.2), specially improved for a reliable treatment of the low x region. The QCD fits were performed simultaneously on hydrogen and deuterium F_2 measurements with

- full NLO singlet + non-singlet treatment;
- correlations of the main systematic errors included in the fit, the most important one being the relative normalisation uncertainty between the data sets corresponding to 90 and 280 GeV beam energy (see 1.3.5);
- adequately versatile parton distributions parametrisations corresponding to $Q_0^2 = 7\,\text{GeV}^2$, the gluon distribution containing 3 to 4 free parameters;
- *both* no inclusion of higher twist terms *or* inclusion of twist-4 terms assumed to be equal to the ones extracted in the QCD analysis of SLAC + BCDMS data for $x > 0.20$ (see 2.3.2, Fig. 9) and extrapolated as a constant value ($-0.04\,\text{GeV}^2$) to lower x (other extrapolations were also tried);
- target mass corrections, their effect on the results of the fits being very small;
- constraint from the momentum sum rule *imposed* (!); here this constraint becomes much more legitimate (see 2.2.3) because of the narrowness of the "unseen" low x region that corresponds to only a few percents of the nucleon momentum ($\int_0^{0.008} x(q^{SI} + G)dx \simeq 0.05$);
- $R(x, Q^2)$ taken to be equal to a parametrisation [29] compatible, at large Q^2 (and also at low x), with R_{QCD};
- inclusion of the charm threshold in the splitting functions (P_{cG}) and in the Q^2 evolution of α_S, directly computed from the RGE (see 2.2.6);
- separate computation of the non-singlet Q^2 evolution of the small $s-c$ component of F_2^D (very small effect).

As shown in Fig. 21 for the deuterium measurements, the data are well described by the QCD fit. The main results can be summarised as follows.

On the *experimental* side, an offset of the relative normalisation between 90 and 280 GeV data sets of 2.4% is preferred by the fit. This value is within the expected normalisation uncertainties (of the order of 0.8 standard deviations). Furthermore it is already known that this preferred relative normalisation shift is reduced (to about 1.5%) when using the final version of the NMC data, the QCD analysis of which is currently under way.

On the *phenomenological* side, this analysis has yielded the most precise (yet preliminary) determination of the gluon distribution obtained in DIS. This gluon distribution is displayed in Fig. 22 with an error band corresponding to total errors. As a comparison, two previous determinations of this distribution (roughly scaled to correspond to the same value of Q^2, $Q_0^2 = 7\,\text{GeV}^2$), the ones from the SLAC + BCDMS

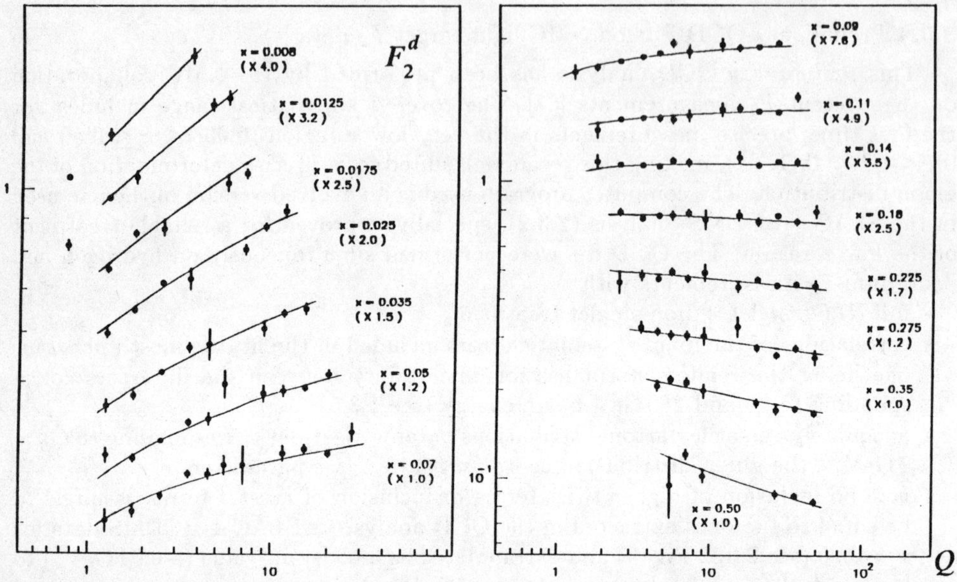

Fig. 21. Preliminary QCD fit on the NMC F_2 measurements on deuterium (see text).

analysis (2.3.2) and from the CDHSW analysis (2.3.5), are also shown. The determination from the NMC data extends to much lower x and is more precise. The chosen parametrisation for this gluon distribution is of the form $x\tilde{G}(x) = A(1-x)^{\eta}P(x)$ with $P(x)$ a two (or three) parameter function constructed such that it can only be different from unity for $x < 0.10$ (the fitted central value for the exponent η is 8.9). This result on the gluon distribution does not change much whether the value of α_S is fixed in the fit (e.g. to the value obtained in the SLAC + BCDMS analysis) or left free, provided the constraint from the momentum sum rule is kept.

It does not change much, either, whether higher twist terms (see above) are included or not in the fit, while the quality of the fit (the χ^2 value) is slightly improved by the inclusion of such terms. The predicted x dependence of the scaling violations, obtained with this gluon distribution, is in good agreement with the measured one, as shown in Fig. 23. The shaded area illustrates the large contribution from the gluons to the (positive) scaling violations at low x, i.e. the sensitivity of the F_2 measurement to the gluon distribution. Another result from this analysis concerns the measurement of the *Gottfried* sum that can be derived from the fitted effective quark distributions, $\int_0^1 (F_2^p(x) - F_2^n(x))\mathrm{d}x/x = 1/3\ (0.73 \pm 0.05 \pm 0.03)$. This result is in agreement with a previous measurement of the *Gottfried* sum from F_2^n/F_2^p data [65] and its signifi-

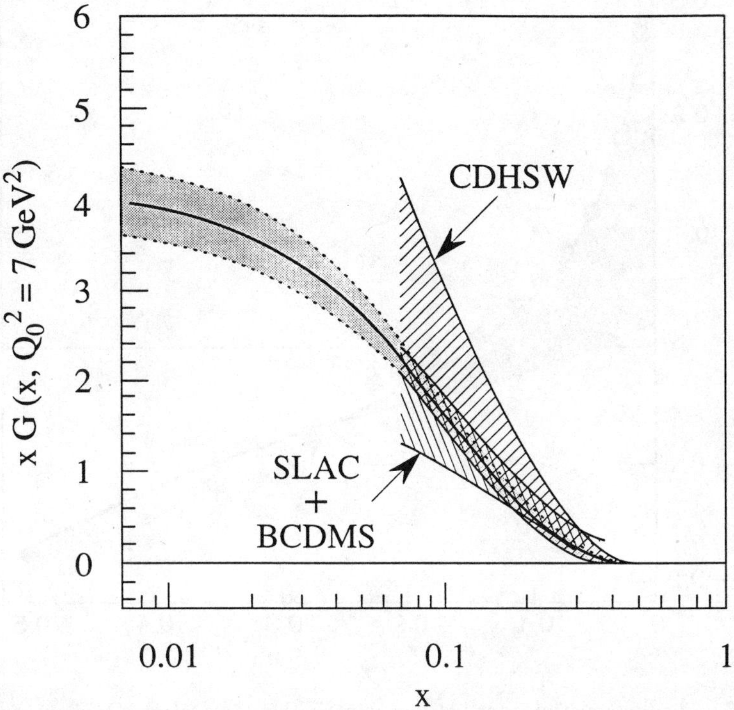

Fig. 22. The gluon distribution determined in the preliminary QCD analysis of the NMC F_2 data, compared to previous determinations in DIS.

cant departure from the value 1/3 may be explained by a non-flavour-symmetric sea ($\bar{u} \neq \bar{d}$, see 2.1.4).

On the *theoretical* side, although these NMC F_2 data are less precise for $x >$ 0.3, they allow a significant determination of α_S through the constraint given by the momentum sum rule (see 2.2.3). The preliminary result, extrapolated to M_Z^2 is $\alpha_S(M_Z^2) = 0.117 \pm 0.005$, where the error is experimental only (statistical + systematic). This result corresponds to a fit performed without any higher twist terms. The inclusion of such terms (see above) shifts the central value to $\sim$0.114 .

3.3.5 QCD Fits on CDHSW xF_3 Data

The xF_3 measurements of CDHSW cover roughly the same kinematical domain as their F_2 measurements ($0.015 \leq x \leq 0.65$ and $\sim 0.3 < Q^2 < 200\,\mathrm{GeV}^2$). In their final publication [24], the CDHSW collaboration has presented a QCD analysis of these measurements of the purely non-singlet structure function xF_3. Different computer programs were used, that all lead to similar results. The QCD fits were performed with

- full NLO non-singlet treatment;

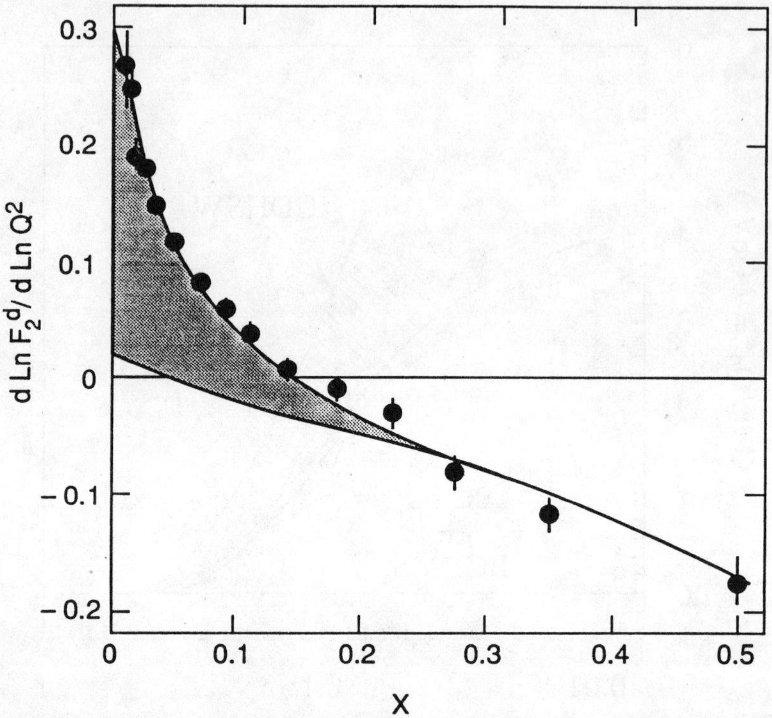

Fig. 23. The logarithmic derivatives $\mathrm{d}\ln F_2/\mathrm{d}\ln Q^2$ of the NMC hydrogen+deuterium data versus x. The upper (lower) curve correspond to the QCD prediction with (without) the contribution from gluons.

- no correlations of any systematic errors included in the fit;
- a non-singlet parton distributions parametrisation $\tilde{q}^{NS}(x)$ (at $Q_0^2 = 5\,\mathrm{GeV}^2$) containing (only) 4 free parameters;
- no inclusion of higher twist terms but various Q^2 cuts tried (from 2 to $30\,\mathrm{GeV}^2$);
- target mass corrections.

Of course, any consideration about momentum sum rule, $R(x, Q^2)$ or P_{cG} is irrelevant about the xF_3 structure function. As already stated, the result of these fits is that the data do *not* appear to show good agreement with the pQCD predictions. This is most obvious in Fig. 24 where the measured scaling violations of xF_3 (i.e. in practice the high Q^2 averaged logarithmic slopes $\mathrm{d}\ln xF_3/\mathrm{d}\ln Q^2$) are compared to the predicted ones for different values of $\Lambda_{\overline{\mathrm{MS}}}^{(4)}$ (i.e. of α_S). We recall that, as xF_3 is purely non-singlet, these scaling violations are predicted in pQCD to depend *only* on α_S over the *whole* x range. In Fig. 24, the measured Q^2 slopes are displayed for two different sets of Q^2 and ν cuts. As the predicted Q^2 slopes depend on kinematical cuts, this figure should thus contain, instead of one, two sets of curves corresponding to the different cuts; the difference between the two sets of curves are, however, likely to be small compared to the experimental errors. For $x < 0.25$, the measured Q^2 slopes are clearly above the pQCD predictions which, in this low x region, depend

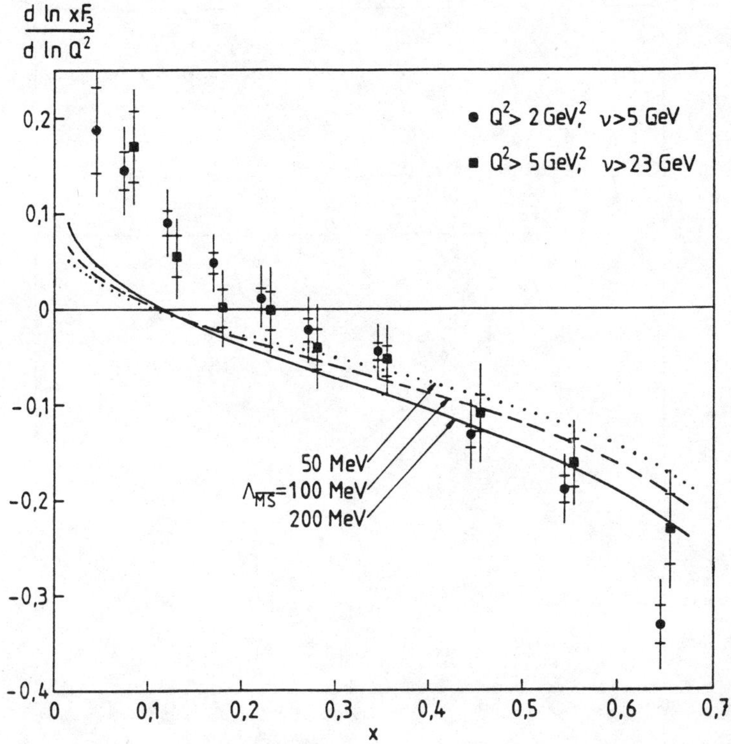

Fig. 24. The logarithmic derivatives $\mathrm{d}\ln xF_3/\mathrm{d}\ln Q^2$ of the CDHSW data (as published) compared to QCD predictions.

only weakly on α_S. As the outer error bars shown here are supposed to be total errors, this difference may appear very significant. However, this visual impression overestimates the significance of a possible discrepancy because the systematic errors on the measured slopes are greatly correlated along x. As stated in their publication, the CDHSW collaboration "*...do(es) not consider these differences to be significant*" because "*systematic errors dominate and are strongly correlated between neighbouring points...*" and they quote, as their α_S measurement, the result of fits performed over the whole x range but for $Q^2 > 10\,\mathrm{GeV}^2$ and $W^2 > 11\,\mathrm{GeV}^2$, $\Lambda^{(4)}_{\overline{\mathrm{MS}}} = 100^{+90}_{-60}\,\mathrm{MeV}$ which corresponds to $\alpha_S(M_Z^2) = 0.098 \pm 0.010$ (experimental errors only).

It seems, however, that in the region where the data agree well with the pQCD predictions, i.e. at large x, a higher value of α_S is favoured. In any case, one may wonder what would be the results of fits taking explicitly into account the high correlations in the dominant systematic errors that are invoked to explain the differences between the data and the pQCD predictions. It happens that such fits have actually been performed (in 1989) by a small subset of the CDHSW collaboration [62]. This "parallel" analysis allowed to show that the observed differences may be essentially explained, for example, by systematic offsets of 1.0 to 1.5 standard deviation on the as-

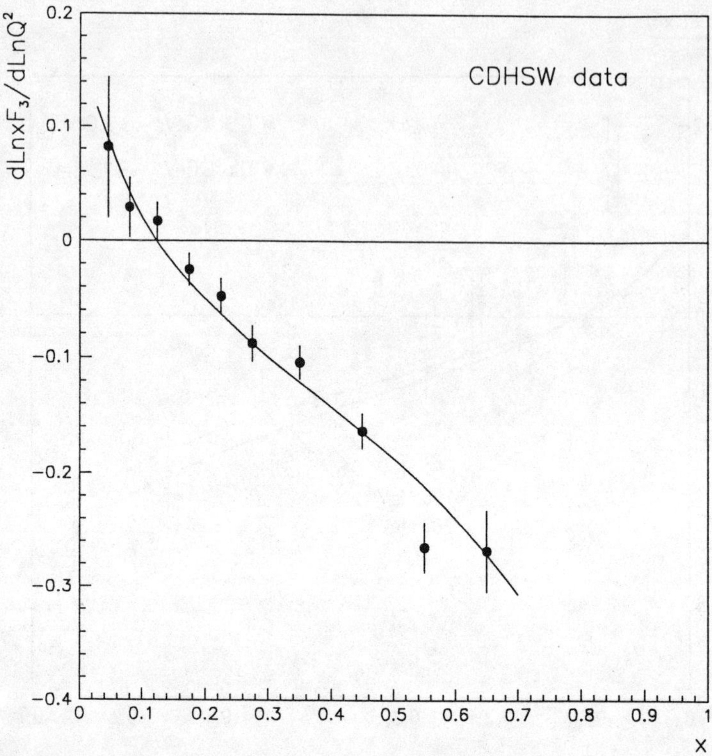

Fig. 25. The logarithmic derivatives $\mathrm{d}\ln xF_3/\mathrm{d}\ln Q^2$ of the CDHSW data resulting from an analysis taking into account the correlations in the dominant systematic errors (see text).

sumed total cross sections of both neutrinos and antineutrinos, by no means excluded (even favoured) by later determination of the total cross sections. When correlations of all dominant systematic errors (uncertainties on total cross sections, calorimeter calibration, magnetic field calibration,...) are taken into account in the fit, one gets a good agreement between data and predictions within less than 1.5 standard deviation of each systematic error, even for Q^2 cuts as low as $2\,\mathrm{GeV}^2$. This is illustrated in Fig. 25. It is interesting to remark that the offset on the calorimeter calibration (of about 1 standard deviation of its uncertainty) that is preferred in these xF_3 fits, may also account (at least partly) for the differences observed at low Q^2 between the CDHSW measurement of F_2 and the reference fit in Fig. 8.

The values of $\Lambda^{(4)}_{\overline{\mathrm{MS}}}$ obtained with such fits are much more stable against Q^2 (or x) cuts than the ones obtained with the published procedure. These values are constantly around 300 MeV for $Q^2 > 5\,\mathrm{GeV}^2$. Similar values are even obtained with lower Q^2 cuts, when simple higher twist terms are included in the fits. More precisely, the higher twists considered in this parallel analysis are twist-4 terms (see 2.2.4) with a two parameter parametrisation of the x dependence of the twist-4 coefficient, $C(x) = h_1 x^{h_2}/(1-x)$ (see Eq. 24). This parallel analysis is much more convincing, in our view, than the published one. Its results can be summarised as follows. The xF_3 data of CDHSW are in good agreement, over a wide kinematical domain ($Q^2 > 2\,\text{GeV}^2$ and $W^2 > 2\,\text{GeV}^2$), with the pQCD predictions within less than 1 to 1.5 standard deviations of their main systematic errors. They allow a determination of α_S, stable against kinematical cuts, that corresponds to $\Lambda^{(4)}_{\overline{\text{MS}}} = 300 \pm 100\,\text{MeV}$ or $\alpha_S(M_Z^2) = 0.115 \pm 0.006$, i.e. to a (more precise) result, closer to other DIS determinations (see preceding sections).

In the published analysis, an interesting result is derived from the xF_3 data in the form of a measurement, for different values of ν, of the *Gross–Llewellyn-Smith* sum (see 2.1.4). This sum is estimated at fixed values of ν instead of Q^2 because the xF_3 measurements cover the whole accessible range of x at any fixed value of ν but not at fixed values of Q^2. At high values of ν, i.e. in the region with less systematic problems, the measurement of this sum turns out to be compatible with the predicted value ($\sim$2.5 for $\Lambda^{(4)}_{\overline{\text{MS}}} \simeq 300\,\text{MeV}$ - different from the one of Eq. 21 estimated at fixed value of Q^2); it is below the prediction for lower ν.

To end this section, one can add that attempts to determine the gluon distribution (and α_S), through direct QCD fits of the differential cross sections (instead of F_2), were made in both the published and the parallel analyses. But these fits were performed at LO only and have yielded rather unprecise results for the gluon distribution (see, e.g. Fig. 22).

3.3.6 QCD Fits on CCFR xF_3 (and F_2) data

The recent and precise xF_3 measurements of CCFR cover, as for CDHSW, roughly the same kinematical domain as their F_2 measurements ($0.045 \leq x \leq 0.65$ and $1 < Q^2 < 400\,\text{GeV}^2$). The CCFR collaboration has recently presented results from a QCD analysis of these xF_3 measurements as well as from a *preliminary* analysis of the F_2 data [27]. The computer program used for the non-singlet fit of the xF_3 results included

- full NLO non-singlet treatment;
- no correlations of systematic errors included in the fit;
- a non-singlet parton distribution parametrisation $\tilde{q}^{NS}(x)$ probably adequate;
- no inclusion of higher twist terms but Q^2 cuts applied (from 5 to $15\,\text{GeV}^2$);
- target mass corrections.

The result of the fit is quite satisfactory. The data are globally well described by the result of the fit, as shown in Fig. 26 (for the fit corresponding to $Q^2 > 5\,\text{GeV}^2$) and the measured x dependence of the scaling violations is in good agreement with the specific behaviour predicted by pQCD, as shown in Fig. 27 for the fit corresponding to $Q^2 > 15\,\text{GeV}^2$. (For lower Q^2 cuts, the measured Q^2 slopes are more precise at low x but the agreement with the predicted slopes remains very good.) A particular feature of the x dependence of non-singlet Q^2 slopes predicted by pQCD is that there exists an x value for which the predicted Q^2 slopes have the same value whatever the value of α_S (i.e. do not depend on α_S - see, e.g. Fig. 24). So, for this particular x value (~ 0.12), there is a kind of "zero parameter" absolute QCD prediction on the value of the Q^2 slope (close to 0). This prediction is nicely verified by the CCFR xF_3 measurements while it was much less obvious for the CDHSW results (Fig. 24). Although such a good agreement is observed between the CCFR data and the QCD

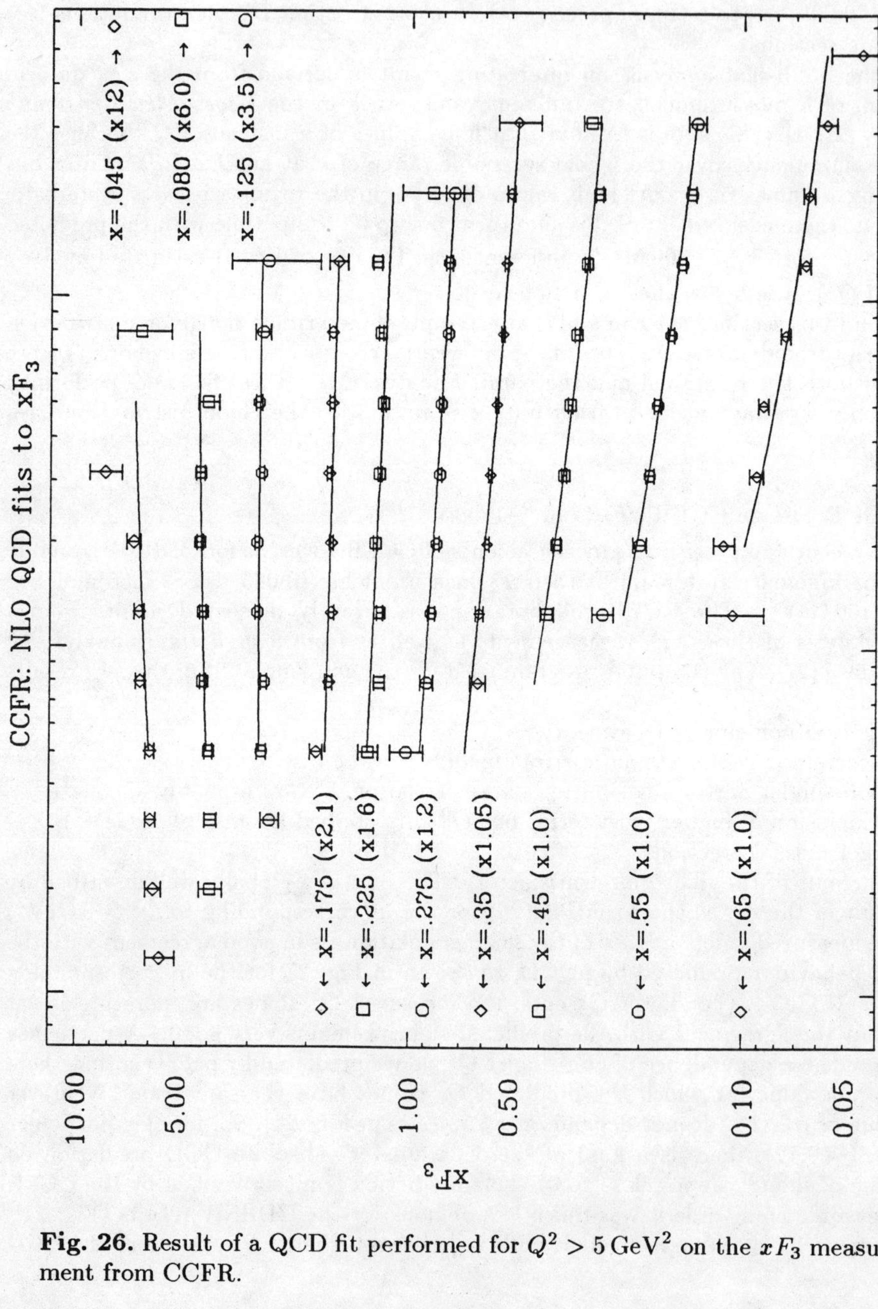

Fig. 26. Result of a QCD fit performed for $Q^2 > 5\,\mathrm{GeV}^2$ on the xF_3 measurement from CCFR.

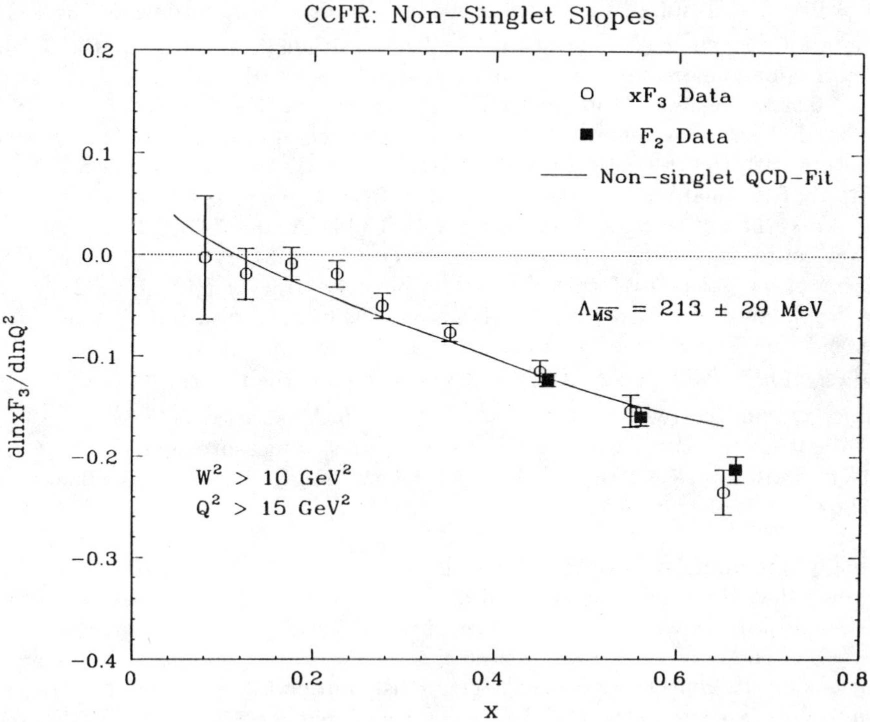

Fig. 27. The logarithmic derivatives $\mathrm{d}\ln xF_3/\mathrm{d}\ln Q^2$ of the CCFR data (open circles) compared to the QCD predictions (solid line). The F_2 derivatives (squares) are also shown for $x > 0.4$.

predictions without taking into account any correlation in the systematic errors, it would be better, e.g. for the determination of α_S, to include such correlations in the fits. The magnitude of the systematic errors, however, seems to be rather small (significantly smaller than in the case of CDHSW) so that one does not expect their correlations to have a large effect on the fitted quantities. In any case, the value of $\Lambda^{(4)}_{\overline{\mathrm{MS}}}$ obtained in these xF_3 fits is rather stable, around 180 MeV, for different Q^2 cuts.

There is a way, however, to boost the precision of this α_S measurement through non-singlet fits. Indeed, one can take advantage of the fact that the structure functions xF_3 and F_2 are expected to be identical if the sea quarks (and the gluons) can be neglected, i.e. in the large x region (see, e.g. 1.1.2, Eqs. 4 and 5). So, for $x > x_{cut}$, one can substitute the more precisely measured F_2 for xF_3 in the non-singlet fits. After different trials for x_{cut} between 0.4 and 0.6, the CCFR collaboration has chosen to quote as their result the $\Lambda^{(4)}_{\overline{\mathrm{MS}}}$ value obtained for $Q^2 > 15\,\mathrm{GeV}^2$ and with the F_2 substitution for $x > 0.5$ because this is where their measured antiquark distribution becomes totally negligible and the F_2 and xF_3 measurements totally compatible. This value, $\Lambda^{(4)}_{\overline{\mathrm{MS}}} = 213 \pm 29$ (stat.) ± 41 (syst.) MeV corresponds to the predicted Q^2 slopes in Fig. 27. The measured F_2 slopes are displayed in this figure on top of the xF_3 ones for $x > 0.4$. This result, $\Lambda^{(4)}_{\overline{\rm MS}} = 213 \pm 50\,{\rm MeV}$, translates, according to the CCFR collaboration [66], to $\alpha_S(M_Z^2) = 0.111 \pm 0.003$ (although in our view the 50 MeV error corresponds more to an error on α_S of 0.004). So, with an assumed theoretical error of 0.004, the final result of CCFR is $\alpha_S(M_Z^2) = 0.111 \pm 0.005$, where the quoted global error corresponds to the total experimental error of 0.003 combined in quadrature with the theoretical error of 0.004.

The CCFR collaboration has also derived from their xF_3 data a measurement of the *Gross–Llewellyn-Smith* sum (see 2.1.4). Unlike in the CDHSW analysis, this sum was estimated, here, at a fixed value of Q^2, $Q_0^2 = 3\,{\rm GeV}^2$. Fits of the Q^2 dependence of xF_3 were performed in each x bin to extract $xF_3(x, Q_0^2)$. The resulting x dependence of xF_3 at $Q_0^2 = 3\,{\rm GeV}^2$ was then fitted with different functional forms and the *Gross–Llewellyn-Smith* sum estimated from these fits. The result is $\int_0^1 F_3(x, Q_0^2){\rm d}x = 2.50 \pm 0.02$ (stat.) ± 0.08 (syst.) while the theoretical prediction, at the same Q_0^2 and corresponding to $\Lambda^{(4)}_{\overline{\rm MS}} = 213 \pm 50\,{\rm MeV}$ is considered to be equal to 2.66 ± 0.04. So, the CCFR collaboration claims a measurement of the *Gross–Llewellyn-Smith* sum 1.8 standard deviation below the pQCD predicted value. In our view, however, the error on the predicted value (0.04) is underestimated (Q_0^2 is rather small !).

Finally, preliminary singlet QCD fits have also been performed on the F_2 data. It appears that these data can be well described by the pQCD predictions. This is illustrated in Fig. 28 where a good agreement is observed between the measured and predicted Q^2 slopes of F_2. However, these fits seem to be *very* preliminary and no unambiguous results concerning α_S or the gluon distribution are extracted. It seems nevertheless already clear, in view of the impressive agreement between the CCFR measurements of F_2 and the result of the QCD fit performed on the SLAC + BCDMS F_2 data (see Fig. 11), that any QCD fit on these CCFR F_2 data will yield results very close to those of the SLAC + BCDMS fit (see 2.3.2).

3.3.7 Summary of the Most Significant α_S Measurements in DIS

We have seen (2.3.2) that the comprehensive QCD analysis of the SLAC and BCDMS measurements of F_2 on hydrogen and deuterium targets has yielded the result,

$$\alpha_S(M_Z^2) = 0.113 \pm 0.003\,({\rm exp.}) \pm 0.004\,({\rm theo.}) \; .$$

The result of the (published) QCD analysis of the F_2 measurement on a carbon target from BCDMS (2.3.3) is $\alpha_S(M_Z^2) = 0.111 \pm 0.005$, (exp.). However, if one applies to these carbon data the constraint on the BCDMS dominant systematic uncertainty that allowed the later appearance of the precise reanalysed F_2 data from SLAC, this result becomes $\alpha_S(M_Z^2) = 0.113 \pm 0.004$ (exp.). So, to avoid either to ignore this fact or to give too much weight to it, we choose to quote, in this summary, an average for the central value and to retain the largest experimental error, i.e.

$$\alpha_S(M_Z^2) = 0.112 \pm 0.005\,({\rm exp.}) \pm \; ? \quad ({\rm theo.}) \; .$$

The *preliminary* result of the QCD analysis of the F_2 measurements on light targets from NMC (2.3.4) is $\alpha_S(M_Z^2) = 0.117 \pm 0.005$ (exp.). However, because these data extend down to rather low Q^2, one may worry about possible higher twist effects.

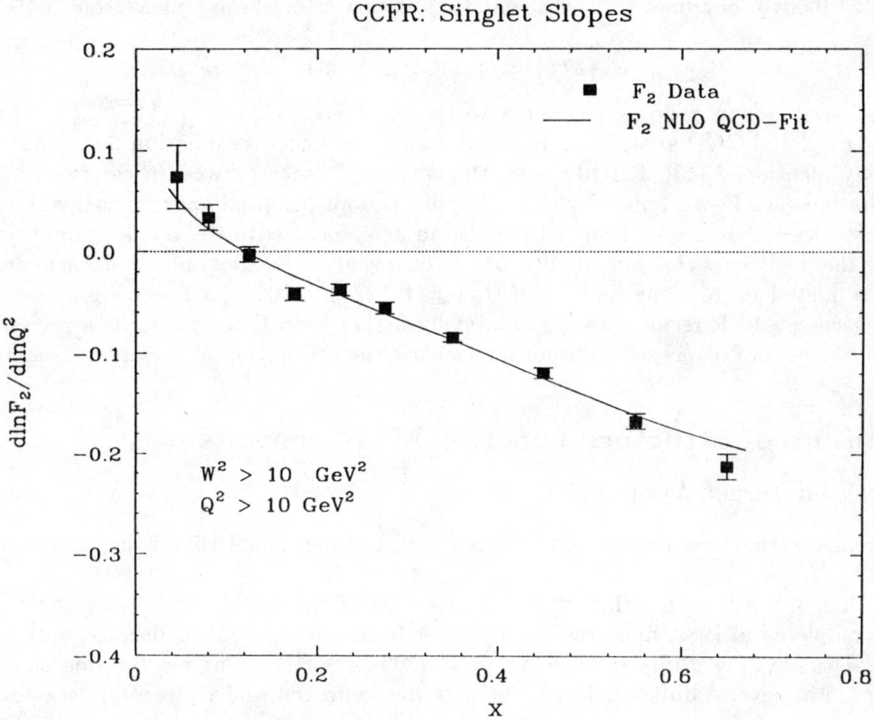

Fig. 28. The logarithmic derivatives $\mathrm{d}\ln F_2/\mathrm{d}\ln Q^2$ of the CCFR data compared to the QCD predictions from a singlet fit.

When such effects are taken into account, with a magnitude extrapolated from other measurements, the central value for α_S is shifted down to ~ 0.114. So, as for the previous result, it is reasonable to quote here a (rounded) average of these two values, with an experimental error enlarged to take into account higher twist uncertainties (of the order of 0.003),

$$\alpha_S(M_Z^2) = 0.116 \pm 0.006\,(\text{exp.}) \pm \; ? \quad (\text{theo.}) \; .$$

The result of the published QCD analysis of the xF_3 measurements from CDHSW is $\alpha_S(M_Z^2) = 0.098 \pm 0.010$ (exp.) but the result we prefer to quote here comes from another analysis that seems definitely more convincing than the published one (see 2.3.5),

$$\alpha_S(M_Z^2) = 0.115 \pm 0.006\,(\text{exp.}) \pm \; ? \quad (\text{theo.}) \; .$$

Finally, the result of the QCD analysis of the xF_3 measurements from CCFR is probably not too much sensitive to higher twists because of the Q^2 and W^2 cuts applied in the analysis. It is

$$\alpha_S(M_Z^2) = 0.111 \pm 0.003\,(\text{exp.}) \pm 0.004\,(\text{theo.}) \; .$$

Making a weighted average of all these results with the experimental errors only one gets $\alpha_S(M_Z^2) = 0.1126 \pm 0.0018$ (exp.). So, assuming a common theoretical error of 0.0040, our final (slightly rounded) estimate for the α_S measurement from DIS is

$$\alpha_S(M_Z^2) = 0.1125 \pm 0.0045 \ .$$

This value is in good agreement with the result $\alpha_S(M_Z^2) = 0.111^{+0.004}_{-0.005}$ coming from the global QCD analysis, performed mainly on structure function data, that we already mentioned [45]. Furthermore, the small difference between these two global results can even be explained by the different assumptions (and approximations) used in both cases. This result from DIS is also in agreement with the α_S determinations from the LEP experiments mainly based on event topologies and τ decays; from values quoted in [67], one finds $\alpha_S(M_Z^2) = 0.122 \pm 0.006$. (A few people wonder, however, why LEP results are systematically higher than DIS ones.) For a review of measurements of α_S as well as another view on the estimation of the errors, see [68].

4 Future of Structure Function Measurements

4.1 Fixed Target Approved Program

There are rather few approved DIS fixed target experiments that will provide new data.

In muon scattering, although the data taking of the NMC experiment at CERN was completed in 1989, more results will come from this experiment because only part of its data have been fully analysed up to now. These forthcoming results concern data recorded at several different beam energies (on hydrogen and deuterium targets) so that, on top of an extension of the measured kinematical range, measurements of the ratio $R(x, Q^2)$ are expected. The only other approved experiments are the one of the SMC collaboration which will specifically measure spin structure functions at CERN and the one from the E665 collaboration at Fermilab, which has chosen to concentrate on hadronic and nuclear studies.

In neutrino scattering, besides the result from the final QCD analysis of the CCFR F_2 measurements that is not yet available, the only new data for some time will come from an extension of the CCFR experiment recently approved, but which aims essentially at precise electroweak tests.

In electron scattering, the SLAC – E140 experiment was approved for an extended running in 1991; the main goal is to improve the measurement of R.

4.2 HERA

Indisputably, the main experimental effort in DIS will be pursued in the forthcoming years at the big electron-proton collider HERA in which the very first deep inelastic interactions have been obtained at the time of this conference.

HERA is a collider of 30 GeV electrons against 820 GeV protons where two big experiments (H1 and ZEUS) are currently starting to take data that will eventually include DIS interactions at Q^2 values up to 40000 GeV^2 (see [2]). For QCD studies, one of the main interests of this collider is that, at such high values of Q^2, one can be confident that all non-perturbative effects are totally negligible. More generally, the high centre-of-mass energy may also allow the discovery of new phenomena like the "leptoquarks" or a quark substructure. On the other hand, at these high Q^2, the cross sections are rather small and, as collider luminosities are limited compared to fixed target ones, the structure function measurements will be limited by statistical errors in part of the accessible kinematic range. This is illustrated in Fig. 29, in which is shown a simulation of the expected HERA F_2 measurements after typically two years of running at the maximum energy with 100% data-taking efficiency (luminosity of $200\,\text{pb}^{-1}$). This prediction is based on a QCD-inspired phenomenological parametrisation of $F_2(x, Q^2)$ which is also represented in the figure together with the SLAC, BCDMS and NMC hydrogen data on which it has been fitted. For the HERA simulated measurement, the errors shown are purely statistical and the kinematic range corresponds to the region where F_2 is extracted from the sole measurement of the scattered electron with systematic errors expected to remain below 10% ($0.1 < y < 0.9$). This range can be extended (to lower Q^2 values) by using also the measurement of the hadronic shower. It is nevertheless clear from Fig. 29 that, although F_2 measurements at HERA will reach amazingly low x values ($\sim 10^{-4}$), they will only be precise at low x (say for $x < 0.20$). In this region, QCD studies are complicated because the Q^2 dependence of F_2 results from an interplay of many features: α_S, the R structure function (which can be rather large), the gluon distribution (which is essentially unknown at very low x), the radiative corrections (that can be as large as 100%) ... Moreover, at very low x values ($x <\sim 0.001$), the validity of perturbative QCD is questionable. On the other hand, within pQCD, specific low x asymptotic behaviours are predicted [69] that will be most interesting to check. Actually, the sensitivity to all these effects proves that this region is very rich and as it is, for $x < 0.005$, an *uncharted* domain, the HERA data will be of the greatest interest (see also [2,3]]). (For a connection to present fixed target data, it would be worthwhile operating HERA at a reduced centre-of-mass energy.)

4.3 Desirable Extensions

Complementary to HERA, it would also be extremely valuable to use the high energy muon or neutrino beams, like the ones existing at CERN and at Fermilab or the even more energetic ones that may exist one day, e.g. at LHC, to measure the nucleon structure functions in the x range from ~ 0.003 to 1 and at Q^2 values up to $\sim 700\,\text{GeV}^2$ (or $\sim 4000\,\text{GeV}^2$ in the case of LHC) with new fixed target experiments, especially designed to lower the systematic uncertainties by one order of magnitude compared to the accuracy of present data. Different projects exist. A new generation neutrino experiment at Fermilab [70] could take advantage of the big increase in neutrino beam intensity due to the main injector improvement. Actually a first phase of such an experiment (based on the CCFR detector) has been already approved, but essentially for electroweak tests. A new muon experiment based on a superconducting air-core toroidal magnet [71] could use any existing or future muon beam to measure structure functions over a wide kinematic range with systematic errors reduced by typically a factor 10 compared to present measurements.

On top of possible improvements of QCD tests, a new high statistics neutrino experiment using hydrogen and deuterium targets would have a specific interest concerning the extraction of the separate distributions of quarks of different flavours in free or quasi-free nucleons. These important distributions are at the moment rather

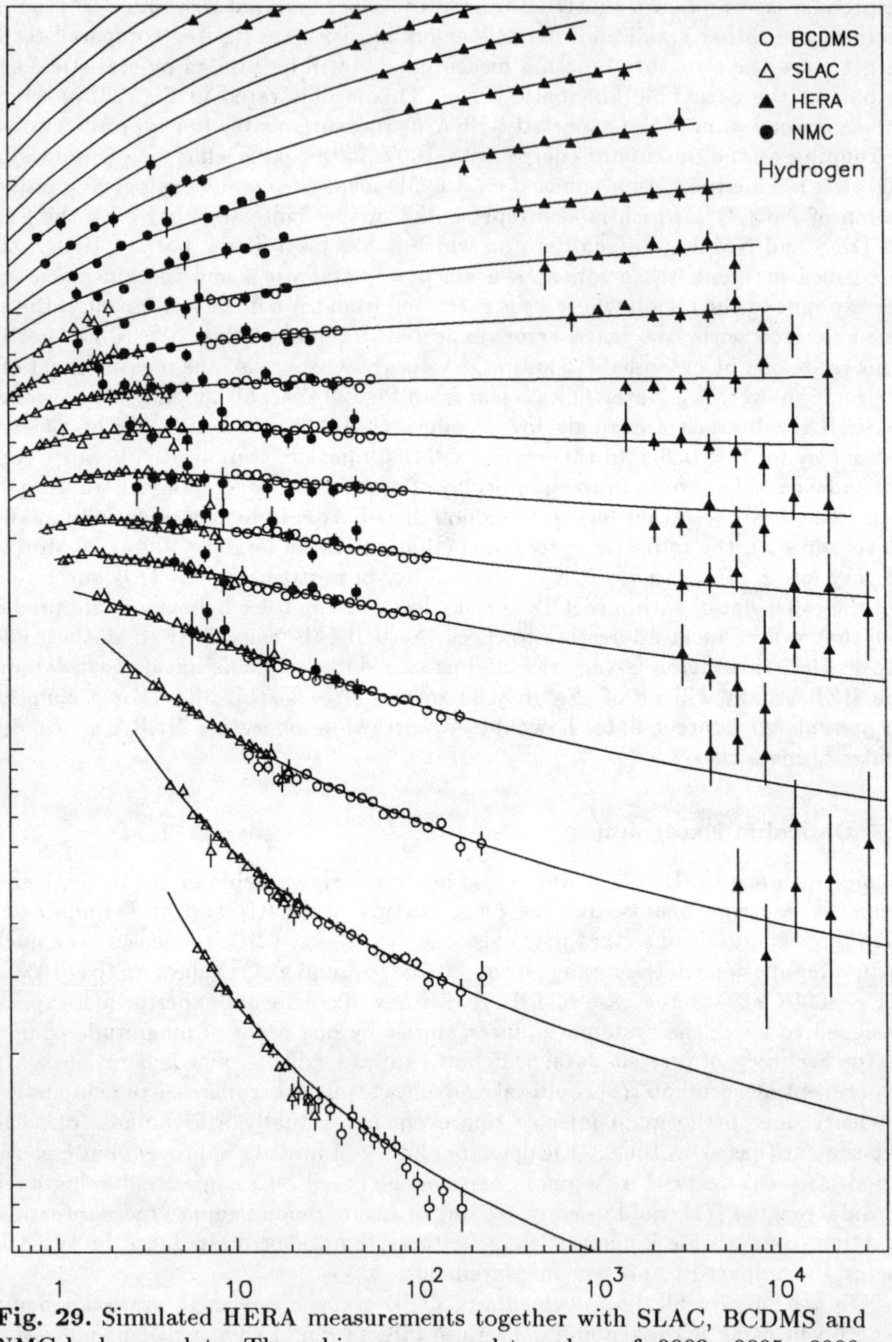

Fig. 29. Simulated HERA measurements together with SLAC, BCDMS and NMC hydrogen data. The x bins correspond to values of .0003, .0005, .0008, .0015, .003, .005, .008, .015, .03, .05, .07, .10, .14, .18, .225, .275, .35, .45, .55, .65, .75 . For $x \leq 0.005$ many HERA data points are above the figure limit.

badly known and the flavour separation can only be obtained through neutrino scattering. At present the total statistics (summed over all the experiments) of recorded events corresponding to neutrino interactions on light targets is very low (below 100 Kevents).

For QCD studies, the high accuracy needed gives an advantage to muon scattering because of the higher statistics of events it allows to gather. Among muon experiments, the principal advantages of an air-core toroid compared to other DIS experiments concern the luminosity (it can use targets more than 30 m long), the crucial high Q^2 acceptance (for example, its acceptance remains flat along Q^2 up to $\sim 700\,\mathrm{GeV}^2$ in the Fermilab muon beam) and the even more crucial control of systematics (see [71]).

Fig. 30 gives an idea of the potentialities of such an experiment, by displaying a simulation of the F_2 measurements that it can provide in one year of running with a hydrogen target. Compared to the expected HERA data, the covered kinematic range is complementary in x and roughly comparable in Q^2 if one restricts the HERA range to where the bulk of the statistics is ($x < 0.10$). The already impressive accuracy of the data over the whole x range can even be improved if one uses a denser target like carbon and, even in this case which corresponds to a huge statistical power, the apparatus is designed to keep the systematic errors at the level of the statistical ones.

In summary, by complementing the HERA data, such an experiment would extend (for $\sim$3% of the HERA cost) the very accurate measurement of the nucleon structure functions to an extremely wide kinematic range (x from 0.0001 to 1 and Q^2 up to 1000 - 5000 GeV^2). This could provide an unmatched opportunity of detailed studies of QCD.

5 Conclusions

The experimental situation concerning structure function measurements is now clear. The discrepancies that used to exist between results of different experiments, mostly EMC and BCDMS, have been traced back to specific experimental problems with the former. All the other precise measurements of structure functions, chronologically the ones from SLAC, BCDMS, NMC, and CCFR, obtained with various lepton beams (electrons, muons, neutrinos) on various targets (hydrogen, deuterium, iron, carbon), are in very good agreement (see e.g. Fig. 12!)

This consistent set of precise structure function measurements corresponds to a large kinematical domain ($0.008 \leq x \leq 0.85$, $0.5 \leq Q^2 \leq 350\,\mathrm{GeV}^2$) and allows

- → an improved determination of the various parton distributions (the numerous obsolete parametrisations, fitted on incorrect and/or less precise data, should no longer be used),
- → stringent quantitative tests of the validity of perturbative QCD, free of any uncertainty from the hadronisation processes,
- → some of the most precise determinations of the strong interaction coupling constant, typically $\alpha_S(M_Z^2) = 0.113 \pm 0.005$, in agreement with other determinations obtained at LEP from completely different processes (jet topologies, Z^0 width, τ decays). In fact, the experimental accuracy of DIS experiments has reached such a level that the errors on the α_S determinations in DIS are now

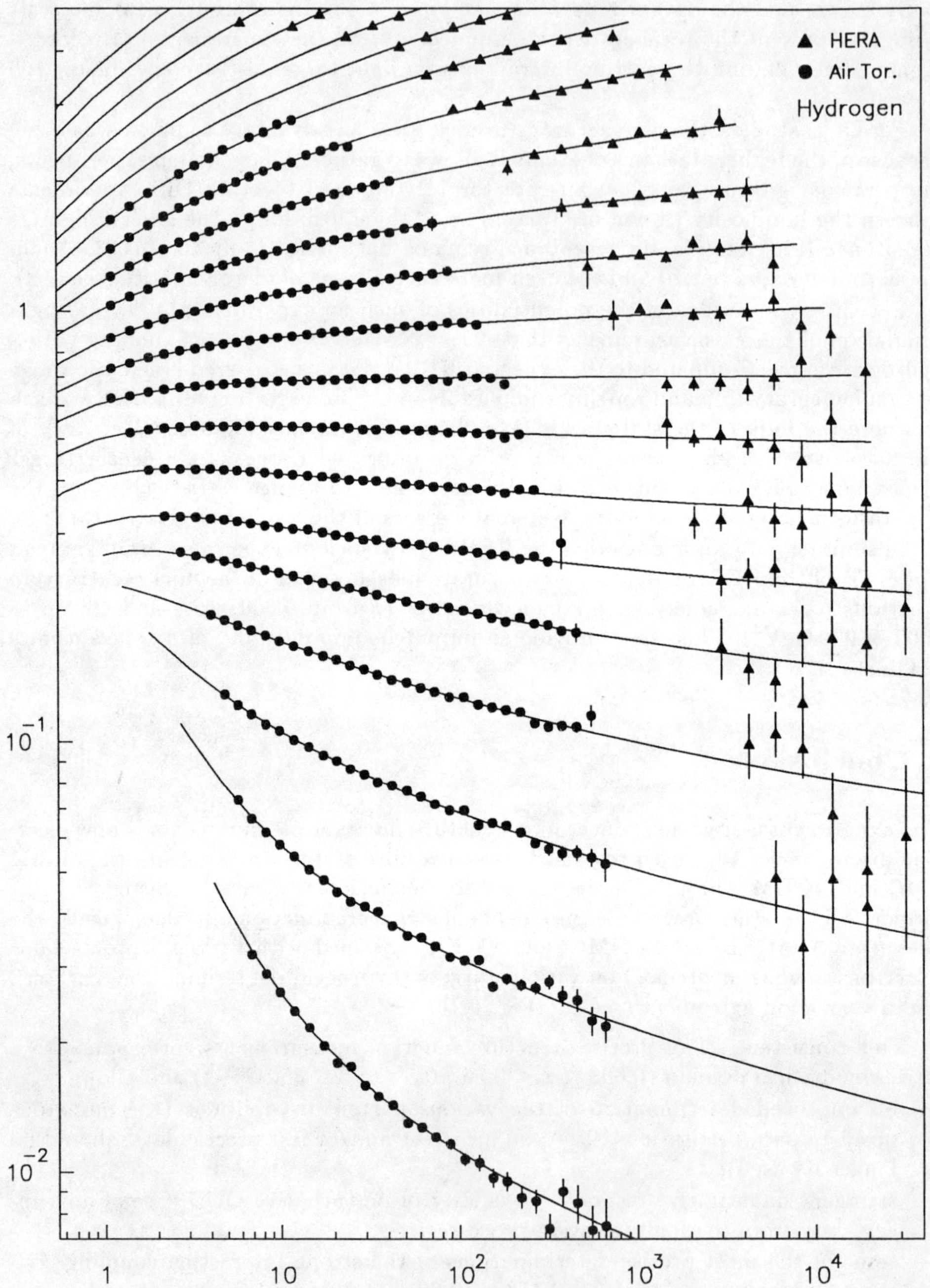

Fig. 30. Simulated measurements from HERA and from a possible new fixed target experiment (air-core toroid) in the Fermilab muon beam. The x bins are the same as in the previous figure.

dominated by theoretical uncertainties. The computation of the NNLO corrections to the splitting functions would be welcome.

In connection to those points, we think that **the description of the scaling violations of structure functions is one of the most wonderful successes of QCD**.

With the start of HERA that opens an exciting new domain at very low x and high Q^2 where, on top of many interesting measurements of expected phenomena, surprises may happen, and with (maybe) a new generation of fixed target experiments that could boost the measurement accuracy by one order of magnitude over a wide kinematical range, the future of deep inelastic scattering may be as rich as its glorious past.

Acknowledgment

It is a pleasure to thank the organisers for a very profitable and enjoyable meeting.

References

0. J. Stirling, *these proceedings.*
1. E. Reya, *these proceedings*, and preprint DO-TH-92-17 (Sep 1992).
2. G. Wolf, *these proceedings.*
3. J. Kwieciński, *these proceedings.*
4. A very nice account of this discovery is found in: R. E. Taylor, *Rev. Mod. Phys.* **63** (1991) 573; H. W. Kendall, *ibid.* 597; J. I. Friedmann, *ibid.* 615.
5. C. Chang et al., *Phys. Rev. Lett.* **35** (1975) 901;
E.M. Riordan et al., *Phys. Lett.* **B 52** (1974) 249.
6. CHARM, F. Bergsma et al., *Phys. Lett.* **B 123** (1983) 269;
CHARM, F. Bergsma et al., *Phys. Lett.* **B 141** (1984) 129.
7. see e.g. CHARM-2, A. Staude, in *Proceedings of the XXVI-th International HEP Conference* (Dallas, 1992).
8. CDHS, H. Abramowicz et al., *Z. Phys.* **C 17** (1983) 283.
9. BFP, P. D. Meyers et al., *Phys. Rev.* **D 34** (1986) 1265.
10. CHIO, B. A. Gordon et al., *Phys. Rev.* **D 20** (1979) 2645.
11. BEBC-WA25, D. Allasia et al., *Phys. Lett.* **B 135** (1984) 231;
BEBC-WA25, D. Allasia et al., *Z. Phys.* **C 28** (1985) 321.
12. BEBC-WA21, G. T. Jones et al., *Z. Phys.* **C 44** (1989) 379.
13. EMC, O. C. Allkofer et al., *Nucl. Instr. Meth.* **179** (1981) 445.
14. EMC-NA2, J. J. Aubert et al., *Nucl. Phys.* **B 272** (1986) 158.
15. EMC-NA2, J. J. Aubert et al., *Nucl. Phys.* **B 259** (1985) 189.
16. EMC-NA2, J. J. Aubert et al., *Nucl. Phys.* **B 293** (1987) 740.
17. K. Bazizi and S. J. Wimpenny, preprint UCR/DIS/91-02 (July 1991).
18. EMC-NA28, M. Arneodo et al., *Nucl. Phys.* **B 333** (1990) 1.

19. BCDMS, D. Bollini et al., *Nucl. Instr. Meth.* **204** (1983) 333;
BCDMS, A. C. Benvenuti et al., *Nucl. Instr. Meth.* **226** (1984) 330.
20. BCDMS, A.C. Benvenuti et al., *Phys. Lett.* **B 195** (1987) 91.
21. BCDMS, A.C. Benvenuti et al., *Phys. Lett.* **B 223** (1989) 485.
22. BCDMS, A.C. Benvenuti et al., *Phys. Lett.* **B 237** (1990) 592.
23. CDHS, M. Holder et al., *Nucl. Instr. Meth.* **151** (1978) 69;
CDHSW, W. von Rüden et al., *IEEE Trans. Nucl. Sci.* **NS-29** (1982) 360.
24. CDHSW, P. Berge et al., *Z. Phys.* **C 49** (1991) 187.
25. CCFRR, B. C. Barish et al., *IEEE Trans. Nucl. Sci.* **NS-25** (1978) 232;
CCFR, W. K. Sakumoto et al., *Nucl. Instr. Meth.* **A 294** (1990) 179;
CCFR, B. J. King et al., *Nucl. Instr. Meth.* **A 302** (1991) 254.
26. CCFRR, D. B. MacFarlane et al., *Z. Phys.* **C 26** (1984) 1;
CCFR, E. Oltman et al., *Z. Phys.* **C 53** (1992) 51.
27. CCFR, S. R. Mishra et al., preprint NEVIS-1465 (Jan. 1992).
28. L. W. Whitlow et al., *Phys. Lett.* **B 282** (1992) 475.
29. L. W. Whitlow et al., *Phys. Lett.* **B 250** (1990) 193.
30. NMC, P. Amaudruz et al., preprint CERN-PPE/92-124 (Jul 1992), to appear in *Phys. Lett. B.*
31. A. Milsztajn, in *Proceedings of the EPS Conference* (Uppsala, 1987);
T. Sloan, *ibid.*; R. Voss, in *Proceedings of the Lepton-Photon conference* (Hamburg, 1987).
32. A. Milsztajn et al., *Z. Phys.* **C 49** (1991) 527.
33. See Fig. 42 of: CDHSW, P. Berge et al., preprint CERN-EP/89-103 (Aug. 1989).
34. J. J. Aubert et al., *Phys. Lett.* **B 123** (1983) 275.
35. A recent review can be found in : M. Arneodo, preprint CERN-PPE/92-113 (June 1992), to appear in *Phys. Rep.* (1993).
36. L. W. Mo and Y. S. Tsai, *Rev. Mod. Phys.* **41** (1969) 205;
Y. S. Tsai, SLAC-PUB-848 (1971).
37. A. A. Akhundov et al., *Sov. J. Nucl. Phys.* **26** (1977) 660; *ibid.* **44** (1986) 988; JINR-Dubna preprints E2-10147 (1976), E2-10205 (1976), E2-86-104 (1986); D. Bardin and N. Shumeiko, *Sov. J. Nucl. Phys.* **29** (1979) 499.
38. SLAC-E140, S. Dasu et al., *Phys. Rev. Lett.* **60** (1988) 2591; *ibid.* **61** (1988) 1061.
39. B. Badelek, D. Yu. Bardin, K. Kurek and C. Scholz, in preparation, and private communications (1992).
40. G. Altarelli and G. Martinelli, *Phys. Lett.* **B 76** (1978) 89.
41. R. M. Barnett, *Phys. Rev.* **D 14** (1976) 70; see also [56].
42. B. Vallage, private communication (1992).
43. M. Virchaux and A. Milsztajn, *Phys. Lett.* **B 274** (1992) 221.
44. BCDMS, A. C. Benvenuti et al., preprint CERN-EP/87-100 (June 1987).
45. A. D. Martin, W. J. Stirling and R. G. Roberts, preprint RAL-92-021 (April 1992).
46. G. Altarelli and G. Parisi, *Nucl. Phys.* **B 126** (1977) 298;
the first reference to evolution equations is found in L. N. Lipatov, *Sov. J. Nucl. Phys.* **20** (1975) 94; see also V. N. Gribov and L. N. Lipatov, *Sov. J. Nucl. Phys.* **15** (1972) 438.

47. see, e.g. G. Altarelli, *Phys. Rep.* **81** (1982) 1.
48. G. Curci, W. Furmański and R. Petronzio, *Nucl. Phys.* **B 175** (1980) 27; W. Furmański and R. Petronzio, *Phys. Lett.* **B 97** (1980) 437; *Z. Phys.* **C 11** (1982) 293.
49. W. L. van Neerven and E.B. Zijlstra, Leiden preprint INLO-PUB-3/91; E.B. Zijlstra and W. L. van Neerven, *Phys. Lett.* **B 273** (1991) 476; *Nucl. Phys.* **B 383** (1992) 525.
50. A very clear discussion is given in : W. J. Marciano, *Nucl. Phys.* B (Proc. Suppl.) **11** (1989) 5.
51. W. J. Marciano, *Phys. Rev.* **D 29** (1984) 580.
52. D. J. Gross and C. H. Llewellyn-Smith, *Nucl. Phys.* **B 14** (1969) 337; W. A. Bardeen et al., *Phys. Rev.* **D 18** (1978) 3998.
53. K. Gottfried, *Phys. Rev. Lett.* **18** (1967) 1174.
54. see, e.g. R. K. Ellis, W. Furmański and R. Petronzio, *Nucl. Phys.* **B 212** (1983) 29.
55. C. Llewellyn-Smith, private communication (1984).
56. H. Georgi and D. Politzer, *Phys. Rev.* **D 14** (1976) 1829.
57. M. Glück, E. Hoffmann and E. Reya, *Z. Phys.* **C 13** (1982) 119.
58. BCDMS, A.C. Benvenuti et al., *Phys. Lett.* **B 195** (1987) 97.
59. BCDMS, A.C. Benvenuti et al., *Phys. Lett.* **B 223** (1989) 490.
60. BCDMS, A.C. Benvenuti et al., *Phys. Lett.* **B 237** (1990) 599.
61. K. Bazizi and S. J. Wimpenny, preprint UCR/DIS/91-03 (July 1991).
62. C. Guyot, P. Perez and B. Vallage, DAPNIA/SPP preprint, in preparation.
63. NMC, M. Botje, preprint NIKHEF H+K 92-01, and *Proceedings of the 27th Moriond Meeting "QCD and High Energy Hadronic Interactions"* (Les Arcs, 1992), Ed. Frontières.
64. A. D. Martin, R. G. Roberts and W. J. Stirling, *Phys. Rev.* **D 43** (1991) 3648; *Phys. Lett.* **B 266** (1991) 273.
65. NMC, P. Amaudruz et al., *Nucl. Phys.* **B 371** (1992) 3.
66. CCFR, M. H. Shaevitz, in *Proceedings of the 27th Moriond meeting "QCD and High Energy Hadronic Interactions"* (Les Arcs, 1992), Ed. Frontières.
67. S. Bethke, to appear in the *Proceedings of the XXVI-th International HEP Conference* (Dallas, 1992);
see also S. Bethke and S. Catani, preprint CERN-TH6484/92 (May 1992).
68. G. Altarelli, *these proceedings*, and preprint CERN-TH6623/92.
69. See, e.g. L.V. Gribov, E.M. Levin and M.G. Ryskin, *Phys. Rep.* **100** (1983) 1.
70. S. Mishra, in *"Physics at Fermilab in the 90's"* (Breckenridge, Colorado, 8/1989).
71. C. Guyot et al., *A New Fixed Target Experiment for Precise Tests of QCD (updated version)*, Saclay preprint (unnumbered, March 14, 1989).
72. NMC, P. Amaudruz et al., preprint CERN-PPE/92-134 (Aug. 1992), to appear in *Phys. Lett. B.*

The Standard Electroweak Theory And Its Experimental Tests

G. Altarelli

Theory Division, CERN, CH-1211 Geneva 23, Switzerland

1 Introduction

These lectures on electroweak interactions start with a short summary of the Glashow–Weinberg–Salam theory [1] and then cover in detail the main subjects of present interest in phenomenology: the Higgs sector and the open problem of the experimental investigation on the origin of the Fermi scale of mass $G_F^{-1/2}$; the structure of radiative corrections; the strategy for the experimental verification of the theory; and finally a discussion of the status of LEP physics. The important domain of flavour-changing processes and of CP violation is not reviewed here.

The modern electroweak theory inherits the phenomenological successes of the $(V-A)\otimes(V-A)$ four-fermion low-energy description of weak interactions [2], and provides a well-defined and consistent theoretical framework including weak interactions and quantum electrodynamics in a unified picture.

As an introduction, in the following we recall some salient physical features of the weak interactions. The weak interactions derive their name from their intensity. At low energy the strength of the effective four-fermion interaction of charged currents is determined by the Fermi coupling constant G_F. For example, the effective interaction for muon decay is given by

$$\mathcal{L}_{\rm eff} = (G_F/\sqrt{2})\,[\bar{\nu}_\mu \gamma_\alpha (1-\gamma_5)\mu]\,[\bar{e}\gamma^\alpha (1-\gamma_5)\nu_e] \ , \tag{1}$$

with [3]

$$G_F = 1.16637(2) \times 10^{-5}\ {\rm GeV}^{-2} \ . \tag{2}$$

In natural units $\hbar = c = 1$, G_F has dimensions of $({\rm mass})^{-2}$. As a result, the intensity of weak interactions at low energy is characterized by $G_F E^2$, where E is the energy scale for a given process ($E \approx m_\mu$ for muon decay). Since

$$G_F E^2 = G_F m_p^2 (E/m_p)^2 \simeq 10^{-5} (E/m_p)^2 \ , \tag{3}$$

where m_p is the proton mass, the weak interactions are indeed weak at low energies (of order m_p). The quadratic increase with energy cannot continue for ever, because it would lead to a violation of unitarity. In fact, at large energies the propagator effects can no longer be neglected, and the current–current interaction is resolved into current–W gauge boson vertices connected by a W propagator. The strength of the weak interactions at high energies is then measured by g_W, the $W-\mu\text{-}\nu_\mu$ coupling, or, even better, by $\alpha_W = g_W^2/4\pi$ analogous to the fine-structure constant α of QED. In the standard electroweak theory, we have

$$\alpha_W = \sqrt{2}\, G_F\, m_W^2/\pi = \alpha/\sin^2\theta_W \cong 1/30\ . \tag{4}$$

That is, at high energies the weak interactions are no longer so weak.

The range r_W of weak interactions is very short: it is only with the experimental discovery of the W and Z gauge bosons that it could be demonstrated that r_W is non-vanishing. Now we know that

$$r_W = \hbar/m_W c \simeq 2.5 \times 10^{-16}\ \text{cm}\ , \tag{5}$$

corresponding to $m_W \simeq 80$ GeV. This very large value for the W (or the Z) mass makes a drastic difference, compared with the massless photon and the infinite range of the QED force. The experimental limits on the photon mass [3] are listed in the following. From a laboratory experiment, one obtains $m_\gamma < 10^{-14}$ eV by a method based on the vanishing of the electric field inside a cavity with conducting walls, predicted by Gauss' law. In fact, the exact r^{-2} behaviour of the electric field corresponds to $m_\gamma = 0$. From the observed distribution of planetary magnetic fields (the field should be damped by an extra factor $e^{-m_\gamma r}$ if $m_\gamma \neq 0$) the Pioneer probe to Jupiter obtained $m_\gamma < 6 \times 10^{-16}$ eV. Finally, indirect evidence from galactic magnetic fields indicates that $m_\gamma < 3 \times 10^{-27}$ eV. Thus, on the one hand, there is very good evidence that the photon is massless. On the other hand, the weak bosons are very heavy. A unified theory of electroweak interactions has to face this striking difference.

Another apparent obstacle in the way of electroweak unification is the chiral structure of weak interactions: in the massless limit for fermions, only left-handed quarks and leptons (and right-handed antiquarks and antileptons) are coupled to W's. This clearly implies parity and charge-conjugation violation in weak interactions.

The universality of weak interactions and the algebraic properties of the electromagnetic and weak currents [the conservation of vector currents (CVC), the partial conservation of axial currents (PCAC), the algebra of currents, etc.] have been crucial in pointing to a symmetric role of electromagnetism and weak interactions at a more fundamental level. The old Cabibbo universality for the weak charged current [4]:

$$\begin{aligned} J_\alpha^{\text{weak}} &= \bar{\nu}_\mu \gamma_\alpha (1-\gamma_5)\mu + \bar{\nu}_e \gamma_\alpha (1-\gamma_5) e + \cos\theta_c\, \bar{u}\gamma_\alpha(1-\gamma_5) d\, + \\ &\quad \sin\theta_c\, \bar{u}\gamma_\alpha (1-\gamma_5) s + \dots\ , \end{aligned} \tag{6}$$

suitably extended, is naturally implied by the standard electroweak theory. In this theory the weak gauge bosons couple to all particles with couplings that are proportional to their weak charges, in the same way as the photon couples to all particles in proportion to their electric charges [in Eq. (6), $d' = \cos\theta_c\, d + \sin\theta_c\, s$ is the weak-isospin partner of u in a doublet. The (u, d') doublet has the same couplings as the (ν_e, ℓ) and (ν_μ, μ) doublets].

Another crucial feature is that the charged weak interactions are the only known interactions that can change flavour: charged leptons into neutrinos or up-type quarks into down-type quarks. On the contrary, there are no flavour-changing neutral currents at tree level. This is a remarkable property of the weak neutral current, which is explained by the introduction of the GIM mechanism [5] and has led to the successful prediction of charm.

The natural suppression of flavour-changing neutral currents, the separate conservation of e, μ and τ leptonic flavours, the mechanism of CP violation [6] through the phase in the quark-mixing matrix, are all crucial features of the Standard Model. Many examples of new physics tend to break the selection rules of the standard theory. Thus the experimental study of rare flavour-changing transitions is an important window on possible new physics.

In the following sections we shall see how these properties of weak interactions fit into the standard electroweak theory.

2 Gauge Theories

In this section we summarize the definition and the structure of a gauge Yang–Mills theory [7],[8]. We will list here the general rules for constructing such a theory. Then in the next section these results will be applied to the electroweak theory.

Consider a Lagrangian density $\mathcal{L}[\phi, \partial_\mu \phi]$ which is invariant under a D dimensional continuous group of transformations:

$$\phi' = U(\theta^A)\phi \qquad (A = 1, 2, ..., D) . \tag{7}$$

For θ^A infinitesimal, $U(\theta^A) = 1 + ig\sum_A \theta^A T^A$, where T^A are the generators of the group Γ of transformations (7) in the (in general reducible) representation of the fields ϕ. Here we restrict ourselves to the case of internal symmetries, so that T^A are matrices that are independent of the space–time coordinates. The generators T^A are normalized in such a way that for the lowest dimensional non-trivial representation of the group Γ (we use t^A to denote the generators in this particular representation) we have

$$\mathrm{tr}(t^A t^B) = 1/2\delta^{AB} . \tag{8}$$

The generators satisfy the commutation relations

$$[T^A, T^B] = iC_{ABC}T^C . \tag{9}$$

In the following, for each quantity V^A we define

$$\mathbf{V} = \sum_A T^A V^A . \tag{10}$$

If we now make the parameters θ^A depend on the space–time coordinates $\theta^A = \theta^A(x_\mu)$, $\mathcal{L}[\phi, \partial_\mu \phi]$ is in general no longer invariant under the gauge transformations $U[\theta^A(x_\mu)]$, because of the derivative terms. Gauge invariance is recovered if the ordinary derivative is replaced by the covariant derivative:

$$D_\mu = \partial_\mu + ig\mathbf{V}_\mu , \tag{11}$$

where V^A_μ are a set of D gauge fields (in one-to-one correspondence with the group generators) with the transformation law

$$\mathbf{V}'_\mu = U\mathbf{V}_\mu U^{-1} - (1/ig)(\partial_\mu U)U^{-1} . \tag{12}$$

For constant θ^A, $\mathbf{V}$ reduces to a tensor of the adjoint (or regular) representation of the group:

$$\mathbf{V}'_\mu = U\mathbf{V}_\mu U^{-1} \simeq \mathbf{V}_\mu + ig[\theta, \mathbf{V}_\mu] , \tag{13}$$

which implies that

$$V'^C_\mu = V^C_\mu - gC_{ABC}\theta^A V^B_\mu , \tag{14}$$

where repeated indices are summed up.

As a consequence of Eqs. (11) and (12), $D_\mu\phi$ has the same transformation properties as ϕ:

$$(D_\mu\phi)' = U(D_\mu\phi) . \tag{15}$$

Thus $\mathcal{L}[\phi, D_\mu\phi]$ is indeed invariant under gauge transformations. In order to construct a gauge-invariant kinetic energy term for the gauge fields V^A, we consider

$$[D_\mu, D_\nu]\phi = ig\{\partial_\mu\mathbf{V}_\nu - \partial_\nu\mathbf{V}_\mu + ig[\mathbf{V}_\mu, \mathbf{V}_\nu]\}\phi \equiv ig\mathbf{F}_{\mu\nu}\phi , \tag{16}$$

which is equivalent to

$$F^A_{\mu\nu} = \partial_\mu V^A_\nu - \partial_\nu V^A_\mu - gC_{ABC}V^B_\mu V^C_\nu . \tag{17}$$

From Eqs. (7), (15) and (16) it follows that the transformation properties of $F^A_{\mu\nu}$ are those of a tensor of the adjoint representation

$$\mathbf{F}'_{\mu\nu} = U\mathbf{F}_{\mu\nu}U^{-1} . \tag{18}$$

The complete Yang–Mills Lagrangian, which is invariant under gauge transformations, can be written in the form

$$\mathcal{L}_{\mathrm{YM}} = -\frac{1}{4}\sum_A F^A_{\mu\nu}F^{A\mu\nu} + \mathcal{L}[\phi, D_\mu\phi] . \tag{19}$$

For an Abelian theory, as for example QED, the gauge transformation reduces to $U[\theta(x)] = \exp[ieQ\theta(x)]$, where Q is the charge generator. The associated gauge field (the photon), according to Eq. (12), transforms as

$$V'_\mu = V_\mu - \partial_\mu\theta(x) . \tag{20}$$

In this case, the $F_{\mu\nu}$ tensor is linear in the gauge field V_μ so that in the absence of matter fields the theory is free. On the other hand, in the non-Abelian case the $F^A_{\mu\nu}$ tensor contains both linear and quadratic terms in V^A_μ, so that the theory is non-trivial even in the absence of matter fields.

3 The Standard Model of the Electroweak Interactions

In this section, we summarize the structure of the standard electroweak Lagrangian and specify the couplings of $W^\pm$ and Z, the intermediate vector bosons (IVBs).

For this discussion we split the Lagrangian into two parts by separating the Higgs boson couplings:

$$\mathcal{L} = \mathcal{L}_{\rm symm} + \mathcal{L}_{\rm Higgs} \,. \tag{21}$$

We start by specifying $\mathcal{L}_{\rm symm}$, which involves only gauge bosons and fermions:

$$\begin{aligned}\mathcal{L}_{\rm symm} = & -\frac{1}{4}\sum_{A=1}^{3} F^A_{\mu\nu}F^{A\mu\nu} - \frac{1}{4}B_{\mu\nu}B^{\mu\nu} + \bar{\psi}_L i\gamma^\mu D_\mu \psi_L \\ & + \bar{\psi}_R i\gamma^\mu D_\mu \psi_R \,. \end{aligned}\tag{22}$$

This is the Yang–Mills Lagrangian for the gauge group $SU(2)\otimes U(1)$ with fermion matter fields. Here

$$B_{\mu\nu} = \partial_\mu B_\nu - \partial_\nu B_\mu \quad \text{and} \quad F^A_{\mu\nu} = \partial_\mu W^A_\nu - \partial_\nu W^A_\mu - g\epsilon_{ABC}\, W^B_\mu W^C_\nu \tag{23}$$

are the gauge antisymmetric tensors constructed out of the gauge field B_μ associated with $U(1)$, and W^A_μ corresponding to the three $SU(2)$ generators; ϵ_{ABC} are the group structure constants [see Eqs. (9)] which, for $SU(2)$, coincide with the totally antisymmetric Levi-Civita tensor (recall the familiar angular momentum commutators). The normalization of the $SU(2)$ gauge coupling g is therefore specified by Eq. (23).

The fermion fields are described through their left-hand and right-hand components:

$$\psi_{L,R} = [(1\mp\gamma_5)/2]\psi, \quad \bar{\psi}_{L,R} = \bar{\psi}[(1\pm\gamma_5)/2] \,, \tag{24}$$

with γ_5 and other Dirac matrices defined as in the book by Bjorken–Drell [9]. In particular, $\gamma_5^2 = 1, \gamma_5^\dagger = \gamma_5$. Note that, as given in Eq. (24),

$$\bar{\psi}_L = \psi_L^\dagger \gamma_0 = \psi^\dagger[(1-\gamma_5)/2]\gamma_0 = \bar{\psi}[\gamma_0(1-\gamma_5)/2]\gamma_0 = \bar{\psi}[(1+\gamma_5)/2] \,.$$

The matrices $P_\pm = (1\pm\gamma_5)/2$ are projectors. They satisfy the relations $P_\pm P_\pm = P_\pm, P_\pm P_\mp = 0, P_+ + P_- = 1$.

The sixteen linearly independent Dirac matrices can be divided into γ_5-even and γ_5-odd according to whether they commute or anticommute with γ_5. For the γ_5-even, we have

$$\bar{\psi}\Gamma_E\psi = \bar{\psi}_L\Gamma_E\psi_R + \bar{\psi}_R\Gamma_E\psi_L \qquad (\Gamma_E \equiv 1, i\gamma_5, \sigma_{\mu\nu}) \,, \tag{25}$$

whilst for the γ_5-odd,

$$\bar{\psi}\Gamma_O\psi = \bar{\psi}_L\Gamma_O\psi_L + \bar{\psi}_R\Gamma_O\psi_R \qquad (\Gamma_O \equiv \gamma_\mu, \gamma_\mu\gamma_5) \,. \tag{26}$$

In the Standard Model the left and right fermions have different transformation properties under the gauge group. Thus, mass terms for fermions (of the form $\bar{\psi}_L\psi_R$ + h.c.) are forbidden in the symmetric limit. In particular, all ψ_R are singlets in the minimal Standard Model. But for the moment, by ψ_R we mean a column vector, including all fermions in the theory that span a generic reducible representation of $SU(2)\otimes U(1)$. The standard electroweak theory is a chiral theory, in the sense that ψ_L

and ψ_R behave differently under the gauge group. In the absence of mass terms, there are only vector and axial vector interactions in the Lagrangian that have the property of not mixing ψ_L and ψ_R. Fermion masses will be introduced, together with $W^\pm$ and Z masses, by the mechanism of symmetry breaking. The covariant derivatives $D_\mu\psi_{L,R}$ are explicitly given by

$$D_\mu\psi_{L,R} = \left[\partial_\mu + ig\sum_{A=1}^{3} t^A_{L,R}W^A_\mu + ig'\frac{1}{2}Y_{L,R}B_\mu\right]\psi_{L,R} \,, \tag{27}$$

where $t^A_{L,R}$ and $1/2Y_{L,R}$ are the $SU(2)$ and $U(1)$ generators, respectively, in the reducible representations $\psi_{L,R}$. The commutation relations of the $SU(2)$ generators are given by

$$[t^A_L, t^B_L] = i\,\epsilon_{ABC}t^C_L \quad \text{and} \quad [t^A_R, t^B_R] = i\epsilon_{ABC}t^C_R \,. \tag{28}$$

We use the normalization (8) [in the fundamental representation of $SU(2)$]. The electric charge generator Q (in units of e, the positron charge) is given by

$$Q = t^3_L + 1/2\,Y_L = t^3_R + 1/2\,Y_R \,. \tag{29}$$

Note that the normalization of the $U(1)$ gauge coupling g' in (27) is now specified as a consequence of (29).

All fermion couplings to the gauge bosons can be derived directly from Eqs. (22) and (27). The charged-current (CC) couplings are the simplest. From

$$\begin{aligned} g(t^1W^1_\mu + t^2W^2_\mu) &= g\left\{[(t^1+it^2)/\sqrt{2}](W^1_\mu - iW^2_\mu)/\sqrt{2}] + \text{h.c.}\right\} \\ &= g\left\{[(t^+W^-_\mu)/\sqrt{2}] + \text{h.c.}\right\} \,, \end{aligned} \tag{30}$$

where $t^\pm = t^1 \pm it^2$ and $W^\pm = (W^1 \pm iW^2)/\sqrt{2}$, we obtain the vertex

$$V_{\bar\psi\psi W} = g\bar\psi\gamma_\mu\left[(t^+_L/\sqrt{2})(1-\gamma_5)/2 + (t^+_R/\sqrt{2})(1+\gamma_5)/2\right]\psi W^-_\mu + \text{h.c.} \tag{31}$$

In the neutral-current (NC) sector, the photon A_μ and the mediator Z_μ of the weak NC are orthogonal and normalized linear combinations of B_μ and W^3_μ:

$$\begin{aligned} A_\mu &= \cos\theta_W B_\mu + \sin\theta_W W^3_\mu \,, \\ Z_\mu &= -\sin\theta_W B_\mu + \cos\theta_W W^3_\mu \,. \end{aligned} \tag{32}$$

Equations (32) define the weak mixing angle θ_W. The photon is characterized by equal couplings to left and right fermions with a strength equal to the electric charge. Recalling Eq. (29) for the charge matrix Q, we immediately obtain

$$g\,\sin\theta_W = g'\cos\theta_W = e \,, \tag{33}$$

or equivalently,

$$\text{tg}\,\theta_W = g'/g. \tag{34}$$

Once θ_W has been fixed by the photon couplings, it is a simple matter of algebra to derive the Z couplings, with the result

$$\Gamma_{\bar\psi\psi Z} = g/(2\,\cos\theta_W)\bar\psi\gamma_\mu[t^3_L(1-\gamma_5) + t^3_R(1+\gamma_5) - 2Q\sin^2\theta_W]\psi Z^\mu \,, \tag{35}$$

where $\Gamma_{\bar{\psi}\psi Z}$ is a notation for the vertex. In the minimal Standard Model, $t^3_R = 0$ and $t^3_L = \pm 1/2$.

In order to derive the effective four-fermion interactions that are equivalent, at low energies, to the CC and NC couplings given in Eqs. (31) and (35), we anticipate that large masses, as experimentally observed, are provided for $W^\pm$ and Z by $\mathcal{L}_{\mathrm{Higgs}}$. For left–left CC couplings, when the momentum transfer squared can be neglected with respect to m^2_W in the propagator of Born diagrams with single W exchange, from Eq. (31) we can write

$$\mathcal{L}^{\mathrm{CC}}_{\mathrm{eff}} \simeq (g^2/8m^2_W)[\bar{\psi}\gamma_\mu(1-\gamma_5)t^+_L\psi][\bar{\psi}\gamma^\mu(1-\gamma_5)t^-_L\psi] \,. \tag{36}$$

By specializing further in the case of doublet fields such as $\nu_e - e^-$ or $\nu_\mu - \mu^-$, we obtain the tree-level relation of g with the Fermi coupling constant G_F measured from μ decay [see Eq. (2)]:

$$G_F/\sqrt{2} = g^2/8m^2_W \,. \tag{37}$$

By recalling that $g \sin\theta_W = e$, we can also cast this relation in the form

$$m_W = \mu_{\mathrm{Born}}/\sin\theta_W \,, \tag{38}$$

with

$$\mu_{\mathrm{Born}} = (\pi\alpha/\sqrt{2}G_F)^{1/2} \simeq 37.2802 \text{ GeV} \,, \tag{39}$$

where α is the fine-structure constant of QED ($\alpha \equiv e^2/4\pi = 1/137.036$).

In the same way, for neutral currents we obtain in Born approximation from Eq. (35) the effective four-fermion interaction given by

$$\mathcal{L}^{\mathrm{NC}}_{\mathrm{eff}} \simeq \sqrt{2}\, G_F \rho_0 \bar{\psi}\gamma_\mu[...]\psi\bar{\psi}\gamma^\mu[...]\psi \,, \tag{40}$$

where

$$[...] \equiv t^3_L(1-\gamma_5) + t^3_R(1+\gamma_5) - 2Q\sin^2\theta_W \tag{41}$$

and

$$\rho_0 = m^2_W/m^2_Z \cos^2\theta_W \,. \tag{42}$$

All couplings given in this section are obtained at tree level and are modified in higher orders of perturbation theory. In particular, the relations between m_W and $\sin\theta_W$ [Eqs. (38) and (39)] and the observed values of ρ ($\rho = \rho_0$ at tree level) in different NC processes, are altered by computable electroweak radiative corrections, as discussed in Section 7.

The gauge-boson self-interactions can be derived from the $F_{\mu\nu}$ term in $\mathcal{L}_{\mathrm{symm}}$, by using Eq. (32) and $W^\pm = (W^1 \pm iW^2)/\sqrt{2}$. Defining the three-gauge-boson vertex as in Fig. 1, we obtain ($V \equiv \gamma, Z$)

$$\Gamma_{W^-W^+V} = ig_{W^-W^+V}[g_{\mu\nu}(q-p)_\lambda + g_{\mu\lambda}(p-r)_\nu + g_{\nu\lambda}(r-q)_\mu] \,, \tag{43}$$

with

$$g_{W^-W^+\gamma} = g \sin\theta_W = e \quad \text{and} \quad g_{W^-W^+Z} = g \cos\theta_W \,. \tag{44}$$

We now turn to the Higgs sector [10] of the electroweak Lagrangian. Here we simply review the formalism of the Higgs mechanism applied to the electroweak theory. In the next section we shall make a more general and detailed discussion of the physics of

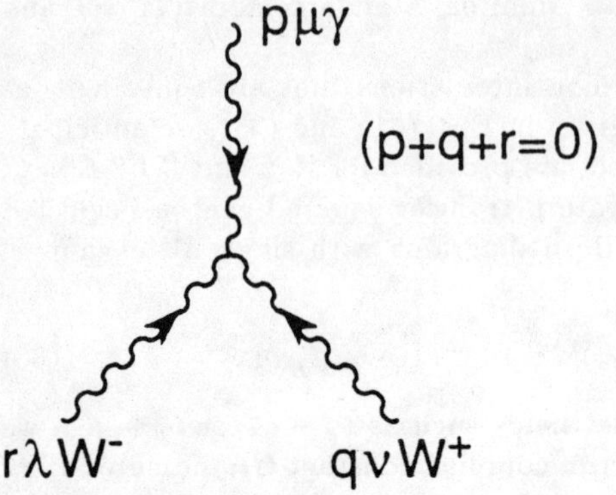

Fig. 1. The three-gauge boson vertex

the electroweak symmetry breaking. The Higgs Lagrangian is specified by the gauge principle and the requirement of renormalizability to be

$$\mathcal{L}_{\text{Higgs}} = (D_\mu\phi)^\dagger(D^\mu\phi) - V(\phi^\dagger\phi) - \bar{\psi}_L\Gamma\psi_R\phi - \bar{\psi}_R\Gamma^\dagger\psi_L\phi^\dagger \ , \tag{45}$$

where ϕ is a column vector including all Higgs fields; it transforms as a reducible representation of the gauge group. The quantities Γ (which include all coupling constants) are matrices that make the Yukawa couplings invariant under the Lorentz and gauge groups. The potential $V(\phi^\dagger\phi)$, symmetric under $SU(2)\otimes U(1)$, contains, at most, quartic terms in ϕ so that the theory is renormalizable. Spontaneous symmetry breaking is induced if the minimum of V which is the classical analogue of the quantum mechanical vacuum state (both are the states of minimum energy) is obtained for non-vanishing ϕ values. Precisely, we denote the vacuum expectation value (VEV) of ϕ, i.e. the position of the minimum, by v:

$$\langle 0|\phi(x)|0\rangle = v \neq 0 \ . \tag{46}$$

The fermion mass matrix is obtained from the Yukawa couplings by replacing $\phi(x)$ by v:

$$M = \bar{\psi}_L\ \mathcal{M}\psi_R + \bar{\psi}_R\mathcal{M}^\dagger\psi_L \ , \tag{47}$$

with

$$\mathcal{M} = \Gamma\cdot v \ . \tag{48}$$

In the minimal Standard Model, where all left fermions ψ_L are doublets and all right fermions ψ_R are singlets, only Higgs doublets can contribute to fermion masses. There are enough free couplings in Γ, so that one single complex Higgs doublet is indeed sufficient to generate the most general fermion mass matrix. It is important to observe that by a suitable change of basis we can always make the matrix $\mathcal{M}$ Hermitian, γ_5-free, and diagonal. In fact, we can make separate unitary transformations on ψ_L and ψ_R according to

$$\psi'_L = U\psi_L, \quad \psi'_R = V\psi_R \tag{49}$$

and consequently

$$\mathcal{M} \to \mathcal{M}' = U^\dagger \mathcal{M} V \, . \tag{50}$$

This transformation does not alter the general structure of the fermion couplings in $\mathcal{L}_{\rm symm}$. For quarks, the Cabibbo–Kobayashi–Maskawa [4],[11] unitary transformation relates the mass eigenstates d, s and b to the CC eigenstates d', s', b', i.e. the states coupled by W emission to u, c, and t, respectively. The NC is then automatically diagonal in flavour at tree level (GIM mechanism [5]). In the case of leptons, if the neutrinos are massless then clearly there is no mixing.

If only one Higgs doublet is present, the change of basis that makes $\mathcal{M}$ diagonal will at the same time diagonalize also the fermion–Higgs Yukawa couplings. Thus, in this case, no flavour-changing neutral Higgs exchanges are present. This is not true, in general, when there are several Higgs doublets. But one Higgs doublet for each electric charge sector i.e. one doublet coupled only to u-type quarks, one doublet to d-type quarks, one doublet to charged leptons would also be all right [12], because the mass matrices of fermions with different charges are diagonalized separately. For several Higgs doublets it is also possible to generate CP violation by complex phases in the Higgs couplings [13]. In the presence of six quark flavours, this CP-violation mechanism is not necessary. In fact, at the moment, the simplest model with only one Higgs doublet seems adequate for describing all observed phenomena.

We recall that the Standard Model, with N fermion families with the observed quantum numbers, is automatically free of γ_5 anomalies [14] owing to cancellation of quarks with lepton loops.

We now consider the gauge-boson masses and their couplings to the Higgs. These effects are induced by the $(D_\mu\phi)^\dagger(D^\mu\phi)$ term in $\mathcal{L}_{\rm Higgs}$ [Eq. (45)], where

$$D_\mu\phi = \left[\partial_\mu + ig\sum_{A=1}^{3} t^A W_\mu^A + ig'(Y/2)B_\mu\right]\phi \, . \tag{51}$$

Here t^A and $1/2Y$ are the $SU(2) \otimes U(1)$ generators in the reducible representation spanned by ϕ. Not only doublets but all non-singlet Higgs representations can contribute to gauge-boson masses. The condition that the photon remains massless is equivalent to the condition that the vacuum is electrically neutral:

$$Q|v\rangle = (t^3 + 1/2Y)|v\rangle = 0 \, . \tag{52}$$

The charged W mass is given by the quadratic terms in the W field arising from $\mathcal{L}_{\rm Higgs}$, when $\phi(x)$ is replaced by v. We obtain

$$m_W^2 W_\mu^+ W^{-\mu} = g^2|(t^+ v/\sqrt{2})|^2 W_\mu^+ W^{-\mu} \, , \tag{53}$$

whilst for the Z mass we get [recalling Eq. (32)]

$$1/2 m_Z^2 Z_\mu Z^\mu = |[g\cos\theta_W t^3 - g'\sin\theta_W (Y/2)]v|^2 Z_\mu Z^\mu \, , \tag{54}$$

where the factor of 1/2 on the left-hand side is the correct normalization for the definition of the mass of a neutral field. By using Eq. (52), relating the action of t^3 and $1/2Y$ on the vacuum v, and Eqs. (34), we obtain

$$1/2 m_Z^2 = (g\cos\theta_W + g'\sin\theta_W)^2 |t^3 v|^2 = (g^2/\cos^2\theta_W)|t^3 v|^2 \, . \tag{55}$$

For Higgs doublets

$$\phi = \begin{pmatrix} \phi^+ \\ \phi^0 \end{pmatrix}, \quad v = \begin{pmatrix} 0 \\ v \end{pmatrix}, \tag{56}$$

we have

$$|t^+ v|^2 = v^2, \quad |t^3 v|^2 = 1/4 v^2 , \tag{57}$$

so that

$$m_W^2 = 1/2 g^2 v^2, \quad m_Z^2 = 1/2 g^2 v^2 / \cos^2 \theta_W . \tag{58}$$

Note that by using Eq. (37) we obtain

$$v = 2^{-3/4} G_F^{-1/2} = 174.1 \text{ GeV} . \tag{59}$$

It is also evident that for Higgs doublets

$$\rho_0 = m_W^2 / m_Z^2 \cos^2 \theta_W = 1 . \tag{60}$$

This relation is typical of one or more Higgs doublets and would be spoiled by the existence of Higgs triplets etc. In general,

$$\rho_0 = \sum_i \left[(t_i)^2 - (t_i^3)^2 + t_i \right] v_i^2 / \sum_i 2 (t_i^3)^2 v_i^2 \tag{61}$$

for several Higgses with VEVs v_i, weak isospin t_i, and z-component t_i^3. These results are valid at the tree level and are modified by calculable electroweak radiative corrections, as discussed in Sections 7 and 8.

If only one Higgs doublet is present, then the fermion–Higgs couplings are in proportion to the fermion masses. In fact, from the Yukawa couplings $g_{\phi \bar{f} f} (\bar{f}_L \phi f_R + h.c.)$, the mass m_f is obtained by replacing ϕ by v, so that $m_f = g_{\phi \bar{f} f} v$.

With only one complex Higgs doublet, three out of the four Hermitian fields are removed from the physical spectrum by the Higgs mechanism and become the longitudinal modes of W^+, W^-, and Z. The fourth neutral Higgs is physical and should be found. If more doublets are present, two more charged and two more neutral Higgs scalars should be around for each additional doublet.

Finally, the couplings of the physical Higgs H to the gauge bosons can be simply obtained from $\mathcal{L}_{\text{Higgs}}$, by the replacement

$$\phi(x) = \begin{pmatrix} \phi^+(x) \\ \phi^0(x) \end{pmatrix} \to \begin{pmatrix} 0 \\ v + (H/\sqrt{2}) \end{pmatrix} , \tag{62}$$

[so that $(D_\mu \phi)^\dagger (D^\mu \phi) = 1/2 (\partial_\mu H)^2 + ...$], with the result

$$\begin{aligned} \mathcal{L}[H, W, Z] = & \; g^2 (v/\sqrt{2}) W_\mu^+ W^{-\mu} H + (g^2/4) W_\mu^+ W^{-\mu} H^2 \\ & + [(g^2 v Z_\mu Z^\mu)/(2\sqrt{2} \cos^2 \theta_W)] H \\ & + [g^2/(8 \cos^2 \theta_W)] Z_\mu Z^\mu H^2 . \end{aligned} \tag{63}$$

We have thus completed our summary of the standard electroweak theory and of the $W^\pm, Z$ couplings.

4 The Higgs Sector

The gauge symmetry of the Standard Model was difficult to discover because it is well hidden in nature. The only observed gauge boson that is massless is the photon. The graviton is still unobserved, even at the classical level of gravitational waves; the gluons are presumed massless but unobservable because of confinement, and the W and Z weak bosons carry a heavy mass. Actually the main difficulty in unifying weak and electromagnetic interactions was the fact that e.m. interactions have infinite range ($m_\gamma = 0$), whilst the weak forces have a very short range, owing to $m_{W,Z} \neq 0$.

The solution of this problem is in the concept of spontaneous symmetry breaking, which was borrowed from statistical mechanics.

Consider a ferromagnet at zero magnetic field in the Landau–Ginzburg approximation. The free energy in terms of the temperature T and the magnetization $\mathbf{M}$ can be written as

$$F(\mathbf{M},T) \simeq F_0(T) + 1/2\, \mu^2(T)\mathbf{M}^2 + 1/4\, \lambda(T)(\mathbf{M}^2)^2 + \dots \,. \tag{64}$$

This is an expansion which is valid at small magnetization. The neglect of terms of higher order in $\mathbf{M}^2$ is the analogue in this context of the renormalizability criterion. Also, $\lambda(T) > 0$ is assumed for stability; F is invariant under rotations, i.e. all directions of $\mathbf{M}$ in space are equivalent. The minimum condition for F reads

$$\partial F/\partial M = 0, \quad [\mu^2(T) + \lambda(T)\mathbf{M}^2]\mathbf{M} = 0 \,. \tag{65}$$

There are two cases. If $\mu^2 > 0$, then the only solution is $\mathbf{M} = 0$, there is no magnetization, and the rotation symmetry is respected. If $\mu^2 < 0$, then another solution appears, which is

$$|\mathbf{M}_0|^2 = -\mu^2/\lambda \,. \tag{66}$$

The direction chosen by the vector $\mathbf{M}_0$ is a breaking of the rotation symmetry. The critical temperature T_{crit} is where $\mu^2(T)$ changes sign:

$$\mu^2(T_{\mathrm{crit}}) = 0 \,. \tag{67}$$

It is simple to realize that the Goldstone theorem holds. It states that when spontaneous symmetry breaking takes place, there is always a zero-mass mode in the spectrum. In a classical context this can be proven as follows. Consider a Lagrangian

$$\mathcal{L} = |\partial_\mu \phi|^2 - V(\phi) \tag{68}$$

symmetric under the infinitesimal transformations

$$\phi \to \phi' = \phi + \delta\phi, \quad \delta\phi_i = i\delta\theta t_{ij}\phi_j \,. \tag{69}$$

The minimum condition on V that identifies the equilibrium position (or the ground state in quantum language) is

$$(\partial V/\partial \phi_i)(\phi_i = \phi_i^0) = 0 \,. \tag{70}$$

The symmetry of V implies that

$$\delta V = (\partial V/\partial \phi_i)\delta\phi_i = i\delta\theta(\partial V/\partial \phi_i)t_{ij}\phi_j = 0 \,. \tag{71}$$

By taking a second derivative at the minimum $\phi_i = \phi_i^0$ of the previous equation, we obtain

$$\partial^2 V/\partial\phi_k\partial\phi_i(\phi_i = \phi_i^0)t_{ij}\phi_i^0 + \frac{\partial V}{\partial\phi_i}(\phi_i = \phi_i^0)t_{ik} = 0 \ . \tag{72}$$

The second term vanishes owing to the minimum condition, Eq. (70). We then find

$$\partial^2 V/\partial\phi_k\partial\phi_i \ (\phi_i = \phi_i^0)t_{ij}\phi_j^0 = 0 \ . \tag{73}$$

The second derivatives $M_{ki}^2 = (\partial^2 V/\partial\phi_k\partial\phi_i)(\phi_i = \phi_i^0)$ define the squared mass matrix. Thus the above equation in matrix notation can be read as

$$M^2 t\phi^0 = 0 \ , \tag{74}$$

which shows that if the vector $(t\phi^0)$ is non-vanishing, i.e. there is some generator that shifts the ground state into some other state with the same energy, then $t\phi^0$ is an eigenstate of the squared mass matrix with zero eigenvalue. Therefore, a massless mode is associated with each broken generator.

When spontaneous symmetry breaking takes place in a gauge theory, the massless Goldstone mode exists, but it is unphysical and disappears from the spectrum. It becomes, in fact, the third helicity state of a gauge boson that takes mass. This is the Higgs mechanism. Consider, for example, the simplest Higgs model described by the Lagrangian

$$\mathcal{L} = -\frac{1}{4} F_{\mu\nu}^2 + |(\partial_\mu - ieA_\mu)\phi|^2 + \frac{1}{2}\mu^2\phi^*\phi - (\lambda/4)(\phi^*\phi)^2 \ . \tag{75}$$

Note the 'wrong' sign in front of the mass term for the scalar field ϕ, which is necessary for the spontaneous symmetry breaking to take place. The above Lagrangian is invariant under the $U(1)$ gauge symmetry

$$A_\mu \to A'_\mu = A_\mu - (1/e)\partial_\mu\theta(x), \quad \phi \to \phi' = \phi \exp[i\theta(x)] \ . \tag{76}$$

Let $\phi^0 = v \neq 0$, with v real, be the ground state that minimizes the potential and induces the spontaneous symmetry breaking. Making use of gauge invariance, we can make the change of variables

$$\begin{aligned} \phi(x) &\to (1/\sqrt{2})[\rho(x) + v] \exp[i\zeta(x)/v] \ , \\ A_\mu(x) &\to A_\mu - (1/ev)\partial_\mu\zeta(x). \end{aligned} \tag{77}$$

Then $\rho = 0$ is the position of the minimum, and the Lagrangian becomes

$$\mathcal{L} = -\frac{1}{4}F_{\mu\nu}^2 + \frac{1}{2}e^2v^2A_\mu^2 + \frac{1}{2}e^2\rho^2A_\mu^2 + e^2\rho v A_\mu^2 + \mathcal{L}(\rho) \ . \tag{78}$$

The field $\zeta(x)$, which corresponds to the would-be Goldstone boson, disappears, whilst the mass term $\frac{1}{2}e^2v^2A_\mu^2$ for A_μ is now present; ρ is the massive Higgs particle.

The Higgs mechanism is realized in well-known physical situations. For a superconductor in the Landau–Ginzburg approximation the free energy can be written as

$$F = F_0 + \frac{1}{2}\mathbf{B}^2 + |(\nabla - 2ie\mathbf{A})\phi|^2/4m - \alpha|\phi|^2 + \beta|\phi|^4 \ . \tag{79}$$

Here **B** is the magnetic field, $|\phi|^2$ is the Cooper pair (e^-e^-) density, $2e$ and $2m$ are the charge and mass of the Cooper pair. The 'wrong' sign of α leads to $\phi \neq 0$ at the minimum. This is precisely the non-relativistic analogue of the Higgs model of the previous example. The Higgs mechanism implies the absence of propagation of massless phonons (states with dispersion relation $\omega = kv$ with constant v). Also the mass term for **A** is manifested by the exponential decrease of **B** inside the superconductor (Meissner effect).

Thus the Higgs effect is not an abstract device, but it is endowed with a precise physical reality. In the electroweak theory it is absolutely necessary in order to give masses to the W's and Z^0 and to the fermions, as well as to ensure the correct high-energy behaviour required by renormalizability. However a more profound physical reality could be hidden behind or accompany the Higgs formalism. In fact, the Higgs mechanism is at present without experimental support. Actually the clarification of the physical origin of the electroweak symmetry breaking is one of the most important problems for experimental particle physics in the next decade. There are arguments indicating that the minimal Standard Model with fundamental Higgs fields cannot be the whole story and that some kind of new physics must necessarily appear near the Fermi scale. The most famous argument of this type is based on the so-called 'hierarchy problem', which we now describe. There is no unification of the fundamental forces in the Standard Model, because a separate gauge group and coupling is introduced for each interaction. On the other hand, the structural unity implied by the common, restrictive property of gauge invariance strongly suggests the possibility that all the observed interactions actually stem from a unified theory at some more fundamental level. The idea of unification at energies of order $m_{\mathrm{GUT}} \approx 10^{15}$ GeV, below the energy scale where quantum gravity becomes effective at masses of the order of the Planck mass, $m_P \approx 10^{19}$ GeV, has been much studied in recent years. However, the question remains whether unification without the inclusion of quantum gravity is really plausible. It is clear that the inclusion of gravity must induce major changes in the physics of the Standard Model at energies of order m_P and possibly even below.

Thus, at least because of the fact that gravity is not included in the Standard Model, new physics must necessarily emerge at some large energy scale Λ (equivalent to some small distance scale). Then the problem is to understand what order of magnitude can reasonably be expected for Λ. In particular, we can ask whether it is natural to expect that Λ may be as large as m_P or m_{GUT}. In other words, is it possible that the Standard Model holds without any new physical input up to the energy scale of quantum gravity? The answer is probably negative, because then we could not naturally explain the enormous value of m_P/m_W, i.e. the ratio between the Planck and the Fermi scales of masses.

To develop this point further, we recall that in the Standard Model the fermion and vector-boson masses are all specified in terms of the VEV of the Higgs field v, according to Eqs. (48) and (58). The value of v is determined by the curvature scale of the Higgs potential V:

$$V(\phi) = -\frac{1}{2}\mu^2\phi^\dagger\phi + (\lambda/4)(\phi^\dagger\phi)^2\,, \tag{80}$$

according to

$$v = \mu/\sqrt{\lambda}\,. \tag{81}$$

The observed values of the masses require for v (and, therefore, roughly for μ as well) that $v \approx 10^2$ GeV [see Eq. (59)].

If $\Lambda \approx (10^{15}\text{–}10^{19})$ GeV, then we face the problem of justifying the presence of two so largely different mass scales in a single theory (the so-called hierarchy problem). In general, if Λ is very large in comparison with μ, then, even if we set by hand a small value for μ at the tree level, the radiative corrections would make μ increase up to nearly the order of Λ. This problem is particularly acute in theories with scalars, as in the Standard Model, because the degree of divergence of mass corrections is quadratic, whilst the same divergences are only logarithmic for spin-1/2 fermions.

One general way out would be that the limit $\mu \to 0$ corresponds to an increase of the symmetry of the theory. In fact, the observed value of μ^2 can be decomposed as

$$\mu^2 = \mu_0^2 + \delta\mu^2 , \qquad (82)$$

where μ_0 is the tree-level value and $\delta\mu^2$ arises from the loop quantum corrections. If no new symmetry is induced when $\mu \to 0$ and no other non-renormalization theorem is operative, then a small value for μ^2/Λ^2 can only arise from an unbelievably precise cancellation between μ_0^2/Λ^2 and $\delta\mu^2/\Lambda^2$. If, however, $\mu = 0$ leads to an additional symmetry, then $\delta\mu^2$ must be proportional to μ_0^2, because for $\mu_0 = 0$ both the tree diagrams and the loop corrections must respect the symmetry. Then, if one starts from a small value of μ_0^2/Λ^2, the radiative corrections would preserve the smallness of μ^2/Λ^2.

For fermions, chiral symmetry is added when $m_f \to 0$, because the axial currents are also conserved in this limit, as their divergence is proportional to the fermion mass. Chiral symmetry and the logarithmic degree of divergence for fermion masses considerably alleviate the hierarchy problem in theories with no fundamental scalars.

In the Standard Model no additional symmetry is gained for $\mu_0 = 0$. This is also seen from the explicit formula for $\delta\mu^2$ at the one-loop level, which shows that $\delta\mu^2$ is not proportional to μ_0^2:

$$\begin{aligned} \mu^2/\Lambda^2 = (\mu_0^2/\Lambda^2) + (1/128\pi^2)(d/dv)^2 \times \\ \times \sum_J (2J+1)(-1)^{2J} m_J^2(v) + \dots , \end{aligned} \qquad (83)$$

where terms which vanish with $\Lambda \to \infty$ are indicated by the dots. The sum over J includes both particles and antiparticles (counted separately) of spin J and mass m_J (expressed as a function of v). Thus we are forced to the conclusion that in the Standard Model the natural value for $\delta\mu^2/\Lambda^2$ is of order one or so. Therefore, as Λ can be interpreted as the energy scale where some essentially new physical ingredient becomes important, we are led to expect that the validity of the present framework cannot be extended beyond $\Lambda \approx (1\text{–}10)$ TeV. Note that here the discussion is on the relation between bare and renormalized masses, where the cut-off dependence is hidden. In the renormalization procedure a physical value is simply assigned to m_H and it is left to the bare mass and the cut-off to adjust to each other. The naturalness problem arises if the divergences are seen as a low-energy effect, to be eventually removed by some new physics at the scale Λ (e.g. by gravity at M_{P}). Then the large momentum cut-off and the scale of new physics can be physically identified.

The problem of explaining the Fermi scale is seen to be closely connected to the Higgs mechanism and to the consequent presence of scalars, which makes the problem of testing the Higgs sector particularly crucial.

One possible solution is that the Higgses are really scalar fundamental fields, but naturality is restored by supersymmetry.

Supersymmetry (SUSY for short) [15] relates bosons and fermions, so that in a multiplet that forms one representation of supersymmetry there is an equal number of bosonic and fermionic degrees of freedom. This implies that SUSY generators are spin-1/2 charges Q_α. SUSY leads to an extension of the Poincaré algebra. Besides the obvious algebraic relations between Q_α and the Poincaré generators, which specify the spinorial transformations of Q_α under Lorentz transformations and its invariance under translations, the essentially new relation is the anticommutator

$$\{Q_\alpha, \bar{Q}_\beta\} = -2(\gamma_\mu)_{\alpha\beta} P^\mu \ , \tag{84}$$

where P^μ is the energy–momentum four-vector, which generates space–time translations. If all fundamental symmetries are gauge symmetries, then also SUSY is presumably a local symmetry. This immediately leads to the realm of gravity. In fact, the product of two local SUSY transformations is a translation with space–time-dependent parameters, as follows from Eq. (84). But a translation with space–time-dependent parameters is a general coordinate transformation. As ordinary gravity can be seen to arise from gauging the Poincaré group, a similar gauging of the Poincaré algebra enlarged by SUSY generators leads to an extended version of gravity, called supergravity. In fact, supersymmetry and supergravity play a crucial role in most present attempts at constructing a sensible theory of quantum gravity, including superstring theories [16] that at present represent the most advanced and promising project for a theory of gravity (and of all particle interactions).

Theorists like SUSY for several reasons. SUSY is the maximum symmetry compatible with a non-trivial S-matrix in a local relativistic field theory. The powerful constraints between couplings and masses implied by SUSY drastically reduce the degree of singularity of field theory as deduced from power counting. In some cases a finite field theory is even obtained. For example, $N = 4$ extended SUSY Yang–Mills theories are finite in four dimensions. In general, more powerful non-renormalization theorems are deduced for SUSY theories. This property may solve many naturalness problems of the Standard Model. In particular, the hierarchy problem can be solved in SUSY theories because the mass divergences of bosons and fermions become the same, and the quadratic divergences of scalars are reduced to logarithmic divergences as for spin-1/2 fermions. Thus in SUSY theories, we automatically obtain

$$(d/dv)^2 \sum_J [(2J+1)(-1)^{2J} m_J^2(v)] = 0 \ , \tag{85}$$

because of a cancellation between bosons and fermions. Of course the cancellation is only exact in the limit of unbroken SUSY. But we know that SUSY must be broken because the SUSY partners of ordinary particles have not been observed. In broken SUSY, the Λ appearing in Eq. (83) can be identified with the mass scale that is typical of SUSY partners of ordinary states. Thus if Λ is of order $G_F^{-1/2}$ or at most $O(1\ \mathrm{TeV})$, then the hierarchy problem would be solved in a natural way. This is the only known way out of the hierarchy problem compatible with fundamental scalar Higgs fields.

We have seen that most theorists working on quantum gravity and superstrings tend to consider SUSY as 'established' at m_P and beyond. For economy, we are naturally led to also try to use SUSY at low energy to solve some of the problems of the Standard Model, including the hierarchy problem. It is therefore very important that it was indeed shown [17] that models where SUSY is softly broken by gravity do offer a viable alternative. We stress again that the supersymmetric way is very appealing to theorists. In fact, it would represent the ultimate triumph of a continuous line of progress obtained by constructing field theories with an increasing degree of exact and/or broken symmetry and applying them to fundamental interactions. Also the value of the ratio of knowledge versus ignorance would be remarkably large in the case of SUSY: the correct degrees of freedom for a description of physics up to gravity would have been identified, the Hamiltonian would be essentially known, and the theory would, to a large extent, be computable up to m_P. In the limit of exact supersymmetry and exact gauge $SU(2) \otimes U(1)$, all particles are massless. When supersymmetry is broken while $SU(2) \otimes U(1)$ is still preserved, ordinary particles remain massless while sparticles become massive. It is important to note that observed particles are precisely those whose mass terms are forbidden in the $SU(2) \otimes U(1)$ limit, while sparticle masses are allowed. For example, quark and lepton masses are forbidden while squark or slepton masses are allowed, the gauge boson masses are forbidden but the gaugino masses are allowed. Thus the fact that all ordinary particles were observed but no sparticles is not unnatural. When finally the $SU(2) \otimes U(1)$ symmetry breaking is switched on, the mass μ^2 in the Higgs potential naturally takes a value of the order of the scale of sparticle masses and all ordinary particles acquire a mass.

The alternative main avenue to physics beyond the Standard Model is compositeness or, more generally, the existence of new strong forces. For example, the electroweak symmetry could be broken by condensates of new fermions attracted by a new force with $\Lambda_{\text{new}} \approx m_F \approx G_F^{-1/2}$($\Lambda_{\text{new}}$ being the analogous quantity of Λ_{QCD}), as in technicolour theories [18]. Recently it has been proposed [19] that a very heavy top mass ($m_t \geq 230$ GeV) could induce the electroweak symmetry breaking. The Higgs would be a sort of $t\bar{t}$ bound state with mass $m_H \simeq O(m_t)$. This model (of the Nambu–Jona–Lasinio type) is non-renormalizable and involves many *ad hoc* four-fermion interactions to fix the fermion masses. Arguments were given indicating that, if m_H is $O(m_t)$, then this view of the Higgs is just a matter of language without great physical significance [20] when compared with the standard picture. Or the Higgs scalar can be a composite of new fermions bound by a new force [21]. In the last two cases there are unsolved problems related to the fermion masses. Or the $SU(2) \otimes U(1)$ gauge symmetry can be a low-energy fake [22]. At high energies $\gtrsim m_F$, the W and Z^0 would be resolved into their constituents. Another possibility is that the Higgs mass becomes large [23] [O(1 TeV)]. Then, as we shall see, the weak interactions become strong and a spectroscopy of tightly bound weak- interaction resonances appears (e.g. WW, WZ, or ZZ states).

However, it is fair to say that the compositeness alternative is not at all so neat and clearly formulated as the supersymmetric option. On the contrary, in many respects the compositeness way is not well defined at all and leads to many unsolved problems.

Of course, the two avenues are not necessarily mutually exclusive, and theoretical frameworks where both appear have been considered [24].

Searching for the standard Higgs particle appears to be a good way to organize the experimental solution of the symmetry-breaking problem.

The search for the Higgs is being pursued at LEP 1 and will continue at LEP 200. Indeed all previous limits on the Higgs mass m_H have been dwarfed by only a few months of LEP operation. For the standard Higgs we have at present the following combined limits from the four LEP experiments [25]:

$$m_H \gtrsim 62.5 \text{ GeV}. \tag{86}$$

This limit is obtained from negative searches of the process $e^+e^- \rightarrow HZ^* (\rightarrow f\bar{f})$.

As is well known, the value of the Higgs mass is not predictable even in the minimal Standard Model with a single Higgs doublet. What is certainly true is that the Higgs boson cannot be too heavy or the perturbative theory becomes sick and breaks down [26]. If $m_H \geq O(1 \text{ TeV})$ the perturbative rates for $VV \rightarrow VV$ scattering ($V \equiv W, Z$) violate the unitarity limit [23] for $\sqrt{s} \gg m_W$. More important than this, in non-asymptotically free gauge theories there are Landau poles where the coupling constant blows up according to the renormalization group improved perturbation theory (unless the renormalized coupling is not vanishing so that the theory is a free theory, i.e. trivial). This phenomenon is also present in QED, but it would only occur beyond the Planck scale of mass, so that the problem can be solved at such large energies by embedding the theory in a larger context (e.g. grand unification). The coupling of the quartic term $\lambda(\phi^+\phi)^2$ in the Higgs potential increases with m_H^2 ($m_H^2 \sim \lambda/G_\mathrm{F}$); see Eqs. (59) and (81). In addition, for a given m_H, λ increases logarithmically with energy because the theory is not asymptotically free in the Higgs sector. Thus the position of the Landau pole depends on m_H. Imposing that the Landau pole be far enough for the theory to make sense up to a scale Λ, gives a bound [27] on the standard Higgs mass which is plotted in Fig. 2, taken from Ref. [28]. We see that for a light Higgs, i.e. $m_H \leq$ 180–200 GeV, the perturbative regime is valid up to M_GUT or M_P. For a heavier Higgs the value of Λ decreases until eventually $\Lambda \sim m_H$: for $m_H \sim 1$ TeV, the theory is valid up to $\Lambda \sim 1$ TeV.

We can understand these results by the following crude simplification [26]. The renormalization group equation for the quartic coupling λ, in the limit of neglecting gauge and Yukawa couplings, becomes:

$$\frac{d\lambda(t)}{dt} = \frac{3}{4\pi^2}\,\lambda^2(t)\,, \tag{87}$$

with $t = \ln \Lambda/v$, where v is the Higgs vacuum expectation value and Λ is the scale where λ is evaluated. The coefficient $\beta_0 = 3/4\pi^2$ is obtained from one-loop corrections to the quartic coupling in the $\lambda(\phi^+\phi)^2$ theory. The normalization of v and λ, in physical terms, is here chosen such that

$$\lambda \equiv \lambda(v) = \sqrt{2}\, G_F\, m_H^2 \tag{88}$$

$$v = (2\sqrt{2}\, G_F)^{-1/2} \simeq 174 \text{ GeV}\,. \tag{89}$$

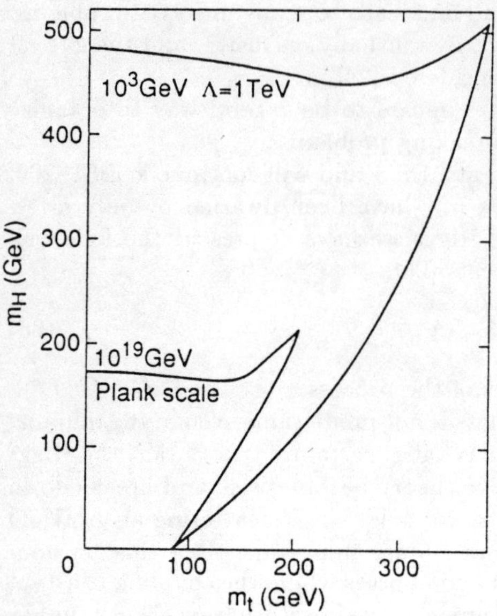

Fig. 2. Combined limits (from Ref. [28]) on m_H and m_t from vacuum stability and avoiding the Landau pole up to a scale Λ

When $\lambda(t)$ is large the gauge and Yukawa couplings can be neglected with the exception of the top Yukawa coupling, which can become large if $m_t \geq v : g_{\text{top}} = m_t/v$. By solving (87) one obtains

$$\lambda(t) = \frac{\lambda}{1 - 3/4\pi^2 \lambda t} \,. \tag{90}$$

The minus sign in the denominator, typical of non-asymptotically free theories, implies the increase of $\lambda(t)$ with the scale Λ up to an infinite value which is obtained for $3/4\pi^2\ \lambda t = 1$. In order to avoid the Landau pole the condition

$$\frac{3}{4\pi^2}\ \lambda t = \frac{3}{4\pi^2}\ \sqrt{2}\, G_F\, m_H^2\ \ln\ \Lambda/v < 1 \tag{91}$$

must be imposed. This condition is equivalent to:

$$m_H \leq \frac{893\ \text{GeV}}{\sqrt{\ln\ \Lambda/174\ \text{GeV}}} \tag{92}$$

or $m_H < 144,\ 165,\ 675$ GeV for $\Lambda = 10^{19}, 10^{15}, 10^3$ GeV, respectively.

We see that this simple model reproduces the quantitative features of the bounds on m_H in Fig. 2 fairly well. The curves in Fig. 2 are obtained by a more refined renormalization group treatment of the problem, with inclusion of gauge and top effects. The obvious criticism to the above approach is that a perturbative evaluation of the β function is not justified in the vicinity of the Landau pole. Thus it is very interesting that the validity of the bound has been confirmed by recent computer simulations of the electroweak theory on the lattice [29]. The precise value of the upper limit on m_H depends on the exact definition of Λ and on where one fixes the line between acceptable and not acceptable. In fact the lattice results nicely extrapolate the perturbative evaluation and impose limits on m_H such that:

$$m_H \leq (8\text{–}10) m_W \simeq 0.6\text{–}0.8 \text{ TeV} . \tag{93}$$

It is thus fair to conclude that, to the best of our knowledge, the internal consistency of the Standard Model demands that the Higgs mass is below 1 TeV.

In Fig. 2 there is also a forbidden region at large m_t and small m_H. This boundary is determined by the requirement of vacuum stability [26],[30].

At tree level the scalar potential can be written in the form:

$$V(\varphi) = -\mu^2 |\varphi|^2 + \frac{\mu^2}{2v^2} |\varphi|^4 . \tag{94}$$

The quantum corrections can be computed by expanding in the number of loops. At one loop one obtains:

$$V(\varphi) = -\mu^2 |\varphi|^2 + \frac{\mu^2}{2v^2} |\varphi|^4 + \gamma |\varphi|^4 \left(\ln \frac{|\varphi|^2}{v^2} - \frac{1}{2} \right) \tag{95}$$

with

$$\gamma = \frac{3 \sum_{\text{vectors}} m_v^4 + \sum_{\text{scalars}} m_s^4 - 4 \sum_{\text{fermions}} m_f^4}{64\pi^2 v^4} . \tag{96}$$

It is simple to check that also in the corrected form v is an extremum of $V(\phi)$. In the minimal Standard Model with one Higgs doublet and three fermion families one obtains

$$\gamma = \frac{3m_Z^4 + 6m_W^4 + m_H^4 - 12m_t^4}{64\pi^2 v^4} . \tag{97}$$

The extra factor of three in front of m_t^4 is of course due to colour.

For the realization of spontaneous symmetry breaking and stability of the theory one requires that i) the extremum at $\varphi = v$ is a minimum, i.e. $V(v) < 0$, and ii) $V(\varphi) \to +\infty$ for $|\varphi| \to \infty$, so that the Hamiltonian is bounded from below.

At small m_t, $m_t < 80$ GeV, the first requirement leads to the Linde–Weinberg limit $m_H^2 > 2\gamma v^2$, or $m_H \geq 7$ GeV. This limit is by now void, because of the experimental lower bounds on both m_t and m_H. For the second requirement to be fulfilled, m_H must increase with m_t in order to prevent γ from becoming too negative [30]. At large $|\varphi|$ the one-loop evaluation of $V(\varphi)$ is not sufficient, and one needs a resummation of the large logarithms $\log \varphi^2/v^2$. The results are shown in Fig. 2 [28]. The above limits are only valid in the minimal Standard Model with one Higgs doublet. Note that in case that there are two or more Higgs doublets the limits refer to some average mass. Thus for the lightest Higgs the lower limit can be easily evaded but the upper limit is *a fortiori* valid. The bottom line is that either the Higgs is found below $\simeq 1$ TeV or new physics beyond the Standard Model should appear. At least one should see the onset of a new non-perturbative regime where the weak interactions become strong.

In conclusion there are solid arguments for new physics near the Fermi scale of mass $G_F^{-1/2}$. Either a fundamental scalar Higgs exists and naturalness is restored by supersymmetry, or new strong forces will manifest themselves, drastically changing the framework of the Standard Model beyond $O(1\ \mathrm{TeV})$. A new non-perturbative regime will set in, with new resonances, and the physics will become less predictable above that energy. An important point is that all conceivable possibilities are very complex. Each of them implies a rich new spectrum of states and phenomena: the whole spectrum of superpartners in SUSY; new hadrons, excited vector bosons, etc., in the composite alternative. The new physics is in all cases distributed over a large interval of energies. The low-lying fringes of the new spectroscopy, or at least their virtual effects, should already be accessible to LEP 1 and LEP 200. A lot of discoveries are expected at the LHC, to be followed by more at the SSC.

5 Basic Relations for Precision Tests

In the standard electroweak theory, there are a number of basic relations that one wants to verify experimentally with the best possible precision. The same quantity $\sin^2\theta_W$ appears in all these relations.

First, $\sin^2\theta_W$ can be measured from the value of m_W. Starting from the tree-level relations [Eqs. (33) and (37), (38)], $\sin^2\theta_W = e^2/g^2$ and $G_F/\sqrt{2} = g^2/8m_W^2$, we obtain

$$\begin{aligned}\sin^2\theta_W &= (\pi\alpha/\sqrt{2}\,G_F)/[m_W^2(1-\Delta r)] \\ &= (37.2802\ \mathrm{GeV})^2/[m_W^2(1-\Delta r)]\ , \end{aligned} \tag{98}$$

where $\Delta r \neq 0$ owing to the effect of radiative corrections.

Then $\sin^2\theta_W$ is also related to the ratio of the vector boson masses. At tree level [see Eq. (42)],

$$\sin^2\theta_W = 1 - m_W^2/\rho_0 m_Z^2\ , \tag{99}$$

where $\rho_0 = 1$ in the Standard Model with only doublets of Higgs bosons. In general, this relation is also modified by radiative corrections:

$$\sin^2\theta_W = 1 - m_W^2/\rho_{\mathrm{mass}} m_Z^2\ , \tag{100}$$

with $\rho_{\mathrm{mass}} = \rho_0(1+\delta\rho_{\mathrm{mass}})$. However, as we shall discuss in detail later, Eq. (99) with $\rho_0 = 1$ is often adopted as a definition of $\sin^2\theta_W$ at all orders. Clearly in this case, $\rho_{\mathrm{mass}} = 1$ by definition.

Finally, $\sin^2\theta_W$ can be obtained for neutral-current couplings. At tree level, the four-fermion interaction from Z exchange is given [see Eqs. (41) and (42)] by

$$\begin{aligned}M_{if} &= [\sqrt{2}\,G_F m_Z^2/D(s)] \\ &\quad \rho_0(J_3^i - 2\sin^2\theta_W J_{em}^i)(J_3^f - 2\sin^2\theta_W J_{em}^f)\ , \end{aligned} \tag{101}$$

where $D(s)$ is the Z propagator, and J_3^f and J_{em}^f are the weak isospin and electromagnetic currents for the fermion f. Excluding pure QED corrections, electroweak radiative corrections modify M_{if} according to

$$\begin{aligned}M_{if} &= [\sqrt{2}\,G_F m_Z^2/D(s)] \\ &\quad \rho_{if}(J_3^i - 2k_i\ \sin^2\theta_W J_{em}^i)(J_3^f - 2k_f\ \sin^2\theta_W J_{em}^f) + \dots\ , \end{aligned} \tag{102}$$

where $\rho_{if} = \rho_0(1 + \delta\rho_{if})$; $k_a = 1 + \delta k_a (a = i, f)$ are in general different for different fermions and depend on the scheme adopted (for example, δk_a depend on the definition of $\sin^2\theta_W$). The ellipse indicates possible additional small non-factorizable terms.

6 Input Parameters

For LEP physics [31], a self-imposing set of input parameters is given by $\alpha_s, \alpha, G_F, m_Z$, m_f and m_H. Clearly the Fermi coupling $G_F = 1.16637(22) \times 10^{-5}$ GeV^{-2} is conceptually more complicated than $\alpha_{weak} = \frac{g^2}{4\pi}$ (which would more naturally accompany $\alpha = 1/137.036$ and α_s) or $\sin^2\theta_W$ or m_W, but is preferred for practical reasons because it is known with all the desirable accuracy. Similarly, m_Z has now been measured at LEP with remarkable precision. This preliminary task of LEP in view of precision tests of the Standard Model has been accomplished to a nearly final degree of accuracy (see Section 10 for details on the line-shape measurement).

The LEP results on m_Z, as summarized at the La Thuile and Moriond '93 conferences [32],[33] are reported in Table 1 [1]. The resulting relative precision is impressive: $\delta m_Z/m_Z = 8 \times 10^{-5}$.

Table 1: Results on m_Z from LEP

Experiment	m_Z (GeV)
ALEPH	91.187 ± 0.007
DELPHI	91.186 ± 0.007
L3	91.196 ± 0.007
OPAL	91.181 ± 0.007
Average	91.187 ± 0.0033(stat.) ± 0.0063 (LEP)
	$\simeq 91.187 \pm 0.007$

Among the quark and lepton masses, m_f, the main unknown is the top quark mass. Our ignorance of m_t is at present a serious limitation for precise tests of the electroweak theory because the radiative corrections are relatively large for large m_t and depend quadratically on m_t [31]. This fact can be used to put stringent constraints on m_t from the existing electroweak measurements, in particular an upper bound on m_t, to be discussed in detail later. As for lower bounds on m_t the best results arise from the failure to observe the t quark at e^+e^- and hadron colliders. LEP and SLC lead to a model-independent bound $m_t \gtrsim 45$ GeV. From CDF [34] one learns that $m_t \gtrsim 108$ GeV, provided that the t quark production rate and semi-leptonic branching ratio are as predicted by the Standard Model.

The mass of light quarks is in general an ambiguous concept. Fortunately, in the present context what matters is the contribution of light quarks (including c and b) to the vacuum polarization loop. The relevant polarization function can be related via a dispersion relation to the $e^+e^- \rightarrow$ hadrons cross section and computed from the actual data. The value described in the next section is obtained as a result. The error

[1] When writing these notes I decided to update the experimental data by also including results that were not yet available when the lectures were delivered.

is a non-negligible source of ignorance and could only be improved by more precise data, especially in the region between the J/ψ and the Υ.

The Higgs mass m_H is largely unknown. One of the most impressive performances of LEP up to now has been the dwarfing of all previous lower bounds on m_H. The LEP bound is summarized in Eq. (86). The basis for the theoretical upper bound on m_H was reviewed in Section 4. As a result the calculation of radiative corrections is now done for 60 GeV $< m_H <$ 1 TeV. The sensitivity of the radiative corrections to variations of m_H in the range 60 GeV $< m_H <$ 1 TeV is not large. In a sense, this level of accuracy fixes the goal for precision tests of the Standard Model because the clarification of the symmetry-breaking sector of the theory is the main target of present-day experiments.

Finally, for electroweak calculations involving hadrons, the value of the QCD coupling α_s must also be specified. The best value of α_s at the Z mass, obtained from all existing experiments is given by [35] $\alpha_s(m_Z) = 0.118 \pm 0.007$. (Note that I am conservative on errors, which are dominated by theoretical uncertainties.) The QCD corrections to processes involving quarks are typically of order α_s/π. As a consequence the stated error on α_s leads to a few per mille relative uncertainty on the corresponding predictions.

7 Large Contributions to Radiative Corrections

A set of important quantitative contributions to the radiative corrections arise from large logarithms [e.g. terms of the form $(\alpha/\pi \ln(m_Z/m_{f_\ell}))^n$ where f_ℓ is a light fermion], and from quadratic terms in m_t, i.e. terms proportional to $G_F m_t^2$.

The sequences of leading and close-to-leading logarithms are fixed [36] by well-known and consolidated techniques (β functions, anomalous dimensions, penguin-like diagrams, etc.). For example, large logarithms dominate the running of α from m_e, the electron mass, up to m_Z, with the result that [37]

$$\alpha(m_Z)/\alpha = 1/(1 - \delta\alpha) \; . \tag{103}$$

At present, the best value of $\delta\alpha$, obtained by extracting the relevant effective light-quark masses from the data on $e^+e^- \to$ hadrons, is given by [37]

$$\delta\alpha = 0.0596 \pm 0.0009 \text{ or } \alpha(m_Z)^{-1} = 128.87 \pm 0.12. \tag{104}$$

Large logarithms of the form $[\alpha/\pi \ln(m_Z/\mu)]^n$ also enter, for example, in the relation between $\sin^2\theta_W$ at the scales m_Z (LEP, SLC) and μ (e.g. the scale of low-energy neutral-current experiments).

The quadratic dependence on m_t [38] (and on other possible widely broken isospin multiplets from new physics) arises because, in spontaneously broken gauge theories, heavy loops do not decouple. On the contrary, in QED or QCD, the running of α and α_s at a scale Q is not affected by heavy quarks with mass $M \gg Q$. According to an intuitive decoupling theorem [39], diagrams with heavy virtual particles of mass M can be ignored at $Q \ll M$ provided that the couplings do not grow with M and that the theory with no heavy particles is still renormalizable. In spontaneously broken gauge theories, one important difference is in the longitudinal modes of weak gauge bosons. These modes are generated by the Higgs mechanism, and their couplings grow with masses (as is also the case for the physical Higgs couplings). The upper limit on m_t from radiative corrections arises from this phenomenon. Other subtler sources of non-decoupling are related, for example, to the presence of chiral anomalies [14] (which may not be completely canceled if heavy particles are removed). Another very important consequence is that precision tests of the electroweak theory may be sensitive to new physics even if the new particles are too heavy for their direct production. With the value of m_t being continuously pushed up by experiment, the quantitative importance of the terms of order $G_F m_t^2$ is increasingly large. Both the large logarithms and the $G_F m_t^2$ terms have a simple structure and are to a large extent universal, i.e. common to a wide class of processes. Their study is important for an understanding of the pattern of radiative corrections. One can also derive approximate formulae (e.g. improved Born approximations, see Section 9), which can be useful in cases where a limited precision may be adequate.

8 Determinations of $\sin^2\theta_W$ from m_Z

Once $\alpha, \alpha_s, G_F, m_Z, m_f$ and m_H have been chosen as input parameters, $\sin^2\theta_W$ is a derived quantity. We now consider the calculation of $\sin^2\theta_W$ beyond the tree approximation. A precise definition of $\sin^2\theta_W$ must be specified before its value can be computed.

At tree level the relation between $\sin^2\theta_W$ and m_Z is obtained from Eqs. (98) and (99) (with $\Delta r = 0$):

$$\sin^2\theta_W \, \cos^2\theta_W = (\pi\alpha/\sqrt{2}\, G_F)/(\rho_0 m_Z^2) \, . \tag{105}$$

Beyond the tree level, the quantity Δr [40] is introduced by radiative corrections:

$$\sin^2\theta_W \, \cos^2\theta_W = (\pi\alpha/\sqrt{2}\, G_F)/[\rho_0 m_Z^2(1-\Delta r)] \, , \tag{106}$$

where $\Delta r \equiv \Delta r(\alpha, \alpha_s, G_F, m_Z, m_f, m_H)$ is of course different for different definitions of $\sin^2\theta_W$. In the following, some particularly interesting definitions of $\sin^2\theta_W$ will be discussed. We now set $\rho_0 = 1$. Let us first consider the usual definition [41]:

$$\sin^2\theta_W \equiv s_W^2 = 1 - (m_W^2/m_Z^2) \, . \tag{107}$$

From now on, the symbol s_W^2 will always refer to this specific definition of $\sin^2\theta_W$. Clearly, given m_Z from LEP, s_W^2 is directly equivalent to m_W. In this case Δr specifies the relation between m_Z and m_W:

$$(1 - m_W^2/m_Z^2)m_W^2 = (\pi\alpha/\sqrt{2}\, G_F)/(1-\Delta r) \, . \tag{108}$$

The value of Δr as a function of the input parameters has been studied in great detail. In particular, the following results are obtained [37], [40], [42]:

$$\begin{aligned} 1/(1-\Delta r) &= \alpha(m_Z)/\alpha \cdot 1/[1 + (c_W^2/s_W^2)\delta\rho] + \text{`small'} \\ &= 1/(1-\delta\alpha) \cdot 1/[1 + (c_W^2/s_W^2)\delta\rho] + \text{`small'} \, , \end{aligned} \tag{109}$$

where $c_W^2 = 1 - s_W^2$; $\delta\alpha$ was defined in Eqs. (103) and (104), and $\delta\rho \to \delta\rho_t$, for large m_t with $\delta\rho_t$ given by [37], [43]

$$\delta\rho_t = (3G_F m_t^2/8\pi^2\sqrt{2}) + (3G_F m_t^2/8\pi^2\sqrt{2})^2(19 - 2\pi^2)/3 + O[(G_F m_t^2)^3] \, . \qquad (110)$$

By 'small' in Eq. (109), we mean terms (which at one-loop accuracy are known) without large logs and/or leading powers of $G_F m_t^2$; $\delta\rho_t$ is the dominant term, for large m_t, of the famous ρ-parameter first studied in Refs. [38]. The two-loop term was obtained in Ref. [43] and the geometric series resummation was advocated in Ref. [42].

Going back to the Z-exchange amplitude near the resonance and the parameters ρ_{if}, k_i, and k_f defined in Eq. (102) with the definition of $\sin^2\theta_W \equiv s_W^2$, we obtain [37]

$$\rho_{if} = 1 + \delta\rho + \text{'small'} \qquad (111)$$

and

$$k_f = 1 + (c_W^2/s_W^2)\delta\rho + \text{'small'} \qquad (112)$$

(here $f \neq b$; the b-quark will be reconsidered in the following), where the 'small' terms are non-universal (i.e. process-dependent). Note that k_f contains additional 'large' terms with respect to those included in Δr and ρ_{if}. As these 'large' terms are also universal, this suggests that ks_W^2 could be a better effective $\sin^2\theta_W$ than s_W^2 for physics at the Z pole. Before going into this matter, we will add some comments on $\delta\rho$. For any weak isospin fermion doublet (e.g. new heavy quarks or leptons), the quantity '$3m_t^2$' in $\delta\rho_t$ (at one loop) becomes [37],[40]

$$\text{'}3m_t^2\text{'} = N_c[m_u^2 + m_d^2 - 2m_u^2 m_d^2/(m_u^2 - m_d^2)\ln\, m_u^2/m_d^2] \, , \qquad (113)$$

where N_c is the number of colours. For negligible $m_{d,u}$, '$3m_t^2$' $\to N_c m_{u,d}^2$, whilst for $m_u = m_d + \epsilon$: '$3m_t^2$' $\to N_c \, \frac{4}{3} \, \epsilon^2$ to leading order in ϵ (the result vanishes for unsplit doublets). Similarly, many more kinds of broken weak isospin multiplets can contribute to '$3m_t^2$' [44] (squarks and sleptons [45], charged Higgses [46], etc.). The upper limit on 'm_t' $\lesssim 200$ GeV (see Section 11) together with the direct lower limit on the t-quark mass, $m_t \gtrsim 108$ GeV, leave little space for heavy multiplets, except for nearly degenerate ones.

The contribution of the neutral Higgs mass to $\delta\rho$ is only logarithmic at one loop, (see, e.g., Eq. (158) in Section 12). There are no m_H^2 terms but only logs, because the 'custodial' $SU(2)$ symmetry is not broken in the Higgs sector [23]. Power terms appear only at two loops [47], but their effect is sizable only for $m_H \gtrsim 1$ TeV.

We now consider a different class of definitions of $\sin^2\theta_W$. Assume that we fix $k_f = 1$ [defined by Eq. (102)] for one given Z vertex (e.g. $k_f = 1$ in $Z \to e^+e^-$ for on-shell Z). Then it follows from the previous discussion that δk is 'small' for all neutral-current processes (for $f \neq b$) near the Z pole. Let us introduce the notation

$$\sin^2\theta_W(\text{from } Z \to e^+e^-) \equiv \bar{s}_W^2 \, . \qquad (114)$$

In this case, Eq. (102) becomes

$$M_{fi} = [\sqrt{2}\, G_F m_Z^2/D(s)] \, \rho_{if}(J_3^i - 2\bar{s}_W^2 J_{em}^i)(J_3^f - 2\bar{s}_W^2 J_{em}^f) + \text{'small'} \, , \qquad (115)$$

where ρ_{if} is given by Eq. (110) in terms of $\delta\rho$, which is specified in Eqs. (110). In the present case,

$$\bar{s}_W^2 \bar{c}_W^2 = (\pi\alpha/\sqrt{2}G_F)/[m_Z^2(1-\Delta\bar{r})] , \tag{116}$$

and we find [37]

$$\bar{s}_W^2 \simeq s_W^2 + c_W^2 \delta\rho + \text{`small'} . \tag{117}$$

In other words, apart from 'small' terms, we have

$$\bar{s}_W^2 \simeq 1 - (m_W^2/\rho m_Z^2) \tag{118}$$

with $\rho \simeq 1 + \delta\rho$. It is interesting to note that $\bar{s}_W^2$ is less dependent than s_W^2 on m_t^2. In fact [37],[40]

$$\Delta r \simeq \delta\alpha - (c_W^2/s_W^2)\delta\rho , \tag{119}$$

$$\Delta\bar{r} \simeq \delta\alpha - \delta\rho , \tag{120}$$

so that the amplifying factor c_W^2/s_W^2 in front of $\delta\rho$ is missing in $\Delta\bar{r}$. Note that given m_Z, $\bar{s}_W^2$ is known with an accuracy of about ± 0.002 when m_t varies between 60 and 210 GeV (see Fig. 3). There is a class of definitions of $\sin^2\theta_W$ in the literature that correspond to $\bar{s}_W^2$ if 'small' terms are neglected: $\bar{s}_W^2$ of Hollik [48]; $s^{*2}(m_Z^2)$ of Lynn and Kennedy [49]; $(\sin^2\theta_W)_{\overline{MS}}(m_Z^2)$ [50],[51], etc. Note that because of Eqs. (103) and (120) one can also write:

$$\bar{s}_W^2 \bar{c}_W^2 = [\pi\alpha(m_Z)/\sqrt{2}G_F]/\rho m_Z^2 . \tag{121}$$

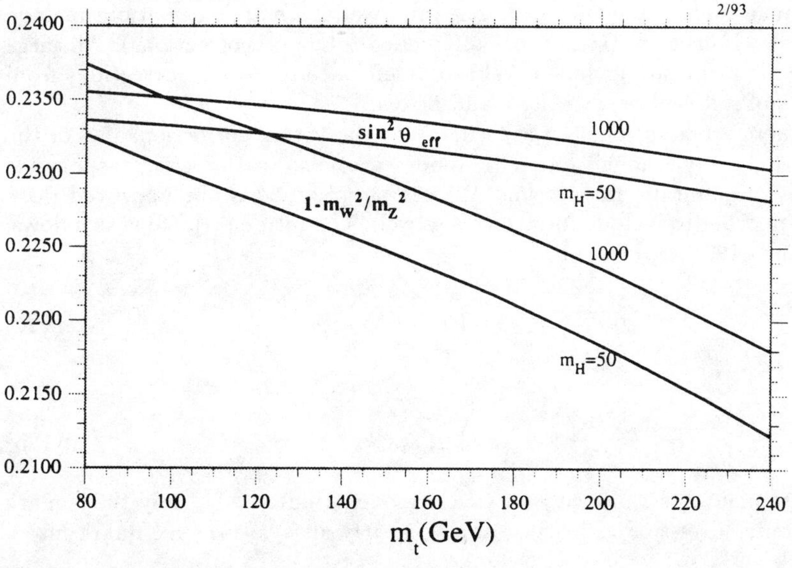

Fig. 3. The behaviour of s_W^2 and $\bar{s}_W^2$ [defined by Eqs. (107) and (114), respectively] as functions of m_t, for $m_Z = 91.187$ GeV, $m_H = 50$–1000 GeV.

9 Improved Born Approximation

For precision tests of the electroweak theory, the complete one-loop radiative corrections (plus higher-order and possibly exponentiated purely photonic corrections) are mandatory and are indeed available for all processes of practical relevance. However, in many cases a less accurate estimate can be enough. For this purpose, it is useful to know [37] that formulae as simple as those of the Born approximation can be written in a way that takes all 'large' corrections into account. As a result, for $e^+e^- \to f\bar{f}$, with $f \neq e, b$, the amplitude for γ and Z exchange near the resonance can be written in the form

$$\begin{aligned} M_{f\bar{f}} = {} & Q_e Q_f [4\pi\alpha(m_Z)/s] J^e_{em} J^f_{em} \\ & + \sqrt{2}\, G_F \rho m_Z^2/(s - m_Z^2 + is\,\Gamma_Z/m_Z) \bar{J}^e \bar{J}^f \ , \end{aligned} \tag{122}$$

where $\rho = 1 + \delta\rho, \delta\rho$ being given by Eq. (110), and

$$J^f_{em} = \gamma_\mu \ , \quad \bar{J}^f = \gamma_\mu [I_3^f(1-\gamma_5) - 2Q_f \bar{s}^2_W] \ , \tag{123}$$

with Q_f and I_3^f being respectively the electric charge and the third component of the weak isospin (e.g. for $f = \mu^-, Q = -1, I_3 = -1/2$). The important features are the replacement of α-fixed with α-running, the inclusion of the s-dependence for the total width Γ_Z in the resonant denominator (see Section 10), the presence of the factor of ρ multiplying G_F, and the use in the Z couplings of the effective $\sin^2\theta_W = \bar{s}^2_W$, introduced in the previous section. Clearly for $f = e$, i.e. for Bhabha scattering, the t-channel exchange is also to be included. The improved Born approximations include the real parts of self-energies (sometimes [49] called 'oblique' corrections). All large logs and all $G_F m_t^2$ terms are included. What are left out are 'small' corrections from imaginary parts of self-energies, vertices, and boxes.

For $f = b$ there are additional large terms from the vertex corrections [52] of the type shown in Fig. 4. The longitudinal W modes are, also in this case, responsible for the presence of quadratic mass terms. We can simply modify the improved Born approximation in order to include these terms as well. The recipe [37],[53] is as follows: in Eqs. (122) and (123), replace ρ by

$$\sqrt{\rho\rho_b}, \quad \rho_b \equiv \rho\left(1 - \frac{4}{3}\delta\rho\right) \ , \tag{124}$$

and $\bar{s}^2_W$ by

$$\bar{s}^2_W \left(1 + \frac{2}{3}\delta\rho\right) \equiv \bar{s}^2_W\ k_b \ . \tag{125}$$

Note that whilst many sorts of heavy particles can contribute to $\delta\rho$, only the t-quark (or a new t') contributes to $\delta\rho_b$, because the W nearly always turns a b-quark into a t (or a t') quark.

As a final example, the improved Born approximation for the inclusive widths $\Gamma[Z \to f\bar{f}(\gamma, g)]$ (including photons and gluons in the final state), is given by

$$\Gamma[Z \to f\bar{f}(\gamma, g)] = N_c (G_F \rho m_Z^3/24\pi\sqrt{2})[1 + (1 - 4|Q_F|\bar{s}^2_W)^2] \ , \tag{126}$$

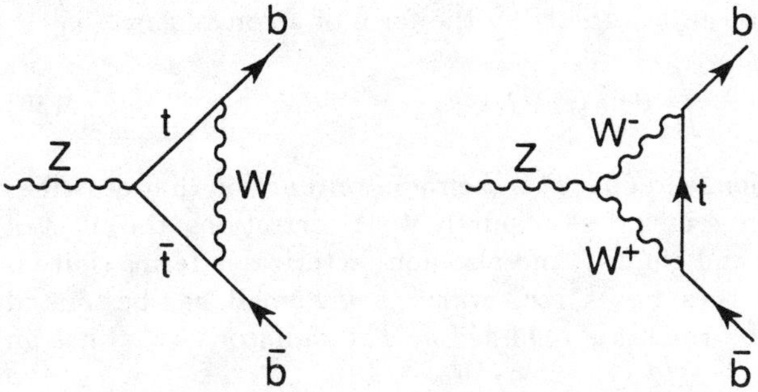

Fig. 4. Vertex corrections that contain large $G_F m_t^2$ terms

where

$$N_c = \begin{cases} 1 & [1+(3\alpha/4\pi)Q_f^2] \qquad\qquad\qquad\qquad\quad \text{(leptons)} \\ 3 & [1+(3\alpha/4\pi)Q_f^2]\,[1+\alpha_s(m_Z)/\pi+...]\ \text{(quarks)} \end{cases} \tag{127}$$

In the same approximation the total width is given simply by

$$\Gamma_Z = \sum_f \Gamma\,[Z \to f\bar{f}(\gamma, g)] + \text{rare decays}\ , \tag{128}$$

where the rare decays [54] (e.g. $Z \to H\mu^+\mu^-$ for a relatively light Higgs) are numerically unimportant in the Standard Model. For $\Gamma(Z \to b\bar{b})$, replace ρ by ρ_b and $\bar{s}_W^2$ by $\bar{s}_W^2 k_b$, where ρ_b and k_b are given in Eqs. (124) and (125). Note that in the Standard Model for $f \neq b$, Eq. (126) can also be written in the form [by using Eq. (121)]:

$$\Gamma[Z \to f\bar{f}(\gamma, g)] = N_c[\alpha(m_Z)m_Z/(48\ \bar{s}_W^2\bar{c}_W^2)][1+(1-4|Q_F|\bar{s}_W^2)^2]\ . \tag{129}$$

Actually, the 'large' $G_F m_t^2$ corrections are empirically not so large (i.e. there is a strong upper limit on m_t). As a consequence, as will be discussed in Section 12, an improved Born approximation based on the Standard Model at tree level plus pure QED corrections is well supported by the data.

10 The Z Line-Shape

A considerable amount of work has deservedly been devoted to the theoretical study of the Z line-shape [55]. The present experimental accuracy on m_Z obtained at LEP is $\delta m_Z = \pm 7$ MeV (recall Table 1, Section 6). This small error was obtained by a precise calibration of the LEP energy scale achieved in '91 by taking advantage of the transverse polarization of the beams and implementing a sophisticated resonant spin depolarization method [56]. Similarly, a measurement of the total width to an accuracy $\delta\Gamma = \pm 7$ MeV has by now been achieved. The prediction of the Z line-shape in the Standard Model to such an accuracy has posed a formidable challenge to theory, which has been successfully met. For the inclusive process $e^+e^- \to f\bar{f}X$, with $f \neq e$ (for simplicity, we leave Bhabha scattering aside) and X including γ's and gluons, the physical cross section can be written in the form of a convolution [55]:

$$\sigma(s) = \int_{z_0}^{1} dz\ \hat{\sigma}(zs)G(z,s)\ , \tag{130}$$

where $\hat{\sigma}$ is the reduced cross section, and $G(z,s)$ is the radiator function that describes the effect of initial-state radiation; $\hat{\sigma}$ includes the purely weak corrections, the effect of final-state radiation (of both γ's and gluons), and also non-factorizable terms (initial- and final-state radiation interferences, boxes, etc.) which, being small, can be treated in lowest order and effectively absorbed in a modified $\hat{\sigma}$. The radiator $G(z,s)$ has an expansion of the form [55]

$$\begin{aligned} G(z,s) &= \delta(1-z) + \alpha/\pi(a_{11}L + a_{10}) + (\alpha/\pi)^2(a_{22}L^2 + a_{11}L + a_{20}) \\ &+ \ldots + (\alpha/\pi)^n \sum_{i=0}^{n} a_{ni}L^i\ , \end{aligned} \tag{131}$$

where $L = \ln s/m_e^2 \simeq 24.2$ for $\sqrt{s} \simeq m_Z$. All first- and second-order terms are known exactly. The sequence of leading and next-to-leading logs can be exponentiated (closely following [57] the formalism of structure functions in QCD). For $m_Z \approx$ 91 GeV, the convolution displaces the peak by +110 MeV, and reduces it by a factor of about 0.74. The exponentiation is important in that it amounts to a shift of about 14 MeV in the peak position.

A model-independent analysis [58] of the reduced cross section $\hat{\sigma}$ leads to the following general expression (here $m \equiv m_Z$):

$$\begin{aligned} \hat{\sigma}(s) &= [12\pi\Gamma_e\Gamma_f/|D(s)|^2][s/m^2 + R_f(s-m^2)/m^2\ +\ \Gamma/m\ I_f + \ldots] \\ &+[4\pi\alpha^2(m^2)Q_f^2 N_c/3s]\ , \end{aligned} \tag{132}$$

with N_c given by Eq. (127) and

$$D(s) = s - m^2 + im\Gamma[s/m^2 + \epsilon(s-m^2)/m^2] + \ldots\ . \tag{133}$$

This form of the resonant term was obtained by starting from a general renormalizable field theory (the Standard Model being a particular case). Near the resonance, $(s-m^2)/m^2 \approx \Gamma/m$ is of order α (or α_W). As only a perturbative calculation of $\hat{\sigma}$ is possible in α or α_W, at the same level of accuracy we can expand vertices, propagators, etc., in $(s-m^2)/m^2$ near the resonance. For example, for the inverse propagator, we can write the expansion

$$\begin{aligned} D(s) &= s - m^2 + \Pi(s) \\ &= s - m^2 + \text{Re}\ \Pi(m^2) + (s-m^2)\text{Re}\ \Pi'(m^2) + i\ \text{Im}\ \Pi(m^2) + \\ &\quad i(s-m^2)\text{Im}\ \Pi'(m^2) + \ldots \\ &= [1 + \text{Re}\ \Pi'(m^2)]\{s - m^2 + im\ \Gamma[s/m^2 + \epsilon(s-m^2)/m^2 + \ldots]\}\ , \end{aligned} \tag{134}$$

with

$$m\Gamma = \text{Im}\ \Pi(m^2)/[1 + \text{Re}\ \Pi'(m^2)]\ , \tag{135}$$

and Re $\Pi(m^2) = 0$ because of the specific definition of $m \equiv m_Z$ that is adopted. (Recently a different definition of m_Z was discussed [59]. The present prescription would lead to problems related to gauge invariance at the two-loop level. Within the one-loop framework it is perfectly all right and is generally adopted because of its calculational simplicity.) The overall factor 1+Re $\Pi'(m^2)$ is reabsorbed in the numerator. The parameter ϵ measures the deviation from the scaling behaviour of the s-dependent width; ϵ is noticeably different from zero only if, in the final state of Z decays, there are important channels with massive particles (by now we know that this is ruled out). For example, for $Z \to A\bar{A}, \epsilon_A \approx (4m_A^2/m^2)B(Z \to A\bar{A})$. The following identity is valid:

$$\begin{aligned} D(s) &= s - m^2 + im\,\Gamma[s/m^2 + \epsilon(s - m^2)/m^2 + ...] \\ &= [1 + i\gamma(1+\epsilon)](s - \bar{m}^2 + i\bar{m}\bar{\Gamma}) , \end{aligned} \tag{136}$$

with $\gamma = \Gamma/m$ and

$$\bar{m} = m[1 - (\gamma^2/2)(1+\epsilon) + ...] , \tag{137}$$

$$\bar{\Gamma} = \Gamma[1 - (\gamma^2/2)(1+3\epsilon) + ...] . \tag{138}$$

As the factor $1+i\gamma(1+\epsilon) \simeq \exp i\gamma(1+\epsilon)$ cannot be observed from the absolute square of $D(s)$, we see that, on the one hand, at $\epsilon = 0$ the replacement of the constant Γ by the s-dependent width $s\Gamma/m^2$ leads to a variation of the apparent mass by

$$\delta m = -\frac{1}{2}\gamma^2 m \simeq -34 \text{ MeV} , \tag{139}$$

which is clearly an important effect [60],[61]. On the other hand, the additional effect from ϵ,

$$\delta m_\epsilon = -\frac{1}{2}\gamma^2 \epsilon m \simeq -34\epsilon \text{ MeV} , \tag{140}$$

is certainly small because ϵ cannot exceed several per mille or so at most. Note that the effect of ϵ cannot be disentangled from m in the fit, so that it leads to an ambiguity in the determination of m. In conclusion, ϵ can be safely neglected.

Of the two parameters R_f and I_f which appear in Eq. (132) for $\hat{\sigma}$ (in addition to the inclusive partial widths Γ_e and Γ_f), I_f, like ϵ, depends only on the spectrum of particles below the Z. In fact, I_f is determined by absorptive parts of vertices and boxes. Thus large deviations from the Standard Model value cannot occur for I_f. In the Standard Model, I_f is very small [for $\mu^+\mu^-, I_\mu \simeq -(1\text{–}2) \times 10^{-2}$], and as it is multiplied by Γ/m, it can be neglected. The main contribution to R_f is already present at the Born level and arises from γ–Z interference. Higher-order corrections are relatively important, especially for muons. New physics, for example a heavy Z', can modify R_f because the non-resonating background is changed. In the Standard Model, R_f is small [for muons, $R_\mu \simeq (4.5\text{–}6) \times 10^{-2}$, for hadrons $R_h \simeq (0.75\text{–}1) \times 10^{-1}$]. Determining R_f from a fit is difficult. In first approximation, it can be fixed at its Standard Model value. Realistic deviations from the Standard Model would not appreciably affect the determination of m and Γ.

In conclusion, a model-independent analysis of the reduced cross-section proves that a modified Breit–Wigner with s-dependent width, plus photon exchange and interference is a perfectly adequate basis for the experimental study of the line-shape.

Whilst in the Minimal Standard Model all the parameters $\Gamma, \Gamma_e, \Gamma_f, R_f$ and I_f can be computed from m_Z, given m_t and m_H, the general expression of $\hat{\sigma}$ will, in principle, allow a model-independent measurement of the Z parameters. Starting from Eq. (132) for $\hat{\sigma}$, approximate analytic solutions of the convolution integral can be found [58], [62]–[64]. The resulting compact analytic expressions are sufficiently precise for most applications.

11 Precision Tests of the Standard Electroweak Theory

It is clear that the set of input parameters specified in Section 6 can be separated into two parts. On the one hand, $\alpha, G_F, m_Z, m_{f_{light}}$ are well known and the ambiguities associated with these quantities on the radiative corrections are quite small ($m_{f_{light}}$ is actually replaced by $\alpha(m_Z)$ computed via the experimental data on $e^+e^- \to$ hadrons, as discussed in Section 7). We can add α_s to this class, in that, if it is true that the experimental error on α_s is relatively large, it only enters as a small correction to electroweak processes involving hadrons and is practically irrelevant for purely leptonic processes. Also, by fitting the electroweak data one cannot obtain a better value for $\alpha_s(m_Z)$ than that derived from QCD tests. On the other hand m_t and m_H are largely unknown. Thus, for each relevant observable, one can only express the prediction of the Standard Model as a function of m_t and m_H, obtained by using the best available calculations of radiative corrections, with $\alpha, \alpha_s, G_F, m_Z$ and $m_{f_{light}}$ fixed at their experimental values with the corresponding errors. By comparing these predictions with experiment, one can check their mutual consistency and derive constraints on m_t and m_H.

Actually the sensitivity on m_H is so small that for all the measured quantities the ambiguity due to varying m_H in the range 60 GeV $< m_H <$ 1 TeV is below the present experimental error, so that for practical purposes, at the present stage of accuracy, the relevant predictions can be plotted as functions of m_t in the form of a band of values determined by $\delta m_H, \delta m_Z(\delta\alpha_s)$ (see Figs. 5 to 14).

Note that from this point of view $\sin^2\theta_W$ is not a primary quantity. It is not part of the set of input parameters. It is a derived quantity that one could even decide not to introduce at all. I stress this point in order to make it clear that all disputes over which is the better definition of $\sin^2\theta_W$ beyond the tree level are completely secondary. First of all it is always true that physical results are independent of definitions. Differences in physical results obtained from a different definition of input parameters (scheme dependence) can at most occur by terms of higher order, due to the truncation of the perturbative series at a given order. But, for $\sin^2\theta_W$, its precise definition is only necessary to compute it from the input parameters, but cannot matter for the prediction of observables because, with the choice specified above, $\sin^2\theta_W$ is not taken as an input parameter of the theory. And in fact, the emphasis on $\sin^2\theta_W$ is fading away with the improved refinement of precision tests of the Standard Model.

The widespread use of expressing the experimental result for each given observable in terms of the corresponding value of $\sin^2\theta_W$ (within a specified definition for it) is no longer adequate at the present level of sophistication. For comparing the constraining power of different experiments it would be more appropriate to quote the range of m_t implied by each of them [65]. In fact, $\sin^2\theta_W$ is just one particular observable

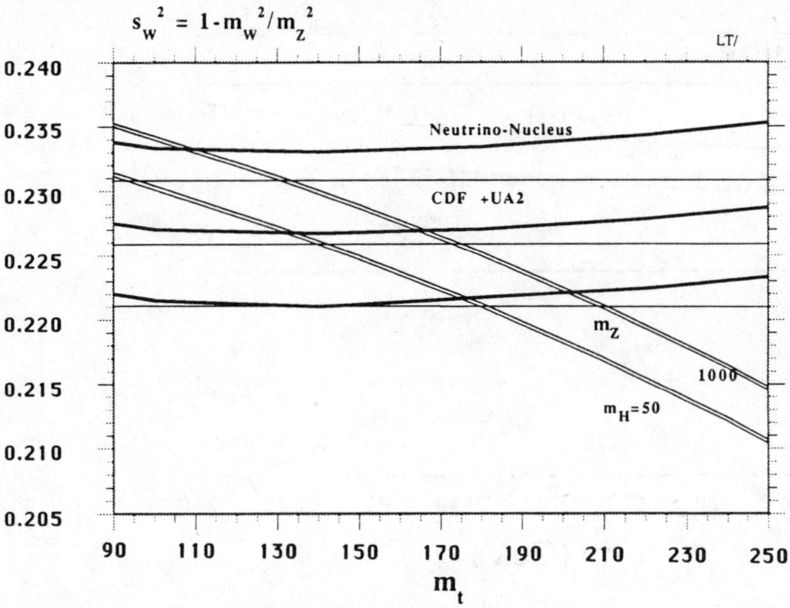

Fig. 5. Labeled by m_Z is the theoretical prediction for s_W^2 obtained from m_Z as a function of m_t for m_H = 50–1000 GeV. The error bands implied by the CDF and UA2 measurements of m_W/m_Z and by the data on R_ν are also shown.

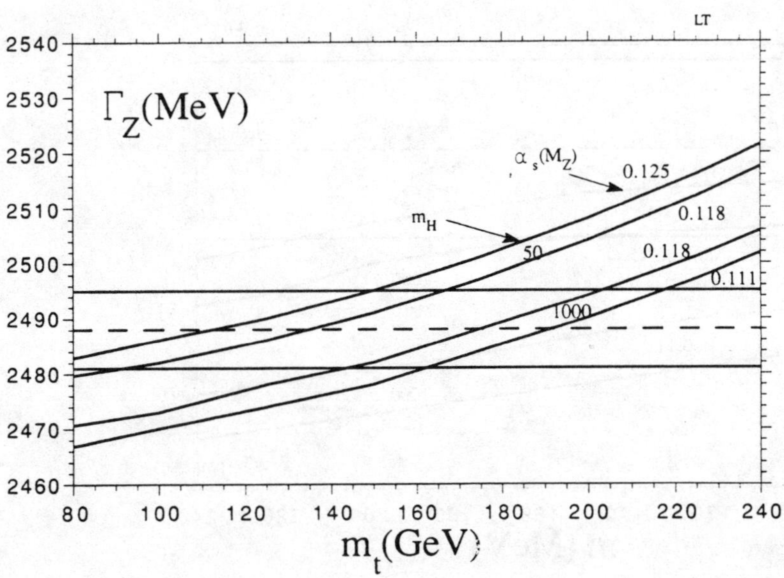

Fig. 6. Γ_Z vs. m_t as predicted by the Standard Model for m_H = 50–1000 GeV and $\alpha_s(m_Z)$ = 0.111–0.125 compared with the LEP result

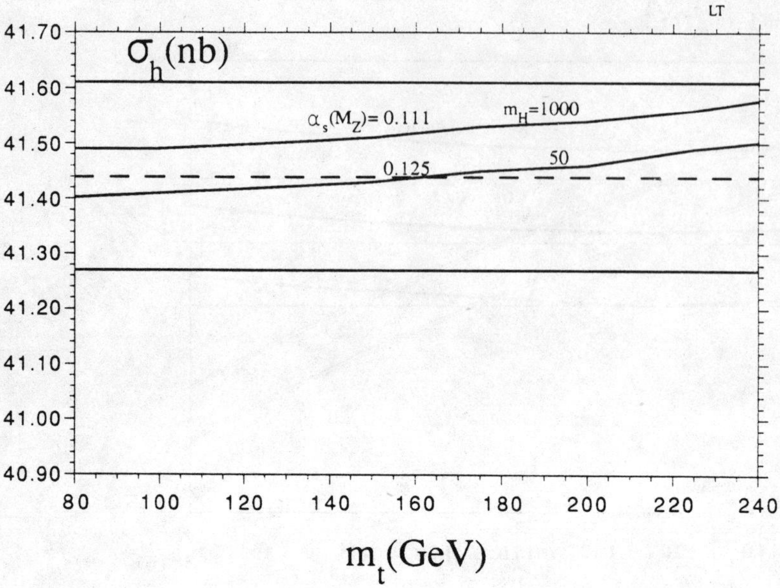

Fig. 7. σ_h vs. m_t as predicted by the Standard Model for m_H = 50–1000 GeV and $\alpha_s(m_Z)$ = 0.111–0.125 compared with the LEP result

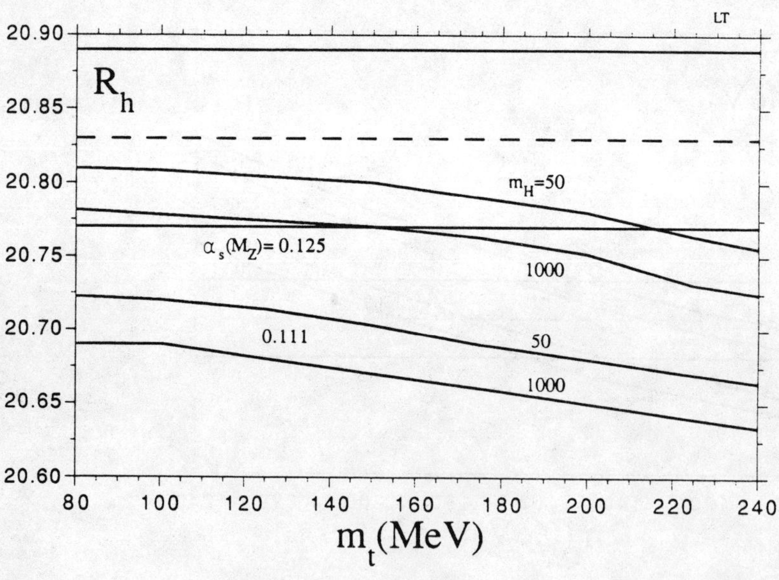

Fig. 8. R_h vs. m_t as predicted by the Standard Model for m_H = 50–1000 GeV and $\alpha_s(m_Z)$ = 0.111–0.125 compared with the LEP result

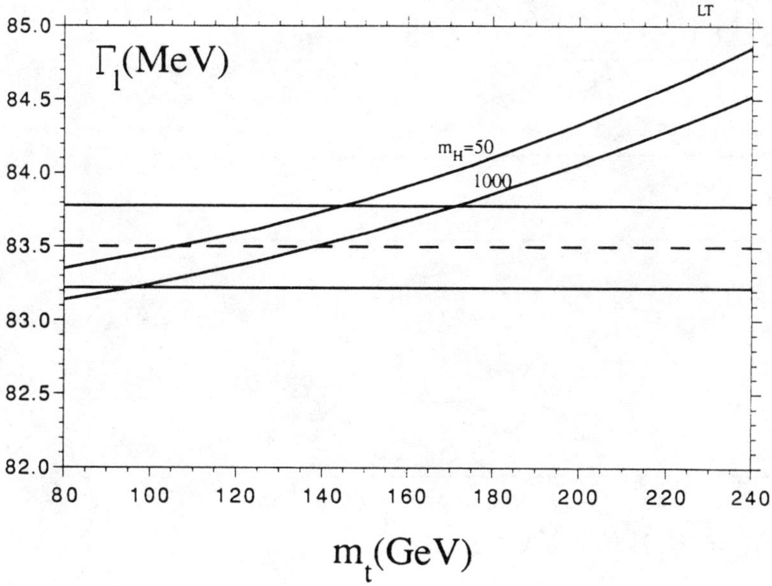

Fig. 9. Γ_ℓ vs. m_t as predicted by the Standard Model for m_H = 50–1000 GeV compared with the LEP result

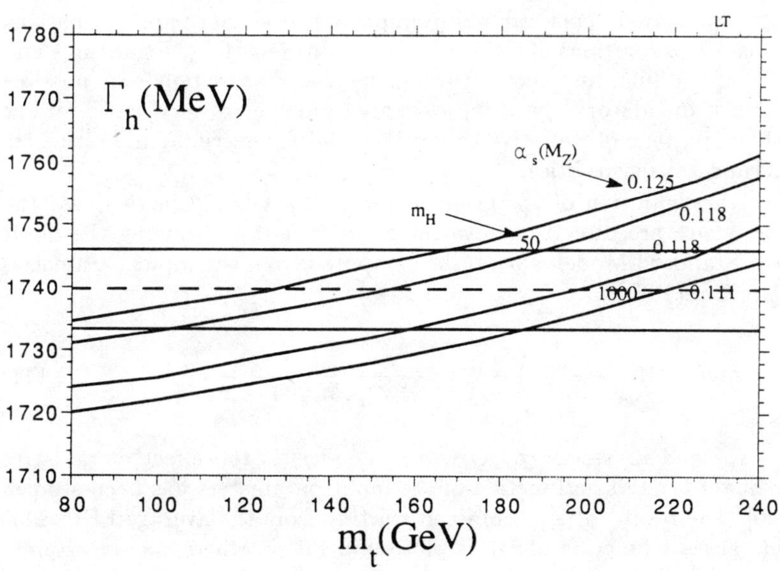

Fig. 10. Γ_h vs. m_t as predicted by the Standard Model for m_H = 50–1000 GeV and $\alpha_s(m_Z)$ = 0.111–0.125 compared with the LEP result

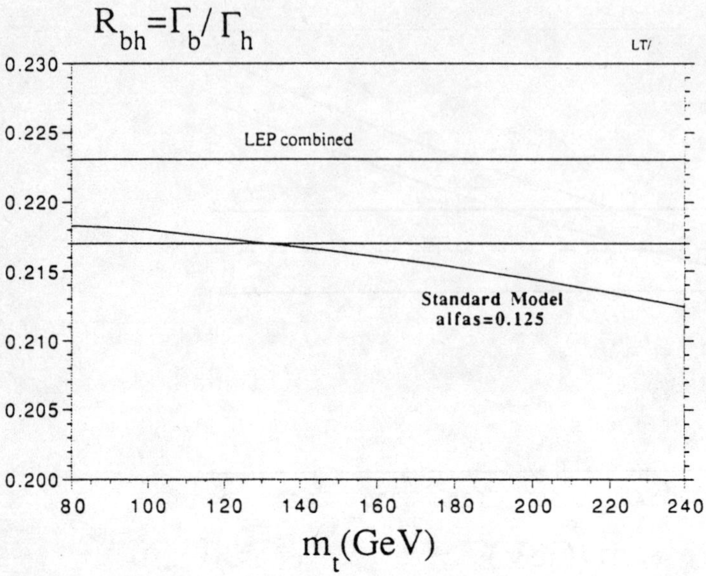

Fig. 11. R_{bh} vs. m_t as predicted by the Standard Model with $\alpha_s(m_Z) = 0.125$. (The dependence on α_s is very mild; $R_{bh} \simeq \left(1 + 0.15\frac{\alpha_s}{\pi}\right)$.)

of the theory. Outside the domain of precision tests, with appropriate definitions of $\sin^2\theta_W$, as was seen in Section 9, one can write simple improved Born approximations that include the main contributions of radiative corrections (e.g. large logarithms and terms of order $G_F m_t^2$). While, for precision tests, the use of as complete as possible radiative corrections is mandatory, these approximate formulae are very useful for our understanding of the pattern of radiative corrections and for everyday-life estimates of rates and experimental sensitivities.

We go back to the definition of s_W^2 given in Eq. (107). Clearly in this case the observables s_W^2 and m_W are directly equivalent given that m_Z is among the input parameters. In the Standard Model, s_W^2 can be computed from the input parameters by the relation [see Eq. (108)]:

$$s_W^2 c_W^2 \equiv \left(1 - \frac{m_W^2}{m_Z^2}\right)\frac{m_W^2}{m_Z^2} = \frac{\pi\alpha}{\sqrt{2}G_F}\,\frac{1}{m_Z^2}\,\frac{1}{1-\Delta r} \tag{141}$$

where $c_W^2 = 1 - s_W^2$ and $\Delta r \equiv \Delta r(\alpha_s, G_F, m_Z, m_f, m_H)$ is the effect of radiative corrections. The quantity Δr as a function of the input parameters has been studied in great detail [40]. The result for s_W^2, obtained starting from the average LEP value for m_Z (see Table 1), as a function of m_t, is plotted in Fig. 5, where the uncertainty for 50 GeV $< m_H <$ 1 TeV is also visible. We see that m_t is the main unknown in the calculation of m_W/m_Z from m_Z, followed in importance by the ambiguities from varying the Higgs mass in the above range and the value of $\alpha(m_Z)$, while the remaining uncertainty from the experimental error on m_Z is very small.

When the available direct experimental information on m_W/m_Z is added, the sensitivity of s_W^2 to m_t provides a very strong constraint on m_t. The ratio m_W/m_Z is directly measured at hadron colliders but can also be indirectly obtained (assuming the validity of the Standard Model) from the ratio $R_\nu = \sigma^{NC}/\sigma^{CC}$ of neutral current (NC) to charged current (CC) cross-sections in neutrino–nucleus deep inelastic scattering. The value of m_W/m_Z has been measured at hadron colliders [66]. From CDF and UA2 we have the results reported in Table 2, where the PDG'92 value is also shown (which includes in addition UA1 data on $m_W - m_Z$). We use the latter value in the following.

Table 2

Experiment	m_W/m_Z	$s_W^2 = 1 - m_W^2/m_Z^2$
CDF	0.8763±0.0043	0.2320±0.0075
UA2	0.8813±0.0041	0.2234 ± 0.0072
Average	0.8789±0.0030	0.2275±0.0052
PDG'92	0.8798±0.0028	0.2259±0.0049

By combining m_W/m_Z from PDG'92 [3] with the LEP value for m_Z one obtains $m_W = 80.23 \pm 0.26$ GeV. The corresponding average value of s_W^2 is also shown in Fig. 5 as a horizontal band, obviously independent of m_t, in the s_W^2–m_t plane.

As is well known, the value of s_W^2 extracted from R_ν is also nearly independent of m_t in the interesting range of values for the top mass. This fact arises from a largely accidental cancellation [67], specific to this process and to the Standard Model, between two different sources of m_t dependence, as discussed in the following.

The ratio $R_\nu = \sigma_{NC}/\sigma_{CC}$ for ν–N scattering is given in terms of s_W^2 by [see Eqs. (101) and (102)]:

$$R_\nu = \rho_{\nu N}^2 \left(\frac{1}{2} - k_{\nu N} s_W^2 + \frac{5}{9}(k_{\nu N} s_W^2)^2(1+r)\right) + \ldots \,, \tag{142}$$

where $r = (\sigma^{\bar\nu}/\sigma^\nu)_{CC} \simeq 0.4$ is also measured. The tree approximation (with $\rho_0 = 1$) is recovered for $\rho_{\nu N} = k_{\nu N} = 1$. Some large logarithms from the radiative corrections to σ^{CC} are also included in $\rho_{\nu N}$. But for the sake of this argument we are only considering the $G_F m_t^2$ terms. For fixed $R_\nu =$ (the experimental value) and $s_W^2 \sim 0.23$ there is a strong cancellation in the Standard Model between the m_t dependence of $\rho_{\nu N} \simeq 1 + \delta\rho$ and of $k_{\nu N} \simeq 1 + c_W^2/s_W^2 \delta\rho$ [Eqs. (111),(112)], so that as a result $\delta s_W^2 \simeq 0.2\delta\rho$, where $\delta\rho$ is given in Eq. (110). For realistic values of m_t the resulting contribution of the quadratic m_t terms is no longer dominant.

The most precise experimental results on R_ν were obtained by the CHARM [68] and CDHS [69] collaborations at CERN. Recently, a new result from the CCFR Collaboration at FNAL has been published. The original results on $\sin^2\theta_W$ were given for fixed m_t and m_H. CHARM obtained $s_W^2 = 0.236 \pm 0.005$ (exp) $\pm$ 0.005 (th) for $m_t = 45$ GeV and $m_H = 100$ GeV, while the CDHS result was $s_W^2 = 0.2275 \pm 0.005$ (exp) $\pm$ 0.005 (th) for $m_t = 60$ GeV and $m_H = 100$ GeV. The theoretical error arises from hadronic uncertainties and the effect of the charm threshold. An average at

$m_t = 60$ GeV and $m_H = 100$ GeV gives $s_W^2 = 0.232 \pm 0.006$ (where the error 6×10^{-3} is obtained as $6 \times 10^{-3} = \sqrt{(5/\sqrt{2})^2 + 5^2} \times 10^{-3}$). The new result from CCFR [70] quoted for $m_t = 150$ GeV and $m_H = 100$ GeV, $s_W^2 = 0.2222 \pm 0.057$ lowers the average at $m_t = 60$ GeV to something like 0.229 ± 0.006. The corresponding combined result at different values of m_t and m_H can also be obtained from the known form of the radiative corrections. The result is shown in Fig. 5 [71].

There are many more, less precise experimental results on s_W^2 from low-energy neutral current data, most of them being well known since a long time [72],[73]. These additional data are all consistent among them and with the results in Fig. 5.

We now consider the implications for the standard electroweak theory of the LEP results on the Z partial widths and asymmetries. About 2 million Z events were collected by the four LEP experiments in the '89–'91 runs and 3 million more were stored on tape in '92. The final results from the analysis of the '89–'91 data plus some preliminary results from '92 run are now available [32],[33] and will be discussed in the following. From the shape of the resonance, measured by energy scanning, one obtains m_Z and Γ_Z. Since in '92 there has been no scanning, i.e. all data were collected at the Z peak, there is essentially no progress on m_Z and Γ_Z with respect to the '91 results. At the peak the basic measurements are those of the hadronic peak cross-section σ_h (deconvoluted from all QED effects: $\sigma_h = \frac{12\pi}{m_Z^2}\frac{\Gamma_e\Gamma_h}{\Gamma_Z^2}$), the ratio R_h of hadrons to leptons: $R_h = \Gamma_h/\Gamma_\ell$ (either using e-μ-τ universality, which is supported by the data within the present accuracy, or separately for each charged lepton type) and the ratio $R_{bh} = \Gamma_b/\Gamma_h$ of the $Z \to b\bar{b}$ to the hadronic widths. From the above primary measurements one can also derive the experimental values of $\Gamma_h, \Gamma_\ell, \Gamma_b$ and Γ_{inv} (or equivalently N_ν the number of light neutrinos, which is now $N_\nu = 2.99 \pm 0.03$) given in Table 3. In Figs. 6 to 11, we compare the data on the Z widths (collected in Table 3) with the predictions of the Standard Model, obtained with the programme ZFITTER [74] which includes a state-of-the-art set of electroweak radiative corrections.

Table 3: Summary of combined data used in the text

M_Z	91.187 ± 0.007
m_W/m_Z	0.8798±0.0028
Γ_T (MeV)	2488 ± 7
$R = \Gamma_h/\Gamma_\ell$	20.83±0.06
$\sigma_h = \frac{12\pi\Gamma_e\Gamma_h}{m_Z^2\Gamma_T^2}$(nb)	41.44 ± 0.17
Γ_ℓ(MeV)	83.52 ± 0.28
Γ_h (MeV)	1739.7 ± 6.3
Γ_b (MeV)	383 ± 6
$R_{bh} = \Gamma_b/\Gamma_h$	0.220 ± 0.003
A_{FB}^ℓ	0.0164 ± 0.0021
A_{pol}^τ	0.142 ± 0.017
A^e	0.130 ± 0.025
A_{FB}^b	0.098 ± 0.09
g_v/g_a (all asymmetries)	0.0725 ± 0.0033

Additional important information is provided by the measurement of a number of asymmetries. In particular we refer to the forward–backward asymmetries for charged leptons $\left(A_{FB}^{\ell}\right)$ and for the b-quark $\left(A_{FB}^{b}\right)$ and the τ polarization asymmetry $\left(A_{pol}^{\tau}\right)$. The value of those asymmetries, combined over the LEP experiments, are given in the following [32],[33]. For A_{FB}^{ℓ} deconvoluted for all QED effects, one has:

$$A_{FB}^{\ell}(\sqrt{s}=m_Z)=0.0164\pm 0.0021 \tag{143}$$

which corresponds to $x=g_V/g_A=0.0743\pm 0.0048$ via the relation:

$$A_{FB}^{\ell}=\frac{3x^2}{(1+x)^2}. \tag{144}$$

For A_{FB}^{b} the result, after correction for the $B-\bar{B}$ mixing effect and for QCD effects [75],[76], is given by

$$A_{FB}^{b}(\sqrt{s}=m_Z)=0.098\pm 0.09\ . \tag{145}$$

For A_{pol}^{τ} one has:

$$A_{\mathrm{pol}}^{\tau}(\sqrt{s}=m_Z)=\frac{2x}{1+x^2}=0.142\pm 0.017\ . \tag{146}$$

Actually, from the angular dependence of the τ polarization, the analogue of A_τ for the electron, A_e, can be measured. LEP finds:

$$A_\ell=0.130\pm 0.025\ . \tag{147}$$

Assuming universality the values in Eqs.(146), (147) can be combined:

$$A_\tau=0.138\pm 0.014 \tag{148}$$

which is the value shown in Fig. 14. It is important to mention that recently the first results on A_{LR} with polarized e^- beams were published by the SLD collaboration at SLAC [77]. The measured value is $A_{LR}=0.100\pm 0.044$. The combined value of $x=g_V/g_A$ from all asymmetries measured at LEP is shown in Table 3. The experimental results on the asymmetries are compared with the Standard Model predictions in Figs. 12–14.

All measurements are in good agreement with the Standard Model and a consistent range of m_t values is indicated. When the data are fitted in the Standard Model (taking the experimental correlations also into account, see Ref. [32] with $\alpha_s(m_Z)=0.118\pm 0.007$ fixed, one obtains

$$\begin{aligned} m_t &= 132\pm 22 \text{ GeV for } m_H\sim 60 \text{ GeV}\\ &= 174\pm 17 \text{ GeV for } m_H\sim 1 \text{ TeV}. \end{aligned}$$

When also the data on m_W/m_Z [66], R_ν [69],[70] and atomic parity violation on Cs [78] are added, the corresponding results are

$$\begin{aligned} m_t &= 135\pm 18 \text{ GeV for } m_H\sim 60 \text{ GeV}\\ &= 174\pm 15 \text{ GeV for } m_H\sim 1 \text{ TeV}. \end{aligned}$$

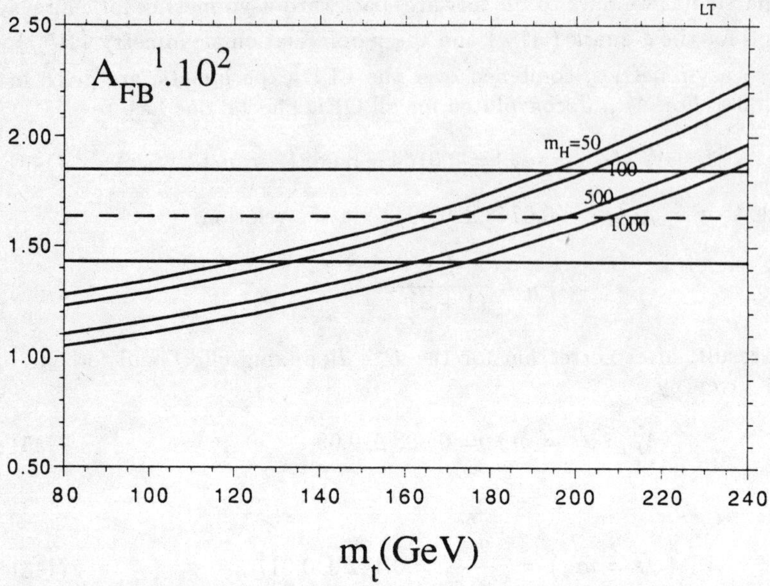

Fig. 12. $A^{\ell}_{FB} \times 10^2$ vs. m_t as predicted the Standard Model for $m_H = 50$–1000 GeV compared with the LEP result

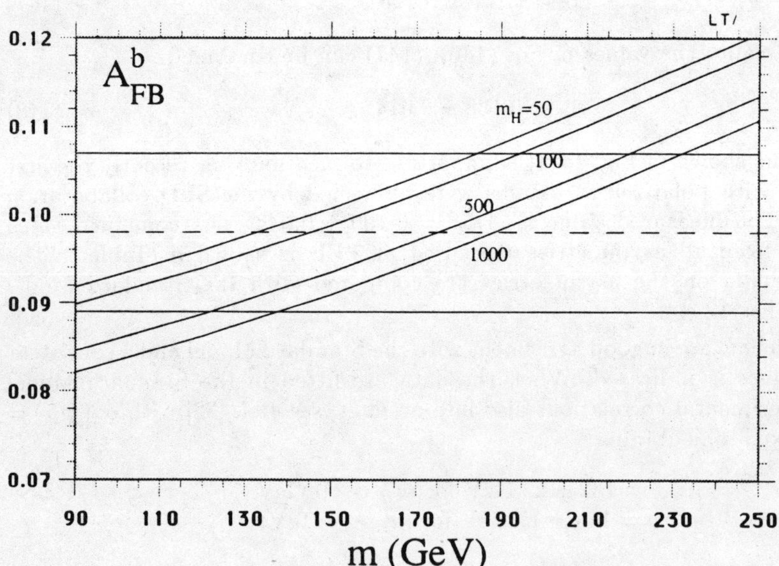

Fig. 13. A^{b}_{FB} vs. m_t as predicted by the Standard Model for $m_H = 50$–1000 GeV compared with the LEP result

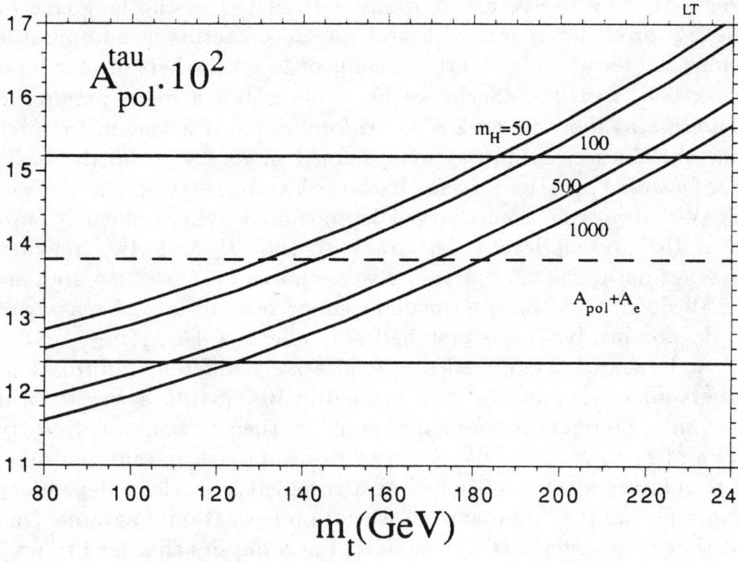

Fig. 14. A^{τ}_{pol} vs. m_t as predicted by the Standard Model for $m_H = 50$–1000 GeV compared with the LEP result

With the progress of LEP the direct impact of other electroweak tests on m_t is becoming marginal. The χ^2 value prefers small m_H, but within 2σ value of m_H as large as 1 TeV or more are allowed. If $\alpha_s(m_z)$ is left free, the fit leads to the value [2]

$$\alpha_s(m_Z) = 0.128 \pm 0.008$$

while the central values of m_t are shifted down by ~ 2 GeV.

Finally, it is interesting to note that the observed magnitude of the radiative corrections implies a lower bound on m_t as well, i.e., $m_t > 80$ GeV at 95% c.l.

12 Toward a Model-Independent Analysis of the Data

Recently much attention has been devoted to develop a strategy for a model independent analysis of precision electroweak data [79]–[84]. The general framework consists in defining a set of variables describing possible deviations from the Standard Model. These variables are chosen in such a way as to minimize the negative impact of our ignorance of the top quark mass m_t and to be as much as possible sensitive to new physics in particular problematic sectors of the theory, such as the Higgs sector. A very common approach [79]–[81], in terms of the variables S, T and U, is based on the assumption that the dominant effects of new physics only appear through vacuum

[2] Note that from the experimental value of R_h alone, shown in Table 3, which is often used by QCD experts, one would obtain $\alpha_s(m_Z) = 0.131 \pm 0.010$ by using $(R_h)_0 = 19.945 \pm 0.03$, the updated value of the QCD-uncorrected prediction [32].

polarization diagrams [85]. Alternative assumptions, formulated in the language of effective Lagrangians [82] or of dominance of lowest dimension terms in an operator expansion [83] or in physical terms [86], lead to the same or to similar sets of variables.

In a previous work [84], with S. Jadach, we have advocated a more pragmatic approach where no specific assumptions are *a priori* formulated. Rather, in the first paper of Ref. [84] the variables ϵ_1, ϵ_2 and ϵ_3 were defined in general from the basic observables $m_W/m_Z, \Gamma_\ell$ and A^ℓ_{FB} (the forward-backward asymmetry at the Z-peak for charged leptons). We always assume charged lepton universality, which is supported by the data at the present level of accuracy, so that Γ_ℓ and A^ℓ_{FB} refer to the corresponding average data. These observables were chosen because they are particularly simple, are all defined at the intermediate gauge boson scale of mass, are precisely measured, do not involve important hadronic effects (like $\alpha_s(m_Z)$ or the $Z \to b\bar{b}$ vertex [87]) and each of them carries a qualitatively different information. In terms of these observables ϵ_1, ϵ_2 and ϵ_3 were defined in Ref. [84] in a way that, if the assumption of vacuum polarization dominance is made, then ϵ_1 would be linearly related to T, ϵ_2 to U and ϵ_3 to S. But one virtue of this approach is that the most interesting physical results are already obtained at a completely model independent level, without assumptions like the dominance of vacuum polarization diagrams. One example is the result that present data favour models of new physics that lead to negative corrections to ϵ_3 with respect to the Standard Model, like suitable versions of extended gauge models [88] or the Minimal Supersymmetric Standard Model [89],[90], while simple technicolour models [91],[92] are disfavoured.

Of course some assumptions are needed when additional observables other than $m_W/m_Z, \Gamma_\ell$ and A^ℓ_{FB} are introduced and one wants to relate them to the same set of epsilons in order to make their determination from the data more precise. One can formulate a hierarchy of simple and rather general assumptions valid in large classes of models which are needed in order to relate the epsilons to a progressively larger set of observables. First, very mild assumptions are required in order that all other observables connected to charged leptons at the Z pole, e.g. the τ-polarization asymmetry $A^\tau_{\rm pol}$ and the left-right asymmetry A_{LR}, are uniquely determined by the ϵ_i. This is true in all models where the contributions of new physics only occur either through vacuum polarization terms and/or in $Z \to \ell^+\ell^-$ vertex corrections of the form $\Delta V_\mu(Z \to \ell^+\ell^-) = \bar{u}(\Delta g_A \gamma_5 + \Delta g_V)\gamma_\mu u$ with Δg_V and Δg_A real form factors universal for all charged lepton flavours. These two kinds of contributions cannot be disentangled if we only consider on-shell Z properties in the charged lepton sector. Stronger hypotheses about possible forms of new physics are needed if one wants to relate the epsilons to observables involving quarks, like the hadronic widths or asymmetries, or to observables measured at low q^2, far from the Z pole (e.g. deep inelastic neutrino-nucleus scattering [93],[94] , ν-e scattering [95] and atomic parity violation in atomic physics [96],[97]). For example, we could restrict ourselves to the case of new physics in oblique corrections with the assumption that the q^2-dependence of the relevant vacuum polarization amplitudes is weak enough for all second and higher order derivatives in q^2 to be safely neglected [79]–[81]. However, because m_t is unknown, the observables related to the b-quark, e.g. the b partial width and the forward-backward b asymmetry A^b_{FB}, in principle cannot be obtained from the ϵ_i's even if the new physics only occurs in oblique and/or in universal Z-vertex corrections. This is because of the large m_t-dependent Standard Model corrections to the $Z \to b\bar{b}$

vertex [87]. As a consequence the relation between the previous ϵ_i's and Γ_b or A^b_{FB} can only be specified for a given value of m_t. In practice, the sensitivity to the new quadratic terms in m_t^2 from the $Z \rightarrow b\bar{b}$ vertex is only present in Γ_b, and consequently in the hadronic and total widths Γ_h and Γ_t, while it is almost non-existent, at realistic values of m_t, for A^b_{FB}. For this reason and because of the dependence on $\alpha_s(m_Z)$, the precise measurements of Γ_T, Γ_h and Γ_b were not included in the previous analysis [84] (while A^b_{FB} was included).

In the second paper of Ref. [84] we have overcome the above limitation by adding a new parameter, ϵ_b, which describes the m_t-dependent part of the $Z \rightarrow b\bar{b}$ vertex, and by correspondingly enlarging the set of basic observables with the inclusion of Γ_b. From the quantities $m_W/m_Z, \Gamma_\ell, A^\ell_{FB}$ and Γ_b the four epsilons can be defined without need of specifying m_t. The important point is that, in the Standard Model, for all observables at the Z pole, the whole dependence on m_t arising from one-loop diagrams only enters through the epsilons. Furthermore, the same is obviously true for any extension of the Standard Model such that all possible deviations only occur through vacuum polarization diagrams and/or the $Z \rightarrow b\bar{b}$ vertex. At the same time we have improved the previous analysis (first paper of Ref. [84]) by slightly modifying the original definition of the epsilons in order to achieve a number of advantages: the modified variables are exactly zero in the Standard Model in the limit of neglecting all pure electroweak loop corrections (i.e. when only the predictions from the tree level Standard Model plus pure QED corrections are taken into account), are more directly related to the forward-backward asymmetry at the peak as by now precisely specified by the LEP experiments [98] and some small top quark mass dependencies are better taken into account. In the present article we have updated the analysis described in the second paper of Ref. [84] by including the new results presented at the La Thuile and Moriond '93 Conferences [98],[99].

12.1 Basic Definitions

In the Standard Model all the observables can be computed starting from the input parameters, a self-imposing set of which is given by $\alpha, \alpha_s, G_F, m_Z, m_f, m_H$. In the following we use the values $\alpha^{-1} = 137.036, G_F = 1.166372 \times 10^{-5}\ \mathrm{GeV}^2, m_Z = 91.187$ GeV. Among the fermion masses m_f, the main unknown quantity is the top quark mass m_t. The CDF lower limit on $m_t, m_t > 108$ GeV, is valid in the Standard Model [100]. The failure of direct searches at LEP imposes the bound $m_H > 62.5$ GeV on the Higgs mass [101]. Also, the Standard Model is affected with serious pathologies for $m_H > 0.6$-1 TeV, when the Landau pole associated to the $\lambda(\phi^+\phi)^2$ coupling comes too close to the physical region in parameter space [102]. Thus in evaluating the Standard Model predictions we vary m_H in the range 60 GeV-1 TeV. Finally direct measurements of $\alpha_s(m_Z)$ from many different experiments lead to a relatively precise value of $\alpha_s(m_Z)$ [103]. In the present discussion we use $\alpha_s = 0.118 \pm 0.007$ [103].

We start from the basic observables $m_W/m_Z, \Gamma_\ell$ and A^ℓ_{FB} and Γ_b. From these four quantities one can isolate the corresponding dynamically significant corrections $\Delta r_W, \Delta\rho_N, \Delta k_N$ and ϵ_b, which contain the small effects one is trying to disentangle, and are defined in the following. First we introduce Δr_W as obtained from m_W/m_Z

by the relation:

$$\left(1-\frac{m_W^2}{m_Z^2}\right)\frac{m_W^2}{m_Z^2}=\frac{\pi\alpha(m_Z)}{\sqrt{2}\,G_F m_Z^2(1-\Delta r_W)}\,. \tag{149}$$

Here $\alpha(m_Z)=\alpha/(1-\Delta\alpha)$ is fixed to the conventional value 1/128.87 so that the effect of the running of α due to known physics is extracted from $(1-\Delta r)=(1-\Delta\alpha)(1-\Delta r_W)$. In fact, in the Standard Model the value of $1/\alpha(m_Z)$ should be 128.87 ± 0.12 [104]. A possible departure from this value would then be included in Δr_W. The present definition of Δr_W is unchanged with respect to both papers of Ref. [84]. In order to define $\Delta\rho_N$ and Δk_N we consider the effective vector and axial-vector couplings g_V and g_A of the on-shell Z to charged leptons, given by the formulae:

$$\Gamma_\ell=\frac{G_F m_Z^3}{6\pi\sqrt{2}}(g_V^2+g_A^2)\left(1+\frac{3\alpha}{4\pi}\right) \tag{150}$$

$$A_{FB}^\ell(\sqrt{s}=m_Z)=\frac{3g_V^2g_A^2}{(g_V^2+g_A^2)^2}=\frac{3x^2}{(1+x^2)^2}\,. \tag{151}$$

Note that Γ_ℓ stands for the inclusive partial width $\Gamma(Z\to\ell\ell+\text{photons})$. With respect to the definition of g_V and g_A in the first Ref. [84] we stress the following differences, which have been introduced in the second. First, we have extracted from $(g_V^2+g_A^2)$ the factor $(1+3\alpha/4\pi)$ which is induced in Γ_ℓ from final state radiation. Second, by the asymmetry at the peak in Eq. (151) we mean the quantity which is now commonly referred to by the LEP experiments (denoted as A_{FB}^0 in Ref. [98]), which is corrected for all QED effects, including initial and final state radiation and also for the effect of the imaginary part of the γ vacuum polarization diagram. The last effect was instead not corrected for in the first [84]. In terms of g_A and $x\ g_V/g_A$, $\Delta\rho_N$ and Δk_N are given by [84]:

$$\begin{aligned}g_A&=-\frac{\sqrt{\rho_N}}{2}=-\frac{1}{2}\left(1+\frac{\Delta\rho_N}{2}\right)\\ x&=\frac{g_V}{g_A}=1-4\bar{s}_W^2=1-4(1+\Delta k_N)s_0^2\,.\end{aligned} \tag{152}$$

The index N for 'new' is there to recall that the corresponding definition was changed in the second Ref. [84] with respect to the first. In fact the formal relations with the couplings are identical but the definition of g_V and g_A has been modified (it is only for simplicity that we do not append the index N to them too). In Eq. (152) $\bar{s}_W^2$ is an effective $\sin^2\theta_W$ for on-shell Z, defined by this very equation, while s_0^2 is the corresponding quantity before non pure-QED corrections, given by:

$$s_0^2c_0^2=\frac{\pi\alpha(m_Z)}{\sqrt{2}G_F m_Z^2} \tag{153}$$

with $c_0^2=1-s_0^2$ ($s_0^2=0.231184$ for $m_Z=91.187$ GeV).

We now define ϵ_b from Γ_b, the inclusive partial width for $Z\to b\bar{b}$ according to the relation

$$\Gamma_b=\frac{G_F m_Z^3}{6\pi\sqrt{2}}\,\beta\left(\frac{3-\beta^2}{2}\,g_{bV}^2+\beta^2g_{bA}^2\right)N_C R_{QCD}\left(1+\frac{\alpha}{12\pi}\right)\,. \tag{154}$$

where $N_C = 3$ is the number of colours, $\beta = \sqrt{1 - 4m_b^2/m_Z^2}$, with $m_b = 4.8$ GeV, R_{QCD} is the QCD correction factor given by

$$R_{QCD} = 1 + 1.2a - 1.1a^2 - 13a^3 \; ; \; a = \frac{\alpha_S(m_Z)}{\pi} \tag{155}$$

(for $\alpha_s(m_Z) = 0.118, R_{QCD} = 1.0428$) and g_{bV}, g_{bA} are specified as follows

$$\begin{aligned} g_{bA} &= -\frac{1}{2}\left(1 + \frac{\Delta\rho_N}{2}\right)(1+\epsilon_b) \\ \frac{g_{bV}}{g_{bA}} &= \frac{1 - \frac{4}{3}\bar{s}_W^2 + \epsilon_b}{1+\epsilon_b} = \frac{1 - \frac{4}{3}(1+\Delta k_N)s_0^2 + \epsilon_b}{1+\epsilon_b} \end{aligned} \tag{156}$$

with $\Delta\rho_N, \Delta k_N, \bar{s}_W^2$ and s_0^2 as defined in Eqs. (150)–(153). The parameter ϵ_b in the Standard Model is closely related to the quantity $-\text{Re}\{\delta_{b-\text{vertex}}\}$ defined in the last of Refs. [87].

As is well known [105], in the Standard Model, $\Delta r_W, \Delta\rho_N, \Delta k_N$ and ϵ_b, for sufficiently large m_t, are all dominated by quadratic terms in m_t of order $G_F m_t^2$. As new physics can more easily be disentangled if not masked by large conventional m_t effects, it is convenient to keep $\Delta\rho_N$ and ϵ_b while trading Δr_W and Δk_N for two quantities with no contributions of order $G_f m_t^2$. We thus introduce the following linear combinations [80],[84]:

$$\begin{aligned} \epsilon_{N1} &= \Delta\rho_N \\ \epsilon_{N2} &= c_0^2\Delta\rho_N + \frac{s_0^2\Delta r_W}{(c_0^2 - s_0^2)} - 2s_0^2\Delta k_N \\ \epsilon_{N3} &= c_0^2\Delta\rho_N + (c_0^2 - s_0^2)\Delta k_N \; . \end{aligned} \tag{157}$$

In analogy with $\Delta\rho_N$ and Δk_N the new epsilons are labeled ϵ_N. Clearly ϵ_{N2} and ϵ_{N3} no longer contain terms of order $G_F m_t^2$ but only logarithmic terms in m_t. The leading terms for large Higgs mass, which are logarithmic, are mainly contained in ϵ_{N1} but are also present in ϵ_{N3}. In the Standard Model one has the following 'large' asymptotic contributions [87],[106],[107]:

$$\begin{aligned} \epsilon_{N1} &= \frac{3G_F m_t^2}{8\pi^2\sqrt{2}} - \frac{3G_F m_W^2}{4\pi^2\sqrt{2}}\text{tg}^2\theta_W \ln\left(\frac{m_H}{m_Z}\right) + \ldots \\ \epsilon_{N2} &= -\frac{G_F m_W^2}{2\pi^2\sqrt{2}} \ln\left(\frac{m_t}{m_Z}\right) + \ldots \\ \epsilon_{N3} &= \frac{G_F m_W^2}{12\pi^2\sqrt{2}} \ln\left(\frac{m_H}{m_Z}\right) - \frac{G_F m_W^2}{6\pi^2\sqrt{2}} \ln\left(\frac{m_t}{m_Z}\right) + \ldots \\ \epsilon_b &= -\frac{G_F m_t^2}{4\pi^2\sqrt{2}} + \ldots \end{aligned} \tag{158}$$

The relations between the basic observables and the epsilons can be linearised and inverted, leading to the formulae

$$\begin{aligned} \epsilon_{N1} &= -0.9882 + 0.011963\Gamma_\ell\,(\text{MeV}) - 0.1511x \\ \epsilon_{N3} &= -0.7146 + 0.009181\Gamma_\ell\,(\text{MeV}) - 0.69735x \end{aligned}$$

$$\epsilon_b = -0.62\epsilon_{N1} + 0.24\epsilon_{N3} + 0.436\left[\frac{\Gamma_b}{\Gamma_{b0}} - 1\right]$$

$$\epsilon_{N2} = \frac{1}{c_0^2 - s_0^2}\left[c_0^2\epsilon_{N1} - 2s_0^2\epsilon_{N3} + s_0^2\Delta r_W\right] = 1.43\epsilon_{N1} - 0.86\epsilon_{N3} + 0.43\Delta r_W \quad (159)$$

where $x = g_V/g_A$ is defined in Eq. (151) from A^ℓ_{FB}. The quantity Γ_{b0} is the value of Γ_b in the limit when all epsilons are neglected. With $\alpha_s(m_Z) = 0.118, \Gamma_{b0} = 379.4$ MeV [more generally its value is given by Eqs. (154)–(156) and (175)].

Equations (159) are also used to obtain, by definition, the behaviour of the epsilons in the Standard Model as functions of m_t and m_H: by means of the latest version of the programme ZFITTER [108] we compute Γ_ℓ, Γ_b, x and Δr_W as functions of m_t and m_H; we obtain the results for $\epsilon_{N1}, \epsilon_{N2}$ and ϵ_{N3} shown in Fig. 15, while the result for ϵ_b, which is practically independent of m_H, is plotted in Fig. 16 for $\alpha_s(m_Z) = 0.118$. The above results were also checked by using the new code TOPAZ0 [109]. As already mentioned, in the Standard Model the dependence on m_t and m_H of all observables at the Z pole only enters through the epsilons (see Section 12.3). This is also true in any extension which deviates from the Standard Model in vacuum polarizations and/or in the $Z \to b\bar{b}$ vertex. This is an exact statement if one loop corrections are only considered. At two loops, some residual dependence on m_t is induced in the various observables from terms of order α_s^2 [110], which can be reabsorbed in the epsilons only in an approximate but sufficiently accurate way.

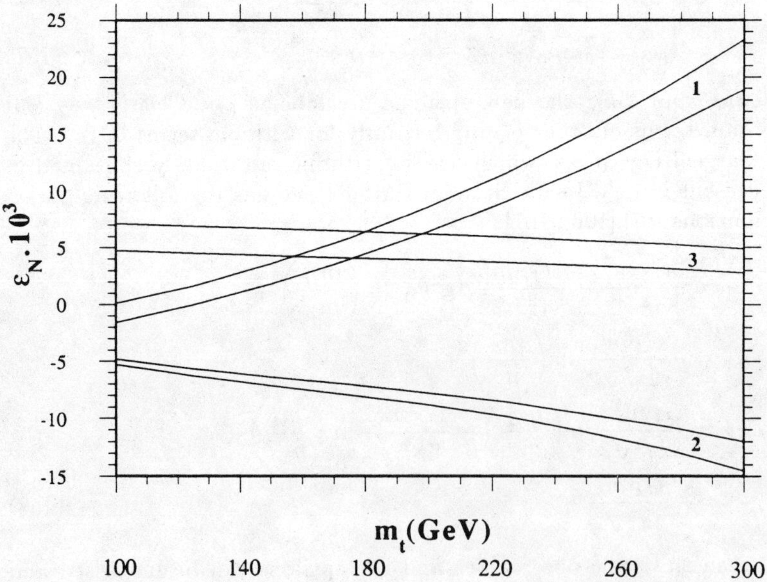

Fig. 15. $\epsilon_{N1}, \epsilon_{N2}$ and ϵ_{N3} vs. m_t as predicted in the Standard Model for $m_H = 50$–1000 GeV

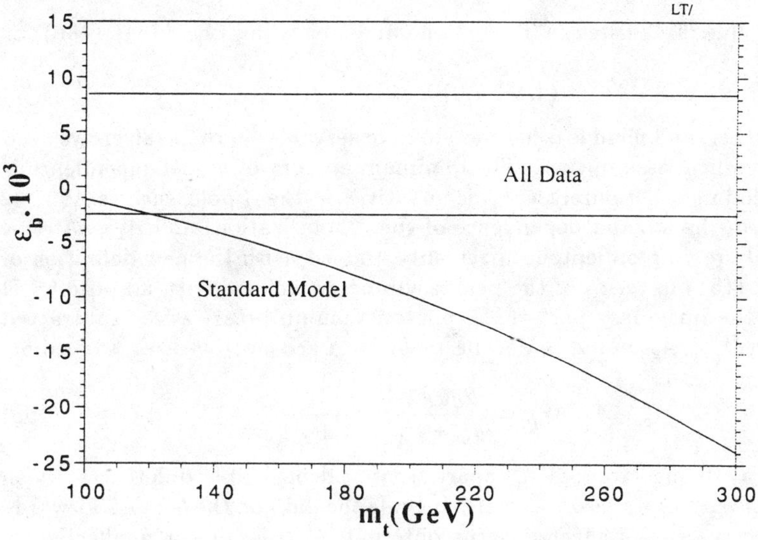

Fig. 16. ϵ_b vs. m_t as predicted by the Standard Model (there is no dependence on m_H). The band is the value fitted in Eqs. (180) from all the data analyzed with $\alpha_s(m_Z) = 0.118 \pm 0.007$.

12.2 Determination of the Epsilons from the Data

The experimental results that will be used in the present analysis are collected in Table 3 (Section 11). We start by deriving from the data on the input observables the values of the epsilons that follow by directly applying their definition, with no dynamical assumptions.

By combining the value of m_W/m_Z [111] with the LEP value for the Z mass [98], both given in Table 3, and using Eq. (149) one obtains:

$$\Delta r_W = (-16 \pm 16) \times 10^{-3} \,. \tag{160}$$

From the LEP results on the charged lepton partial width and the forward-backward asymmetry, assuming e-μ-τ universality, given in Table 3, one finds:

$$g_A^2 = 0.2499 \pm 0.0009 \tag{161}$$

$$x = g_V/g_A = 0.0743 \pm 0.0048 \quad \text{or} \quad \bar{s}_W^2 = 0.2314 \pm 0.0012 \,. \tag{162}$$

One can now use Eqs. (152) to derive the results for $\Delta\rho_N$ and Δk_N:

$$\epsilon_{N1} = \Delta\rho_N = (-0.3 \pm 3.4) \times 10^{-3} \tag{163}$$

$$\Delta k_N = (0.9 \pm 5.0) \times 10^{-3} \,. \tag{164}$$

The corresponding results for ϵ_{N2} and ϵ_{N3} are obtained from Eqs. (157) or (159):

$$\epsilon_{N2} = (-7.6 \pm 7.6) \times 10^{-3} \tag{165}$$

$$\epsilon_{N3} = (0.4 \pm 4.2) \times 10^{-3} \; . \tag{166}$$

Finally from the value of Γ_b listed in Table 3 one finds, by using Eqs. (154)–(156)) or (159):

$$\epsilon_b = (4.4 \pm 7.0) \times 10^{-3} \; . \tag{167}$$

To proceed further, and include other measured observables in the analysis we need to make some dynamical assumptions. The minimum amount of model dependence is introduced by including other purely leptonic quantities at the Z pole, such as $A^{\tau}_{\rm pol}, A_e$ (measured [98] from the angular dependence of the τ polarization) and A_{LR}. At this stage, one is simply relying on lepton universality. Indeed, with the new definition of $x = g_V/g_A$ in Eq. (151) in terms of the peak asymmetry, defined with all pure QED effects, including the imaginary part of the photon vacuum polarization, subtracted away, the data on $A^{\tau}_{\rm pol}, A_e$ and A_{LR} can be taken into account by using the simple formula

$$A^{\tau}_{\rm pol}, A_e, A_{LR} = \frac{2 g_V g_A}{g_V^2 + g_A^2} = \frac{2x}{1+x^2} \tag{168}$$

with the same x as defined from A^{ℓ}_{FB}, apart from still negligible differences. With essentially the same assumptions one can also include the data on the b-quark forward–backward asymmetry A^b_{FB}. In fact it turns out that A^b_{FB} is almost unaffected by the $Z \to b\bar{b}$ vertex correction. Schematically the reason is that $A^b_{FB} = 3\eta_e\eta_b$ with $\eta = \frac{g_V g_A}{(g_V^2 + g_A^2})$, so that $\delta A^b_{FB} = 3(\eta_e \delta\eta_b + \eta_b \delta\eta_e)$. We see that the sensitivity on η_b, which contains the $Z \to b\bar{b}$ vertex correction, is strongly suppressed by the small factor η_e. This is confirmed by an accurate numerical calculation of A^b_{FB}. If we define a new quantity $(\bar{s}^2_W)_b$ by the identity:

$$A^b_{FB} = 3\frac{1 - 4(\bar{s}^2_W)_b}{1 + (1 - 4(\bar{s}^2_W)_b)^2} \, \frac{\beta(1 - \frac{4}{3}(\bar{s}^2_W)_b)}{\beta^2 + \frac{3-\beta^2}{2}(1 - \frac{4}{3}(\bar{s}^2_W)_b)^2} \; , \tag{169}$$

where $\beta = \sqrt{1 - 4m_b^2/m_Z^2}$, we can explicitly evaluate the relation, as a function of m_t and m_H, between $(\bar{s}^2_W)_b$ and the similar quantity $(\bar{s}^2_W)_{FB}$ previously defined from the charged lepton asymmetry A^{ℓ}_{FB} and check that, with the new definition of A^{ℓ}_{FB}, they indeed practically coincide.

As a result of the previous discussion, we can combine the values of $x = g_V/g_A$ obtained from the whole set of asymmetries measured at LEP and we get the value:

$$x = g_V/g_A = 0.0725 \pm 0.0033 \quad \text{or} \quad \bar{s}^2_W = 0.2319 \pm 0.0008 \; , \tag{170}$$

which coincides with the result quoted in Ref. [98]. At this stage the best values of $\epsilon_{N1}, \epsilon_{N2}$ and ϵ_{N3} are modified according to

$$\begin{aligned} \epsilon_{N1} &= \Delta\rho_N = (0.0 \pm 3.4) \times 10^{-3} \\ \epsilon_{N2} &= (-8.3 \pm 7.7) \times 10^{-3} \\ \epsilon_{N3} &= (1.6 \pm 3.5) \times 10^{-3} \\ \epsilon_b &= (-4.5 \pm 7.0) \times 10^{-3} \; . \end{aligned} \tag{171}$$

All observables measured on the Z peak at LEP can be included in the analysis provided that we assume that all deviations from the Standard Model are only contained in vacuum polarization diagrams (without demanding a truncation of the q^2 dependence of the corresponding functions) and/or the $Z \rightarrow b\bar{b}$ vertex. Note that this is true for whatever partition of the new effect between g_{bV} and g_{bA}, because only one combination of them is measured in Γ_b, while, as already mentioned, A^b_{FB} is nearly independent of the $Z \rightarrow b\bar{b}$ vertex. A set of primary measurements at LEP contains $\Gamma_Z, R = \Gamma_h/\Gamma_\ell$, the hadronic peak cross-section σ_h (with a 3×3 correlation matrix given in Ref. [99]), $R_{bh} = \Gamma_b/\Gamma_h$ plus the asymmetries that lead to the value of x in Eq. (170). The relations between these quantities and the epsilons can be written down in the following linearised form:

$$\begin{aligned}\Gamma_Z &= \Gamma_{Z0}[1 + 1.35\epsilon_{N1} - 0.46\epsilon_{N3} + 0.35\epsilon_b] \\ R &= R_0[1 + 0.28\epsilon_{N1} - 0.36\epsilon_{N3} + 0.50\epsilon_b] \\ \sigma_h &= \sigma_{h0}[1 - 0.03\epsilon_{N1} + 0.04\epsilon_{N3} - 0.20\epsilon_b] \\ R_{bh} &= R_{bh0}[1 - 0.06\epsilon_{N1} + 0.07\epsilon_{N3} + 1.79\epsilon_b] \\ x &= x_0[1 + 17.6\epsilon_{N1} - 22.9\epsilon_{N3}]\ .\end{aligned} \tag{172}$$

We stress that, at the one-loop level, all the dependence on m_t and m_H of the quantities in Eqs. (172) enters through the epsilons and not in the various numerical coefficients. This is no longer true for those two-loop effects, unlike those computed in Ref. [112], which are not vacuum polarization or $Z \rightarrow b\bar{b}$ vertex corrections. For example, the dependence on m_t in R_{QCD} introduced by the α_s^2 terms of Ref. [109], was not included in Eq. (155). As a consequence, their effect on Γ_b is, by definition, found in ϵ_b. Their remaining effect on all other observables we consider is numerically irrelevant.

We have made a global fit of the data on $m_W/m_Z, \Gamma_Z, R, \sigma_h, R_{bh}$, given in Table 3 and the results on the asymmetries, summarized by the value of x in Eq. (170). For LEP data, we have taken the correlation matrix for Γ_Z, R and σ_h given in Ref. [99], while we have considered the additional information on R_{bh} and x as independent. The results of our fit are:

$$\begin{aligned}\epsilon_{N1} &= \Delta\rho_N = (-0.3 \pm 3.2) \times 10^{-3} - 0.1\ \delta a_s \\ \epsilon_{N2} &= (-8.5 \pm 7.6) \times 10^{-3} - 0.11\delta a_s + 0.23\delta a \\ \epsilon_{N3} &= (1.5 \pm 3.3) \times 10^{-3} - 0.04\delta a_s - 0.77\delta a \\ \epsilon_b &= (3.1 \pm 4.5) \times 10^{-3} - 1.42\delta a_s\end{aligned} \tag{173}$$

where the dependence on the values of $\alpha_s(m_Z)$ and $\alpha(m_Z)$ has also been indicated. If we neglect possible contributions from δa and add in quadrature the effect of a ± 0.007 error on $\alpha_s(m_Z)$ around the central value 0.118, we obtain:

$$\begin{aligned}\epsilon_{N1} &= \Delta\rho_N = (-0.3 \pm 3.2) \times 10^{-3} \\ \epsilon_{N2} &= (-8.5 \pm 7.6) \times 10^{-3} \\ \epsilon_{N3} &= (1.5 \pm 3.3) \times 10^{-3} \\ \epsilon_b &= (3.1 \pm 6.5) \times 10^{-3}\end{aligned} \tag{174}$$

which shows that the errors induced from $\alpha_s(m_Z)$ are entirely negligible with the exception of the error on ϵ_b.

Provided that all deviations from the Standard Model are only contained in vacuum polarizations and/or in the $Z \to b\bar{b}$ vertex, Eqs. (172),(173) apply. For that to be true, it is important that we focus on observables at the intermediate gauge boson scale of mass. To include in the analysis lower energy observables as well, a stronger hypothesis needs to be made: only vacuum polarization diagrams are allowed to vary from the Standard Model ones and only in their constant and first derivative terms in a q^2-expansion [79]–[81], a likely picture, e.g. in technicolour theories [91],[92]. In such a case, to Eqs. (159) one can add the linearized expressions, in terms of the epsilons, of the ratio R_ν of neutral to charged current processes in deep inelastic neutrino scattering on nuclei [93],[94]

$$R_\nu = 0.3115 \times 0.83\epsilon_{N1} - 0.27\epsilon_{N3} \tag{175}$$

or of the 'weak charge' Q_W measured in atomic parity violation experiments on Cs [96],[97]

$$Q_W = -72.72 \pm 0.13 - 102\epsilon_{N3} \tag{176}$$

(with the error due to the theoretical uncertainties in the atomic wave functions).

It is now possible to include these low energy observables in an overall fit. Using, together with the data in Table 3, the present average value from νN scattering

$$\langle R_\nu \rangle = 0.312 \pm 0.003 \tag{177}$$

(which takes the recent CCFR result [94] into account) and the result from caesium data [96],[97]

$$Q_W = -71.04 \pm 1.81 \ , \tag{178}$$

one obtains the global fit

$$\begin{aligned} \epsilon_{N1} &= \Delta\rho_N = (-0.1 \pm 2.7) \times 10^{-3} \\ \epsilon_{N2} &= (-8.2 \pm 7.5) \times 10^{-3} \\ \epsilon_{N3} &= (1.3 \pm 3.1) \times 10^{-3} \\ \epsilon_b &= (2.0 \pm 5.4) \times 10^{-3} \ . \end{aligned} \tag{179}$$

Here the error induced from $\alpha_s(m_Z)$ was taken into account in the same way as in Eqs. (177).

12.3 Discussion and Conclusion

The results of the various fits are collected in Table 4. The results for ϵ_b and ϵ_{N2} are compared with the Standard Model predictions in Figs. 16 and 17. We also show in Figs. 18–20 the projected ellipses in the planes $\epsilon_{N1} - \epsilon_{N3}, \epsilon_{N1} - \epsilon_b, \epsilon_{N3} - \epsilon_b$ for different fits together with the corresponding Standard Model predictions.

We note that, with the progress of LEP, the effect of low energy data on the epsilons is decreasing in importance. We also observe that all the epsilons are compatible with zero within 1σ (or slightly more for ϵ_{N2}, as a consequence of the fact that Δr_W in Eq. (160) is displaced from zero by 1σ). This means that the tree level Standard Model plus pure QED corrections is quite a good approximation and that the possibility of measuring the pure weak radiative corrections is marginal within the present accuracy.

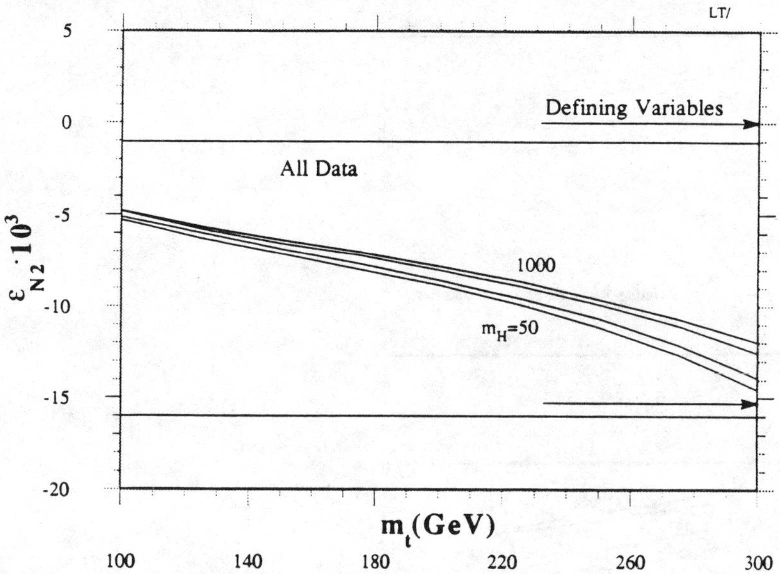

Fig. 17. ϵ_{N2} vs. m_t and m_H as predicted in the Standard Model compared with the value fitted from all the data in Eqs. (179). The arrows indicate the limits of the band if only the defining variables of Eq. (165) are used.

However the a priori size of the purely weak corrections could well be larger than the measured values. As a consequence strong constraints are obtained both in the Standard Model (upper and lower limits on the top mass) and beyond.

Table 4: Summary of the different determinations of the epsilon parameters

	Only defining variables	Defining variables plus all asymmetries	All LEP data	'All data'
$\epsilon_{N1} \cdot 10^3$	-0.3± 3.4.	0.0 ± 3.4	-0.3± 3.2	-0.1± 2.7
$\epsilon_{N2} \cdot 10^3$	-7.6± 7.6	-8.3 ± 7.5	-8.5 ± 7.6	-8.2 ± 7.5
$\epsilon_{N3} \cdot 10^3$	0.4± 4.2	1.6± 3.5	1.5±3.3	1.3± 3.1
$\epsilon_b \cdot 10^3$	4.4 ±7.0	4.5± 7.0	3.1 ± 5.5	2.9±5.4

By comparing the results from the global fit given in Table 4, column 3, with those in column 2, obtained from fitting $m_W/m_Z, \Gamma_\ell, \Gamma_b$ and the value of x derived from all the asymmetries, we observe that the effect on $\epsilon_{N1}, \epsilon_{N2}$, and ϵ_{N3} of the additional information from Γ_Z, R and σ_h is quite modest. Their central values are displaced by less than 1σ and the gain on the errors is very limited. This *a posteriori* supports the procedure proposed in the first paper in Ref. [84], where, for the determination of $\epsilon_{N1}, \epsilon_{N2}$, and ϵ_{N3}, the hadronic quantities were discarded because of the top dependence introduced by the $Z \to b\bar{b}$ vertex and of the need of specifying $\alpha_s(m_Z)$. It is possible to make contact with the approach in terms of the variables S, T and U [79],[81] only under the assumption of dominance of new physics effects through

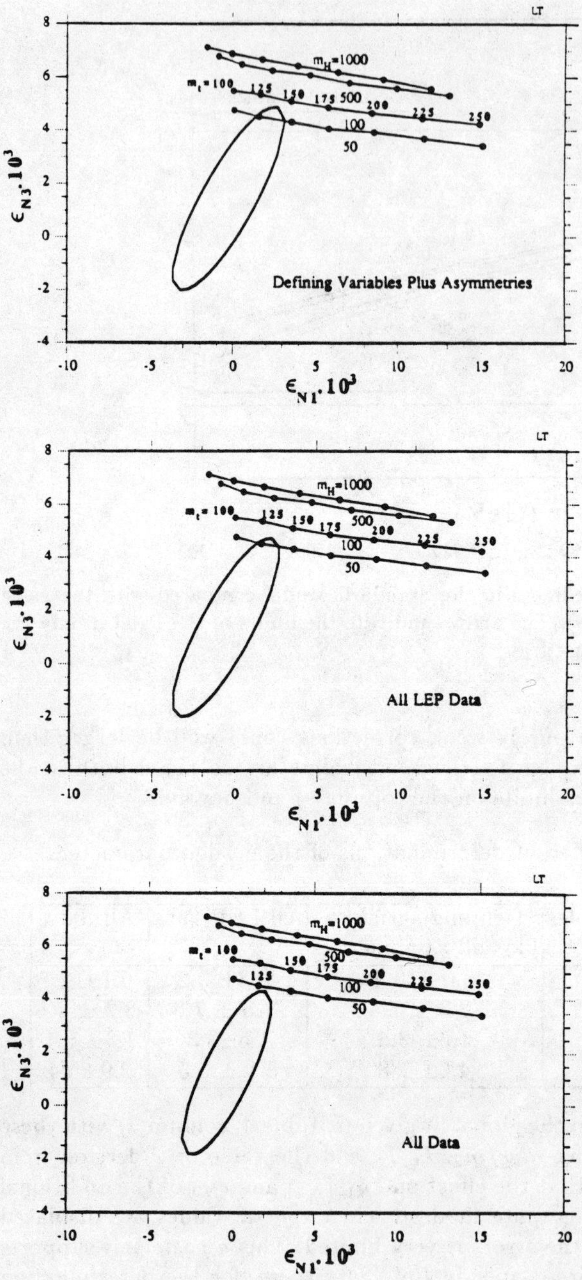

Fig. 18. Plots of ϵ_{N3} vs. ϵ_{N1} for different fits presented in Table 4: a) column 2; b) column 3; c) column 4. The projections on the axis of the ellipses correspond to the 1σ errors displayed in Table 4. The curves refer to the Standard Model predictions for different values of m_t and m_H.

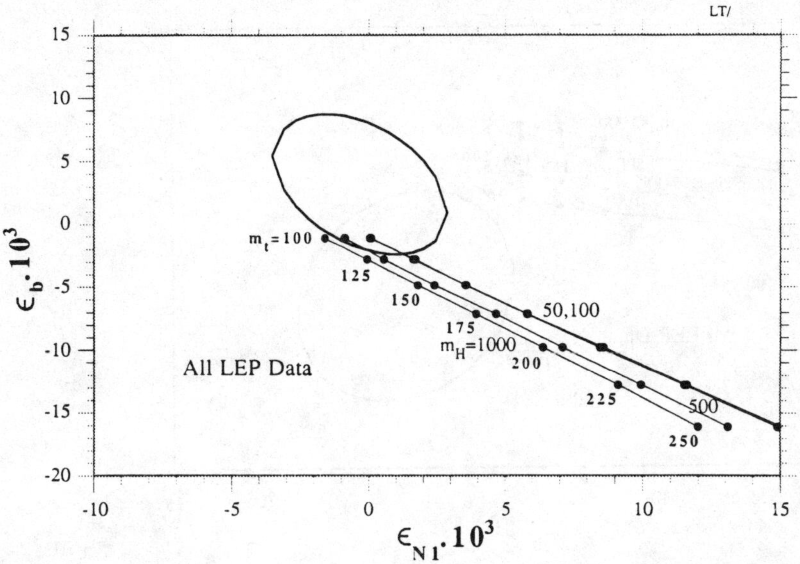

Fig. 19. Plots of ϵ_b vs. ϵ_{N1} for the fit presented in Table 4, column 3 (all LEP data). The projections on the axis of the ellipses correspond to the 1σ errors displayed in Table 4. The curves refer to the Standard Model predictions for different values of m_t and m_H with $\alpha_s(m_Z) = 0.118 \pm 0.007$.

vacuum polarization corrections. In such a case, the precise relation with the epsilons is given by:

$$\begin{aligned} \epsilon_{N1} &= \epsilon_{N10} + \alpha T \\ \epsilon_{N2} &= \epsilon_{N20} - \frac{\alpha}{4s_0^2} U \\ \epsilon_{N3} &= \epsilon_{N30} + \frac{\alpha}{4s_0^2} S \ . \end{aligned} \tag{180}$$

Note that S, T and U have only meaning in terms of a reference set of values for m_t and m_H, while our epsilons are defined from measurable quantities without need of specifying given values of m_t and m_H (because our reference point is the improved Born approximation given by the Standard Model at tree level plus pure QED corrections which is independent of m_t and quite well supported by all present data). The quantities $\epsilon_{N10}, \epsilon_{N20}$ and ϵ_{N30} are the Standard Model values at the reference point, where also ϵ_b would have to be fixed.

It is interesting to observe that the central value of ϵ_b is always positive, while the predictions of the Standard Model are negative for m_t >100 GeV (Fig. 16). Thus, besides ϵ_{N3} we have identified another interesting variable for constraining models of new physics: those extensions of the Standard Model that push the prediction of ϵ_b in the positive direction are favoured. Another interesting consequence of this result is that one can obtain a rather strong limit on m_t, independent of $\Delta\rho_N$ (i.e. ϵ_{N1}).

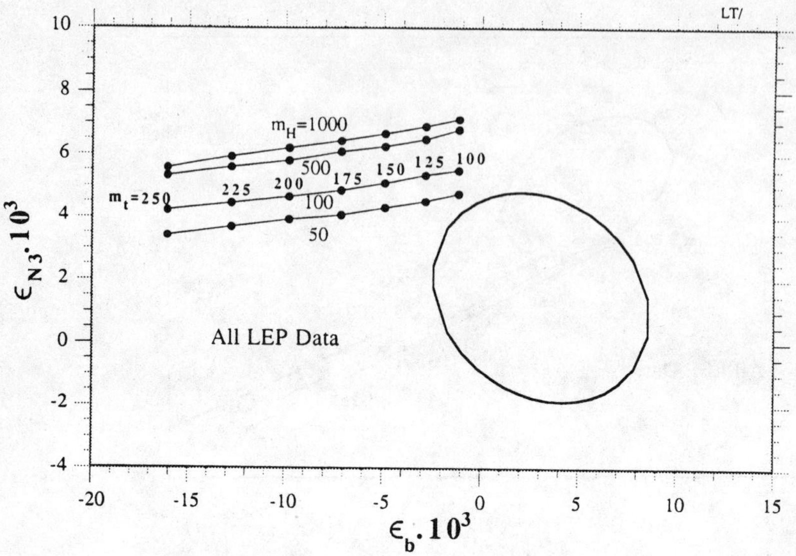

Fig. 20. Plots of ϵ_{N3} vs. ϵ_b for the fit presented in Table 4, column 3 (all LEP data). The projections on the axis of the ellipses correspond to the 1σ errors displayed in Table 4. The curves refer to the Standard Model predictions for different values of m_t and m_H with $\alpha_s(m_Z) = 0.118 \pm 0.007$.

Suppose in fact that one assumes that new physics leads to arbitrary effects in vacuum polarization functions but that all vertices are instead negligibly affected. Then our analysis would apply and, moreover, ϵ_b would be described by the Standard Model prediction as a function of m_t. For $\alpha_s(m_Z) = 0.118 \pm 0.007$, we obtain from ϵ_b the limit

$$m_t < 195 \text{ GeV} \qquad (95\% \text{ c.l.}) . \tag{181}$$

By a similar approach a weaker limit was obtained in Ref. [113] (where only the data on Γ_b and R were taken into account, while also the results on Γ_Z and σ_h have a sizable impact). There may be some question on the meaning of the confidence level assigned to this bound as the data on ϵ_b at 1σ intersect the Standard Model theoretical curve in a region of m_t, which is mostly excluded by the CDF limit: we have taken the ratio of the Gaussian tail above the limit and the area from 108 GeV to infinity to be 5%. If we did not know that the region below 100 GeV is forbidden, the corresponding limit would be $m_t < 160$ GeV. Note that if we perform a complete fit in the Standard Model of the same set of data, with $\alpha_s(m_Z) = 0.118 \pm 0.007$, we obtain the results $m_t = 132 \pm 22$ GeV for $m_H = 50$ GeV and $m_t = 175 \pm 18$ GeV for $m_H = 1$ TeV, with a better χ^2 for the light Higgs case (3.0 vs. 7.1).

It is important to stress that the fitted value of ϵ_b and its excess with respect to the Standard Model prediction for realistic values of m_t depend on the input range of $\alpha_s(m_Z) = 0.118 \pm 0.007$, which has been derived from all existing experiments, assuming the validity of QCD [103]. It is interesting to repeat the fit without constraining $\alpha_s(m_Z)$. By considering all LEP data (as for the results in Eqs. (174) one obtains in this case:

$$\begin{aligned}\epsilon_{N1} &= \Delta\rho_N = (0.03 \pm 3.3) \times 10^{-3} \\ \epsilon_{N3} &= (1.5 \pm 3.4) \times 10^{-3} \\ \epsilon_b &= (4.7 \pm 8.1) \times 10^{-3} \\ \alpha_s(m_Z) &= 0.115 \pm 0.015 \ .\end{aligned} \tag{182}$$

(Note that a fit of the same data in the Standard Model with $\alpha_s(m_Z)$ unconstrained gives $\alpha_s(m_Z) = 0.128 \pm 0.009$ [98]). By comparison with Eqs. (174) we see that, first, ϵ_{N1} and ϵ_{N3} change by only a small fraction of their errors. Second, the central value of ϵ_b becomes even more positive, but the error increases. Third, the value of $\alpha_s(m_Z)$ goes down to a central value, in perfect agreement with the world average. As a matter of fact, when $\alpha_s(m_Z)$ is a free parameter, ϵ_b is essentially determined from Eqs. (172) by R_{bh} alone, which is almost α_s-independent, with ϵ_{N1} and ϵ_{N3} fixed by Γ_ℓ and $x = g_V/g_A$. Thus, in principle, one can experimentally decide if the observed excess of ϵ_b is due to $\alpha_s(m_Z)$ or to the $Z \to b\bar{b}$ vertex, because the pattern of deviations for the various observables is different in the two cases. At present the precision of the data, especially those on R_{bh}, is not sufficient to really decide this issue. The data change rapidly, so the picture could be soon modified, but the present data, taken at face value, suggest that the long-standing result that the central value of $\alpha_s(m_Z)$ comes out a bit too large from the line-shape (or the value of R in Table 3) could be due to an anomaly in the $Z \to b\bar{b}$ vertex.

The decrease of ϵ_{N3} with respect to the Standard Model prediction is possible in many models, for example extended gauge models [88] or the Minimal Supersymmetric Standard Model (MSSM) [89],[90]. In particular, it has been shown in Ref. [91], that in the MSSM a sizable reduction of ϵ_{N3}, of the order of the measured shift, is induced by a Z wave function renormalisation effect if electroweak gaugino-higgsinos ('charginos') are close to their experimental limits. This effect, by itself, would reduce by the same percentage all widths. Then it would not at all affect R_{bh}, so that it would not cure the apparently high value of ϵ_b. In the MSSM there are corrections to the $Z \to b\bar{b}$ vertex from charged Higgs exchange and from chargino and s-top exchange [114],[115]. Charged Higgs exchange goes in the direction of suppressing $\Gamma(Z \to b\bar{b})$ [114]. Chargino and s-top exchange can overcompensate this effect for light charginos and s-top [115]. However, even for charginos near their experimental limit and for a s-top mass $m_{st} = 100$ GeV the resulting effect is small, especially for a not too heavy top quark, and of marginal importance with respect to the experimental error. We also note that, the vertex corrections (other than the $Z \to b\bar{b}$ vertex) being negligible in this case, the determination of the epsilons from all LEP data given in Table 4 applies. On the contrary, due to the strong q^2 dependence of the Z wave function renormalisation in the case of gauginos and higgsinos close to threshold, the low energy data cannot be included.

In conclusion, especially in view of the envisaged accuracy that will be obtained by the electroweak tests at the end of the LEP 1 phase, the study of the epsilons can provide important constraints and possibly also positive clues on new physics beyond the Standard Model. The isolation of different sectors of the theory, as allowed by the epsilons, may help to clarify the physical origin of any deviation from the Standard

Model that will possibly be found or to discard theories that require some specific pattern of radiative effects on the epsilons which turns out to be incompatible with the data.

Finally, I most warmly thank the Organizing Committee of the Schladming School for their kind invitation and for the very nice hospitality received in this magnificent mountain resort.

References

1. S.L. Glashow, *Nucl. Phys.* **22** (1961) 579;
   S. Weinberg, *Phys. Rev. Lett.* **19** (1967) 1264;
   A. Salam, *Proc. 8th Nobel Symposium*, Aspenäsgården, 1968, ed. N. Svartholm (Almqvist and Wiksell, Stockholm, 1968), p. 367.
2. See, for example, the following textbooks:
   I. Aitchison and A. Hey, *Gauge Theories in Particle Physics: A Practical Introduction* (A. Hilger, Bristol, 1982);
   D. Bailin, *Introduction to Gauge Field Theory* (A. Hilger, Bristol, 1986);
   E.D. Commins and P.H. Bucksbaum, *Weak Interactions of Leptons and Quarks* (Cambridge Univ. Press, Cambridge, 1983);
   L.B. Okun, *Leptons and Quarks* (North Holland Publ. Co., Amsterdam, 1982).
3. Particle Data Group, *Phys. Rev.* **D45** (1992) 1.
4. N. Cabibbo, *Phys. Rev. Lett.* **10** (1963) 531.
5. S.L. Glashow, J. Iliopoulos and L. Maiani, *Phys. Rev.* **D2** (1970) 1285.
6. J.H. Christenson, J.W. Cronin, V.L. Fitch and R. Turlay, *Phys. Rev. Lett.* **13** (1964) 138.
7. C.N. Yang and R. Mills, *Phys. Rev.* **96** (1954) 191.
8. E.S. Abers and B.W. Lee, *Phys. Rep.* **9** (1973) 1.
9. J. Bjorken and S. Drell, *Relativistic Quantum Fields* (McGraw-Hill, New York, 1965).
10. P.W. Higgs, *Phys. Lett.* **12** (1964) 132; *Phys. Rev.* **13** (1964) 508; *Phys. Rev. Lett.* **145** (1966) 1156;
    P.W. Anderson, *Phys. Rev.* **130** (1963) 439;
    F. Englert and R. Brout, *Phys. Rev. Lett.* **13** (1964) 321;
    G.S. Guralnik, C.R. Hagen and T.W.B. Kibble, *Phys. Rev. Lett.* **13** (1964) 585.
11. M. Kobayashi and M. Maskawa, *Progr. Theor. Phys.* **49** (1963) 652.
12. S.L. Glashow and S. Weinberg, *Phys. Rev.* **D15** (1977) 1968;
    E.A. Paschos, *Phys. Rev.* **D15** (1977) 1966.
13. T.D. Lee, *Phys. Rep.* **9C** (1974) 143;
    S. Weinberg, *Phys. Rev. Lett.* **37** (1976) 657.
14. S.L. Adler, *Phys. Rev.* **177** (1969) 2426;
    J.S. Bell and R. Jackiw, *Nuovo Cimento* **51** (1969) 47;
    See also, C. Bouchiat, J. Iliopoulos and Ph. Meyer, *Phys. Lett.* **38B** (1972) 519;
    D. Gross and R. Jackiw, *Phys. Rev.* **D6** (1972) 477.
15. For reviews, see for example:
    P. Fayet and S. Ferrara, *Phys. Rep.* **32** (1977) 251;
    P. van Nieuwenhuizen, *Phys. Rep.* **68** (1981) 191;
    J. Wess and J. Bagger, *Supersymmetry and Supergravity* (Princeton Univ. Press, New York, 1983);
    P. West, *Introduction to Supersymmetry and Supergravity* (World Scientific, Singapore, 1986);
    N.P. Nilles, *Phys. Rep.* **110** (1984) 1;
    H.E. Haber and G. Kane, *Phys. Rep.* **117** (1985) 75;
    R. Barbieri, *Riv. N. Cimento* **11** (1988) 1.
16. See, for example, D.J. Gross, Proc. 55th Summer School, Les Houches, 1991, and/or M.B. Green, J.H. Schwarz and E. Witten, *Superstring Theory* (Cambridge Univ. Press, Cambridge, 1987).
17. E. Cremmer et al., *Phys. Lett.* **116B** (1982) 231;
    R. Barbieri, S. Ferrara and A. Savoy, *Phys. Lett.* **119B** (1982) 343.
18. E. Farhi and L. Susskind, *Phys. Rep.* **74** (1981) 277.

19. V.A. Miransky, M. Tanabashi and K. Yamawaki, *Phys. Lett.* **B221** (1989) 177 and *Mod. Phys. Lett.* **A4** (1989) 29;
W.A. Bardeen, C.T. Hill and M. Lindner, *Phys. Rev.* **D41** (1990) 1647;
For the pioneering paper in this direction, see Y. Nambu and G. Jona-Lasinio, *Phys. Rev.* **122** (1961) 345.
20. See, for example, A. Hasenfratz, Proc. LP-HEP 91 Conf. Geneva, 1991.
21. H. Georgi et al., *Phys. Lett.* **143B** (1984) 152;
M.J. Dugan et al., *Nucl. Phys.* **B254** (1985) 299.
22. L.F. Abbott and E. Farhi, *Phys. Lett.* **101B** (1981) 69; *Nucl. Phys.* **B189** (1981) 547.
23. M. Veltman, *Acta Phys. Polon.* **B8** (1977) 475;
B.W. Lee, C. Quigg and H.B. Thacker, *Phys. Rev.* **D16** (1979) 1519.
24. J.C. Pati, *Phys. Lett.* **B228** (1989) 228;
K.S. Babu, J.C. Pati and H. Stremnitzer, *Phys. Lett.* **B256** (1991) 206.
25. V. Innocente, Proceedings of the Rencontres de Moriond, Les Arcs, France 1993).
26. See, for example, M. Sher, *Phys. Rep.* **179** (1989) 273 and L. Maiani, Rome preprint No. 775 (1991), Lectures given at Cargèse Summer School, 1990.
27. M.A.B. Bég et al., *Phys. Rev.* **52** (1984) 883;
D.J. Callaway, *Nucl. Phys.* **B233** (1984) 189;
R. Dashen and H. Neuberger, *Phys. Rev. Lett.* **50** (1983) 189;
K.J. Babu and E. Ma, *Phys. Rev.* **D31** (1984) 2861;
E. Ma, *Phys. Rev.* **D31** (1985) 322.
28. M. Lindner, *Z. Phys.* **C31** (1986) 295.
29. P. Hasenfratz, *Nucl. Phys.* (Proc. Suppl.) **B9** (1989) 3;
J. Kuti, ibid., p. 55;
H. Neuberger, Proc. Symposium on Lattice Field Theory, Capri, 1989, *Nucl. Phys. B* (Proc. Suppl.) **17** (1990);
M. Lüscher and P. Weisz, *Nucl. Phys.* **B290** (1987) 5, **B295** (1988) 65 and **B318** (1989) 705;
S. Sharpe, Univ. Washington preprint DOE-ER-40423-10 P90 (1990), to appear in Proc. PASCOS '90 (see P. Langacker in Ref. [5]).
30. M. Chanowitz, M. Furman and I. Hinchliffe, *Phys. Lett.* **B78** (1978) 285;
N. Cabibbo et al., *Nucl. Phys.* **B158** (1979) 295;
R.A. Flores and M. Sher, *Phys. Rev.* **D27** (1983) 1679.
31. G. Altarelli, R. Kleiss and C. Verzegnassi (eds.), *Z Physics at LEP 1* (CERN 89-08, Geneva, 1989), Vols 1–3.
32. M. Pepe-Altarelli, Proceedings of the Rencontres de la Vallée d'Aoste, La Thuile, Italy, 1993.
33. Talks by C. De Clercq, V. Innocente and R. Tenchini, Proceedings of the Rencontres de Moriond, Les Arcs, France, 1993.
34. R. Raja, Proceedings of the Rencontres de la Vallée d'Aoste, La Thuile, Italy, 1993.
35. G. Altarelli, preprint CERN-TH 6623/92 (1992) in *QCD-20 Years Later*, Aachen, 1992;
S. Bethke, Proceedings of the XXVI International Conference on High Energy Physics, Dallas, 1992.
36. See, for example: F. Antonelli and L. Maiani, *Nucl. Phys.* **B186** (1981) 269;
S. Bellucci, M. Lusignoli and L. Maiani, *Nucl. Phys.* **B189** (1981) 329;
A recent discussion of 'large' terms can be found in G. Gounaris and D. Schildknecht, *Z. Phys.* **C42** (1989) 107.
37. M. Consoli, W. Hollik and F. Jegerlehner, in Ref. [31], Vol. 1, p. 7;
See also F. Jegerlehner, Villigen preprint PSI-PR-91-08 (1991); and
H. Burkhardt, F. Jegerlehner, G. Penso and C. Verzegnassi, *Z. Phys.* **C43** (1989) 497.

38. M. Veltman, *Nucl. Phys.* **B123** (1977) 89;
M.S. Chanowitz, M.A. Furman and I. Hinchliffe, *Phys. Lett.* **78B** (1978) 285.
39. T. Appelquist and J. Carazzone, *Phys. Rev.* **D11** (1975) 2856.
40. G. Burgers et al., in Ref. [31], Vol 1, p. 55.
41. A. Sirlin, *Phys. Rev.* **D22** (1980) 971;
W.J. Marciano and A. Sirlin, *Phys. Rev.* **D22** (1980) 2695 and **D29** (1984) 75, 945.
42. M. Consoli, W. Hollik and F. Jegerlehner, *Phys. Lett.* **B227** (1989) 167.
43. J.J. van der Bij and F. Hoogeveen, *Nucl. Phys.* **B283** (1987) 477.
44. B.W. Lynn, M. Peskin and R.G. Stuart, in *Physics at LEP*, eds. J. Ellis and R. Peccei (CERN 86-02, Geneva, 1986).
45. R. Barbieri and L. Maiani, *Nucl. Phys.* **B224** (1983) 32;
L. Alvarez-Gaumé, J. Polchinski and M. Wise, *Nucl. Phys.* **B221** (1983) 495.
46. W. Hollik, *Z. Phys.* **C37** (1988) 569.
47. J. van der Bij and M. Veltman, *Nucl. Phys.* **B231** (1984) 205.
48. W. Hollik, *Fortschr. Phys.* **38** (1990) 165.
49. D.C. Kennedy et al., *Nucl. Phys.* **B321** (1989) 83;
D.C. Kennedy and B.W. Lynn, *Nucl. Phys.* **B322** (1989) 1;
B.W. Lynn, Stanford report SU-ITP-867 (1989).
50. W.J. Marciano and A. Sirlin, *Phys. Rev. Lett.* **46** (1981) 163;
J. Sarantakos, W.J. Marciano and A. Sirlin, *Nucl. Phys.* **B217** (1983) 84.
51. A. Sirlin, *Phys. Lett.* **B232** (1989) 123.
52. A.A. Akundov et al., *Nucl. Phys.* **B276** (1988) 1;
F. Diakonov and W. Wetzel, HD-THEP-88-4 (1998);
W. Beenakker and W. Hollik, *Z. Phys.* **C40** (1988) 141;
J. Bernabeu, A. Pich and A. Santamaria, *Phys. Lett.* **B200** (1988) 569; *Nucl. Phys.* **B363** (1991) 326;
B.W. Lynn and R.G. Stuart, *Phys. Lett.* **B252** (1990) 676.
53. J.H. Kühn et al., in Ref. [31], Vol. 1, p. 267.
54. E.W.N. Glover et al., in Ref. [31], Vol. 2, p. 1.
55. F.A. Berends et al., in Ref. [31], Vol. 1, p. 89.
56. See, for example, G. Alexander et al. (eds.), *Polarization at LEP* (CERN 88-06, Geneva, 1988);
L. Arnaudon et al., CERN-PPE/92-49 (1992).
57. E.A. Kuraev and V.S. Fadin, *Sov. J. Nucl. Phys.* **41** (1985) 466;
G. Altarelli and G. Martinelli, same Proc. as in Ref. [44], Vol. 1, p. 47.
58. A. Borrelli et al., *Nucl. Phys.* **B333** (1990) 357.
59. R.G. Stuart, *Phys. Lett.* **B262** (1991) 113; **B272** (1991) 353; CERN-TH 6261/91 (1991).
60. W. Wetzel, *Nucl. Phys.* **B227** (1983) 1, and same Proc. as Ref. [44], Vol. 1, p. 40.
61. D.Yu. Bardin et al., *Phys. Lett.* **206B** (1988) 539.
62. R.N. Cahn, *Phys. Rev.* **D36** (1987) 2666.
63. F. Aversa and M. Greco, *Phys. Lett.* **B228** (1989) 134.
64. S.N. Ganguli, A. Gurtu and K. Mazumdar, Tata preprints TIFR-EHEP/89/2 and 89/3 (1989).
65. See, for example, V.A. Novikov, L.B. Okun and M.I. Vysotsky, CERN preprint TH 6053/91 (1991).
66. CDF Collaboration, F. Abe et al., *Phys. Rev. Lett.* **65** (1990) 2243; *Phys. Rev.* **D43** (1991) 2070;
UA2 Collaboration, *Phys. Lett.* **276** (1992) 354.
67. R.G. Stuart, *Z. Phys.* **C34** (1987) 445;
A. Blondel, preprint CERN LEP Phys Works, Geneva 89 EP/89-84 (1989).

68. J.V. Allaby et al., *Phys. Lett.* **B177** (1986) 446, *Z. Phys.* **C36** (1987) 611.
69. H. Abramowicz et al., *Phys. Rev. Lett.* **57** (1986) 298;
A. Blondel et al., *Z. Phys.* **C45** (1990) 361.
70. B. King, Proceedings of the Rencontres de Moriond, Les Arcs, France, 1993.
71. G.L. Fogli, private communication. I am grateful to G.L. Fogli for providing me with these curves.
72. U. Amaldi et al., *Phys. Rev.* **D36** (1987) 1385;
G. Costa et al., *Nucl. Phys.* **B297** (1988) 244.
73. P. Langacker, in Review of Particle Properties, *Phys. Lett.* **239** (1990) 1;
W.J. Marciano, Brookhaven preprint BNL-45999 (1991).
74. D. Bardin et al., CERN preprint TH 6443/92 (1992) and references therein.
75. M. Böhm and W. Hollik, Ref. [31], p. 203;
J.H. Kühn and P.M. Zerwas, ibid., p. 267.
76. G. Altarelli and B. Lampe, *Nucl. Phys.* **B391** (1993) 3.
77. The SLD Collaboration, Stanford preprint SLAC-PUB/6030 (1993);
M. Piccolo, Proceedings of the Rencontres de la Vallée d'Aoste, La Thuile, Italy, 1993.
78. M.C. Noecker et al., *Phys. Rev. Lett.* **61** (1988) 310;
M. Bouchiat, M.A. Proceedings of the 12th International Atomic Physics Conference (1990).
79. M.E. Peskin and T. Takeuchi, *Phys. Rev. Lett.* **65** (1990) 964 and *Phys. Rev.* **D46** (1991) 381.
80. G. Altarelli and R. Barbieri, *Phys. Lett.* **B253** (1990) 161.
81. B. Holdom and J. Terning, *Phys. Lett.* **B247** (1990) 88;
D.C. Kennedy and P. Langacker, *Phys. Rev. Lett.* **65** (1990) 2967 and preprint UPR-0467T;
B. Holdom, Fermilab 90/263-T(1990);
W.J. Marciano, as in Ref. [73];
A. Ali and G. Degrassi, DESY preprint DESY 91-035 (1991);
E. Gates and J. Terning, *Phys. Rev. Lett.* **67** (1991) 1840;
E. Ma and P. Roy, *Phys. Rev. Lett.* **68** (1992) 2879;
G. Bhattacharyya, S. Banerjee and P. Roy, *Phys. Rev.* **D45** (1992) 729.
82. M. Golden and L. Randall, *Nucl. Phys.* **B361** (1991) 3;
M. Dugan and L. Randall, *Phys. Lett.* **B264** (1991) 154;
A. Dobado et al., *Phys. Lett.* **B255** (1991) 405;
R.D. Peccei and S. Peris, *Phys. Rev.* **D44** (1991) 809.
83. B. Grinstein and M. Wise, *Phys. Lett.* **B265** (1991) 326.
84. G. Altarelli, R. Barbieri and S. Jadach, *Nucl. Phys.* **B369** (1992) 3;
G. Altarelli, R. Barbieri and F. Caravaglios, CERN preprints CERN-TH 6770/93 and TH 6859/93 (1993).
85. M. Consoli, S. Lopresti and L. Maiani, *Nucl. Phys.* **B223** (1983) 472;
D.C. Kennedy and B.W. Lynn, *Nucl. Phys.* **B322** (1989) 1;
D.C. Kennedy et al., *Nucl. Phys.* **B231** (1989) 83;
B.W. Lynn, Univ. Stanford preprint SU-ITP-867 (1989);
B. Lynn et al., in Ref. [44], Vol. 1.
86. V.A. Novikov, L.B. Okun and M.I. Vysotsky, CERN preprints TH 6538/92, 6696/92, TH 6715/92 (1992).
87. A.A. Akundov et al., *Nucl. Phys.* **B276** (1988) 1;
F. Diakonov and W. Wetzel, HD-THEP-88-4 (1988);
W. Beenakker and H. Hollik, *Z. Phys.* **C40** (1988) 569;
B.W. Lynn and R.G. Stuart, *Phys. Lett.* **B252** (1990) 676;

J. Bernabeu, A. Pich and A. Santamaria, *Phys. Lett.* **B200** (1988) 569; *Nucl. Phys.* **B363** (1991) 326.
88. G. Altarelli, R. Casalbuoni, D. Dominici, F. Feruglio and R. Gatto, *Mod. Phys. Lett.* **A5** (1990) 495 and *Nucl. Phys.* **B342** (1990) 15;
G. Altarelli, R. Casalbuoni, F. Feruglio and R. Gatto, *Phys. Lett.* **B245** (1990) 669;
J. Layssac, F.M. Renard and C. Verzegnassi, LAPP preprint LAPP-TH-290/90 (1990);
M.C. Gonzales-Garcia and J.W.F. Valle, *Phys. Lett.* **B259** (1991) 365 ;
F. Del Aguila, J.M. Moreno and M. Quiros, preprint CERN-TH 6184/91 and *Nucl. Phys.* **B361** (1991) 45;
G. Bhattacharyya et al., TIFR/EHEP 91-6 (1991);
G. Altarelli et al., *Phys.Lett.* **B263** (1991) 459.
89. H.P. Nilles, *Phys. Rep.* **C110** (1984) 1;
H.E. Haber and G.L. Kane, *Phys. Rep.* **C117** (1985) 75;
R. Barbieri, *Riv. N. Cim.* **11** (1988) 1.
90. R. Barbieri, F. Caravaglios and M. Frigeni, *Phys. Lett.* **B279** (1992) 169.
91. E. Farhi and L. Susskind, *Phys. Rep.* **74** (1981) 277.
92. R. Casalbuoni et al., *Phys. Lett.* **B258** (1991) 161;
R.N. Cahn and M. Suzuki, Berkeley preprint LBL-30351 (1991);
C. Roisnel and Tran N. Truong, *Phys. Lett.* **B253** (1991) 439;
T. Appelquist and G. Triantaphyllou, Yale Univ. preprint YCTP-P49-91;
see also the detailed analysis by J. Ellis, G.L. Fogli and E. Lisi, preprint CERN-TH 6383/92 (1992).
93. U. Amaldi et al., *Phys. Rev.* **D26** (1987) 1385;
A. Blondel et al., CDHSW Collaboration, *Z. Phys.* **C45** (1990) 361;
J.V. Allaby et al., CHARM Collaboration, *Z. Phys.* **C36** (1987) 611;
P.G. Reutens et al., CCFR Collaboration, *Z. Phys.* **C45** (1990) 539;
T.S. Mattison et al., FMM Collaboration, *Phys. Rev.* **D42** (1990) 1311.
94. T. Bolton, CCFR Collaboration, Proceedings of the XXVI International Conference on High Energy Physics, Dallas, 1992.
95. D. Geigerat et al., CHARM II Collaboration, *Phys. Lett.* **B232** (1989) 539.
96. W.J. Marciano and J.L. Rosner, *Phys. Rev. Lett.* **65** (1990) 2963.
97. M.C. Noecker et al., *Phys. Rev. Lett.* **61** (1988) 310;
M. Bouchiat, M.A. Proceedings of the 12th International Atomic Physics Conference (1990).
98. M. Pepe-Altarelli, Proceedings of the Recontres de la Vallée d'Aoste, La Thuile, Italy, 1993;
Talks by C. De Clercq and V. Innocente, Proceedings of the Rencontres de Moriond, Les Arcs, France, 1993.
99. In the basis Γ_Z, σ_h and R we use the correlation matrix elements $C_{12} = C_{21} = -0.14, C_{13} = C_{31} = 0.007, C_{32} = C_{23} = 0.13$ (the LEP Collaborations, private communication from M. Pepe-Altarelli).
100. R. Raja, CDF Collaboration, Proceedings of the Recontres de la Vallée, d'Aoste, La Thuile, Italy, 1993.
101. V. Innocente, Proceedings of the Rencontres de Moriond, Les Arcs, France, 1993.
102. For reviews, see, e.g., M. Sher, *Phys. Rep.* **179** (1989) 275;
L. Maiani, in Z^0 *Physics*, Cargèse, 1990, eds. M. Levy et al. (Plenum Press, New York, 1991) p. 237;
see also S. Cortese, E. Pallante and R. Petronzio, *Phys. Lett.* **B301** (1993) 203.
103. G. Altarelli, in *QCD- 20 Years Later*, Aachen 1992, CERN-TH.6623/92; S. Bethke, Proceedings of the XXVI International Conference on High Energy Physics, Dallas, 1992.

104. F. Jegerlehner, Villigen preprint PSI-PR-91-08 (1991);
H. Burkhardt, F. Jegerlehner, G. Penso and C. Verzegnassi, *Z. Phys.* **C43** (1989) 497.
105. For reviews, see, for example, G. Altarelli, preprint CERN-TH 6305/91, Proc. 55th Summer School, Les Houches, 1991;
R. Barbieri, preprint CERN-TH 6659/92, Proceedings of the Cargèse Summer School, 1992, .
106. M. Consoli et al., Ref. [31].
107. M. Veltman, *Nucl. Phys.* **B123** (1977) 89;
M.S. Chanowitz et al., *Phys. Lett.* **78B** (1978) 285;
G. Passarino and M. Veltman, *Nucl. Phys.* **B160** (1979) 151.
108. D. Bardin et al., CERN preprint TH 6443/92 (1992) and references therein.
109. G.Montagna et al., University of Pavia preprints FNT/T-93/02 (to appear in *Nucl. Phys. B*) and FNT/T-93/05 (to appear in *Computer Physics Comm.*).
110. B. A. Kniehl and J.H. Kühn, *Phys. Lett.* **B224** (1989) 229; *Nucl. Phys.* **B329** (1990) 547.
111. CDF Collaboration, F. Abe et al., *Phys. Rev. Lett.* **65** (1990) 2243; *Phys. Rev.* **D43** (1991) 2070;
UA2 Collaboration, *Phys. Lett.* **B276** (1992) 354; the value shown in Table 3 is from the Particle Data Group 1992: *Phys. Rev.* **D45** (1992).
112. R. Barbieri et al., CERN preprint CERN-TH 6713/92 (1992).
113. A. Blondel, A. Djouadi and C. Verzegnassi, preprint DESY 92-112 (1992).
114. W. Hollik, *Mod. Phys. Lett.* **A5** (1990) 1909.
115. A. Djouadi et al., *Nucl. Phys.* **B349** (1991) 48;
M. Boulware and D. Finnell, *Phys. Rev.* **D44** (1991) 2054. The sign discrepancy between these two papers appears now to be solved in favour of the second one.
116. The SLD Collaboration, SLAC-PUB-6030 (1993);
M. Piccolo, Proceedings of the Recontres de la Vallèc d'Aoste, La Thuile, Italy, 1993.

First Results from HERA

G. Wolf

Deutsches Elektronen Synchrotron DESY, D-22603 Hamburg, Germany

Abstract. The new electron-proton collider HERA at DESY has been put into operation in 1992 together with the experiments H1 and ZEUS. With HERA, high energy physics has entered a new territory far beyond the reach of previous lepton-nucleon scattering experiments. The very first run has already given new results on several topics. Deep inelastic neutral current scattering has been extended by two orders of magnitude towards very small x values and multijet production as expected from higher order QCD processes has been detected. For photoproduction the total cross section has been measured and a large excess of events from hard scattering has been observed. The general characteristics of these hard scatters agrees with the expectations for direct and resolved photon interactions.

1 Introduction

In May of last year, the two HERA experiments H1 and ZEUS observed for the first time electron-proton collisions in their central detectors. This feat was the culmination of eight years of prototyping, construction and commissioning of the machine [1] and the detectors H1 [2] and ZEUS [3]. In parallel with the construction of the machine and detectors, the horizon of HERA physics has been substantially expanded as documented in several workshops. The most recent and complete proceedings are those from 1987 [4] and 1991 [5]. These lectures focus on the results obtained by H1 and ZEUS in the two running periods of 1992.

2 The HERA Collider

2.1 Layout

The layout of HERA is shown in Fig. 1. Two separate magnet systems guide the e and p beams around the 6.3 km long ring. DESY and PETRA serve as injectors. There are four interaction regions, two of which are occupied by H1 and ZEUS. A third region has been allocated to the experiment HERMES [6] which has been approved recently for the measurement of the nucleon spin structure (see the lectures by E. Reya [7]) by colliding the polarized electron beam with a gas jet of polarized nucleons. There is a fourth experiment (HERA-B) under discussion to study CP-violation in the $B\bar{B}$ system produced by scattering of protons from the beam halo on a wire target.

Table 1 shows some of the parameters of the collider.

Table 1. HERA design parameters.

	electron ring		proton ring
circumference		6336 m	
energy	30 GeV		820 GeV
$e-p$ c.m. energy		314 GeV	
magnetic bending field	0.164 T		4.682 T
bending radius of dipole	610 m		584 m
circulating current	60 mA		160 mA
number of bunch buckets	220		220
number of bunches	210		210
time between beam crossings		96 ns	
luminosity		$1.5 \cdot 10^{31}\,\mathrm{cm}^{-2}\mathrm{s}^{-1}$	
polarization time at $E_e = 30$ GeV	25 min		

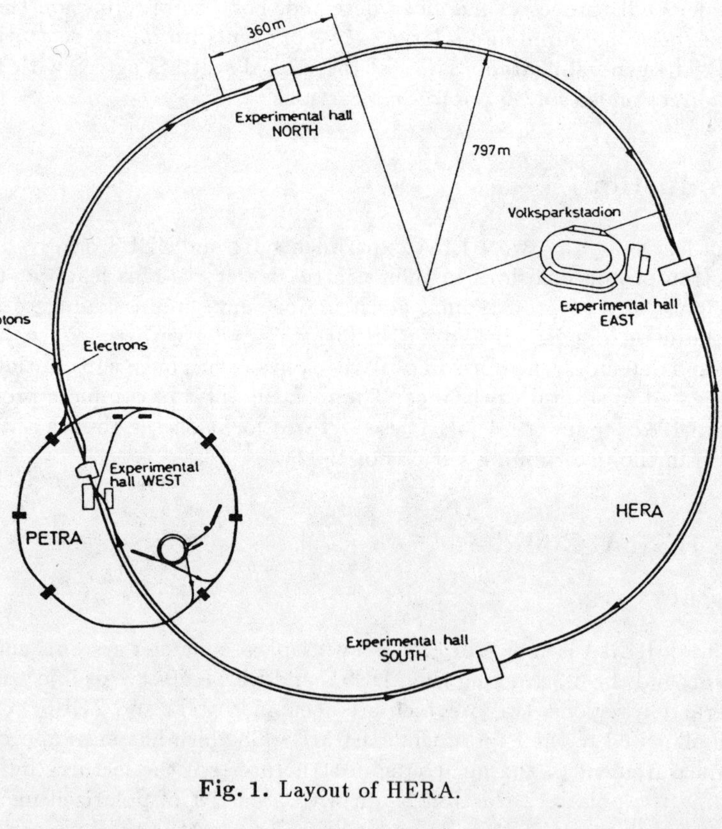

Fig. 1. Layout of HERA.

2.2 Commissioning Phase

Construction of HERA started in April 1984. In August of 1987 the 6.3 km long tunnel was completed. A year later the electron ring made of normal conducting magnets was put into operation. The proton ring was closed in fall of 1990. Most of the proton magnets are superconducting and are operated at 4.4 K providing a bending field of 4.68 T. The first protons were stored in April 1991 at the injection energy of 40 GeV. Having learned how to accelerate protons and to operate proton and electron beams simultaneously the machine crew began with collision studies. Luminosity was observed for the first time on October 19, 1991 by colliding one electron bunch of 12 GeV with one proton bunch of 480 GeV. The measured luminosity was about $L = 10^{26}\mathrm{cm}^{-2}\mathrm{s}^{-1}$. In the following weeks the electron energy, the number of bunches and the beam currents were increased. After reaching $L = 2 \cdot 10^{28}\mathrm{cm}^{-2}\mathrm{s}^{-1}$ with 10 bunches per beam at beam energies $E_e = 26.6$ GeV and $E_p = 480$ GeV, HERA operations were interrupted and the experiments H1 and ZEUS were installed in the interaction regions during a 4 months shut down.

In April of 1992 HERA resumed operation. By mid-April electrons were stored at 26.6 GeV; one month later protons were accelerated to the nominal energy of 820 GeV and on May 31st the two experiments registered the first collisions between the two beams. The H1 and ZEUS detectors were brought on-line concurrently with the beam tests of HERA: no special running period was provided for running-in or tuning of the detectors.

2.3 Performance

Data taking started at the end of June with 10 bunch operation, maximum beam currents of 1 – 2 mA and maximum luminosities around $6 \cdot 10^{28}\mathrm{cm}^{-2}\mathrm{s}^{-1}$. In a first

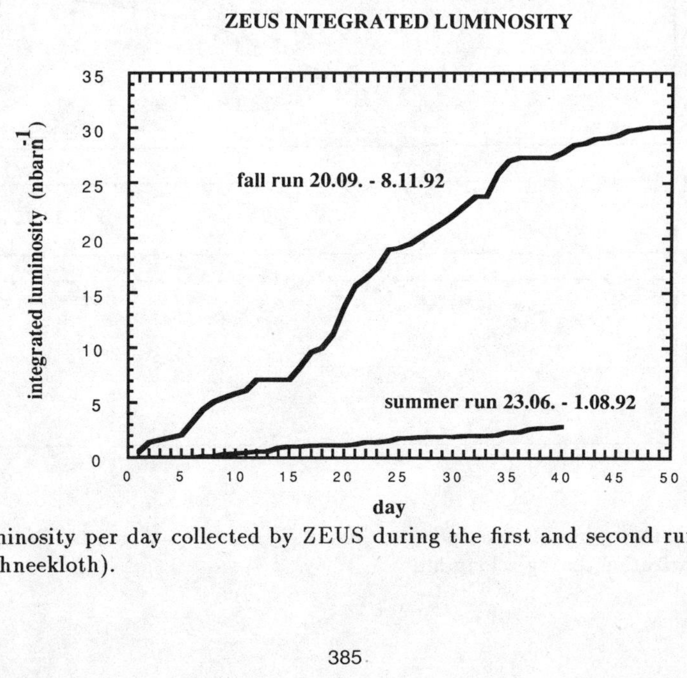

Fig. 2. Luminosity per day collected by ZEUS during the first and second running periods (from U. Schneekloth).

one month long period, data were recorded for integrated luminosities of $2-3$ nb^{-1}. Physics results from this running period were reported a week after data taking had finished [8-10]. After a shut down and machine tests a second data taking period from mid-September until beginning of November followed with maximum luminosities of $\sim 3\cdot 10^{29}\text{cm}^{-2}\text{s}^{-1}$ and a seven-fold increase in luminosity per day providing integrated luminosities of about 30 nb^{-1} for the two experiments (see Fig.2).

After coasting for some time the electrons become polarized with spins being antiparallel to the direction of the bending field as a result of the Sokholov-Ternov effect [11]. The build-up time for the polarization is determined by the synchrotron radiation and is given by $P(t) = P_0\ (1 - \exp(-t/t_p))$ with $P_0 = 92\%$ and $t_p = 98\ r^2RE^{-5}$ (t_p in s, bending radius r in m, average radius R in m and electron beam energy E in GeV). This prediction holds for a perfect machine. Already small imperfections, e.g. in the magnet lattice may produce depolarization effects and make the depolarization time shorter than the build-up time thereby destroying the polarization. Only a few shifts were devoted so far to the study of beam polarization. An electron energy near 26.67 GeV was chosen. Beam polarization in HERA was observed for the first time in fall of 1991 at the level of $P = 5 - 9\%$. After realignment of magnets P increased to 18% by April 1992. After the summer shutdown and with some tuning suggested by tracking programs a polarization of $58 \pm 5\%$ was measured [12], see Fig.3.

For particle physics, instead of transverse polarization electrons such of definite helicity (left – or right handed) are needed. This can be achieved with the help of a pair of spin rotators installed at the interaction regions. A pair of spin rotators was designed and built [13] and is available for installation and testing in HERA.

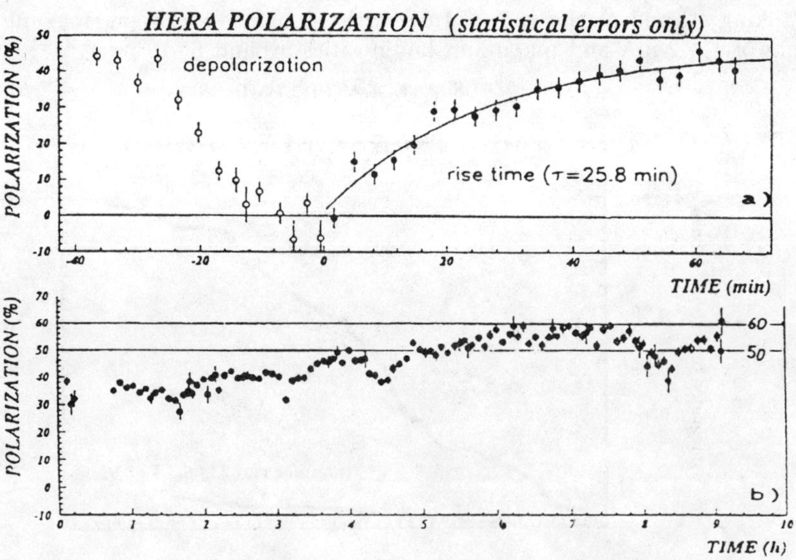

Fig. 3. a) The transverse electron polarization at $E_e = 26.7$ GeV as a function of time. b) Beam polarization during a long fill.

3 The Experiments

3.1 Detector Challenges

HERA produces a large variety of reactions with widely differing energy flows. This feature together with the desire for detecting and identifying the constituents such as electron, photon, quarks and gluons which participate in these reactions places different requirements on the detector. The large momentum imbalance between incident electrons and protons and the nature of space-like processes send most particles into a narrow cone around the proton direction. The observation of deep inelastic (DIS) neutral current (NC) scattering, $ep \to eX$ (Fig. 4a), is fairly straightforward. It produces a high energy electron whose transverse momentum is balanced by the current jet. The remnants from the breakup of the proton escape mostly unseen down the beam pipe. The variables x and Q^2 which describe the process (see below) can be determined from the energy and angle of either the electron or the current jet. This requires a precise electromagnetic and hadronic calorimeter with the calibration known at the $1-2\%$ level.

In charged current (CC) scattering, $ep \to \nu X$ (Fig. 4b), only the current jet can be observed. The identification of such events is based on the observation of missing transverse momentum carried away by the neutrino. It requires a hermetic calorimeter which covers the full solid angle such that e.g. photons, neutrons or K_2^0 cannot escape undetected. The variables x and Q^2 are measured from the current jet.

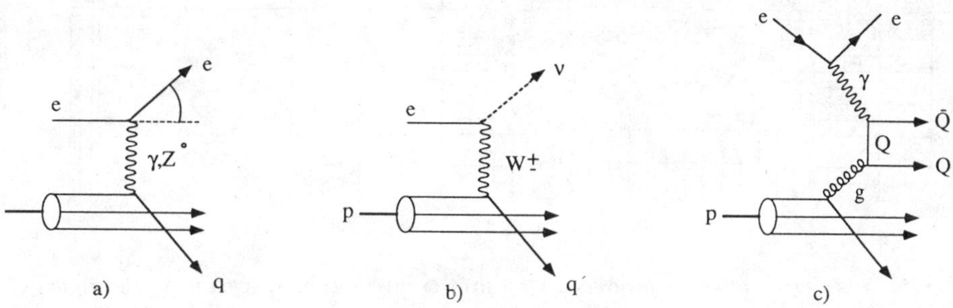

Fig. 4. Diagrams for NC and CC scattering and for photon-gluon fusion.

The observation of processes at low Q^2, in particular from scattering on soft partons or from production of pairs of charm or bottom mesons (Fig. 4c) is more difficult. The energy deposited in the calorimeter often is only a few GeV. Additional information from tracking detectors which surround the interaction point is necessary for their identification.

Background presents another challenge. The number of events from $e-p$ interactions is tiny $(10^{-3}-10^{-5})$ compared to the background events produced for instance by beam protons on the beam pipe wall or in the residual gas. What is worse, this type of background deposits a large amount of energy in the detector. A typical background event is shown in Fig. 5 where 225 GeV are observed in the calorimeter and many tracks in the tracking detector. At proton design current the background rate is expected to be around $10-100$ kHz. The detector must be able to discriminate quickly – within a few microseconds – against background events although both beams cross each other every 96 ns. The high background rates combined with the short bunch crossing interval forced the HERA experiments to develop novel concepts of electronic readout and triggering, concepts which are suitable also for detectors at the next generation of *pp* colliders, SSC and LHC.

The H1 and ZEUS detectors are driven by their choice of calorimeter. ZEUS uses a compensating uranium-scintillator calorimeter which provides the best possible energy resolution for hadrons. Compensation means that electromagnetic particles (electrons, photons) and hadrons of the same energy yield the same pulse height, $e/h = 1$. The radioactivity of the (depleted) uranium provides an extremely stable calibration signal, the mean life time of ^{238}U being $6.5 \cdot 10^9$ years. The H1 calorimeter uses liquid argon for readout which promises a very stable and precise energy calibration and allows a high transverse and longitudinal segmentation. The calorimeter is noncompensating, $e/h = 1.1 - 1.2$. However, due to the high segmentation and using software weighting with the observed shower profile equal signals for electrons and hadrons can be obtained.

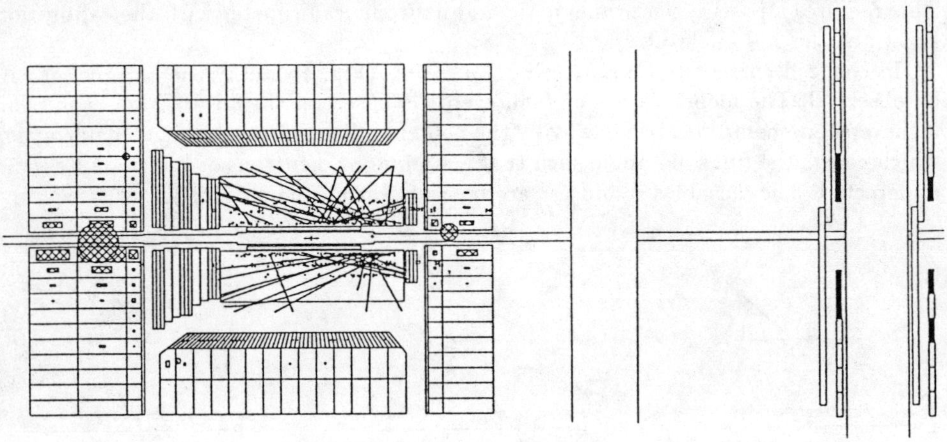

Fig. 5. A background event produced by a proton interaction upstream (to the right) of the ZEUS detector.

3.2 The H1 Detector

The H1 detector [2] is displayed in Fig. 6. The liquid-argon calorimeter (LA) covers the angular region $4^0 < \Theta < 155^0$ (the forward direction, $\Theta = 0^0$, is given by the direction of the proton beam). The calorimeter is longitudinally subdivided into an electromagnetic section with lead plates and a hadronic section with stainless steel plates as absorbers. The total depth varies between 8 and 4.5 absorption lengths. The calorimeter was optimized for a precise measurement and identification of electrons and for a stable calibration for electrons and hadrons. The energy resolution σ/E for electrons is $12\%/\sqrt{E} \oplus 1\%$ ($\oplus$ means quadratic addition) and $45\%/\sqrt{E} \oplus 1\%$ for hadrons (after weighting) as measured with test beams (Fig. 7). The calorimeter has been in operation since April 1991. Since then the charge collection has changed by less than 0.2% and the number of dead or problematic readout channels is below 0.1%.

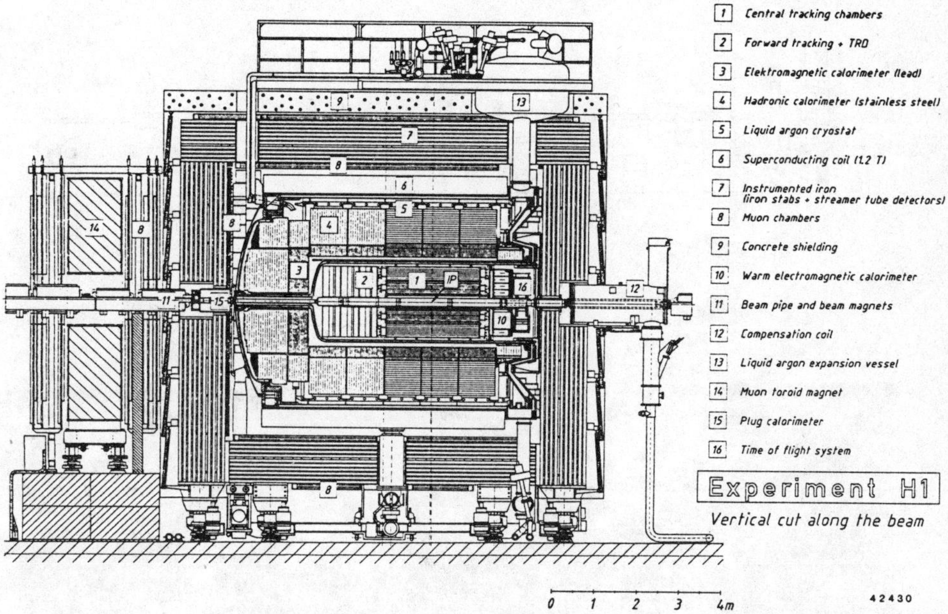

Fig. 6. Schematic view of the H1 detector.

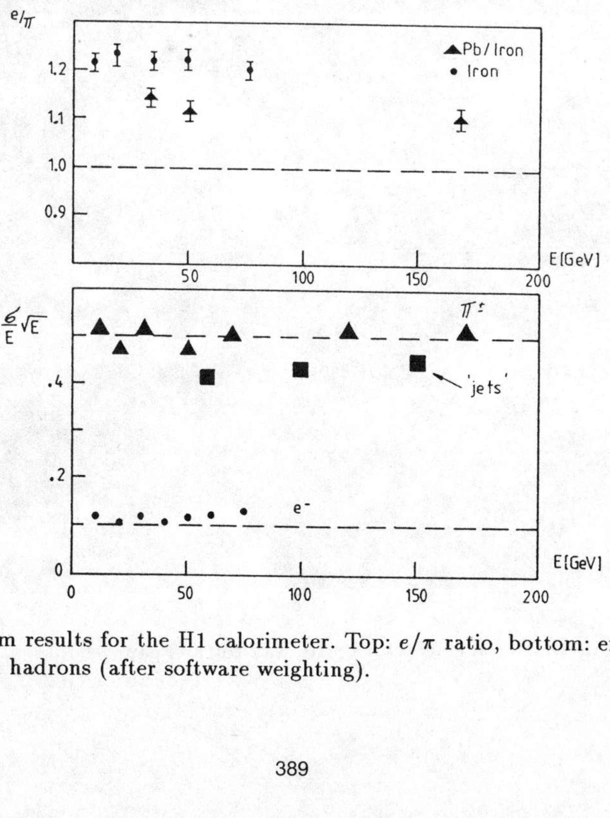

Fig. 7. Test beam results for the H1 calorimeter. Top: e/π ratio, bottom: energy resolution for electrons and hadrons (after software weighting).

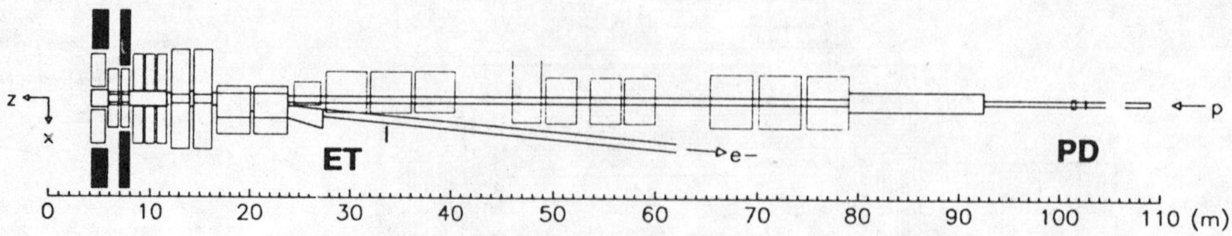

Fig. 8. Layout of the H1 luminosity detector.

Fig. 9. The H1 detector on its way to the beam position.

In the backward direction the calorimeter is supplemented by an electromagnetic lead-scintillator calorimeter (BEMC) followed by time-of-flight counters (TOF-VETO). In the forward direction the plug calorimeter with copper plates and silicon diode readout extends the energy measurement for hadrons down to angles of 0.7^0.

Charged particles are tracked in a magnetic field of 1.2 T which is produced by a superconducting solenoid that surrounds the calorimeter. The tracking system consists of cylindrical jet- and z-drift chambers in the central region, and of three radial and three planar drift chambers in the forward direction. The drift chambers are interleaved with proportional wire chambers for a fast trigger selection. The backward direction is covered by a four layer proportional wire chamber providing space points up to a scattering angle of 175^0. In forward direction a transition radiation detector (TRD) enhances the electron identification.

The magnet yoke is made of 10 layers of 7.5 cm thick iron plates. The gaps are instrumented with limited streamer tube (LST) chambers for measuring energy which has not been fully absorbed in the liquid argon calorimeter and for tracking of muons. Large area LST chambers in front and behind the iron yoke and an iron toroid magnet plus 6 layers of drift chambers in forward direction complete the muon detection system.

The luminosity is measured by observing the bremsstrahlung process $ep \longrightarrow ep\gamma$ at very small angles to the electron beam direction. The final state electron and photon are detected in coincidence in electromagnetic calorimeters of the luminosity detector LUMI positioned at 33 m (electron tagger) and 100 m (photon tagger) upstream (in proton direction) of the central detector (Fig. 8). At nominal luminosity the rate of luminosity events is between 50 – 100 kHz depending on the selection criteria.

Fig. 9 shows the H1 detector on its way into the interaction region. The H1 collaboration at present consists of 320 physicists from 32 institutes and 11 countries.

3.3 The ZEUS detector

A cross section of the ZEUS detector along the beams is shown in Fig. 10. The main component is the uranium-scintillator calorimeter (CAL) subdivided mechanically into the forward (FCAL), barrel (BCAL) and rear (RCAL) calorimeters. The CAL covers polar angles from 2.6^0 to 176.1^0 and 99.7% of the total solid angle. It consists of a total of 80 modules. Every module is constructed with up to 180 layers of 3.3 mm thick depleted uranium plates plus 2.6 mm thick scintillator plates.

Wave-length shifter bars transport the light to photomultipliers. The modules are subdivided longitudinally into an electromagnetic and two (one) hadronic sections in FCAL, BCAL (RCAL) presenting a total depth of 7 to 4 absorption lengths. The scintillator plates form 5×20 cm^2 (10×20 cm^2) cells in the electromagnetic section and 20×20 cm^2 in the hadronic sections of FCAL, BCAL (RCAL).

The calibration of the photomultipliers is being monitored with the signal (UNO) from the radioactivity of the uranium to a precision of $< 0.2\%$. The pulse heights of electrons and hadrons (Fig. 11a) are equal to within 3%, i.e. $e/h = 1.0 \pm 0.03$ (Fig. 11b), for momenta above 3 GeV/c. The energy resolution as measured in the test beam is for electrons $\sigma/E = 18\%/\sqrt{E}$ (E in GeV) and for hadrons $35\%/\sqrt{E}$ (Fig. 11c). The calorimeter yields also an accurate time measurement. The time resolution of a calorimeter cell is $\sigma = 1.5/\sqrt{E} \sim 0.5$ ns or < 1 ns above 3 GeV.

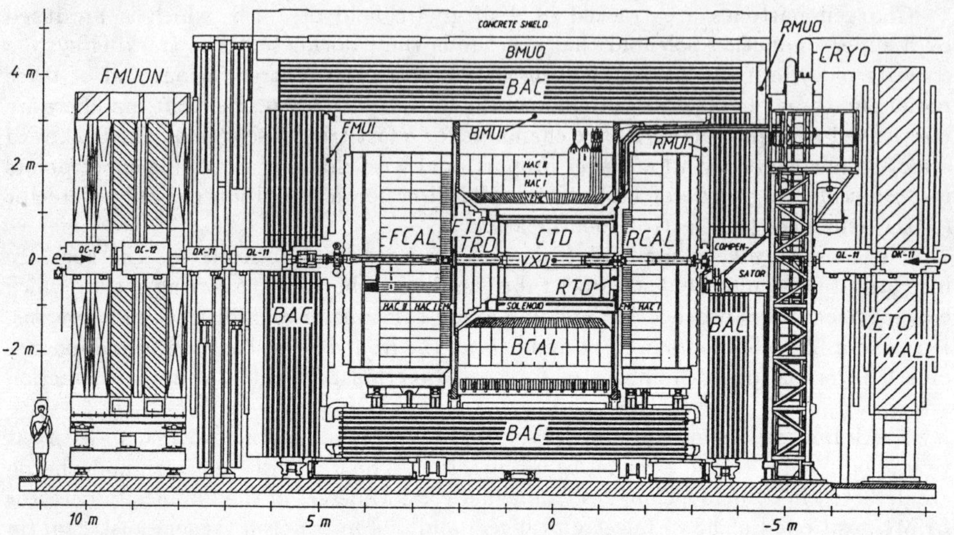

Fig. 10. Cross section of the ZEUS detector transverse to the beams.

In the course of an upgrade program, the transverse segmentation of the forward and rear parts of the calorimeter is being increased by inserting a plane of 3×3 cm^2 silicon diodes after the first 3 radiation lengths.

Charged particles coming from the interaction point are detected in the tracking system. It consists of a vertex detector, a cylindrical jet-type drift chamber and planar drift chambers in the forward and backward directions. In the forward direction planar transition radiation chambers are used for enhanced electron identification. The tracking detectors are surrounded by a thin-walled solenoid which produces a magnetic field of up to 1.8 T.

The iron yoke serves as the absorber for the backing calorimeter and as a muon filter. It is made of 7.5 cm thick iron plates and instrumented with proportional tube chambers for measuring the energy not absorbed in the uranium calorimeter. For identification and momentum measurement of muons the yoke is magnetized to 1.6 T with copper coils. Large area LST chambers measure the position and direction of muons in front and behind the iron yoke. In the forward direction a spectrometer of two iron toroids and drift- and LST chambers identifies muons and measures their momenta up to $100 - 150$ GeV$/c$.

For luminosity measurement the same reaction and a setup similar to that of H1 is used.

Very forward scattered protons (transverse momenta < 1 GeV$/c$) are measured in the leading proton spectrometer which uses the proton ring magnets for momentum analysis and detects the scattered protons in 6 stations with silicon strip detectors

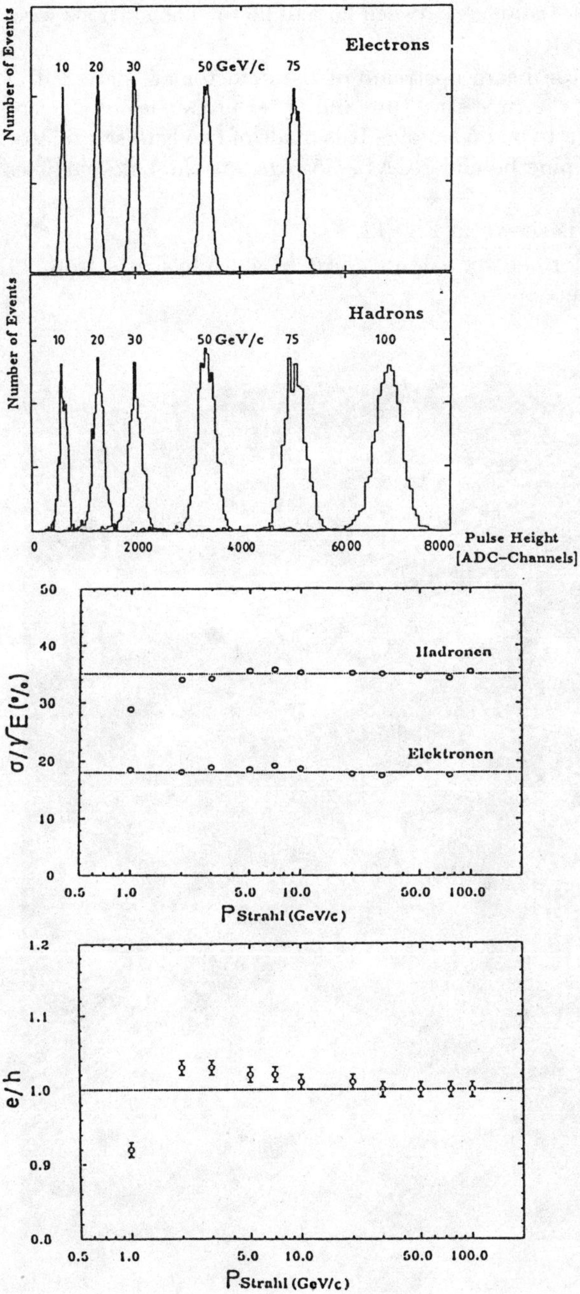

Fig. 11. Measurements with the ZEUS prototype calorimeter:
(a) Pulse height distributions for electrons and hadrons;
(b) Energy resolution for electrons and hadrons;
(c) e/h ratio as a function of momentum.

mounted very close to the beam at distances between 26 and 96 m. The stations have been installed, the detectors not yet.

Particles produced by the proton beam upstream of the detector are detected in the VETOWALL. For monitoring the time structure and other properties of the two beams a ring counter C5 has proven to be invaluable. It is made of two lead-scintillator layers and mounted on the beam pipe behind RCAL. C5 registers the halo particles accompanying both beams.

The central detector of ZEUS is shown in Fig. 12.

The ZEUS experiment is performed by a joint effort of 450 physicists from 11 countries and more than 50 institutes.

Fig. 12. View of the ZEUS detector: in the foreground the uranium-scintillator calorimeter.

3.4 Trigger-Selection

The selection of interesting events during data acquisition proceeds in four (three) trigger steps for H1 (ZEUS). The selection at trigger level one is made after 2.4 μs (H1) and 4.4 μs (ZEUS), respectively. Up to this point, information from 200 000 to 300 000 electronic channels are stored dead time free in analog or digital pipelines for 25 (H1) and 46 (ZEUS) consecutive beam crossings, respectively. Global information from various components like the calorimeter energy sums obtained by summing over specific regions of the calorimeter are stored in trigger pipelines. In case an interesting event is detected, the signal pipelines are stopped and the data for the bunch crossing(s) in question are digitized. The digitized data are used on the next trigger level(s) for a more restrictive event selection.

The final event selection is done in computer farms. At this point the complete digitized information for the event is available and a first reconstruction of the event is performed. The filter farms consist of a large number of fast processors with computing power of $\sim$ 300 (H1) to 1000 MIPS (ZEUS), each processor processing one event at a time. The rate of accepted events varies in both experiments between about 3 and 7 Hz, typical event sizes are 60 kByte (H1) to 140 kByte (ZEUS). The accepted events are reconstructed off-line in processor farms with sufficient computing power to have the reconstructed events available for analysis within a few hours after data taking.

4 Running Conditions

The typical bunch configuration of HERA during the first weeks of data taking is sketched in Fig. 13: 10 consecutive proton bunches (1-10) and 9 electron bunches (1-9) are filled. The 10th proton bunch is unpaired; it has no electron partner and is used for studying proton beam induced background. Similarly, the electron bunch 19 is unpaired and is used to measure electron beam induced background. Typical beam currents were $1-3$ mA.

The luminosity was measured by detecting the process $ep \rightarrow e'p\gamma$ as mentioned before. Fig. 14 shows the scatter plot of the energies E'_e, E_γ for the final state

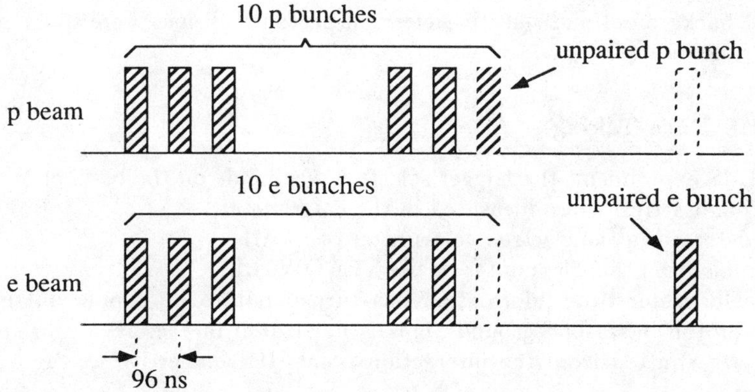

Fig. 13. Sketch of the HERA bunch configuration.

electron and photon as determined by the luminosity detector. There is a well isolated band of events for which the sum of the two energies is equal to the energy of the electron beam, $E_e' + E_\gamma = E_e$, as expected for the luminosity reaction. However, bremsstrahlung of the electron beam on the residual gas in the beam pipe, $eA \rightarrow e'A'\gamma$, satisfies the same condition. The subtraction of this background was done by measuring the gas bremsstrahlung with the unpaired electron bunch and scaling with the currents of the unpaired electron bunch and the total electron beam (see Fig. 15). The achieved precision of the luminosity measurement is at present 10 – 15%.

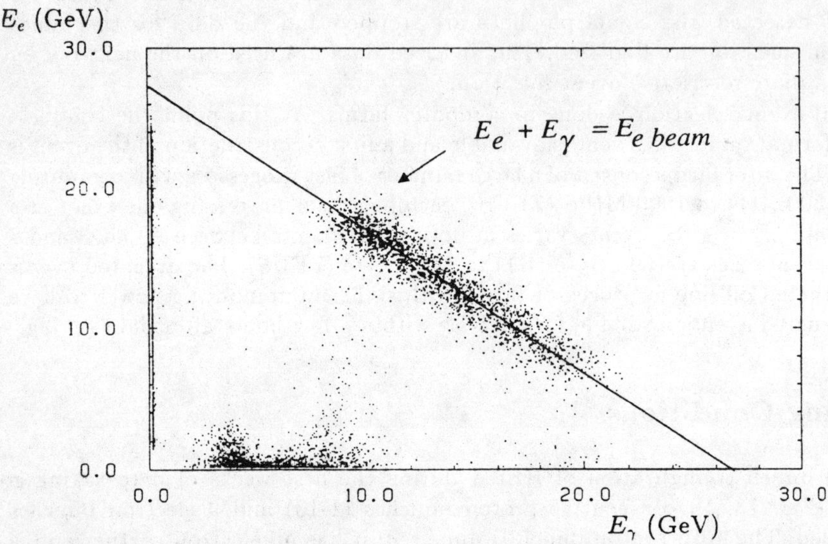

Fig. 14. Distribution of the electron energy versus the photon energy measured in the ZEUS luminosity detector.

Suppression of background events at the trigger stage to a manageable rate was not a particularly difficult task, mostly because the beam currents were only a few percent of their nominal values. By far the most copious background was produced by proton interactions upstream of the detector (see Fig. 5). The strategies for suppressing unwanted background and selecting electron-proton collisions were different for the two experiments.

4.1 ZEUS Data Taking

In the ZEUS experiment, the trigger selection was made on the basis of
- energies and arrival times measured in the calorimeter,
- energy detected in the electron calorimeter of LUMI,
- veto signals from the C5 counter or the VETOWALL.

The calorimeter time information has turned out to be a powerful handle for rejecting proton beam background. This is illustrated in Fig. 16. In an *ep* collision particles are emitted from the interaction point, IP, and arrive at the calorimeter

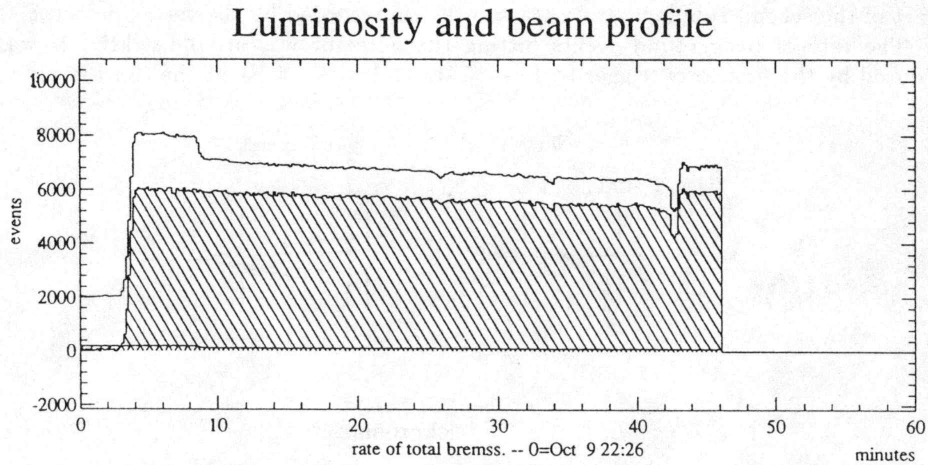

Fig. 15. The total bremsstrahlung rate (open curve) and the luminosity after subtraction of beam gas bremsstrahlung (shaded curve) as a function of time (from ZEUS).

cells at times $t = 0$ while a proton interaction upstream of the detector such as shown in Fig. 5 deposits energy in the RCAL about 10 ns earlier. The 10 ns difference corresponds to twice the distance between RCAL and the IP. Of course, in FCAL also the proton induced background arrives at $t = 0$. The measured distribution of FCAL (t_{FCAL}) versus RCAl (t_{RCAL}) times is shown in Fig. 16c for events with more than 1 GeV deposited in a calorimeter cell in both FCAL and RCAL. The *ep* events with $t_{FCAL} \sim t_{RCAL} \sim 0$ are well separated from the background which clusters around $t_{RCAL} = -10$ ns, $t_{FCAL} - t_{RCAL} = 10$ ns. Note, there are about 1000 times more background than *ep* events.

Samples of event pictures are shown in Fig. 17. The first event stems from neutral current scattering at $Q^2 \approx 2550$ GeV^2, $x \approx 0.07$, with an electron seen in BCAL and a high energy jet in FCAL (Fig. 17a). The jet near the proton beam is presumably produced by the proton remnants. The high energy jet and the electron are back-to-back in the transverse plane and balance transverse momentum as expected for an NC event. The interaction point is marked by tracks detected in the cylindrical drift chamber (CTD). The second event (Fig. 17b) shows a low Q^2, low x event ($Q^2 = 6$ GeV^2, $x = 0.0004$). The electron is produced very close to the beam and is only seen in RCAL. The third event (Fig. 17c) is due to quasireal photoproduction: it has energy deposition and charged tracks in the central detector plus an electron of 10.5 GeV in the LUMI electron calorimeter. The c.m. energy of the photon-proton system for this event is 230 GeV; the equivalent proton energy for a stationary proton would be 29 TeV.

The event pictures were produced by including every calorimeter cell with an energy more than 60 (100) MeV in the electromagnetic (hadronic) section. The calorimeter is seen to be very clean. The information on charged particle tracking is still limited due to missing digitizing electronics which is scheduled for installation later this year and during 1994. For the moment the CTD provides only z and $r - phi$ coordinates from the z-by-timing readout of 16 wire layers in superlayers 1, 3 and 5. Since the start of the second running period tracks are also recorded by the vertex detector.

The rate of background events hitting the detector was around a kHz. It was reduced by the first level trigger to 10 – 15 Hz and to 3 – 5 Hz at the third level

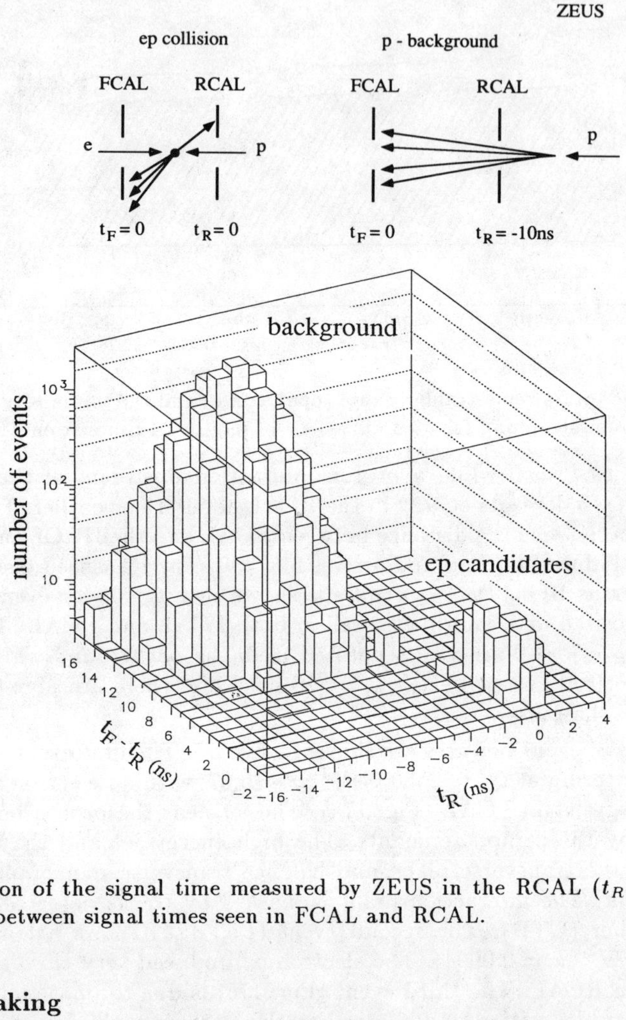

Fig. 16. Distribution of the signal time measured by ZEUS in the RCAL (t_R) versus the difference $t_F - t_R$ between signal times seen in FCAL and RCAL.

4.2 H1 Data Taking

The suppression of background at the trigger stage was accomplished in the H1 experiment with the help of

- the scintillator hodoscope in the backward direction (TOF-Veto),
- the proportional wire chambers of the central tracker requiring at least one ray pointing to the vertex region,
- the liquid-argon (LA) and backward electromagnetic calorimeters
- (BEMC) requiring clusters of $> 8(4)$ GeV in the first (second) component.
- the electron calorimeter of LUMI.

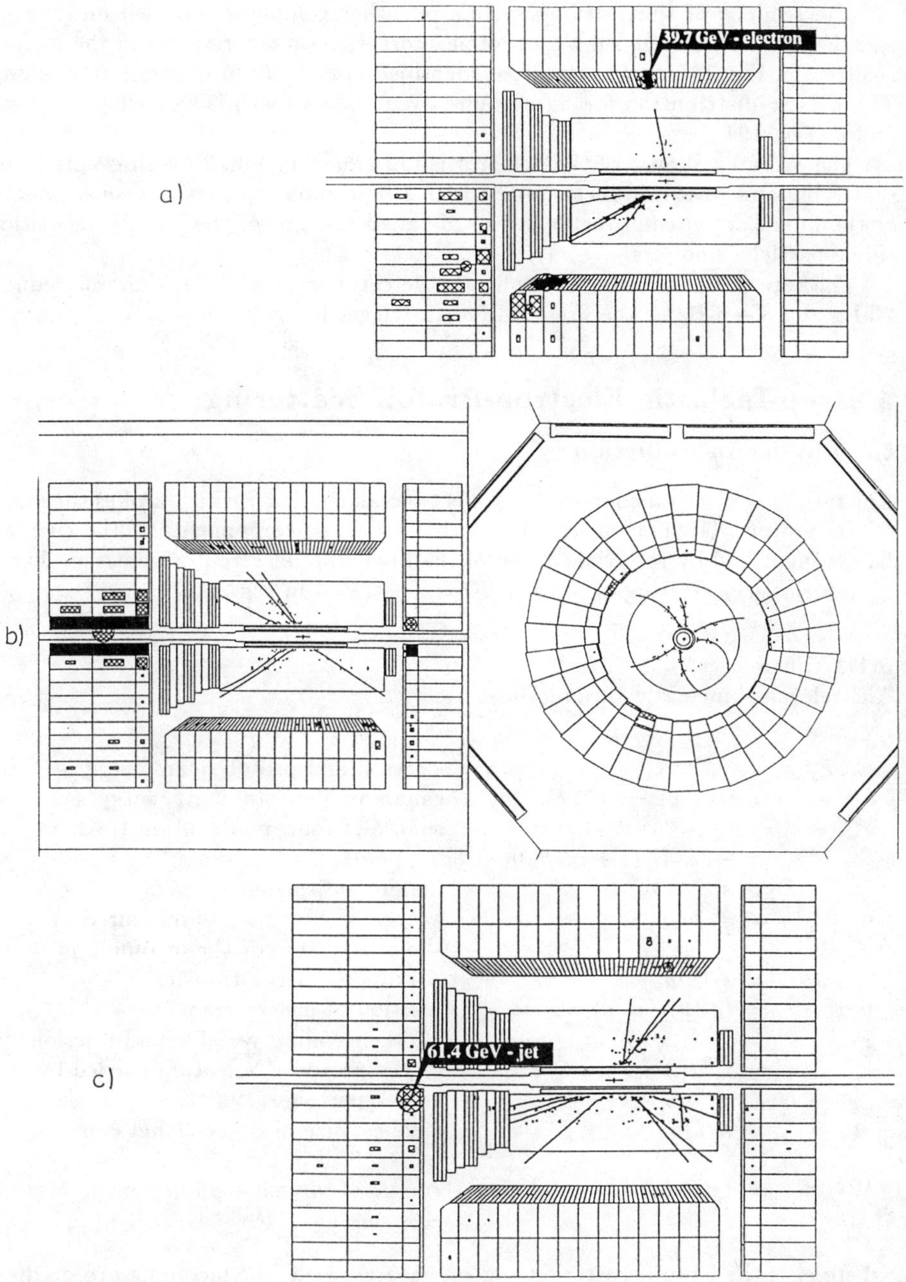

Fig. 17. Event pictures observed by ZEUS:
a) Deep inelastic NC scattering at $Q^2 \approx 2550$ GeV^2, $x \approx 0.07$
b) Deep inelastic NC scattering at $Q^2 \approx 6$ GeV^2, $x \approx 0.0004$
c) Photoproduction at a total c.m. energy of 230 GeV.

The readout of the tracking system is rather complete and well understood. A spectacular event produced by a proton interaction on the rest gas in the beam pipe is shown in Fig. 18. It has 21 protons identified via dE/dx in the central jet chamber. The dE/dx distribution for background events shows well isolated bands of π, K, p and d (Fig. 19).

The overall response of the detector is illustrated in Fig. 20 with events from NC scattering and photoproduction. The high longitudinal and transverse segmentation of the liquid argon calorimeter gives a detailed account of the energy deposition for single particles and jets.

The rate of background events in the detector was ~ 3 kHz; this was reduced to 20 Hz and $3-5$ Hz at the first and fourth trigger levels, respectively.

5 Deep Inelastic Electron-Proton Scattering

5.1 Physics Introduction

The incoming electron couples to the electroweak current j which probes the structure of the proton. The neutral (NC) and charged (CC) components of the current can be distinguished by the observation of the final state electron or neutrino. The basic deep inelastic scattering process (DIS) is illustrated in Fig.21.

5.1.1 Kinematics

The relevant kinematic variables are

E_e, E_p		electron and proton beam energies
s	$= (e+p)^2 = 4E_eE_p$	square of the total c.m. energy
q^2	$= (e-e')^2 = -Q^2 =$	square of four moumentum transfer
	$= -2E_eE'_e(1+cos\theta'_e)$	
Q^2_{max}	$= s$	maximum possible Q^2value
ν	$= q \cdot p/m_p$	energy of current j as measured in the rest system of the incoming proton
ν_{max}	$= s/(2m_p)$	maximum energy transfer
y	$= (q\cdot p)/(p\cdot e) = \nu/\nu_{max}$	fraction of energy transfer
x	$= Q^2/(2q\cdot p) =$	Bjorken scaling variable and fraction
	$= Q^2/(2m_p\nu) = Q^2/(ys)$	of the proton momentum carried by the struck parton
Δ	$= \hbar/Q$	smallest size of objects that can be resolved in the proton
W^2	$= Q^2(1/x-1)+m_p^2$	square of virtual – photon proton c.m. energy

where q'_e is the lepton scattering angle measured w.r.t. the incoming proton direction and the electron and proton masses, m_e, m_p, have been neglected. For NC events the kinematic properties of DIS events can be determined either from the variables of the electron or of the final state hadron system, or from a mixture of both. For CC events only the hadron system is accessible for measurement.

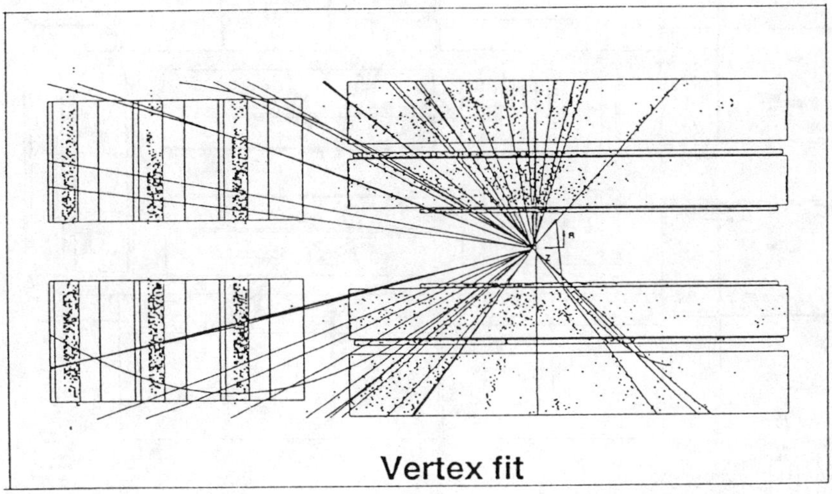

Fig. 18. Example of a proton-gas interaction in the H1 detector with 21 final state protons identified by dE/dx.

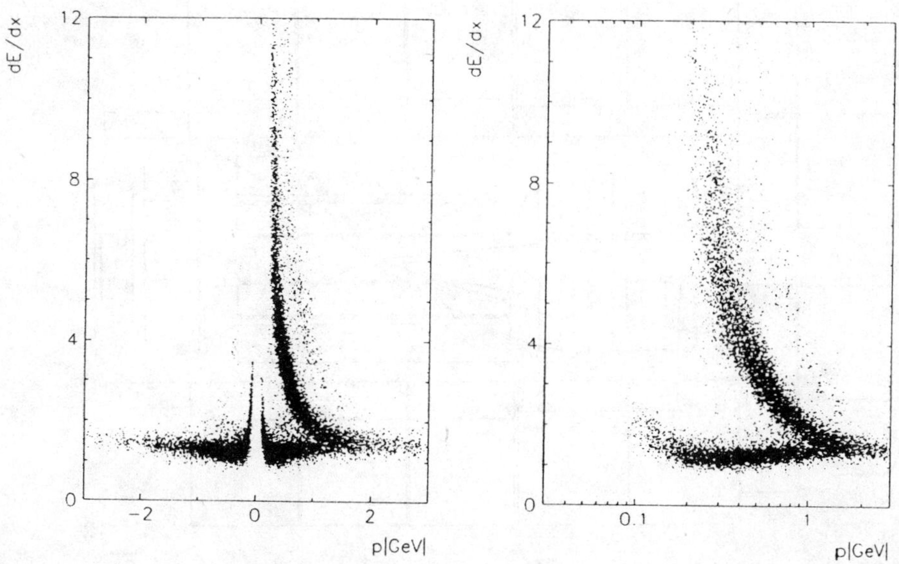

Fig. 19. dE/dx spectra observed by the central tracker of H1 for negative and positive particles.

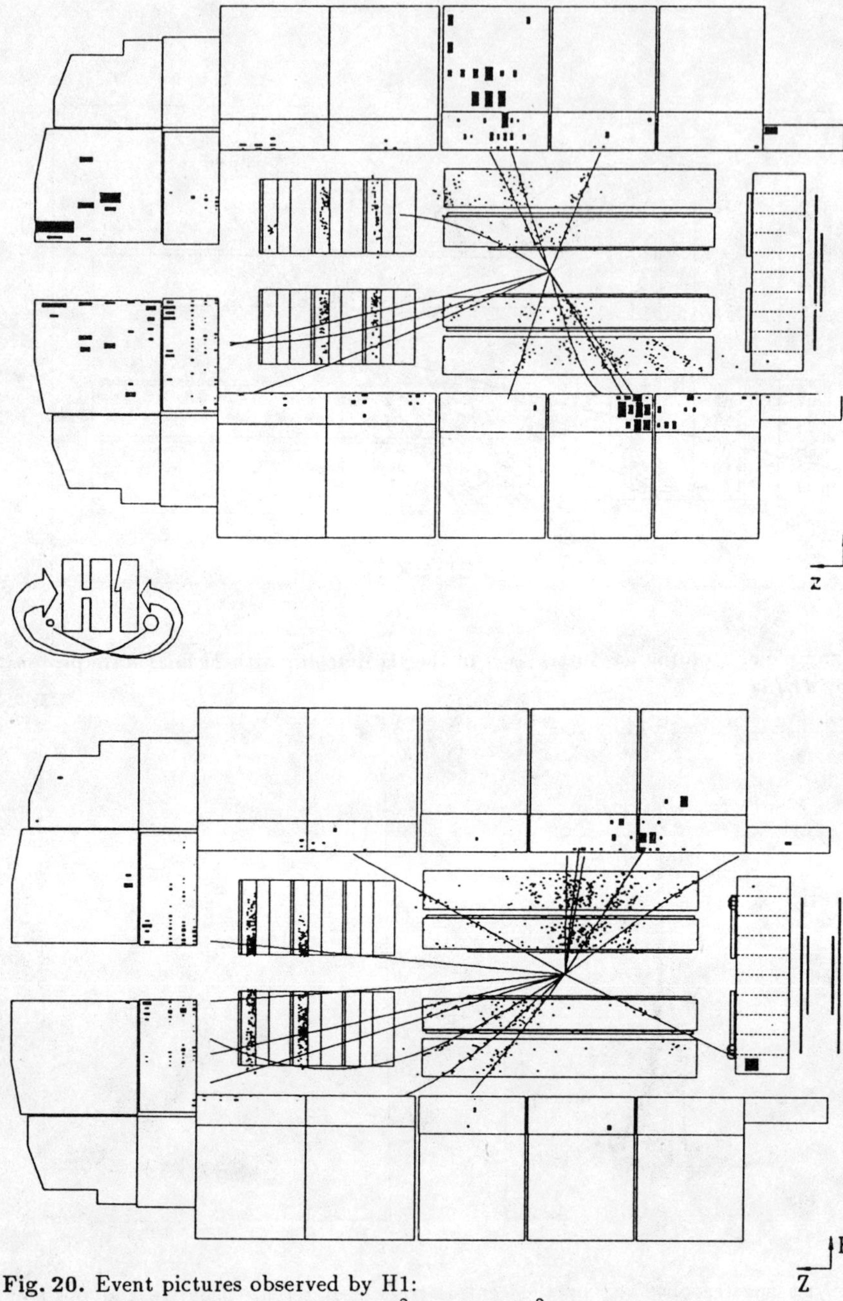

Fig. 20. Event pictures observed by H1:
top: Deep inelastic NC scattering at $Q^2 = 103\ \mathrm{GeV}^2, x = 0.004$
bottom: Photoproduction with two jets in the final state.

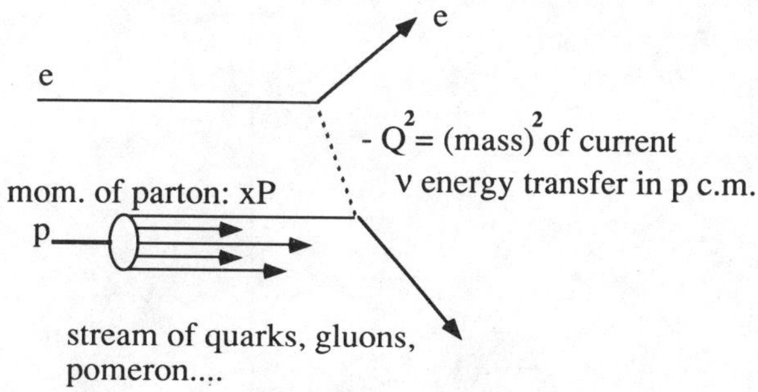

Fig. 21. Diagram for *ep* scattering.

From the electron:

$$Q_e^2 = 2E_e E_e'(1 + cos\theta_e') \qquad y_e = 1 - (E_e'/E_e)(1 - cos\theta_e')/2.$$

From the hadron system, using the method of Jacquet-Blondel [14] and summing the energies (E_h), transverse (P_{Th}) and longitudinal momenta (P_{zh}) of all final state hadrons:

$$Q_{JB}^2 = \sum_h (P_{Th}^2)/(2E_e) \qquad y_{JB} = \sum_h (E_h - P_{zh})/(2E_e).$$

From the electron scattering angle and the angle γ_h which characterizes the longitudinal and transverse momentum flow of the hadron system [15] (in the naive parton model γ_h is the scattering angle of the struck quark):

$$cos\gamma_h = \frac{(\sum_h P_x)^2 + (\sum_h P_y)^2 - (\sum_h (E_h - P_{zh}))^2}{(\sum_h P_x)^2 + (\sum_h P_y)^2 + (\sum_h (E_h - P_{zh}))^2}$$

$$Q_{\gamma\theta}^2 = 4E_e^2 sin\gamma_h(1 + cos\theta_e')/[sin\gamma_h + sin\theta_e' - sin(\gamma_h + \theta_e')]$$

$$y_{\gamma\theta} = sin\theta_e'(1 - cos\gamma_h)/[sin\gamma_h + sin\theta_e' - sin(\gamma_h + \theta_e')].$$

For NC events the precision of x, y and Q^2 can be improved by a simultaneous fit to all measured variables [16].

Table 2 compares the kinematic range accessible at HERA and in previous lepton-nucleon scattering experiments. The maximum energy transfer is increased by a factor of $\sim$ 100: HERA is equivalent to a fixed target experiment with an incident electron beam of 52 TeV. The Q^2 domain over which lepton nucleon scattering can be measured is also increased by two orders of magnitude. Since the typical Q values in DIS are much larger than the proton mass the electron interacts with one of the partons (quarks, gluons, $\cdots$) rather than with the proton as a whole: HERA is in reality an electron-quark (or gluon) collider.

For the analysis of *ep* interactions at HERA an understanding of the event kinematics is helpful. It is complicated by the fact that the c.m. system is not at rest. The large momentum excess of proton over electron beam pushes most final state particles into the proton direction. The correlations between energy and angle of electron and current jet (ignoring gluon emission) are shown in Fig. 22 in the $x - Q^2$ plane.

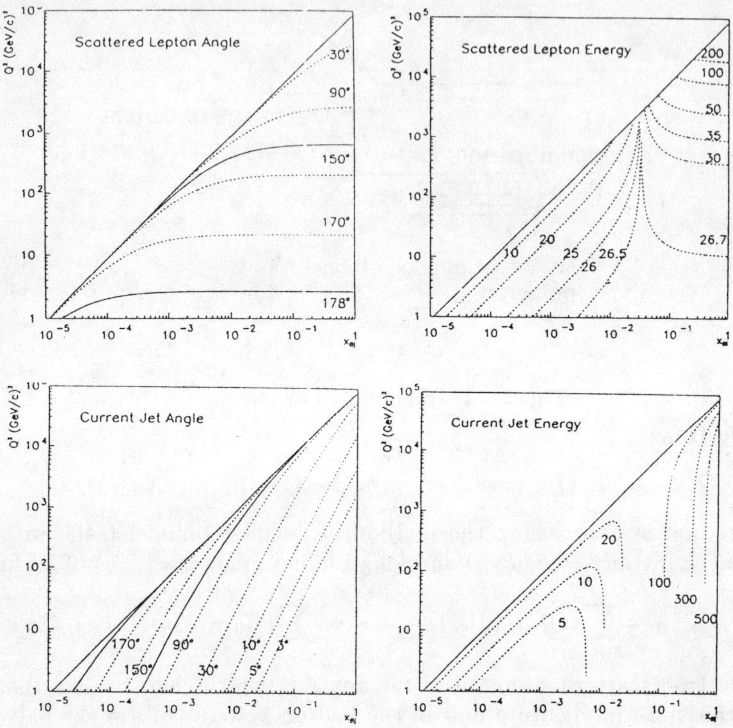

Fig. 22. The $x - Q^2$ dependence of the angle and energy of the scattered electron and the current jet (ignoring gluon radiation) for beam energies of $E_e \times E_p = 30$ GeV $\times$ 820 GeV; the angles are measured w.r.t. the proton beam direction.

Table 2. Kinematic region accessible at HERA ($E_e = 30$ GeV, $E_p = 820$ GeV) and in previous experiments.

	HERA	pre-HERA
s (GeV^2)	10^5	10^3
maximum attainable Q^2 (GeV^2)	40 000	400
Δ (cm)	$1 \cdot 10^{-16}$	$1 \cdot 10^{-15}$
ν_{max} (GeV)	52 000	500
minimum x at $Q^2 = 10$ GeV^2	$1 \cdot 10^{-4}$	$1 \cdot 10^{-2}$

5.1.2 Cross sections

The cross sections for NC and CC scattering are related to structure functions F_i of the proton:

$$NC,\ ep \to eX:$$
$$\frac{d^2\sigma(\gamma+Z^0)}{dxdy} = \frac{4\pi a^2}{sx^2y^2}\left[(1-y+y^2/2)F_2(x,Q^2) - y^2/2F_L(x,Q^2)\pm (y^2/2-y)xF_3(x,Q^2)\right]$$

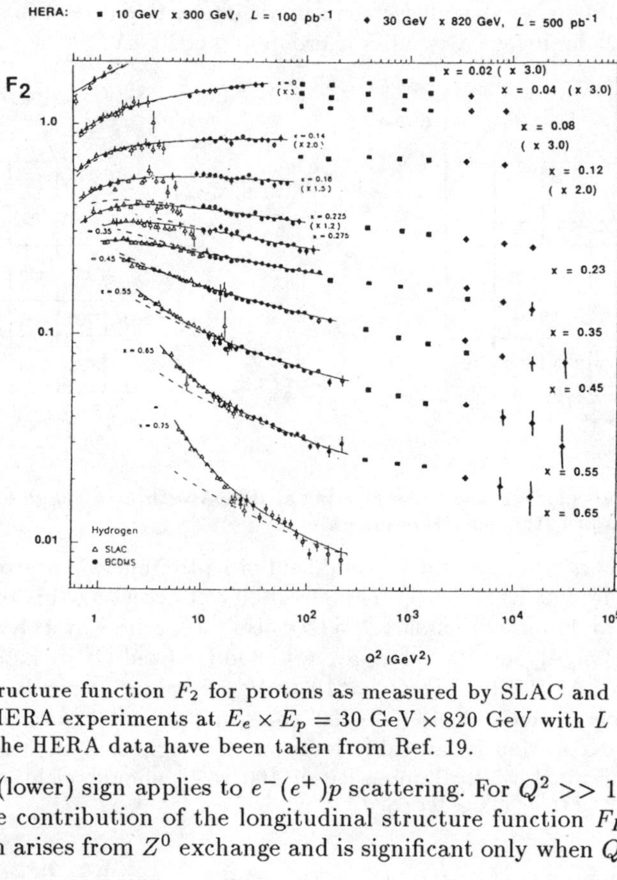

Fig. 23. The structure function F_2 for protons as measured by SLAC and BCDMS and as expected from HERA experiments at $E_e \times E_p = 30$ GeV $\times$ 820 GeV with $L = 500$ nb^{-1}, the predictions for the HERA data have been taken from Ref. 19.

The upper (lower) sign applies to $e^-(e^+)p$ scattering. For $Q^2 >> 1$ GeV^2 and not too small x the contribution of the longitudinal structure function F_L is small. The F_3 contribution arises from Z^0 exchange and is significant only when Q is comparable to the Z mass.

$$CC,\ ep \to \nu X:$$
$$\frac{d^2\sigma(W)}{dxdy} = \frac{G_F^2 s}{2\pi}\frac{1}{(1+Q^2/M_W^2)^2}\left[(1-y+y^2/2)F_2(x,Q^2) \pm (y^2/2-y)xF_3(x,Q^2)\right]$$

where G_F is the Fermi coupling constant, $G_F = 1.02 \cdot 10^{-5}/m_p^2$. As before, the upper (lower) sign applies to $e^-(e^+)p$ scattering. The structure functions F_i from NC and CC scattering are in principle independent. They can be related via the quark parton model, however.

The structure functions measured in previous experiments (see the lectures by A. Milsztain [17]) have been extrapolated by QCD evolution [18] to the HERA regime (see Fig. 23). The number of events expected from NC and CC scattering are shown in Fig. 24 for 500 pb^{-1}. The large NC rates at low Q^2 stem from photon exchange. At $Q^2 > M_Z^2$ contributions from Z-exchange become equally important. The requirement of a minimum of 100 events leads to a maximum Q^2 value of about 35 000 GeV^2 up to which NC measurements are feasible. The event rate for CC scattering at low Q^2 is much smaller. However, for Q^2 around M_W^2 the CC and NC cross sections become equal (see Fig. 25), i.e. weak and electromagnetic interactions are of the same strength. The practical Q^2 limit for CC studies is around 40 000 GeV^2.

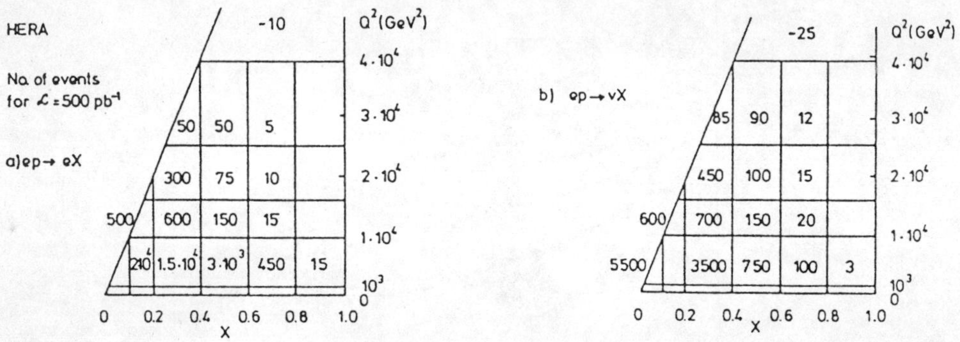

Fig. 24. Event rates for NC and CC scattering at HERA with $L = 500$ pb^{-1} calculated with LUND-LEPTO and EHLQ structure functions.

Small x physics is a new and exciting field of lepton-nucleon scattering pioneered by Gribov, Levin and Ryskin [20]. The possibility of accessing this region at HERA has stimulated an intense discussion [21] (see also the lectures by J. Kwieczinski [22]). Since $x = Q^2/(2m_p\nu)$, small x-values are attained for fixed Q^2 by making the energy transfer ν large. At HERA, for $Q^2 = 10$ GeV^2, x-values as small as 10^{-4} can be reached which is a factor of 100 smaller than in previous experiments (see Table 2). The NC cross section is favorably large in this regime as shown in Fig. 26: for instance, the nominal yearly luminosity of 100 pb^{-1} should yield 10^6 events with $10^{-4} < x < 10^{-3}$, $10 < Q^2 < 20$ GeV^2.

5.2 Results from H1 and ZEUS on the Total Cross Section

For both experiments the prime goal in this first analysis was to establish that events from deep inelastic NC scattering can be isolated from background and have the expected characteristics and cross section. The description of the event selection will be rather sketchy; details can be found in Ref. 23, 24.

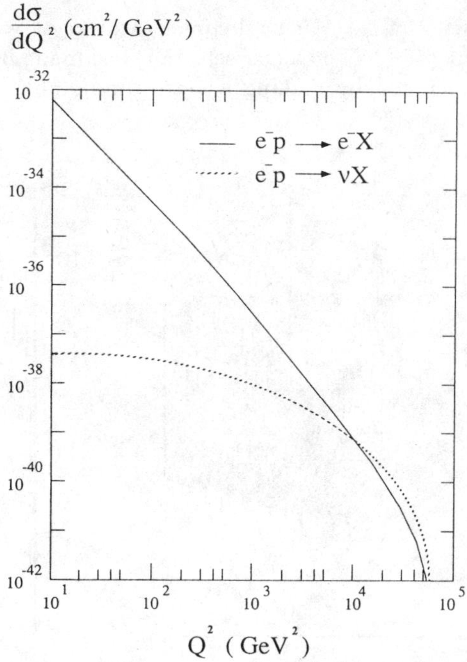

Fig. 25. The cross sections for NC and CC scattering as a function of Q^2.

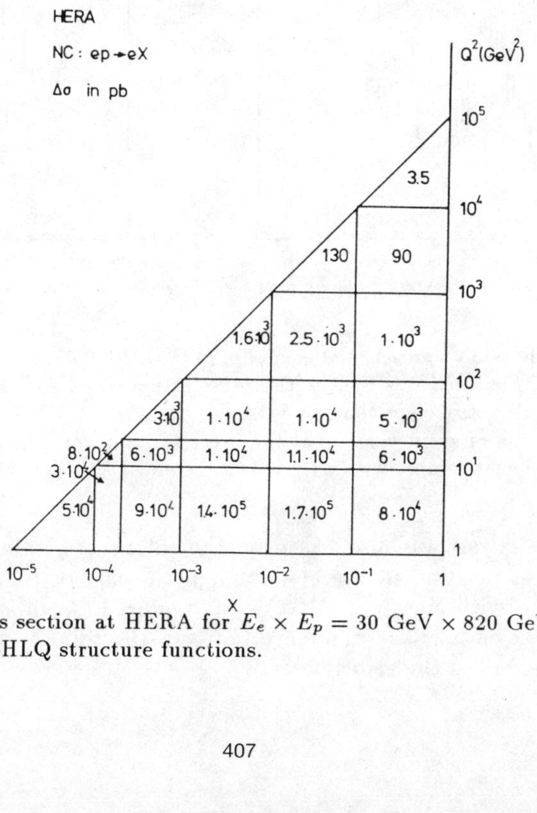

Fig. 26. The NC cross section at HERA for $E_e \times E_p = 30$ GeV $\times$ 820 GeV calculated with LUND-LEPTO and EHLQ structure functions.

ZEUS analysis (Ref. 24): Neutral current scattering was studied with an integrated luminosity of 2.2 nb^{-1}. The trigger selection was made by requiring energy deposition in the calorimeter and by making a loose timing cut. This led to a total of $1 \cdot 10^6$ accepted events.

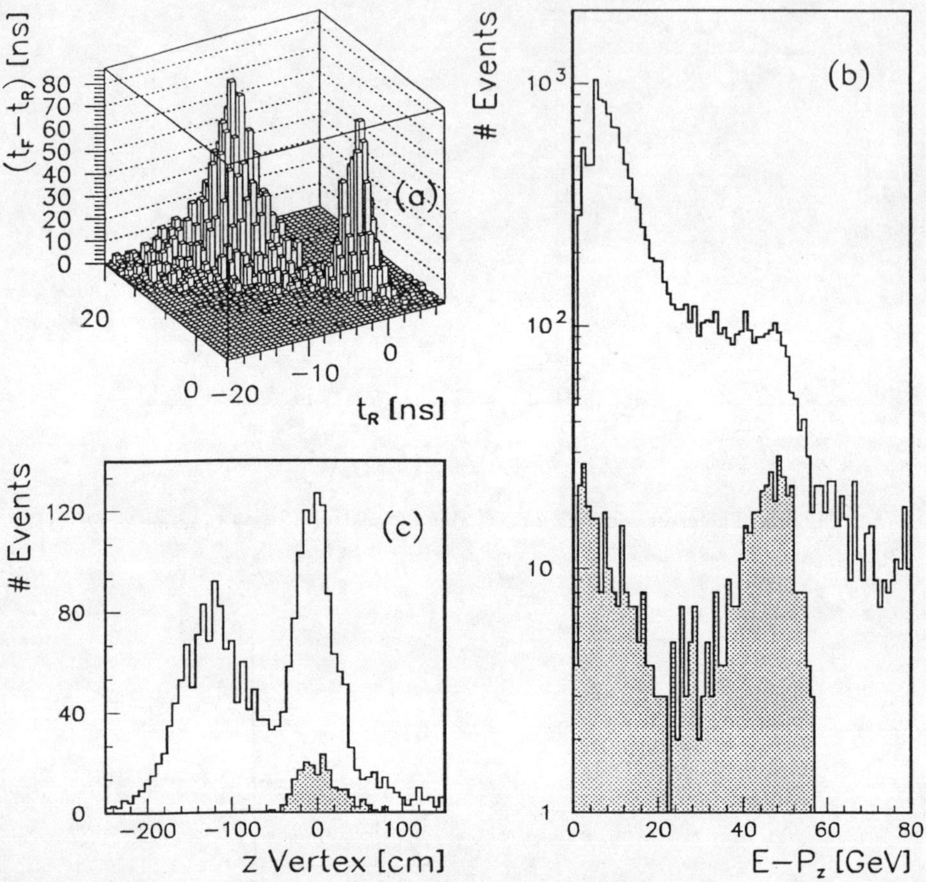

Fig. 27. a) Distribution of preselected events in (FCAL-RCAL) and RCAL times.
b) Distribution of $\delta = \sum_i (E - P_z)_i$ at the preselection level. The shaded area shows the distribution after all cuts except that on δ itself.
c) The Z-distribution of event vertices at the preselection level. The shaded area shows the same for the final event sample with $y_{JB} > 0.02$. From ZEUS, Ref. 27.

The off-line analysis was also based on the calorimeter information. In the *preselection* energy deposition in the electromagnetic part of RCAL or BCAL, or an identified electron with energy > 4 GeV was required. An important variable that characterizes the events is the difference between the total energy and longitudinal momentum measured in the calorimeter,

$$\delta = \sum_i (E - P_z)_i.$$

For ep events $\delta = 2E_e$ if all final state particles are measured. Undetected particles which are emitted into the forward beampipe region give a negligible loss in δ since $E_i \approx P_{zi}$. Initial state radiation moves events to lower δ values and measurement errors lead to a broadening of the δ distribution. Different final states produce different δ distributions, so that δ is a good discriminator against background processes. For instance, photoproduction events, in which the scattered electron escapes in the rear part of the beam pipe, will predominantly give low values of δ. Background events coming from interactions of the proton beam also have small values of δ. The distribution of δ is shown in Fig. 27b. The preselected sample (open histogram) shows a broad shoulder near $\delta = 2E_e$ superimposed on a large falling background.

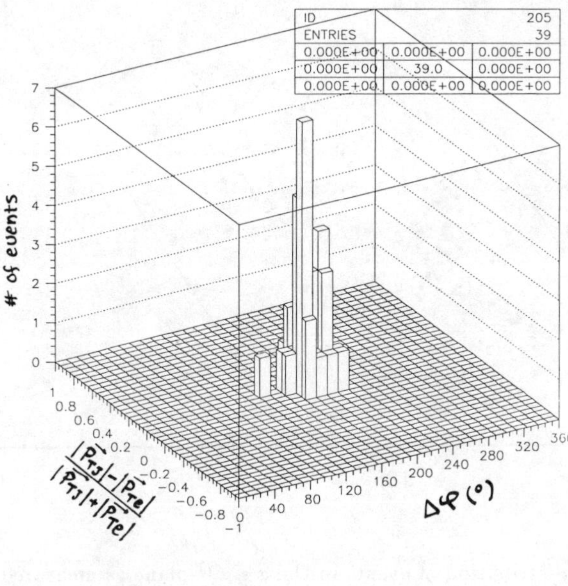

Fig. 28. Distribution of the normalized difference of transverse momenta and azimuth angles of electron and jet for NC events from ZEUS with a scattered electron of $E_e' > 10$ GeV, $\theta_e < 172^0$ and a current jet of $E_{jet} > 15$ GeV.

The time distribution for events with energy deposition in FCAL and RCAL shows a clean separation between events from ep collisions and from background (Fig. 27a). Therefore, the *final* selection was based on a tighter timing cut. It also required an electron with $E_e' > 5$ GeV and 35 GeV $< \delta <$ 60 GeV. The vertex distribution along the beam for the *preselected* event shows a peak concentrated around $Z = 0$ on top of a large background (Fig. 27c) while the events of the *final* sample (shaded distribution in Fig. 27c) are basically background free. The background in the final sample from photoproduction was estimated at $2 \pm 1\%$.

The transverse momentum vectors of the electron, P_{Te}, and of the hadron system, p_{Th}, are back-to-back and of equal magnitude as expected for NC events (see Fig.28).

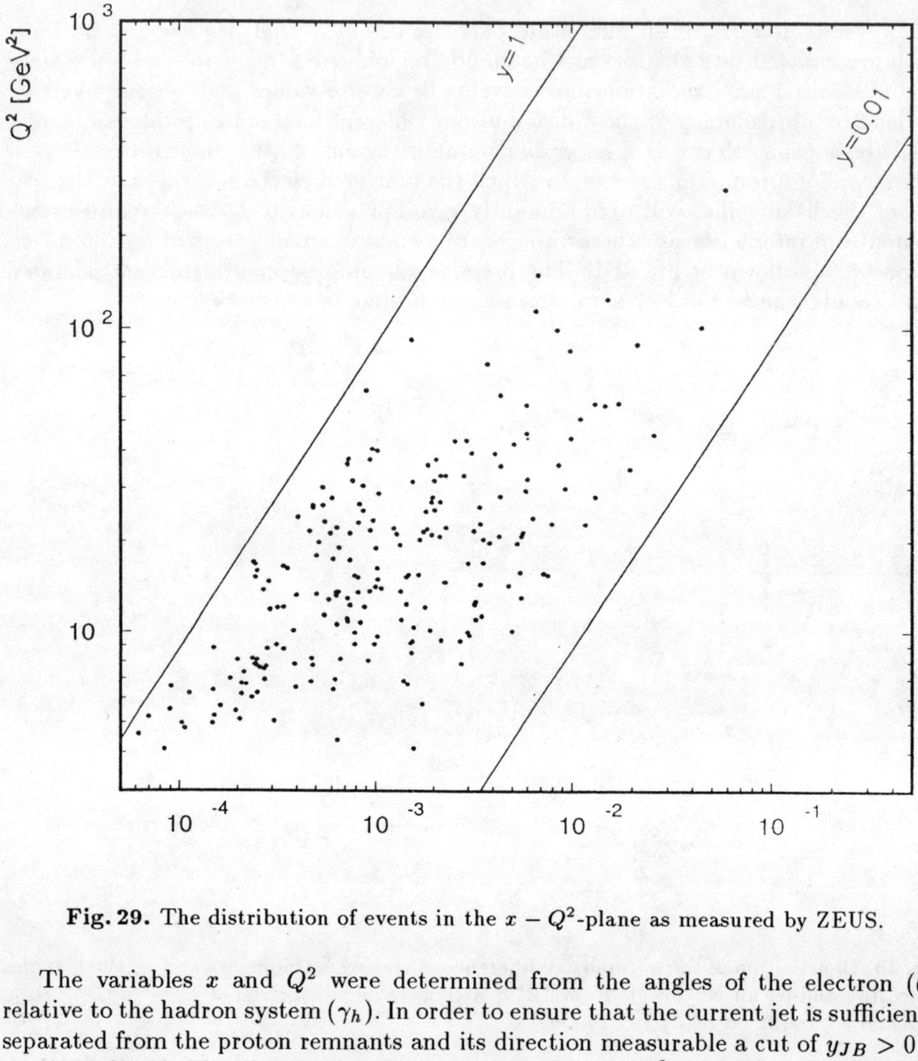

Fig. 29. The distribution of events in the $x - Q^2$-plane as measured by ZEUS.

The variables x and Q^2 were determined from the angles of the electron (θ_e) relative to the hadron system (γ_h). In order to ensure that the current jet is sufficiently separated from the proton remnants and its direction measurable a cut of $y_{JB} > 0.02$ was introduced. The distribution of the events in the (x, Q^2) plane are shown in Fig. 29. Most events lie in the Q^2 region between 5 and 100 GeV^2. In the x-direction the events populate mainly the region of very small x-values, $x < 10^{-2}$, with a sizable number of events between $x = 10^{-3}$ and 10^{-4}. The projections of the data in x and Q^2 (Fig. 30) are in reasonable agreement with Monte Carlo simulations performed with the set D_0 of Ref. 25. The total cross sections, corrected for acceptance, measurement uncertainties and QED radiative effects, are shown in Fig. 31. The data are consistent with all three structure function sets D_0, MTB1 (Ref. 26) – these two assume a constant gluon structure function, xG, as $x \to 0$ – and MTB2 (diverging xG as $x \to 0$).

H1 analysis (Ref. 23). NC candidates were selected from data corresponding to an integrated luminosity of 1.3 nb^{-1}. The trigger required a localized energy deposition or a cluster of more than 4 GeV in the BEMC calorimeter. Out-of-time events were rejected.

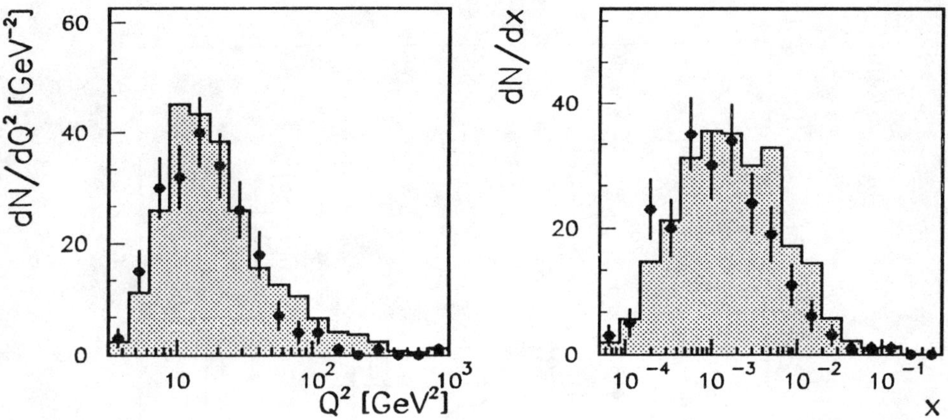

Fig. 30. The Q^2 and x distribution of the events as measured by ZEUS. The Monte Carlo predictions (shaded histograms) were calculated from the extrapolation of the MRSDO parametrization.

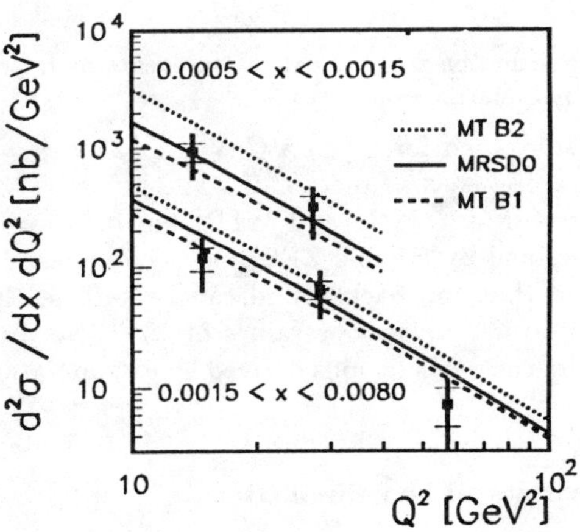

Fig. 31. The differential Born cross section in the $x - Q^2$ ranges indicated as measured by ZEUS. The lines show the extrapolations of the MTB1, MTB2 and MRSDO parametrizations.

Off-line selection of DIS events from the $6 \cdot 10^4$ events accepted by the trigger was done by requiring a correlation between the BEMC cluster and hits from the tracking chambers and performing a vertex cut. In order to suppress further beam background and contributions from photoproduction, events with $y_e > 0.6$ and $y_{JB} < y_e/2$ were rejected. The energy spectrum of the scattered electron for the 148 accepted events is shown in Fig. 32. The calorimeter was recalibrated by fitting the Monte Carlo prediction to the events with $E'_e > 22$ GeV. The background from photoproduction is estimated to be $< 50\%$ for $E'_e < 10$ GeV, $< 20\%$ for $10 < E'_e < 14$ GeV and negligible above; beam induced background is $< 5\%$ for $E'_e > 10$ GeV.

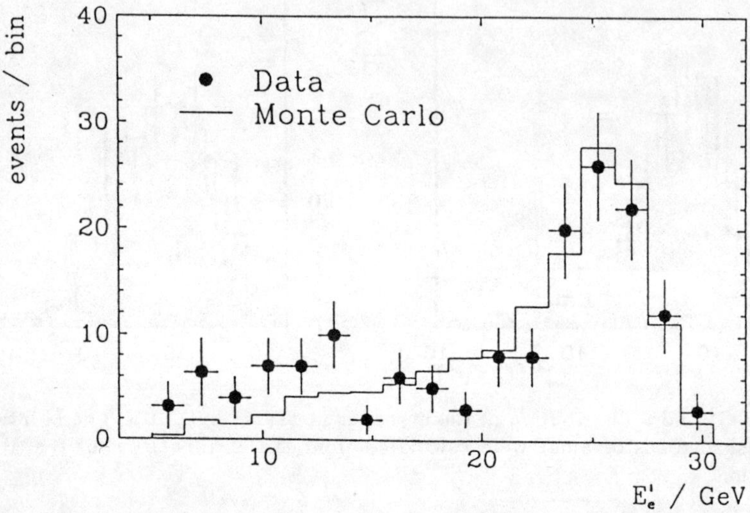

Fig. 32. Electron energy distribution for NC candidate events of H1 compared with the prediction for the MRSDO parametrization.

Cross sections were determined for $Q^2 > 5$ GeV^2, $0.025 < y_{JB} < 0.6$ and $Q'_e < 174^0$ without radiative corrections as a function of x and Q^2 (Fig. 33). The data agree broadly with the predictions from D_0, D_- and MTB1 but disfavor MTB2 somewhat.

These early results obtained by H1 and ZEUS on deep inelastic NC scattering in their very first run showed that the background can be sufficiently suppressed and measurements can be performed down to x-values of 10^{-4}. No dramatic deviations have yet been observed from the expectations derived by extrapolating the fixed target results to the HERA regime.

5.3 General Characteristics of the Final State

H1 performed a first analysis of the final states of NC events [27]. Events were selected with $E'_e > 14$ GeV and hadronic c.m. energies squared $W^2 > 3000$ GeV^2. A total of 88 events were retained. Their average values $< Q^2 >$ and $< W^2 >$ were 15 GeV^2 and 10^4 GeV^2, respectively. The hadronic final states were studied using the information on charged tracks produced with polar angles $22^0 < \theta < 160^0$ and transverse momenta (w.r.t. the beam direction) $p_T > 0.1$ GeV/c.

The results are summarized in Figs. 34, 35. The transverse energy flow of charged particles in terms of the pseudorapidity $\eta = -\log\ (\tan\theta/2)$ and the azimuthal angle

ϕ w.r.t. the scattered electron in the plane transverse to the beams are shown in Fig. 34. Here, all quantities were calculated in the lab. system. The sea-gull plot was determined in the hadronic c.m. system: Fig. 35 shows the average transverse momentum squared as a function of Feynman $x_F = p_z/p_z(max)$.

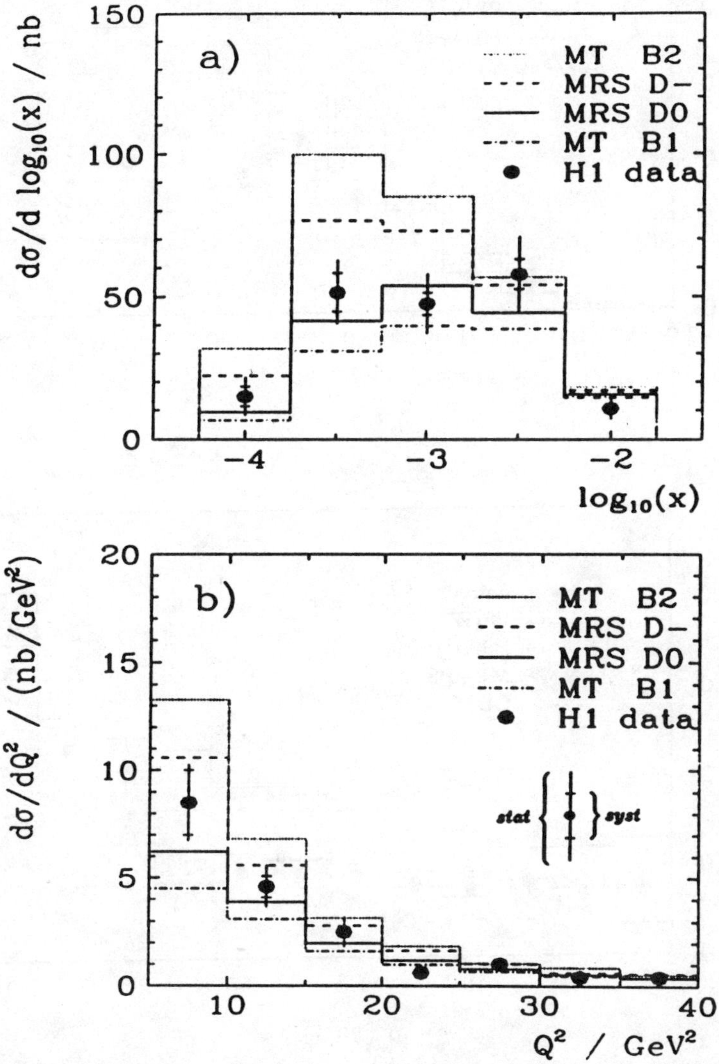

Fig. 33. Differential cross sections in log$x$ and Q^2 in the range $0.6 > y > 0.025$, $Q^2 > 5\ \mathrm{GeV}^2$ and $\theta_e < 174^0$ as measured by H1. The cross sections are not corrected for radiative effects. The full error bars correspond to the statistical errors; the smaller systematic errors are also indicated. The lines show cross section calculations for different parton density parametrizations.

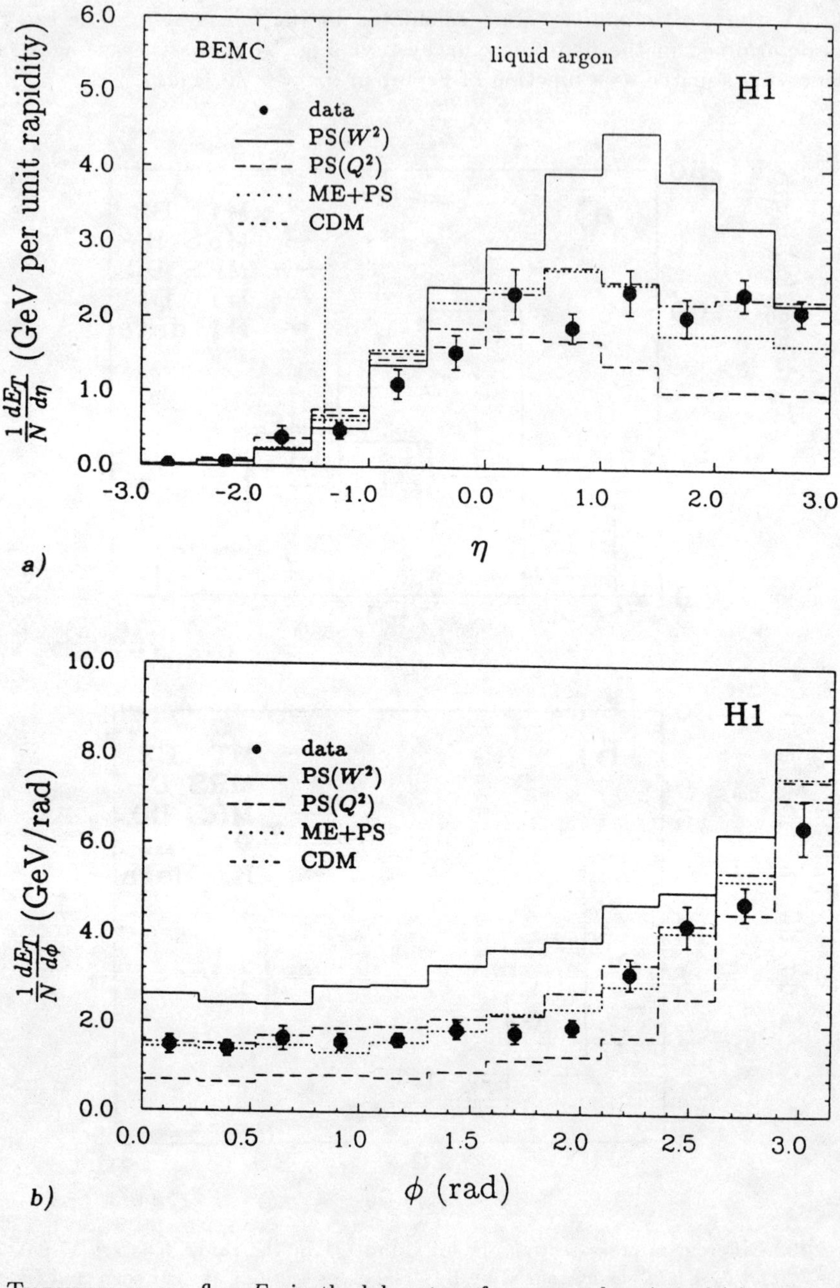

Fig. 34. Transverse energy flow E_T in the laboratory frame as a function of (a) pseudorapidity and (b) of the azimuthal angle with respect to the scattered electron direction in the plane transverse to the beam direction. The model predictions are also shown. From H1.

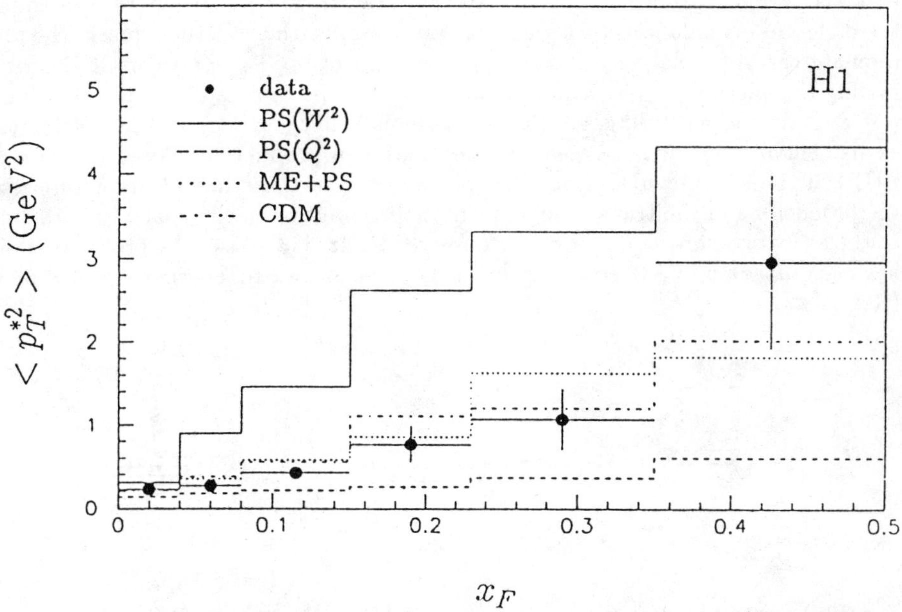

Fig. 35. Distribution of the average transverse momentum squared as a function of Feynman $x_F = p_z/p_z(max)$ in the hadronic c.m. system, as measured by H1.

The data were compared with three different prescriptions for simulating QCD effects using for the soft parton fragmentation the Lund string model as implemented in JETSET [28]. The three prescriptions are described briefly:

(a) Leading log parton showers (PS): parton cascades can start from the initial parton in the proton or from the scattered parton. The amount of gluon radiation depends on the scale of virtuality which can be chosen to be either Q^2 or W^2 (simulation denoted by $PS(Q^2)$ or $PS(W^2)$), or a function of both. Since $< Q^2 >$ is much smaller than $< W^2 >$ in this experiment, $PS(W^2)$ will predict much more gluon radiation than $PS(Q^2)$. The simulation was performed with HERACLES [29] for the electroweak interaction and LEPTO [30] for the QCD processes.

(b) Color dipole model (CDM): Gluon emission is described by a chain of radiating color dipoles between the scattered parton and the proton remnant. As implemented in ARIADNE [31], the virtuality scale is given by the p_T^2 of the radiated gluon and its maximum is proportional to $W^{4/3}$.

(c) $O(\alpha_s)$ matrix element and parton showers (ME + PS): Exact $O(\alpha_s)$ matrix elements are used for simulating the gluon radiation and photon gluon fusion processes while softer gluon emissions are added using the parton shower model. Using LEPTO [32] the maximum virtuality is related to the first order matrix element.

The comparison of the data with these models shows that CDM and ME + PS reproduce the measured results well while the leading log model fails if either Q^2 or W^2 is chosen as the scale of virtuality.

5.4 Jet Production in NC Scattering

In lowest order lepton-proton NC scattering, the transverse momentum of the scattered electron is balanced by a single jet associated with the struck quark, the proton remnant carrying relatively little transverse momentum. Higher order QCD processes modify this picture. In particular, a hard gluon can be radiated from the struck quark (QCD Compton scattering, QCDC), or a gluon from the proton can interact with the exchanged boson giving rise to quark-antiquark pair production (Boson Gluon Fusion, BGF), as illustrated in Fig. 36. One consequence of these higher order processes is the broadening of the transverse momentum distribution of the particles with respect to the exchanged boson direction, as observed before [33]. Also, the production of jets has been reported in a fixed-target muon experiment at c.m. energies up to ~ 30 GeV (Ref. 34).

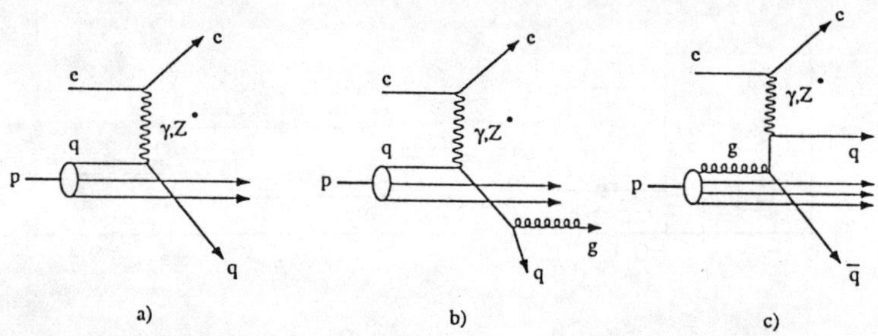

Fig. 36. Diagrams for deep inelastic scattering: (a) Born term, (b) QCD Compton scattering, (c) boson-gluon fusion.

By virtue of the large (virtual) photon proton ($\gamma^* p$) c.m. energies, W, available at HERA the contributions from these higher order QCD diagrams become directly visible as multijet events. This was demonstrated by ZEUS [35].

The analysis was performed for an integrated luminosity of 27 nb^{-1}. The selection criteria mentioned before led to a total of 3808 events with $Q^2 > 4$ GeV^2, $x \geq 10^{-4}$ and $30 < W < 280$ GeV, the average being 110 GeV (Fig. 37).

The transverse momentum distribution was determined in the $\gamma^* p$ rest system with the help of condensates. Condensates are contiguous energy deposits in the calorimeter defined as follows: the calorimeter cell with the highest energy is used as the seed and the cells adjacent to the seed cell are merged. According to simulations, for polar angles above 9^0, the condensate multiplicity agrees with that of the final state particles to 90% and about 80% of the generated stable particles are associated with only one condensate. In the analysis condensates were treated as massless particles.

Multihadron final states were selected by requiring at least 4 condensates with polar angles above 9^0 and the total transverse energy to be larger than 3.5 GeV. In total 3272 events were obtained. From the condensates the sphericity axis was determined in the $\gamma^* p$ rest system. Fig. 38 shows the distribution of the square of the transverse momentum, p_T^2 of the condensates w.r.t. the sphericity axis. The data are restricted to the hemisphere around the γ^* direction, which should contain the struck quark, and where one expects the gluon radiation effects to be most visible. The measured distribution is clearly broader than the QPM expectations. The tail extends up to ~ 20 GeV^2 while the QPM predicts no events above 5 GeV^2.

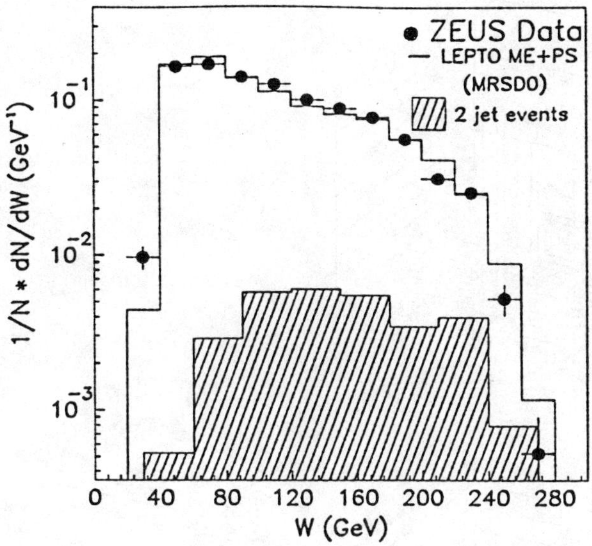

Fig. 37. The distribution of the mass of the hadronic system (W) compared with the LEPTO (ME + PS) prediction. The shaded distribution is for two-jet events. From ZEUS.

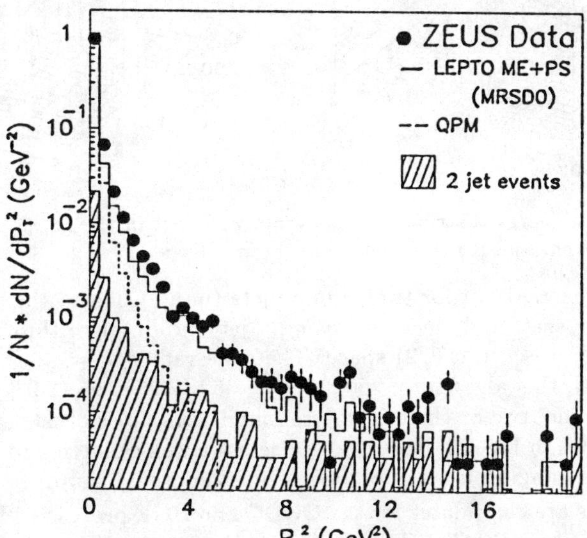

Fig. 38. The square transverse momentum distribution of condensates in the $\gamma^* p$ c.m. system, measured w.r.t. the sphericity axis, together with the predictions of QPM and $ME + PS$. The shaded distribution is for two-jet events. From ZEUS.

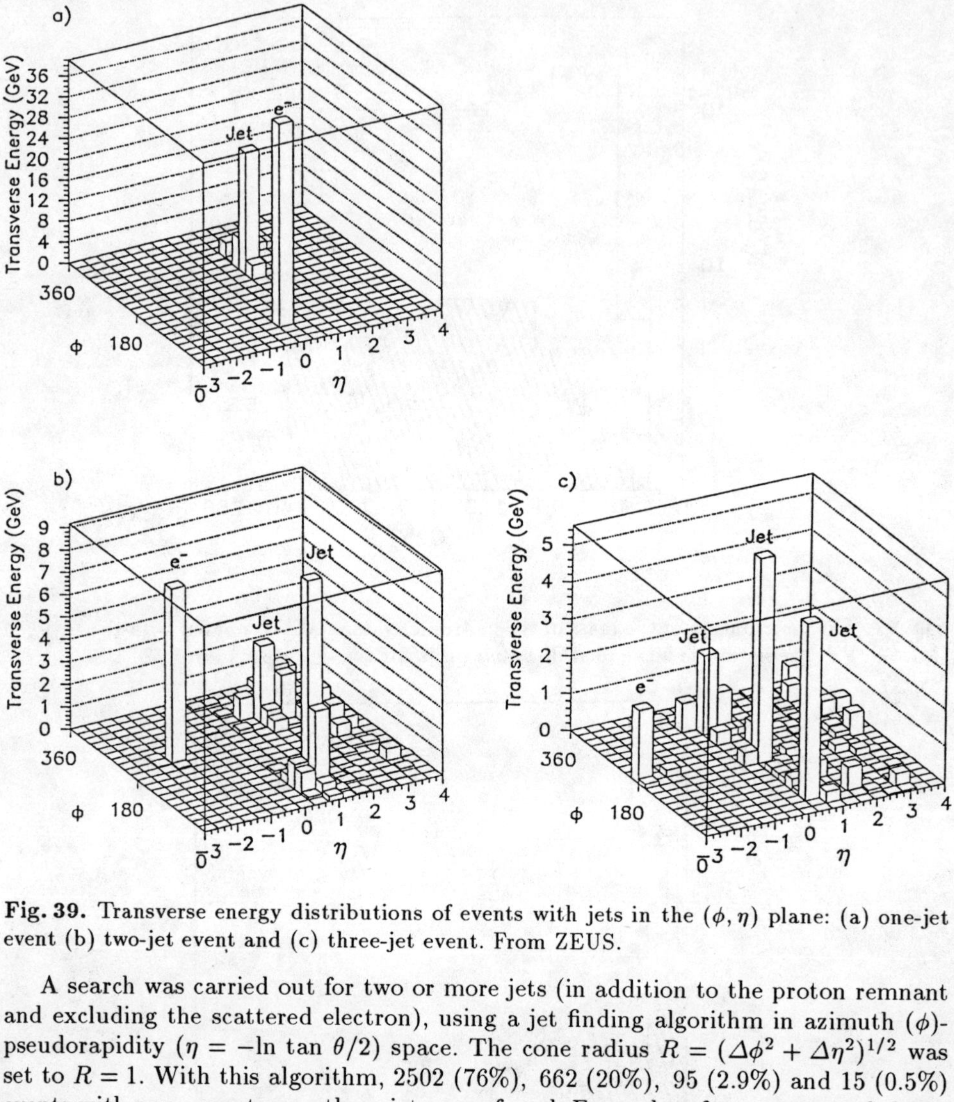

Fig. 39. Transverse energy distributions of events with jets in the (ϕ, η) plane: (a) one-jet event (b) two-jet event and (c) three-jet event. From ZEUS.

A search was carried out for two or more jets (in addition to the proton remnant and excluding the scattered electron), using a jet finding algorithm in azimuth (ϕ)-pseudorapidity ($\eta = -\ln \tan \theta/2$) space. The cone radius $R = (\Delta\phi^2 + \Delta\eta^2)^{1/2}$ was set to $R = 1$. With this algorithm, 2502 (76%), 662 (20%), 95 (2.9%) and 15 (0.5%) events with zero, one, two or three jets were found. Examples of one-, two- and three-jet events are shown in Fig. 39. Note that the jets are well isolated and their transverse energies are large, ranging from 5 to 30 GeV for the events shown.

Two-jet events are candidates for the QCDC and BGF processes. The shaded area in Figs. 37 and 38 show the distributions for two-jet events. They tend to have higher W's (average 150 GeV) and larger p_T^2. The jet energy profile, which is sensitive to fragmentation effects, is shown in Fig. 40 where the transverse energy flow into cells within a cone radius of $R = 2$ around the jet axis was computed. The jet width is $0.5 - 0.6$ (FWHM) in ϕ and η.

The simple QPM fails to describe the data. In contrast, first order QCD matrix elements plus higher order QCD corrections (ME + PS) are in accord with the data: they reproduce the jet rates, describe the W and p_T^2 distributions as well as the jet energy profile. It is noteworthy that the calculation based on exact first order matrix elements only (ME), i.e. without parton showers, predicts a narrower jet profile than observed (Fig. 40). The agreement with the QCD predictions suggests that two-jet production with large transverse jet energies is dominated by the QCD Compton and BGF processes.

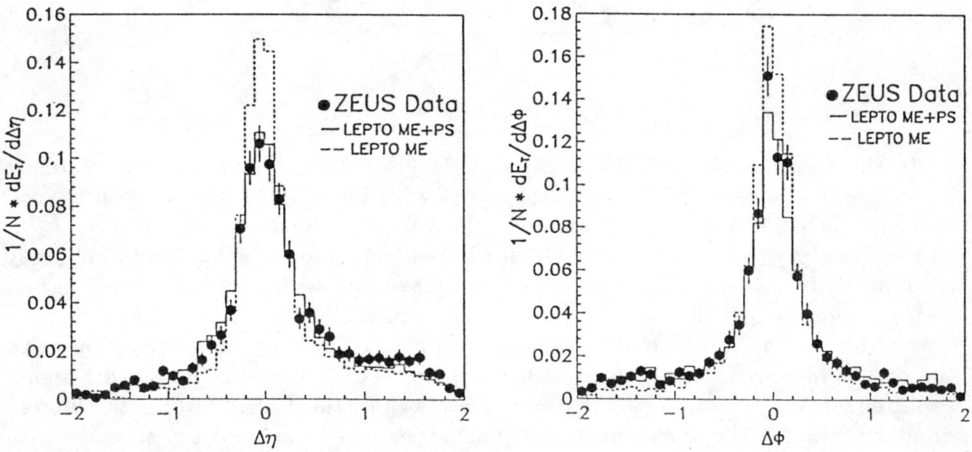

Fig. 40. The profile of jets from the two- and three-jet sample with the expectations from ME and $ME + PS$. From ZEUS.

6 Leptoquarks

Leptoquarks (LQ) are expected in many models. Production of leptoquarks in ep collisions can proceed in several ways depending on the couplings and the leptoquark mass. The most favorable process for HERA is s-channel production (Fig. 41): the electron with energy E_e and a quark with energy $x \cdot E_p$ form a leptoquark of mass squared $M_{LQ}^2 = 4xE_e\ E_P = xs$.

In the $x - Q^2$ distribution leptoquark production will populate a narrow band with fixed x. For scalar leptoquarks the population will be uniform in y. For narrow leptoquarks, the production cross section and the branching ratio into electron $+X$ are given by [36]:

$$\sigma_{LQ} = \pi/(4s) \cdot (g_L^2 + g_R^2) \cdot u(x = M_{LQ}^2/s)$$

$$BR(LQ \to e + u) = (g_L^2 + g_R^2)/(2g_L^2 + g_R^2)$$

where g_L and g_R are the left and right handed coupling constants and $u(x)$ is the u-quark density in the proton. The coupling strengths are unknown. For $g = 0.3$ the coupling would have the same strength as the electromagnetic coupling, $g^2/4\pi = \alpha_{em}$.

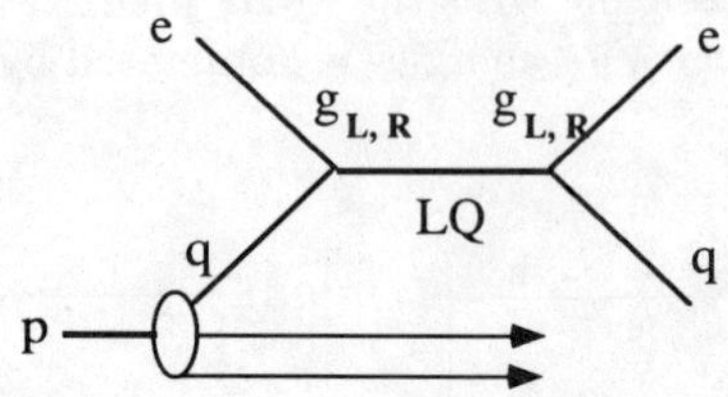

Fig. 41. Diagram for leptoquark production.

In the analyses of ZEUS (Ref. 37) and H1 (Ref. 38) leptoquarks were searched for in the NC ($ep \rightarrow eX$) and CC-like ($ep \rightarrow \nu X$) final states. For NC candidates, ZEUS applied the standard selection criteria with the additional requirements $E'_e >$ 10 GeV and $y_{JB} > 0.1$ resulting in 1659 eX candidates for an integrated luminosity of 26 nb^{-1}. Fig. 42 shows the x distribution for events with $x > 0.001$. The dashed histogram represents the event rate predicted with the MTB1 structure function set normalized to the integrated luminosity. For $x > 0.01$ reasonable agreement with the data is observed. No narrow width resonance is observed. The leptoquark signal expected for $M_{LQ} = 150$ GeV and $g = g_{em}$ is shown by the dashed histogram centered around $x = 0.26$. The search in the νX final state yielded in total 2 candidates with $(x, Q^2) = (0.014,\ 250\ \text{GeV}^2)$ and $(0.44,\ 14\,000\ \text{GeV}^2)$.

The H1 analysis was performed for an integrated luminosity of 24 nb^{-1}. The main selection for NC type candidates required an isolated electron with a large transverse energy, $E_T^e > 10$ GeV, as measured in the calorimeter, and $|y_e - y_{JB}| < 0.3$ providing 43 candidates with $M_{LQ} > 35$ GeV. The M_{LQ} distribution is shown in Fig. 43. By requiring $y_e > 0.25$ no event is left for $M_{LQ} > 80$ GeV. The measured M distribution agrees well with the expectation for deep inelastic NC scattering. The search in the CC-mode required a missing transverse momentum of $p_T^{miss} >$ 20 GeV. One candidate was found with $p_T^{miss} = 37$ GeV and $M_{LQ} = 37$ GeV. The absence of any significant signals in either experiment led to the lower mass limits for scalar leptoquarks displayed in Table 3. For couplings equal to that of electroweak interactions, the mass limits for left-handed coupling to $(e^- u)$ states is 181 GeV, and for right-handed coupling is 178 GeV. The limits exceed those obtained previously at LEP [39] (44 GeV), by UA2 [39] (74 GeV) and CDF [40] (113 GeV).

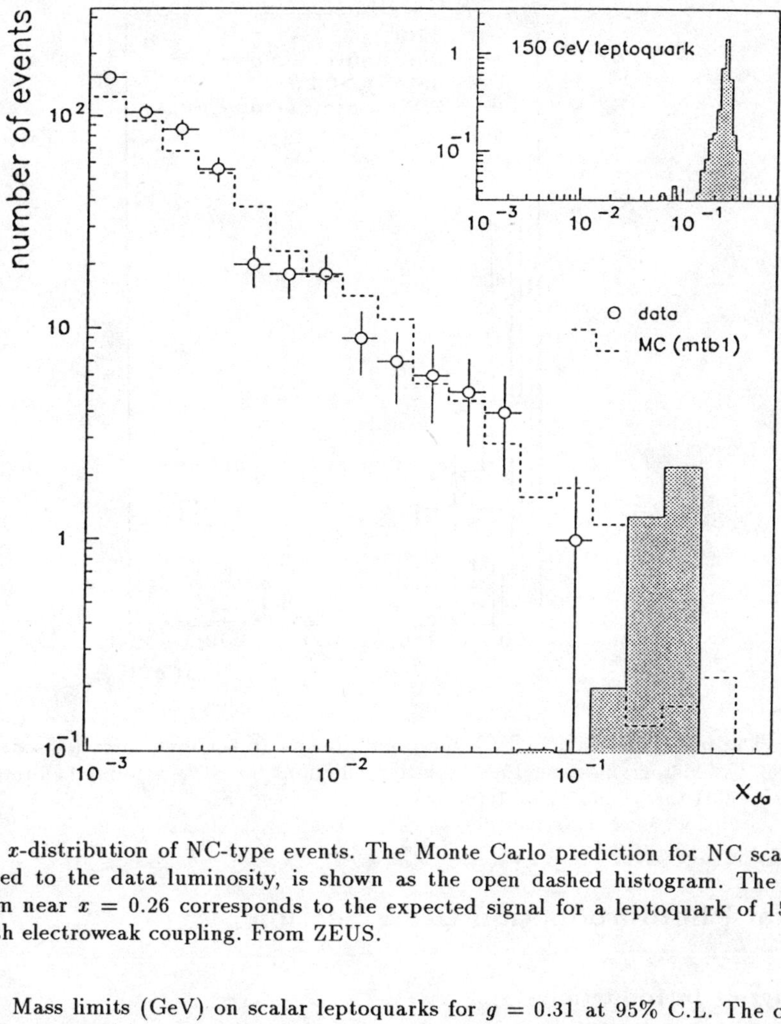

Fig. 42. x-distribution of NC-type events. The Monte Carlo prediction for NC scattering, normalized to the data luminosity, is shown as the open dashed histogram. The shaded histogram near $x = 0.26$ corresponds to the expected signal for a leptoquark of 150 GeV mass with electroweak coupling. From ZEUS.

Table 3. Mass limits (GeV) on scalar leptoquarks for $g = 0.31$ at 95% C.L. The columns provide the charge (q), fermion number (F), weak isospin (I_{wk}), and the flavor.

LQ type	q	F	I_{wk}	left - handed quark limit			right - handed quark limit		
					H1	ZEUS		H1	ZEUS
S_0	$-1/3$	-2	0	u	181	168	u	178	176
$\bar{S}_0$	$-4/3$	-2	0	-	-	-	d	145	146
S_1	$-1/3, -4/3$	-2	1	u, d	192	184	-	-	-
$\bar{S}_{1/2}$	$-5/3, -2/3$	0	1/2	$\bar{u}, -$	97	92	$\bar{u}, \bar{d}$	112	108
$\bar{S}_{1/2}$	$-2/3$	0	1/2	$\bar{d}, -$	97	92	–	–	–

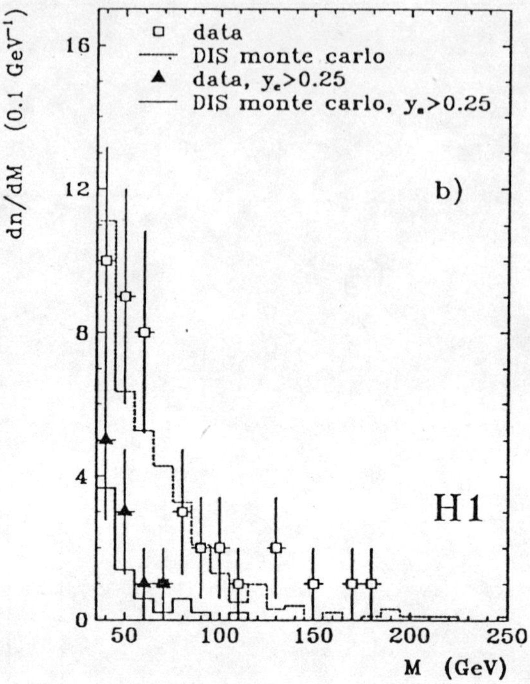

Fig. 43. The mass spectrum of NC-type events before (open) and after (closed) the cut $y_e > 0.25$. The histograms show the absolute predictions for NC scattering before (dashed) and after (solid) the y_e cut. From H1.

7 Total Photoproduction Cross Section

7.1 Physics Introduction

For nonafficionados a measurement of the total photoproduction cross section at high energies may seem rather dull. A closer look at the underlying physics, however, shows that the subject has very interesting aspects. It has recently been reviewed by G. Schuler [41] and the following discussion draws heavily on his report.

Next to $p\bar{p}$ interactions, γp scattering is the only other hadronic or hadron-like reaction for which the total cross section can currently be measured at c.m. energies of several hundreds of GeV. In the VDM picture the photon couples to vector mesons $V(\rho, \omega, \phi)$ which in turn interact with the proton. As a result $\sigma_{tot}(\gamma p)$ is related to the cross section for Vp, and via the quark-parton model, for $\pi^{\pm}p$ and $K^{\pm}p$ scattering. Like the purely hadronic reactions, $\sigma_{tot}(\gamma p)$ in this model has three pieces coming from pomeron exchange, nonpomeron exchange and from partonic scattering in hadron-hadron interactions (Fig. 44a-d). The behavior of $\sigma_{tot}(\gamma p)$ at high energies is therefore expected to be very similar to that of $\sigma_{tot}(p\bar{p})$ (although the hard scattering part of V_P is presumably larger than for $p\bar{p}$ due to a harder x-distribution of quarks).

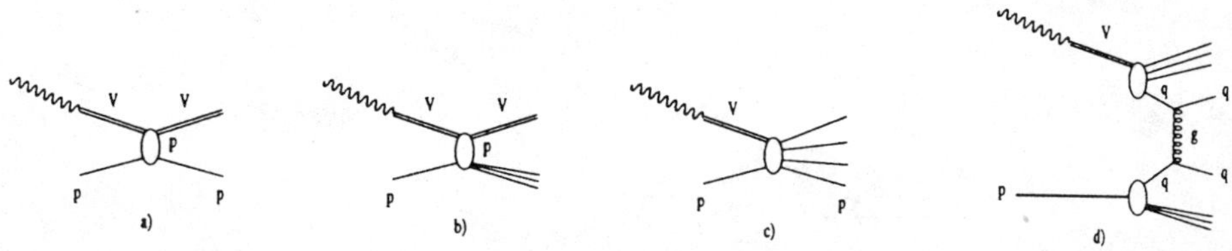

Fig. 44. Diagrams for photoproduction of vector mesons V by elastic (a) and inelastic diffractive scattering (b) and nondiffractive soft (c) and semihard processes (d) via VDM.

The photon, in addition to its hadronic features, however, possesses a property which makes it distinctly different from hadrons: it couples to quarks and the coupling is pointlike. This leads to additional hard scattering processes which become prominent at high energies and which are not present in hadron-hadron interactions. They are represented by two types of diagrams: the first ("direct photon process") results from photon-gluon fusion into a quark-antiquark pair (Fig. 45a) and from photon scattering off a quark in the proton under the emission of a gluon ("QCD Compton process"), see Fig. 45b. But the quarks coupling to the photon can also emit gluons (e.g. Fig. 45c), and either a quark or a gluon may participate in the hard scattering (Fig. 45d,e). Together with the hard scattering of the hadronic (VDM) photon the hard scattering due to the anomalous quark and gluon content of the photon constitute the "resolved photon processes". They are summed in the photon structure functions F^{γ} which describe the quark and gluon content of the photon (note: part if not all of the hard scattering contributions from VDM and from the anomalous quark and gluon content of the photon are presumably identical and care must be taken to avoid double counting). The hard photon processes – direct or resolved – give rise to quark and gluon jets (sometimes called "minijets") with large transverse momenta. Their cross sections have been calculated and found to depend critically on the minimum transfer $p_{T min}$ down to which the integration is performed. As will be seen below, at HERA energies hard scattering originating from photon constituents is clearly visible.

The behavior of $\sigma_{tot}(\gamma p)$ at high energies is closely linked to the density of soft or low-x partons in the proton. The relation between the proton structure function F_2 and the total transverse and longitudinal photon proton cross sections is given by

$$F_2(x,Q^2) = \frac{(1-x)\nu^2 Q^2}{4\pi^2\alpha(Q^2+\nu^2)}(\sigma_{trans} + \sigma_{long})$$

where ν is the photon energy for a stationary proton. As $Q^2 \to 0 : \sigma_{long} << \sigma_{trans} = \sigma_{tot}(\gamma p)$. For small $x << 1$ and high energies, $\nu^2 >> Q^2$, the result for $Q^2 \to 0$ is:

$$F_2(x,Q^2) = \frac{Q^2}{4\pi^2\alpha}\sigma_{tot}(\gamma p)$$

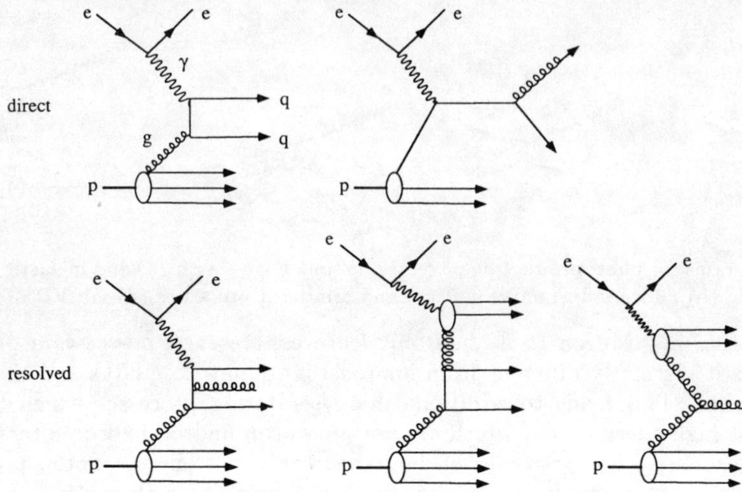

Fig. 45. Diagrams for direct (a,b) and resolved photon processes (c-e).

Under the assumption that for very small x and Q^2, F_2 is of the form

$$F_2(x, Q^2) \sim x^{-\gamma}(Q^2)^{\delta}$$

it follows that

$$F_2(x, Q^2) \sim (Q^2)^{-\gamma+\delta}(2m_p\nu)^{\gamma} \approx (Q^2)^{-\gamma+\delta}W^{2\gamma}$$

where W is the total γp c.m. energy. Continuity for F_2 is preserved as $Q^2 \to 0$ if $F_2 \sim Q^2$ and therefore

$$\sigma_{tot}(\gamma p) \sim W^{2\gamma} \ .$$

The energy dependence of the total cross section at high energies is driven by the parton density of the proton at low x.

The definition for F_2 used here is the standard one. A "microscopic" treatment might separate F_2 into a term describing the "proton proper" (whatever this is) and a term due to resolved photons. In this case F_2 is expected to be dominated by the x distribution of the gluon in the resolved photon structure function [41].

Observations of "diffractive" air showers [42] and an excess of muons in very energetic cosmic air showers [43] have suggested $\sigma_{tot}(\gamma p)$ to rise much faster than $\sigma_{tot}(p\bar{p})$ at beam energies above $\sim$ 100 TeV. This rise has been attributed to the semi-hard scattering arising from the resolved photon contributions. Since gluon initiated processes dominate, the rise is driven by the x-dependence of the gluon distribution for photons and protons [44].

7.2 H1 and ZEUS Results

Photoproduction at HERA is part of neutral current scattering, $ep \rightarrow eX$, with the exchanged photon being almost real, $Q^2 \approx 0$. For events of this type the electron is scattered in the direction of the electron beam. It can be tagged in the electron calorimeter of the luminosity monitor (Fig. 8) and its energy, E'_e, measured. This determines the energy k of the exchanged photon in the HERA system, $k = E_e - E'_e$, and the photon proton c.m. energy W : $W^2 = 4kE_p$. For a minimum energy of the scattered electron of 4 GeV c.m. energies up to 300 GeV can be measured at HERA which is equivalent to a photon of energy $\nu = W^2/(2m_p) \approx 47$ TeV striking a stationary proton.

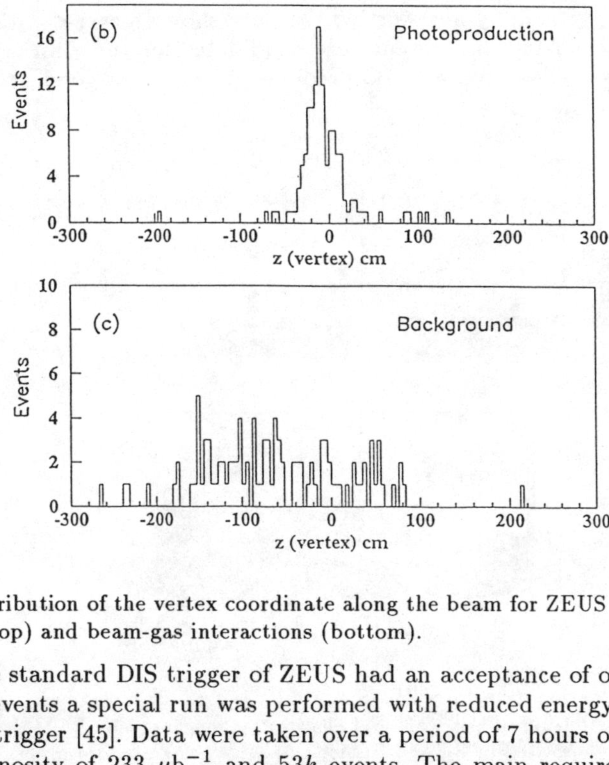

Fig. 46. Distribution of the vertex coordinate along the beam for ZEUS events from photoproduction (top) and beam-gas interactions (bottom).

Since the standard DIS trigger of ZEUS had an acceptance of only 4% for photoproduction events a special run was performed with reduced energy thresholds of the calorimeter trigger [45]. Data were taken over a period of 7 hours of running yielding a total luminosity of 233 μb^{-1} and $53k$ events. The main requirements for photoproduction events were an electron tagged in LUMI (scattering angle < 6 mrad), a minimum energy of 1.1 GeV in the rear calorimeter (RCAL) and proper calorimeter timing. In total 212 events satisfied these conditions of which 182 were accepted after a visual scan. The event vertex was reconstructed from the tracking information for 72% of the events. The vertex position peaks close to the origin in contrast to the uniform distribution characteristic of beam gas events (Fig. 46). For the final sample of 97 events the energy of the scattered electron was required to lie in the range $10 - 16$ GeV ($W = 186 - 233$ GeV) resulting in maximum and average Q^2 values of 0.02 and 0.0006 GeV^2, respectively. Hence the contribution from longitudinal photons can safely be neglected and the measured process is scattering of (almost) real photons on protons.

The main difficulty in the extraction of $\sigma_{tot}(\gamma p)$ lies in the fact that the trigger acceptance A depends strongly on the type of process. It is low for the "elastic" reaction $\gamma p \to \rho p$ (Fig. 44a) and sensitive to the details of the model, $A = 10-27\%$. For these events the proton disappears mostly undetected in the forward direction and the $\pi^{\pm}$ from ρ decay escape often unseen in the backward direction. For inelastic diffractive processes (Fig. 44b) $A = 40-50\%$ while for low$-p_t$ reactions (Fig. 44c) $A = 70\%$; for hard processes such as direct and resolved photon contributions $A = 80\%$. The final acceptance calculation was made assuming 20% from elastic plus diffractive processes and 80% from low p_t processes. The resulting predictions for the energy spectra seen in the three sections of the calorimeter give a reasonable description of the data (Fig. 47). One point is particularly noteworthy: the data show 18 events with zero energy deposition in the FCAL. These events are candidates for the elastic and inelastic diffractive processes. The Monte Carlo predicts 16 events, in good agreement with the data.

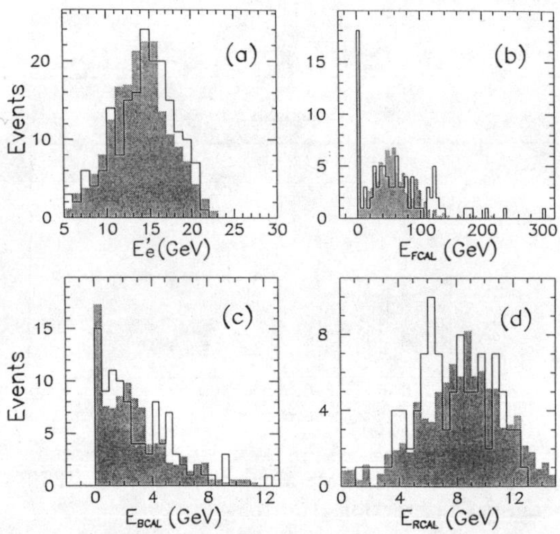

Fig. 47. Photoproduction data from ZEUS:
(a) The energy distribution of the scattered electron as measured by LUMI;
(b)-(d) The energy deposited in FCAL, BCAL and RCAL for photoproduction events with electron energies between 10 and 16 GeV; the shaded distributions are the predictions from the Monte Carlo.

The photoproduction candidates in the H1 experiment were obtained mainly by requiring a tagged electron ($E'_e = 10-19$ GeV) and a track in the central tracking detector pointing to the vertex region [46]. Most of the remaining background was due to proton gas interactions. It was statistically subtracted using the energy spectrum of the tagged electron which is substantially different for the signal and background events (Fig. 48). The final sample consisted of 917 ± 38 events from a luminosity of $1.5\,\mu b^{-1}$. The acceptance was calculated assuming 26% elastic plus inelastic diffractive events, 55% low p_t and 19% hard scattering events as suggested by an extrapolation of low-energy photoproduction data.

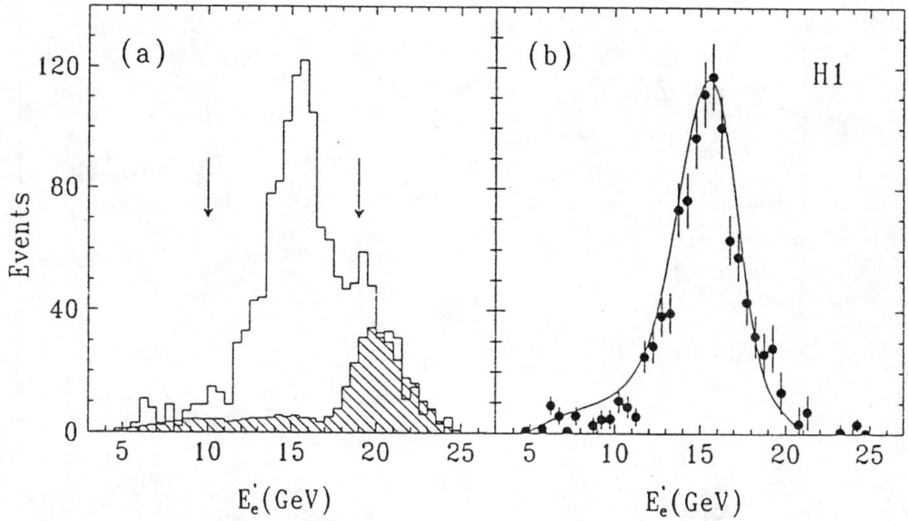

Fig. 48. Energy spectrum from H1 of tagged electrons for background events from the unpaired electron bunch (left) and for tagged events (right), compared with the tagger acceptance (curves).

The total cross section values obtained by the two experiments are

$$\begin{array}{lcllllcll} \sigma_{tot}(\gamma p) & = & 159 & \pm 7 \text{ (stat)} & \pm 20 \text{ (syst) } \mu\text{b} & <W> & = & 195 \text{ GeV}, & H1 \\ & & 154 & \pm 16 \text{ (stat)} & \pm 32 \text{ (syst) } \mu\text{b} & <W> & = & 210 \text{ GeV}, & ZEUS \end{array}$$

Radiative corrections were calculated to be small. For H1 they increase the measured cross section by to $2.4\%^{+1.4\%}_{-2.5\%}$; the cross section was corrected accordingly. For ZEUS the corresponding increase is 1%. The systematic errors reflect the uncertainties in the determination of the acceptance and the luminosity.

Fig. 49 shows $\sigma_{tot}(\gamma p)$ as a function of W above the resonance region ($W >$ 1.75 GeV) as measured in previous experiments up to $W = 18$ GeV (Ref. 47) and by H1 and ZEUS. No dramatic rise is observed between 18 and 200 GeV.

The curve labeled DL in Fig. 49 is the prediction of a Regge-type fit to hadronic and pre-HERA photon proton total cross sections of the form

$$\sigma_{tot} = Xs^{\epsilon} + Ys^{-\eta}$$

where the first term describes Pomeron contributions and the second one those from ρ, ω, f, a exchange [48]. The values of ϵ, η were obtained from fits to pp and $p\bar{p}$ data alone yielding $\epsilon = 0.0808$ and $\eta = 0.4525$ while the coefficients X, Y were determined from fitting the low energy data on $\sigma_{tot}(\gamma p)$. The curve ALLM is also based on a Regge-type analysis using the low energy photoproduction measurements together with proton structure function data [49]. Both predictions agree rather well with the HERA measurements.

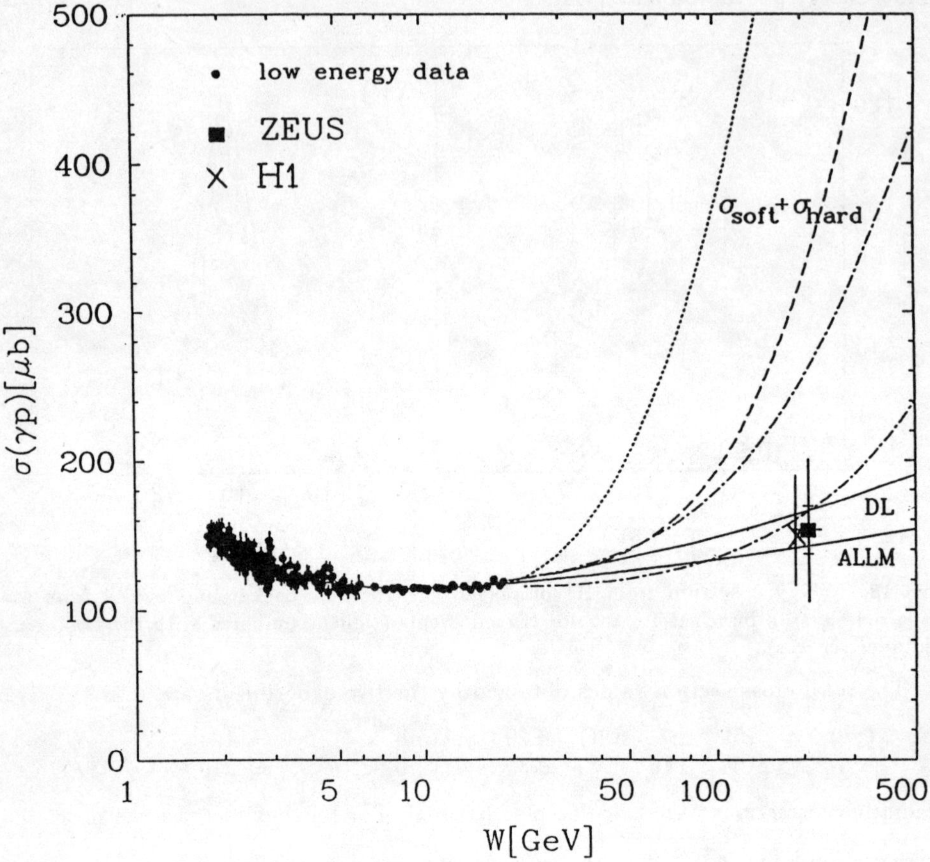

Fig. 49. The total γp cross section as a function of the c.m. energy W as measured below 18 GeV and by H1 and ZEUS near $W = 200$ GeV. The lower solid curve is the prediction of the ALLM [46] parametrization and the higher solid curve is that of DL [45]. The dotted (dashed) line uses the LAC1 parametrization [49] for the photon with $p_{Tmin} = 1.4$ GeV/c (2 GeV/c). The dash-dotted lines use the DG parametrization for the photon [48] with $p_{Tmin} = 1.4$ GeV/c (upper line) and $p_{Tmin} = 2$ GeV/c (lower).

The other curves shown in Fig. 49 are based on the assumption that the total cross section is a sum of a soft part plus the contributions from the direct and resolved photons [50]. They depend critically on the choice of photon structure function F^γ and on the parameter p_T^{min} which is the lower integration limit for the hard processes. Here, p_T^{min} is the transverse momentum cutoff in the c.m. system of the hard subprocess. The dashed-dotted lines use the parametrization of F^γ from Ref. 51 with $p_T^{min} = 2.0$ GeV/c for the lower (1.4 GeV/c for the upper) line. The dashed (dotted) line uses F^γ from Ref. 52 and $p_T^{min} = 2.0$ (1.4 GeV/c). A critical appraisal of the different calculations can be found in Ref. 53. Only the dashed-dotted curve – which makes the lowest cross section prediction – is in agreement with the HERA data. A strong rise of $\sigma_{tot}(\gamma p)$ that might have been expected in the HERA regime from the cosmic ray results and/or from direct and resolved photon contributions is postponed to higher energies, if at all.

8 Hard Scattering in Photoproduction

The presence of direct and resolved photon processes suggests a substantial amount of hard scattering in photoproduction. H1 (Ref. 54) and ZEUS (Ref. 55) analyzed the transverse momentum behavior of events from photoproduction and ep scattering at low Q^2.

8.1 H1 Analysis

The study was made with two event samples recorded for a total luminosity of 0.9 nb^{-1}. The selection criteria for sample I were:

–the scattered electron detected in LUMI
–a track from the interaction region
–$0.35 < y_e < 0.7$ and $y_h > 0.2$
where y is the scaled photon energy $y = \nu/\nu_{max}$
calculated from the scattered electron $y_e = 1 - E'_e/E_e(1 + cos\ \theta'_e)/2$
from the hadrons measured in calorimeter $y_h = \Sigma(E^i - p^i_L)/(2E_e)$
–energy > 5 GeV in the forward region ($\theta < 25^0$).

A total of 330 events satisfied the requirements. The background from beam gas interactions was estimated to be $< 9\%$. Tagging of the scattered electron in LUMI ensures the events to come from photoproduction. The range for the c.m. energy W is $175 - 250$ GeV.

The transverse momentum distribution is shown in Fig. 50. For $p_T^2 < 1\ \text{GeV}^2$ a steeply falling distribution is observed which is well described by a gaussian with an rms of ~ 350 MeV/c. At higher p_T^2 the data show a second component with a long tail of high transverse momenta indicating the presence of hard scattering. A jet search was performed with the calorimeter requiring jets of more than 3 GeV in cones of $\Delta R = (\Delta\phi^2 + \Delta\eta^2)^{1/2} < 1$. For the 19 events with two or more jets the c.m. energy of the two most energetic jets is on average 12.5 GeV.

Since the photon energy is known, $E_\gamma = E_e - E'_e$, the momenta of the two jets are constrained. At the parton level, equating the jets with partons of equal transverse energy E_t and polar angles θ_1 and θ_2, and neglecting initial state radiation:

$$2E_\gamma/E_t = tan\ \theta_1/2 + tan\ \theta_1/2.$$

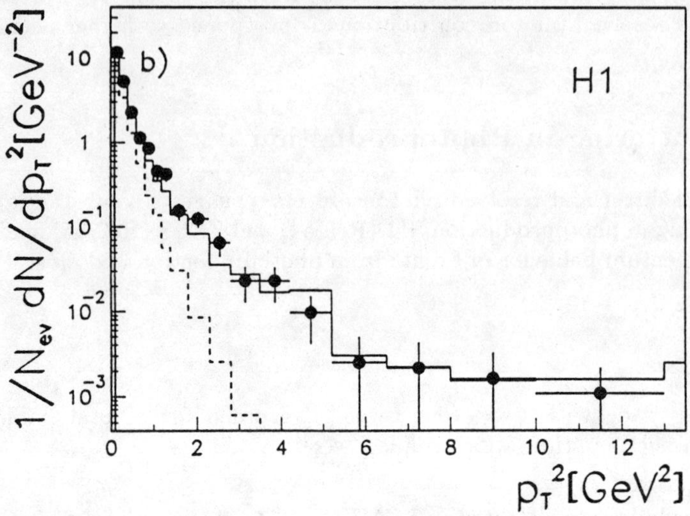

Fig. 50. Squared transverse momentum distribution of charged tracks as observed by H1 from *ep* data compared with 'soft' (dashed line) and 'soft + QCD', Monte Carlo simulation (full line).

Fig. 51 shows the distribution of θ_1 versus θ_2 for the data and the predictions for events from direct plus resolved photons and direct photons alone. There is a clear distinction predicted for direct and resolved processes: the direct process yields rarely events where both jet angles are below 100^0: only 0.9 events should be found in this region. Instead, 9 of the 19 measured events lie in this region in agreement with the predictions for the direct and resolved processes.

The indication that the data require a resolved photon contribution has been further substantiated by a second sample (II) for which an electron tagged in LUMI was not demanded; thus, these events are not necessarily from photoproduction. However, events with a substantial Q^2 were removed by requiring that there be no electron with $E > 10$ GeV in $140^0 < \theta < 176^0$. A total of 51 events were selected in this way.

Fig. 52 shows the energy flow per event as a function of the polar angle for events where the two most energetic jet cones have $\theta < 100^0$. In this case the direct process predicts no energy flow beyond $\theta \sim 130^0$ in marked contrast with the data which show a sizable energy flow up to 180^0. The prediction for the direct *plus* resolved photon processes is in good agreement with the data.

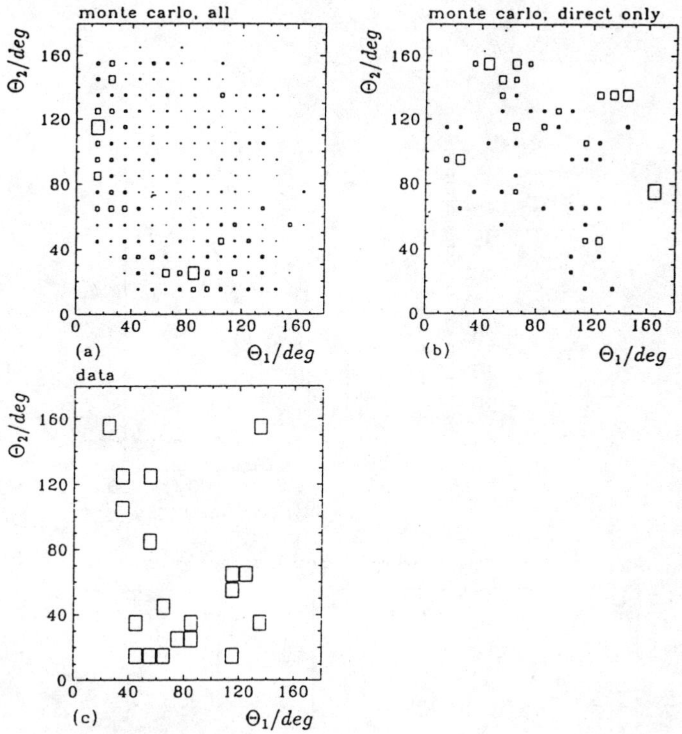

Fig. 51. Correlation of the two highest p_T jets in polar angle for H1 events with an electron tag: (a) Monte Carlo predictions for resolved and direct processes, (b) for direct processes only, (c) correlation for the 19 events with a tagged electron.

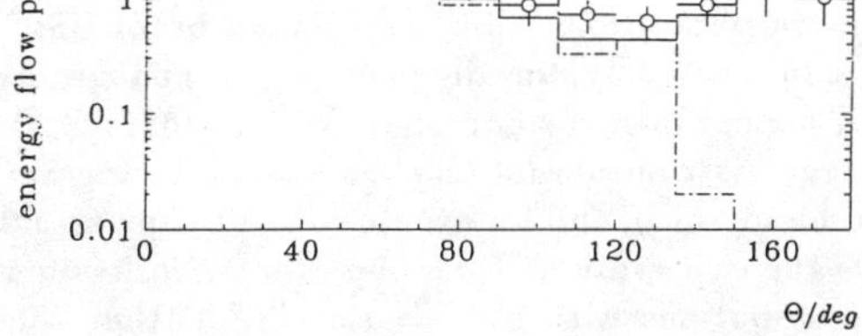

Fig. 52. Energy flow per event versus polar angle as measured by H1; the open points represent the data; the full line is taken from the Monte Carlo simulation for resolved and direct photons; the dotted line gives the prediction for the direct processes alone.

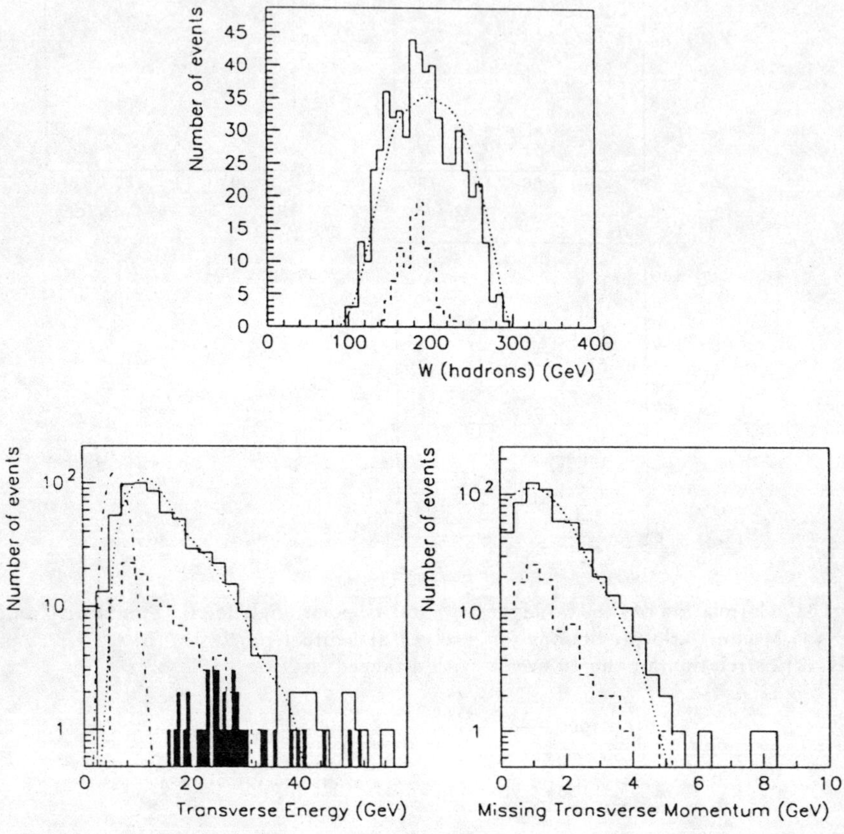

Fig. 53. Hard scattering in photoproduction as analyzed by ZEUS:
(a) Distribution of the c.m. energy W for all events (solid histogram) and for events with a tagged electron (dashed histogram); the curves show the Monte Carlo expectations;
(b) Total transverse energy distribution for all events (solid histogram) and for events with a tagged electron (dashed histogram), and for events with a twojet structure (black histogram); the dashed curve shows the expectations from the Monte Carlo simulation including direct and resolved photon contributions with $p_T^{min} = 1.5$ GeV$/c$; the dashed-dotted curve shows the transverse energy distribution for soft γp interactions;
(c) Missing transverse momentum distribution for all events (solid histogram) and for events with a tagged electron (dashed histogram). The dotted curve shows the Monte Carlo expectations.

8.2 ZEUS Analysis

The events were selected by requiring in addition to the calorimeter trigger

–proper timing observed by the calorimeter,
–more than 10 GeV in FCAL and > 2.5 GeVin RCAL,
or : > 20 GeV in total energy, > 10 GeV in total transverse energy
and $\Sigma(E^i - p_L^i)/(2E_e) > 12$ GeV when summed over all calorimeter cells
–no scattered electron observed in the calorimeter with $y_e < 0.7$.
This cut ensures that the accepted events have $Q^2 < 4$ GeV^2

The cuts were satisfied by 576 events from a total luminosity of 2.2 nb^{-1}. The contamination from beam gas events is $< 3\%$ and $< 2\%$ from deep inelastic events with $Q^2 > 4$ GeV^2.

The data are consistent with originating from photoproduction. This is shown by comparing the full sample with the 96 events for which the scattered electron was detected in LUMI with $5 < E'_e < 22$ GeV. These events have $Q^2 < 0.02$ GeV^2. The ratio of tagged to total number of events agrees well with the Monte Carlo prediction of 20% for pure photoproduction events. The distribution of the γp c.m. energy W is shown in Fig. 53a. For the full sample W is in the range from 100 to 295 GeV; for the tagged electron events it is narrower, $160 < W < 260$ GeV, reflecting the restricted electron energy range. Deep inelastic events would have a completely different shape, peaking sharply at small W.

The distributions of the total transverse energy E_T and the missing transverse momentum p_T^{miss} are shown in Figs. 53b,c. The overall p_T is well balanced as expected for photoproduction, the average p_T^{miss} being 1.5 GeV/c. The E_T distribution exhibits, by contrast, a tail that extends beyond 10 GeV : 391 events have $E_T > 10$ GeV. The tagged electron events show the same behavior (dashed histogram). The tail is much larger than that expected from soft γp interactions which extends to a maximum of 10 GeV (dashed-dotted curve in Fig. 53b). A comparison of the cross sections shows that 20% of all photoproduction events have an $E_T > 10$ GeV.

The events were searched for jets with jet cone $\Delta R < 1$, jet transverse momentum > 4 GeV and $\eta < 2$ ($\theta < 15^0$): 41 events were found with two jets. One of these events is shown in Fig. 54a displaying two clear jets. The distributions of the jet transverse momenta and azimuthal angles are shown in Fig. 55b,c for the two-jet sample. The jets are predominantly back-to-back in the transverse plane. The E_T values for the two-jet events are all above 16 GeV as shown by the black histogram in Fig. 53b.

As discussed before, a major fraction of the high E_T events is expected to come from resolved photon interactions and therefore to show substantial energy flow in the backward (electron beam) direction from the photon remnants. Fig. 55d shows the energy deposited in the RCAL (rear calorimeter) versus the minimum pseudorapidity η of either of the two jets. Note, jets in the forward hemisphere have $\eta > 0$, and in the backward hemisphere $\eta < 0$. If direct photon interactions were the sole origin of these events, sizable energy in RCAL would be expected for events with the minimum $\eta_{min} < -1$, falling essentially to zero as the jets become more distant from the RCAL region ($\eta_{min} > 0$). This trend is indeed observed in Fig. 55d. However, in addition the data show events with as much as $4 - 12$ GeV detected in RCAL, even when both jets are far from the RCAL itself, the nearest jet being as much as three units of rapidity away. These events are interpreted as originating from the resolved photon process, where the two partons go forward and the photon remnants go approximately in the direction of the incident electron, which is close to the direction of the photon. One of the resolved photon candidates is shown in Fig. 54b. It has two jets in the forward direction and a cluster of energy going backward.

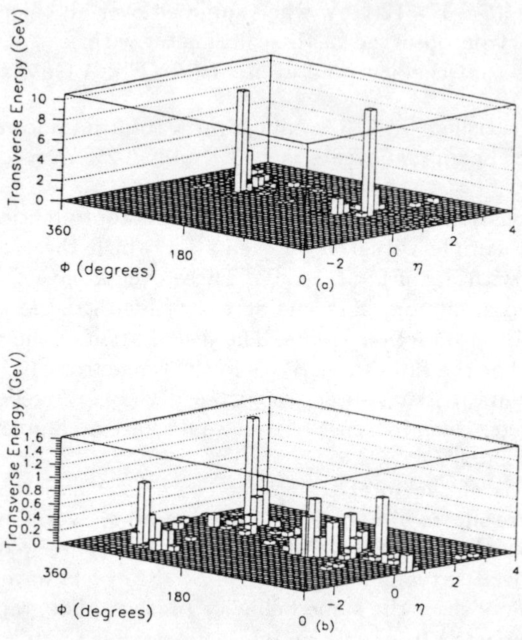

Fig. 54. Examples for events from hard scattering in photoproduction as observed by ZEUS in $\phi - \eta$ space:
(a) An event with a two-jet structure;
(b) An event showing a two-jet structure plus additional energy in the direction of the incident electron associated with the photon remnant.

8.3 Discussion of the Results

The findings of both experiments are consistent. Photoproduction at c.m. energies of 100 – 300 GeV exhibits an excess of events with large transverse momenta. For a fraction of these events large transverse momentum jets have been observed which is evidence for the presence of hard scattering. In events with two jets the jets tend to be back-to-back in the transverse plane. Although the statistics is limited, there is a clear indication that in some of the events where both jets go forward, in addition a substantial amount of energy is emitted in the backward direction, close to the direction of the incident photon. This event topology is expected for resolved photon processes where the photon remnant goes into the direction of the incident photon.

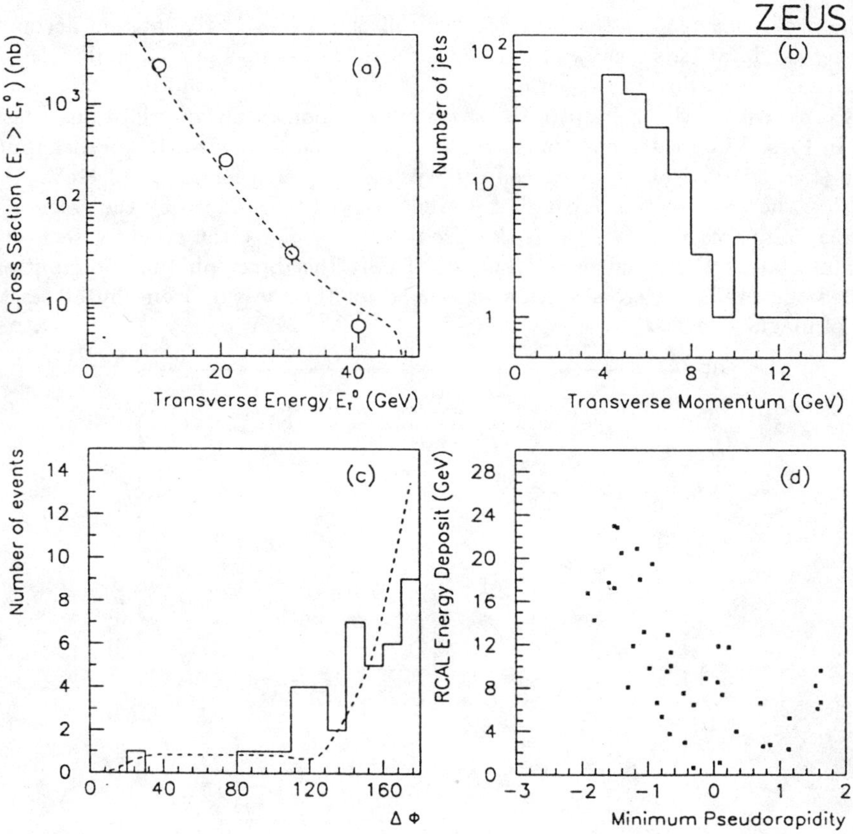

Fig. 55. Hard scattering in photoproduction as analyzed by ZEUS:
(a) Total *ep* cross section for transverse energies $E_T > E_t^o$. The curve is the cross section predicted by HERWIG with $p_T^{min} = 1.5$ GeV/c, including direct and resolved photon contributions.
(b) Transverse momentum distribution of the identified jets;
(c) Distribution of the difference in azimuth between jets for events with two jets; the dotted curve shows the Monte Carlo prediction;
(d) Correlation between energy deposited in the RCAL versus the minimum rapidity of the two jets in the two-jet sample.

Hard scattering in photoproduction was predicted to arise from the direct and resolved photon processes. H1 and ZEUS find that the observed gross features as well as the cross sections can be accounted for by these processes. They were calculated together with the soft contributions by Monte Carlo in full detail, using H1: the PYTHIA generation [56] as well as ZEUS: PYTHIA and HERWIG [57] with the photon structure functions of Ref. 51, the proton structure function MTB1, including initial and final state parton showers, fragmentation into hadrons in the Lund scheme [28], and describing the detector performance within the framework of GEANT [58].

The predictions calculated for a p_T^{min} cutoff of 1.5 GeV/c give a good account of the observed data. This is shown by H1 by the solid curves for the p_T^2 distribution (Fig. 50) and the energy flow in two-jet events (Fig. 52). The corresponding comparison by ZEUS is shown for the E_T and missing transverse momentum distributions (dashed curve in Figs. 53 b,c). H1 has made an absolute comparison with the p_T distribution of jets (Fig. 56) and observes agreement within a factor 2 for $p_T > 10$ GeV. ZEUS finds that the cross section for high E_T events is well reproduced by the calculation: Fig. 55a compares the data with the prediction for the total ep cross section for E_T values larger than a given E_T^0 energy. If only the direct photon contribution is included the predicted cross section falls short of the measured one by at least an order of magnitude.

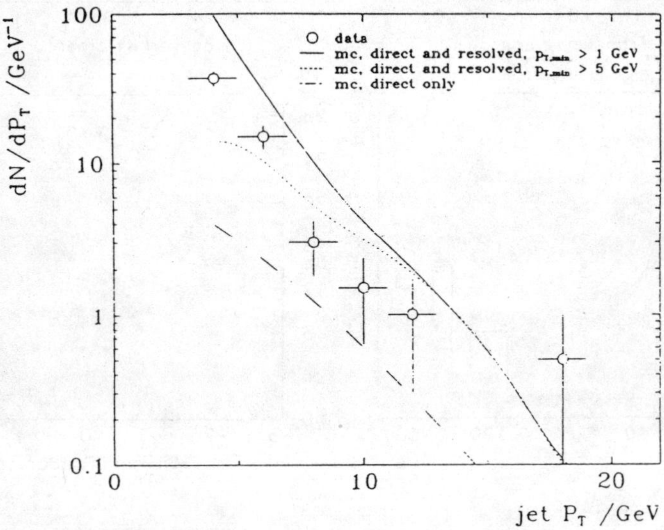

Fig. 56. Hard scattering in photoproduction as analyzed by H1: p_T distribution of jets with $p_T > 3$ GeV. The solid line is the prediction of PYTHIA for $p_T^{min} = 1$ GeV/c, the dashed dotted line corresponds to $p_T^{min} = 5$ GeV/c. The dashed-dotted line is the prediction if resolved photon contributions are excluded ($p_T^{min} = 1$ GeV/c).

In conclusion H1 and ZEUS observed hard scattering in high energy photoproduction at the level of 20% of the total photoproduction cross section. The data are in good qualitative and quantitative agreement with theory which predicted the existence of hard scattering as a result of direct and resolved photon processes. The observed event topologies give an independent indication for the presence of resolved photon contributions.

9 Concluding Remarks

HERA has joined the family of active high energy colliders and the experiments H1 and ZEUS so far have encountered no problem to enter a new territory of electron-proton collisions. The data which led to the results discussed above were obtained in the first few months of data taking. Although the luminosity collected was only a tiny fraction of its nominal value, the experiments presented already exciting results. During the second run period the luminosity rose by a factor of ten and there is good reason for expecting it to rise by several orders of magnitude more in future running. With increasing luminosity more processes will come into the reach of H1 and ZEUS. Examples are deep inelastic neutral scattering where in the beginning every factor of ten increase in luminosity will enlarge the accessible Q^2 range by another order of magnitude, and charged current scattering, for which the first candidates were observed.

Acknowledgments

Prof. Plessas and his colleagues have organized a stimulating and enjoyable meeting. I am grateful to F. Barreiro, H. Kowalski, E. Lohrmann and F. Sciulli for a critical reading of the manuscript, and to G. Schuler for many comments on the current theoretical picture of photoproduction. S. Gharavi has been helpful with the preparation of the drawings.

References

In the following "HERA Workshop 1991" stands for Physics at HERA, Proc. Workshop 1991, ed. by W. Buchmüller and G. Ingelmann, April 1992.

[1] HERA, A Proposal for a Large Electron-Proton Colliding Beam Facility at DESY, DESY HERA 81-10 (1981); B.H. Wiik, Electron-Proton Colliding Beams, The Physics Programme and the Machine, Proc. 10th SLAC Summer Institute, ed. A. Mosher, 1982, p. 233, and Proc. XXVI Int. Conf. High Energy Physics, Dallas, 1992; G.-A. Voss, Proc. First Euro. Acc. Conf., Rome, 1988, p. 7.

[2] H1 Collaboration, Technical Proposal for the H1 Detector (1986).

[3] ZEUS Collaboration, The ZEUS Detector, Technical Proposal (1986), and: The ZEUS Detector, Status Report 1989.

[4] Proc. HERA Workshop 1987, ed. by R. Peccei, August 1988.

[5] Proc. HERA Workshop 1991, ed. by W. Buchmüller and G. Ingelmann 1991.

[6] HERMES Collaboration, A proposal to measure the spin-dependent structure functions of the neutron and the proton at HERA (1990)

[7] E. Reya, lectures presented at this school.

[8] H1 Collaboration, F. Eisele, Proc. XXVI Int. Conf. High Energy Physics, Dallas, 1992.

[9] ZEUS Collaboration, B. Löhr, Proc. XXVI Int. Conf. High Energy Physics, Dallas, 1992.

[10] F. Eisele and G. Wolf, Physikalische Blätter **48** 787 (1992).

[11] A.A. Sokolov and M. Ternov, Sov. Phys. Doklady **8** 1203 (1964).

[12] D.B. Barber et al., DESY Report 92-136 (1992).

[13] J. Buon and K. Steffen, DESY report 85-128 (1985).

[14] F. Jacquet and A. Blondel, Proc. Study of an ep Facility in Europe, ed. U. Amaldi, 79/48 (1979), p. 391.

[15] S. Bentvelsen, J. Engelen and P. Kooijman, HERA Workshop 1991, Vol. 1, p. 23.

[16] H. Chaves, R.J. Seifert and G. Zech, HERA Workshop 1991, Vol. 1, p. 57.

[17] A. Milsztain, lectures presented at this school.
[18] V.N. Gribov and L.N. Lipatov, Sov. J. Nucl. Phys. **15** 438, 675 (1972); G. Altarelli and G. Parisi, Nucl. Phys. **126** 297 (1977).
[19] J. Blumlein et al., Akad. Wissenschaften, Berlin-Zeuthen, Report 88-01 (1988).
[20] L.V. Gribov, E.M. Levin, M.G. Ryskin, Phys. Rep. **100** 1 (1983).
[21] See e.g. J. Bartels and J. Feltesse, HERA Workshop 1991, Vol. 1, p. 131, and references given there.
[22] J. Kwieczinski, lectures presented at this school.
[23] H1 Collaboration, T. Ahmed et al., Phys. Lett. **B299** 385 (1993).
[24] ZEUS Collaboration, M. Derrick et al., Phys. Lett. **B303** 183 (1993).
[25] A.D. Martin, R.G. Roberts and W.J. Stirling, Durham University preprint DTP-92-16 (1992).
[26] J.G. Morfin and W.K. Tung, Z. Phys. **C52** 13 (1992).
[27] H1 Collaboration, T. Ahmed et al., Phys. Lett. **B298** 469 (1993).
[28] T. Sjöstrand, Comp. Phys. Comm. **39** 347 (1986); T. Sjöstrand and M. Bengtsson, ibid. **43** 367 (1987) and CERN-TH 6488/92 (1992).
[29] A. Kwiatkowski, H. Spiesberger and H.-J. Möhring, Comp. Phys. Comm. **69** 155 (1992), and references given there.
[30] G. Ingelman, LEPTO5.2, program manual, unpublished; H. Bengtsson, G. Ingelman and T. Sjöstrand, Nucl. Phys. **B301** 554 (1988).
[31] L. Lönnblad, Comp. Phys. Comm. **71** 15 (1992), and references given there.
[32] G. Ingelman, HERA Workshop 1991, Vol.3, p.1366, and references given there.
[33] See e.g. EMC Collaboration, J.J. Aubert et al., Phys. Lett. **B100** 433 (1981); M. Arneodo et al., Z. Phys. **C36** 527 (1987).
[34] E665 Collaboration, M.R. Adams et al., Phys. Rev. Lett. **69** 1026 (1992).
[35] ZEUS Collaboration, M. Derrick et al., DESY Report 93-030, Phys. Lett., to be published.
[36] See F. Schrempp, HERA Workshop 1991, Vol. 2, p. 1034, and references given there.
[37] ZEUS Collaboration, M. Derrick et al., DESY Report 93-017, Phys. Lett., to be published.
[38] H1 Collaboration, T. Ahmed et al., DESY Report 93-029 (1993).
[39] OPAL Collaboration, M. Akrawy et al., Phys. Lett. **B263** 123 (1991). L3 Collaboration, B. Adeva et al., Phys. Lett. **B261** 169 (1991). DELPHI Collaboration, P. Abreu et al., Phys. Lett. **B275** 222 (1991). ALEPH Collaboration, D. deCamp et al., Phys. Rep. **216** 253 (1992). UA2 Collaboration, J. Alitti et al., Phys. Lett. **B274** 507 (1992).
[40] CDF Collaboration, S. Moulding et al., presentation at the 7th Meeting of the American Physical Society (DPF), Fermilab (November 1992).
[41] G. Schuler, HERA Workshop 1991, Vol. 1, p. 131; see also G. Schuler and T. Sjöstrand, CERN-TH 6796/93 (1983).
[42] Chacaltaya and Pamir Collaborations, contribution to VI Int. Symp. on V.H.E. Cosmic Ray Interactions, ICRR-Rpt-216-90-9, 1990.
[43] G. Yodh, Nucl. Phys. B (Proc. Suppl.) **12** 277 (1990).
[44] M. Drees, F. Halzen, and K. Hikasa, Phys. Rev. **D39** 1310 (1989); T.K. Gaisser et al., Phys. Lett. **243** 444 (1990); R. Ghandi et al., Phys. Rev. **D42** 263 (1990).
[45] ZEUS Collaboration, M. Derrick et al., Phys. Lett. **B293** 465 (1992).
[46] H1 Collaboration, T. Ahmed et al., Phys. Lett. **B299** 374 (1993).
[47] S.I. Alekhin et al., CERN-HERA 87-01 (1987) and references given there.
[48] A. Donnachie and P.V. Landshoff, Nucl. Phys. **B244** 322 (1984).
[49] H. Abramowicz et al., Phys. Lett. **B269** 465 (1991).
[50] M. Drees and K. Grassie, Z. Phys. **C28** 451 (1985); R.S. Fletcher et al., Phys. Rev. **D45** 377 (1992); R. Ghandi and I. Sarcevic, Phys. Rev. **D44** R10 (1991); J.R. Forshaw and

J.K. Storrow, Phys. Lett. **B268** 116 (1991); G. Schuler and J. Terron, HERA Workshop 1991, Vol. 1, p. 599.
[51] M. Drees and K. Grassie, Z. Phys. **C28** 451 (1985).
[52] H. Abramowicz, K. Charchula and A. Levy, Phys. Lett. **B269** 458 (1991).
[53] G. Schuler and J. Terron, HERA Workshop 1991, Vol. 1, p. 599.
[54] H1 Collaboration, T. Ahmed et al. Phys. Lett. **B297** 205 (1992).
[55] ZEUS Collaboration, M. Derrick et al., Phys. Lett. **B297** 404 (1992).
[56] T. Sjöstrand, HERA Workshop 1991, Vol. 3, p. 1405.
[57] G. Marchesini et al., Comp. Phys. Comm. **67** 465 (1992).
[58] GEANT program manual, CERN program library (1992).

Seminars

Several participants contributed interesting seminar talks fitting to the general theme of the School. It was not possible to include them here, but we list the seminar speakers (alphabetically ordered) and the topics covered by them.

J. Adam (Univ. Mainz):
p/m Expansion for Deuteron Elastic Form Factors

B. Bajc (Univ. Ljubljana):
W-loop Versus Ultra-Heavy Fermion Loop in the Reaction $\gamma\gamma \to ZZ$

V. Bernard (Univ. Strasbourg):
Chiral Symmetry and Threshold Pion Electroproduction

S.N. Fedotkin (Inst. for Nuclear Research, Kiev):
Internal Positronium Production from Nuclear Transitions

H.V. von Geramb (Univ. Hamburg):
$p-p$ and $n-p$ Inversion Potentials and Bremsstrahlung

K. Golec-Biernat (Univ. Kraków):
Hot Spots at HERA

G. Karl (Univ. München):
Baryon Magnetic Moments, Axial Currents and the Proton Spin Problem

Ch. Markus (Univ. Bonn):
Studies of Strong and Electroweak Interactions with Direct Photons in Z^0-Decays

R. Raczka (Inst. Nucl. Stud., Warszawa):
Hadronic Standard Model for Strong and Electroweak Interactions

W. Sakuler (Techn. Univ. Wien):
Lattice Study of Hadron-Hadron Interactions with Dynamical Quarks

A. Szczurek (KFA Jülich):
The Role of the Meson Cloud in Deep-Inelastic Scattering of Leptons

Printing: Druckhaus Beltz, Hemsbach
Binding: Buchbinderei Schäffer, Grünstadt